CAMBRIDGE STUDIES IN
ADVANCED MATHEMATICS 87

MULTIDIMENSIONAL REAL ANALYSIS II:

INTEGRATION

Already published; for full details see http://publishing.cambridge.org/stm /mathematics/csam/

11 J.L. Alperin *Local representation theory*
12 P. Koosis *The logarithmic integral I*
13 A. Pietsch *Eigenvalues and s-numbers*
14 S.J. Patterson *An introduction to the theory of the Riemann zeta-function*
15 H.J. Baues *Algebraic homotopy*
16 V.S. Varadarajan *Introduction to harmonic analysis on semisimple Lie groups*
17 W. Dicks & M. Dunwoody *Groups acting on graphs*
18 L.J. Corwin & F.P. Greenleaf *Representations of nilpotent Lie groups and their applications*
19 R. Fritsch & R. Piccinini *Cellular structures in topology*
20 H. Klingen *Introductory lectures on Siegel modular forms*
21 P. Koosis *The logarithmic integral II*
22 M.J. Collins *Representations and characters of finite groups*
24 H. Kunita *Stochastic flows and stochastic differential equations*
25 P. Wojtaszczyk *Banach spaces for analysts*
26 J.E. Gilbert & M.A.M. Murray *Clifford algebras and Dirac operators in harmonic analysis*
27 A. Frohlich & M.J. Taylor *Algebraic number theory*
28 K. Goebel & W.A. Kirk *Topics in metric fixed point theory*
29 J.E. Humphreys *Reflection groups and Coxeter groups*
30 D.J. Benson *Representations and cohomology I*
31 D.J. Benson *Representations and cohomology II*
32 C. Allday & V. Puppe *Cohomological methods in transformation groups*
33 C. Soulé et al *Lectures on Arakelov geometry*
34 A. Ambrosetti & G. Prodi *A primer of nonlinear analysis*
35 J. Palis & F. Takens *Hyperbolicity, stability and chaos at homoclinic bifurcations*
37 Y. Meyer *Wavelets and operators I*
38 C. Weibel *An introduction to homological algebra*
39 W. Bruns & J. Herzog *Cohen-Macaulay rings*
40 V. Snaith *Explicit Brauer induction*
41 G. Laumon *Cohomology of Drinfield modular varieties I*
42 E.B. Davies *Spectral theory and differential operators*
43 J. Diestel, H. Jarchow & A. Tonge *Absolutely summing operators*
44 P. Mattila *Geometry of sets and measures in euclidean spaces*
45 R. Pinsky *Positive harmonic functions and diffusion*
46 G. Tenenbaum *Introduction to analytic and probabilistic number theory*
47 C. Peskine *An algebraic introduction to complex projective geometry I*
48 Y. Meyer & R. Coifman *Wavelets and operators II*
49 R. Stanley *Enumerative combinatorics I*
50 I. Porteous *Clifford algebras and the classical groups*
51 M. Audin *Spinning tops*
52 V. Jurdjevic *Geometric control theory*
53 H. Voelklein *Groups as Galois groups*
54 J. Le Potier *Lectures on vector bundles*
55 D. Bump *Automorphic forms*
56 G. Laumon *Cohomology of Drinfeld modular varieties II*
57 D.M. Clarke & B.A. Davey *Natural dualities for the working algebraist*
59 P. Taylor *Practical foundations of mathematics*
60 M. Brodmann & R. Sharp *Local cohomology*
61 J.D. Dixon, M.P.F. Du Sautoy, A. Mann & D. Segal *Analytic pro-p groups, 2nd edition*
62 R. Stanley *Enumerative combinatorics II*
64 J. Jost & X. Li-Jost *Calculus of variations*
68 Ken-iti Sato *Lévy processes and infinitely divisible distributions*
71 R. Blei *Analysis in integer and fractional dimensions*
72 F. Borceux & G. Janelidze *Galois theories*
73 B. Bollobás *Random graphs*
74 R.M. Dudley *Real analysis and probability*
75 T. Sheil-Small *Complex polynomials*
76 C. Voisin *Hodge theory and complex algebraic geometry I*
77 C. Voisin *Hodge theory and complex algebraic geometry II*
78 V. Paulsen *Completely bounded maps and operator algebra*
79 F. Gesztesy & H. Holden *Soliton equations and their algebro-geometric solutions I*
80 F. Gesztesy & H. Holden *Soliton equations and their algebro-geometric solutions II*

MULTIDIMENSIONAL REAL ANALYSIS II: INTEGRATION

J.J. DUISTERMAAT
J.A.C. KOLK
Utrecht University

Translated from Dutch by J. P. van Braam Houckgeest

CAMBRIDGE UNIVERSITY PRESS
Cambridge, New York, Melbourne, Madrid, Cape Town, Singapore, São Paulo

Cambridge University Press
The Edinburgh Building, Cambridge CB2 8RU, UK

Published in the United States of America by Cambridge University Press, New York

www.cambridge.org
Information on this title: www.cambridge.org/9780521829250

First published 2004

A catalogue record for this publication is available from the British Library

ISBN 978-0-521-82925-0 hardback

Transferred to digital printing 2008

To Saskia and Floortje

With Gratitude and Love

Contents

Preface

I prefer the open landscape under a clear sky with its depth of perspective, where the wealth of sharply defined nearby details gradually fades away towards the horizon.

This book, which is in two parts, provides an introduction to the theory of vector-valued functions on Euclidean space. We focus on four main objects of study and in addition consider the interactions between these. Volume I is devoted to differentiation. Differentiable functions on $\mathbf{R}^n$ come first, in Chapters 1 through 3. Next, differentiable manifolds embedded in $\mathbf{R}^n$ are discussed, in Chapters 4 and 5. In Volume II we take up integration. Chapter 6 deals with the theory of n-dimensional integration over $\mathbf{R}^n$. Finally, in Chapters 7 and 8 lower-dimensional integration over submanifolds of $\mathbf{R}^n$ is developed; particular attention is paid to vector analysis and the theory of differential forms, which are treated independently from each other. Generally speaking, the emphasis is on geometric aspects of analysis rather than on matters belonging to functional analysis.

In presenting the material we have been intentionally concrete, aiming at a thorough understanding of Euclidean space. Once this case is properly understood, it becomes easier to move on to abstract metric spaces or manifolds and to infinite-dimensional function spaces. If the general theory is introduced too soon, the reader might get confused about its relevance and lose motivation. Yet we have tried to organize the book as economically as we could, for instance by making use of linear algebra whenever possible and minimizing the number of ϵ–δ arguments, always without sacrificing rigor. In many cases, a fresh look at old problems, by ourselves and others, led to results or proofs in a form not found in current analysis textbooks. Quite often, similar techniques apply in different parts of mathematics; on the other hand, different techniques may be used to prove the same result. We offer ample illustration of these two principles, in the theory as well as the exercises.

A working knowledge of analysis in one real variable and linear algebra is a prerequisite; furthermore, familiarity with differentiable mappings and submanifolds of $\mathbf{R}^n$, as discussed in volume I, for instance. The main parts of the theory can be used as a text for an introductory course of one semester, as we have been doing for second-year students in Utrecht during the last decade. Sections at the end of many chapters usually contain applications that can be omitted in case of time constraints.

This volume contains 234 exercises, out of a total of 568, offering variations and applications of the main theory, as well as special cases and openings toward applications beyond the scope of this book. Next to routine exercises we tried also to include exercises that represent some mathematical idea. The exercises are independent from each other unless indicated otherwise, and therefore results are sometimes repeated. We have run student seminars based on a selection of the more challenging exercises.

In our experience, interest may be stimulated if from the beginning the student can perceive analysis as a subject intimately connected with many other parts of mathematics and physics: algebra, electromagnetism, geometry, including differential geometry, and topology, Lie groups, mechanics, number theory, partial differential equations, probability, special functions, to name the most important examples. In order to emphasize these relations, many exercises show the way in which results from the aforementioned fields fit in with the present theory; prior knowledge of these subjects is not assumed, however. We hope in this fashion to have created a landscape as preferred by Weyl,[1] thereby contributing to motivation, and facilitating the transition to more advanced treatments and topics.

[1] Weyl, H.: *The Classical Groups.* Princeton University Press, Princeton 1939, p. viii.

Acknowledgments

Since a text like this is deeply rooted in the literature, we have refrained from giving references. Yet we are deeply obliged to many mathematicians for publishing the results that we use freely. Many of our colleagues and friends have made important contributions: E. P. van den Ban, F. Beukers, R. H. Cushman, J. P. Hogendijk, W. L. J. van der Kallen, H. Keers, E. J. N. Looijenga, T. A. Springer, J. Stienstra, and in particular D. Zagier. We were also fortunate to have had the moral support of our special friend V. S. Varadarajan. Numerous small errors and stylistic points were picked up by students who attended our courses; we thank them all.

With regard to the manuscript's technical realization, the help of A. J. de Meijer and F. A. M. van de Wiel has been indispensable, with further contributions coming from K. Barendregt and J. Jaspers. We have to thank R. P. Buitelaar for assistance in preparing some of the illustrations. Without LaTeX, Y&Y TeX and Mathematica this work would never have taken on its present form.

J. P. van Braam Houckgeest translated the manuscript from Dutch into English. We are sincerely grateful to him for his painstaking attention to detail as well as his many suggestions for improvement.

We are indebted to S. J. van Strien to whose encouragement the English version is due; and furthermore to R. Astley and J. Walthoe, our editors, and to F. H. Nex, our copy-editor, for the pleasant collaboration; and to Cambridge University Press for making this work available to a larger audience.

Of course, errors still are bound to occur and we would be grateful to be told of them, at the e-mail address kolk@math.uu.nl. A listing of corrections will be made accessible through http://www.math.uu.nl/people/kolk.

Introduction

Motivation. Analysis came to life in the number space $\mathbf{R}^n$ of dimension n and its complex analog $\mathbf{C}^n$. Developments ever since have consistently shown that further progress and better understanding can be achieved by generalizing the notion of space, for instance to that of a manifold, of a topological vector space, or of a scheme, an algebraic or complex space having infinitesimal neighborhoods, each of these being defined over a field of characteristic which is 0 or positive. The search for unification by continuously reworking old results and blending these with new ones, which is so characteristic of mathematics, nowadays tends to be carried out more and more in these newer contexts, thus bypassing $\mathbf{R}^n$. As a result of this the uninitiated, for whom $\mathbf{R}^n$ is still a difficult object, runs the risk of learning analysis in several real variables in a suboptimal manner. Nevertheless, to quote F. and R. Nevanlinna: "The elimination of coordinates signifies a gain not only in a formal sense. It leads to a greater unity and simplicity in the theory of functions of arbitrarily many variables, the algebraic structure of analysis is clarified, and at the same time the geometric aspects of linear algebra become more prominent, which simplifies one's ability to comprehend the overall structures and promotes the formation of new ideas and methods".[2]

In this text we have tried to strike a balance between the concrete and the abstract: a treatment of integral calculus in the traditional $\mathbf{R}^n$ by efficient methods and using contemporary terminology, providing solid background and adequate preparation for reading more advanced works. The exercises are tightly coordinated with the theory, and most of them have been tried out during practice sessions or exams. Illustrative examples and exercises are offered in order to support and strengthen the reader's intuition.

Organization. This is the second volume, devoted to integration, of a book in two parts; the first volume treats differentiation. The volume at hand uses results from the preceding one, but it should be accessible to the reader who has acquired a working knowledge of differentiable mappings and submanifolds of $\mathbf{R}^n$. Only some of the exercises might require special results from Volume I.

In a subject like this with its many interrelations, the arrangement of the material is more or less determined by the proofs one prefers to or is able to give. Other ways of organizing are possible, but it is our experience that it is not such a simple matter to avoid confusing the reader. In particular, because the Change of Variables Theorem in the present volume is about diffeomorphisms, it is necessary to introduce these initially, in Volume I; a subsequent discussion of the Inverse Function Theorems then is a plausible inference. Next, applications in geometry, to the theory of differentiable manifolds, are natural. This geometry in its turn is indispensable for the description of the boundaries of the open sets that occur in this volume, in the Theorem on Integration of a Total Derivative in $\mathbf{R}^n$, the generalization to $\mathbf{R}^n$ of the

[2]Nevanlinna, F., Nevanlinna, R.: *Absolute Analysis.* Springer-Verlag, Berlin 1973, p. 1.

Fundamental Theorem of Integral Calculus on $\mathbf{R}$. This is why differentiation is treated in the first volume and integration in this second. Moreover, most known proofs of the Change of Variables Theorem require an Inverse Function, or the Implicit Function Theorem, as does our first proof. However, for the benefit of those readers who prefer a discussion of integration at an early stage, we have included a second proof of the Change of Variables Theorem by elementary means.

We have stuck to the (admittedly, old-fashioned) theory of Riemann integration. In our department students take a separate course on Lebesgue integration, where its essential role in establishing completeness in many function spaces is carefully discussed. For the topics in this book, however, the Lebesgue integral is not needed and introducing it would cause an overload. In the applications considered, Arzelà's Dominated Convergence Theorem, for which we give a short proof, is an effective alternative for Lebesgue's Dominated Convergence Theorem.

On some technical points. We have tried hard to reduce the number of ϵ–δ arguments, while maintaining a uniform and high level of rigor.

Even for linear coordinate transformations the Change of Variables Theorem is nontrivial, in contrast to the corresponding result in linear algebra. This stems from the fact that in linear algebra the behavior of volume under invertible linear transformations is usually part of the definition of volume. In analysis the notion of volume relies on the Riemann integral, and for the latter only invariance under translations is an immediate consequence of the definition.

The d-dimensional density on a d-dimensional submanifold in $\mathbf{R}^n$ is considered from two complementary points of view. On the one hand, the tangent space of the manifold can be mapped onto $\mathbf{R}^d \simeq \mathbf{R}^d \times \{0_{\mathbf{R}^{n-d}}\} \subset \mathbf{R}^n$ by means of a suitable orthogonal transformation; pulling back the d-volume on $\mathbf{R}^d$ under this mapping one then finds a d-density on the manifold. On the other hand, one can supplement the basis B_d for the tangent space by a set of mutually perpendicular unit vectors all of which are perpendicular to the tangent space, to form a basis B_n for $\mathbf{R}^n$. Next one defines the d-volume of the span of B_d to be the n-volume of the span of B_n (in other words, area equals volume divided by length). Both ways of thinking lead to the same formalism, which unifies the many different formulae that are in use.

Vector analysis should look familiar to students in physics: therefore we have chosen to center on the notion of vector field initially and on that of differential form only later on. Leitmotiv in our treatment of vector analysis is the generalization of the Fundamental Theorem of Integral Calculus on $\mathbf{R}$ to a theorem on $\mathbf{R}^n$. There are two aspects to the Fundamental Theorem of Integral Calculus on $\mathbf{R}$: the existence of an antiderivative for a continuous function; and the equality of the integral of a derivative of a function over an open set with the integral of the function itself over the boundary of that set. By generalizing the former aspect one arrives at the infinitesimal notions in vector analysis, like grad, curl, div; and at Poincaré's Lemma, and its relation with homotopy. Likewise, the latter aspect leads to the global notions, like the integral theorems, and their relations to homology.

This generalization to $\mathbf{R}^n$ begins with the Theorem on Integration of a Total

Derivative, for which an easy proof is offered, by means of a local substitution of variables that flattens the boundary. All other global theorems are reduced to this theorem.

The existence of an antiderivative (or potential) for a vector field on $\mathbf{R}^n$ with $n > 1$ requires integrability conditions to be satisfied. That is, one needs the vanishing of an obstruction against integrability, viz. of Af, twice the anti-adjoint part of the total derivative Df of the vector field f. In $\mathbf{R}^2$ and $\mathbf{R}^3$, Af essentially is the curl of f. Furthermore, Af approximately equals the sum of the values of f at the vertices of a parallelogram, and that sum in turn is a Riemann sum for a line integral of f along that parallelogram. Globalization of this argument leads to a rudimentary form of Stokes' Integral Theorem: a relation between the circulation of f and a surface integral of Af, i.e. an integral of the obstruction.

Vector analysis in $\mathbf{R}^n$ is not a study of partial derivatives of components of vector-valued functions, leading to a coordinate-dependent formulation and a "débauche d'indices". Rather, it is an investigation of these functions and of their total derivatives in their entirety, which is greatly facilitated by linear algebra, especially by the decomposition of the derivative into self-adjoint and anti-adjoint parts using adjoint linear operators.

The definition of positive orientation of a curve is an infinitesimal one. In concrete examples it is often easy to verify whether it is satisfied without an appeal to geometric intuition. The global definition, which is current in many elementary texts, is less rigorous and may lead to cumbersome formulations and/or proofs, of Green's and Stokes' Integral Theorems in particular.

Although formally the theory of differential forms receives an independent treatment, the stage for it is in fact set by much of the preceding material. The main result in the theory is Stokes' Theorem, and the whole discussion aims at proving that theorem at the earliest possible moment. Therefore we have adopted a definition of exterior derivative whereby we achieve this, and the proof of Stokes' Theorem itself is then presented as a direct generalization of the proof of the rudimentary form mentioned previously. The amount of multilinear algebra required for this has been reduced to a minimum. In particular, the general differential k-form is introduced by means of determinants instead of exterior multiplication of forms of lower order, which usually requires a laborious definition.

Exercises. Quite a few of the exercises are used to develop secondary but interesting themes omitted from the main course of lectures for reasons of time, but which often form the transition to more advanced theories. In many cases, exercises are strung together as projects which, step by easy step, lead the reader to important results. In order to set forth the interdependencies that inevitably arise, we begin an exercise by listing the other ones which (in total or in part only) are prerequisites as well as those exercises that use results from the one under discussion. The reader should not feel obliged to completely cover the preliminaries before setting out to work on subsequent exercises; quite often, only some terminology or minor results are required.

Notational conventions. Our notation is fairly standard, yet we mention the following conventions. Although it will often be convenient to write column vectors as row vectors, the reader should remember that all vectors are in fact column vectors, unless specified otherwise. Mappings always have precisely defined domains and images, thus $f : \mathrm{dom}(f) \to \mathrm{im}(f)$, but if we are unable, or do not wish, to specify the domain we write $f : \mathbf{R}^n \supset\!\to \mathbf{R}^p$ for a mapping that is well-defined on some subset of $\mathbf{R}^n$ and takes values in $\mathbf{R}^p$. We write $\mathbf{N}_0$ for $\{0\} \cup \mathbf{N}$, $\mathbf{N}_\infty$ for $\mathbf{N} \cup \{\infty\}$, and $\mathbf{R}_+$ for $\{ x \in \mathbf{R} \mid x > 0 \}$. The open interval $\{ x \in \mathbf{R} \mid a < x < b \}$ in $\mathbf{R}$ is denoted by $]\, a, b\, [$ and not by (a, b), in order to avoid confusion with the element $(a, b) \in \mathbf{R}^2$.

Making the notation consistent and transparent is difficult; in particular, every way of designating partial derivatives has its flaws. Whenever possible, we write $D_j f$ for the j-th column in a matrix representation of the total derivative Df of a mapping $f : \mathbf{R}^n \to \mathbf{R}^p$. This leads to expressions like $D_j f_i$ instead of Jacobi's classical $\frac{\partial f_i}{\partial x_j}$, etc. The convention just mentioned has not been applied dogmatically; in the case of special coordinate systems like spherical coordinates, Jacobi's notation is the one of preference. As a further complication, D_j is used by many authors, especially in Fourier theory, for the momentum operator $\frac{1}{\sqrt{-1}} \frac{\partial}{\partial x_j}$.

We use the following dictionary of symbols to indicate the ends of various items:

❑ Proof
❍ Definition
✩ Example

Chapter 6

Integration

In this chapter we extend to $\mathbf{R}^n$ the theory of the Riemann integral from the calculus in one real variable. Principal results are a reduction of n-dimensional integration to successive one-dimensional integrations, and the Change of Variables Theorem. For this fundamental theorem we give three proofs: one in the main text and two in the appendix to this chapter. Important technical tools are the theorems from Chapter 3 and partitions of unity over compact sets. As applications we treat Fourier transformation, i.e. the decomposition of arbitrary functions into periodic ones; and dominated convergence, being a sufficient condition for the interchange of limits and integration.

6.1 Rectangles

Definition 6.1.1. An *n-dimensional rectangle B*, parallel to the coordinate axes, is a subset of $\mathbf{R}^n$ of the form

$$B = \{ x \in \mathbf{R}^n \mid a_j \le x_j \le b_j \ \ (1 \le j \le n) \}, \tag{6.1}$$

where it is assumed that $a_j, b_j \in \mathbf{R}$ and $a_j \le b_j$, for $1 \le j \le n$, compare with Definition 1.8.18.

The *n-dimensional volume* of B, notation $\mathrm{vol}_n(B)$, is defined as

$$\mathrm{vol}_n(B) = \prod_{1 \le j \le n} (b_j - a_j).$$

Note that $\mathrm{vol}_n(B) = 0$ if there exists a j with $a_j = b_j$, that is, if B is contained in an $(n-1)$-dimensional hyperplane in $\mathbf{R}^n$, of the form $\{ x \in \mathbf{R}^n \mid x_j = a_j \}$.

A *partition* of a rectangle B is a finite collection $\mathcal{B} = \{ B_i \mid i \in I \}$ (here I is called the *index set* of $\mathcal{B}$) of n-dimensional rectangles B_i such that

$$B = \bigcup_{i \in I} B_i; \qquad B_i \cap B_j = \emptyset \qquad \text{or} \qquad \mathrm{vol}_n(B_i \cap B_j) = 0 \qquad \text{if} \qquad i \neq j. \tag{6.2}$$

Let $\mathcal{B}$ and $\mathcal{B}'$ be partitions of a rectangle B, then $\mathcal{B}'$ is said to be a *refinement* of $\mathcal{B}$ if for every $B_i \in \mathcal{B}$ the $B'_j \in \mathcal{B}'$ with $B'_j \subset B_i$ form a partition of B_i. ❍

Proposition 6.1.2. *Assume* $\{ B_i \mid i \in I \}$ *is a partition of a rectangle* $B \subset \mathbf{R}^n$. *Then*

$$\mathrm{vol}_n(B) = \sum_{i \in I} \mathrm{vol}_n(B_i).$$

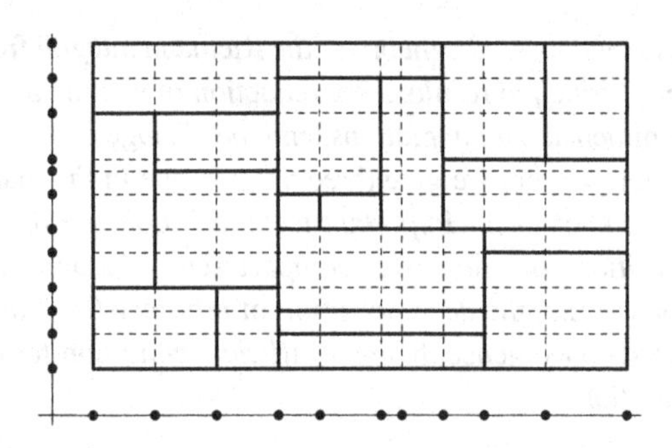

Illustration for the proof of Proposition 6.1.2

Proof. We first prove two auxiliary results.
(i). Assume B as in (6.1); and for $1 \le j \le n$, let $t_j \in [\, a_j, b_j \,]$ be arbitrary. Consider

$$B' = \{ x \in \mathbf{R}^n \mid a_j \le x_j \le t_j, \text{ and } a_k \le x_k \le b_k, \text{ for } k \neq j \},$$
$$B'' = \{ x \in \mathbf{R}^n \mid t_j \le x_j \le b_j, \text{ and } a_k \le x_k \le b_k, \text{ for } k \neq j \}.$$

Because $b_j - a_j = (b_j - t_j) + (t_j - a_j)$, it follows straight away that $\mathrm{vol}_n(B) = \mathrm{vol}_n(B') + \mathrm{vol}_n(B'')$.
(ii). Assume next that for every $1 \le j \le n$ the segment $[\, a_j, b_j \,]$ is subdivided by the intermediate points

$$a_j = t_j^{(0)} \le \cdots \le t_j^{(N(j))} = b_j. \tag{6.3}$$

Then we have, for every n-tuple

$$\alpha = (\alpha(1), \ldots, \alpha(n)) \in \mathbf{N}^n \qquad \text{where} \qquad 1 \le \alpha(j) \le N(j), \tag{6.4}$$

a rectangle

$$B_\alpha = \{ x \in \mathbf{R}^n \mid t_j^{(\alpha(j)-1)} \le x_j \le t_j^{(\alpha(j))} \ (1 \le j \le n) \}.$$

Together the B_α, with α as in (6.4), form a partition of B; and successive application of assertion (i) now gives

$$\mathrm{vol}_n(B) = \sum_\alpha \mathrm{vol}_n(B_\alpha). \tag{6.5}$$

Finally, let $\{B_i\}$ be the given partition of B. For every $1 \le j \le n$, collect the endpoints of the j-th coordinate interval of all rectangles B_i, and arrange these points in increasing order, as in (6.3) (see the illustration). Thus one obtains a partition $\{B_\alpha\}$ of B for which (6.5) holds. Likewise, the B_α with $B_\alpha \subset B_i$ form a partition of B_i for which

$$\mathrm{vol}_n(B_i) = \sum_{\{\alpha \mid B_\alpha \subset B_i\}} \mathrm{vol}_n(B_\alpha).$$

Therefore

$$\sum_{i \in I} \mathrm{vol}_n(B_i) = \sum_{i \in I} \sum_{\{\alpha \mid B_\alpha \subset B_i\}} \mathrm{vol}_n(B_\alpha) = \sum_\alpha \mathrm{vol}_n(B_\alpha) = \mathrm{vol}_n(B).$$

Note that in the second summation $\mathrm{vol}_n(B_\alpha)$ may occur more than once, specifically when $B_\alpha \subset B_i$ and when $B_\alpha \subset B_j$, for $i \neq j$. But this need not concern us, because then $\mathrm{vol}_n(B_\alpha) = 0$ on account of (6.2). ❑

Lemma 6.1.3. *Any two partitions $\mathcal{B}'$ and $\mathcal{B}''$ of a rectangle B possess a common refinement.*

Proof. Analogous to the proof of the foregoing proposition. For every $1 \le j \le n$, collect the endpoints of the j-th coordinate interval of all subrectangles. ❑

6.2 Riemann integrability

Throughout this section we shall assume B to be an n-dimensional rectangle and $f : B \to \mathbf{R}$ to be a bounded function.

Definition 6.2.1. For every partition $\mathcal{B} = \{B_i \mid i \in I\}$ of B we define the *lower sum* and the *upper sum of* f determined by $\mathcal{B}$, notation $\underline{S}(f, \mathcal{B})$ and $\overline{S}(f, \mathcal{B})$, respectively, by

$$\underline{S}(f, \mathcal{B}) = \sum_{i \in I} \inf_{x \in B_i} f(x) \ \mathrm{vol}_n(B_i), \qquad \overline{S}(f, \mathcal{B}) = \sum_{i \in I} \sup_{x \in B_i} f(x) \ \mathrm{vol}_n(B_i). \quad ❍$$

Lemma 6.2.2. (i) *If $\mathcal{B}'$ is a refinement of $\mathcal{B}$, then* $\underline{S}(f, \mathcal{B}) \le \underline{S}(f, \mathcal{B}') \le \overline{S}(f, \mathcal{B}') \le \overline{S}(f, \mathcal{B})$.

(ii) *For any two partitions $\mathcal{B}$ and $\mathcal{B}'$ of B one has* $\underline{S}(f, \mathcal{B}) \le \overline{S}(f, \mathcal{B}')$.

Proof. If $B_j' \subset B_i$, then $\inf_{x\in B_i} f(x) \le \inf_{x\in B_j'} f(x)$. Consequently one has, by Proposition 6.1.2,

$$\begin{aligned}\underline{S}(f, \mathcal{B}) &= \sum_{i\in I} \inf_{x\in B_i} f(x)\, \mathrm{vol}_n(B_i) = \sum_{i\in I} \inf_{x\in B_i} f(x) \sum_{\{\,j\mid B_j'\subset B_i\,\}} \mathrm{vol}_n(B_j') \\ &\le \sum_{i\in I} \sum_{\{\,j\mid B_j'\subset B_i\,\}} \inf_{x\in B_j'} f(x)\, \mathrm{vol}_n(B_j') = \sum_{j\in J} \inf_{x\in B_j'} f(x)\, \mathrm{vol}_n(B_j') \\ &= \underline{S}(f, \mathcal{B}'),\end{aligned}$$

and this proves (i). Assertion (ii) follows from (i) and Lemma 6.1.3. ❑

Definition 6.2.3. The *lower Riemann integral* $\underline{\int}_B f(x)\,dx$ and the *upper Riemann integral* $\overline{\int}_B f(x)\,dx$ *of f over B* are defined by

$$\underline{\int}_B f(x)\,dx = \sup\{\, \underline{S}(f, \mathcal{B}) \mid \mathcal{B} \text{ partition of } B\,\},$$

$$\overline{\int}_B f(x)\,dx = \inf\{\, \overline{S}(f, \mathcal{B}) \mid \mathcal{B} \text{ partition of } B\,\}.$$ ❍

From the preceding lemma follows, for every partition $\mathcal{B}$ of B,

$$\underline{S}(f, \mathcal{B}) \le \underline{\int}_B f(x)\,dx \le \overline{\int}_B f(x)\,dx \le \overline{S}(f, \mathcal{B}).$$

Definition 6.2.4. Let B be an n-dimensional rectangle. A function $f : B \to \mathbf{R}$ is said to be *Riemann integrable over B* if

$$f \text{ is bounded on } B \qquad \text{and} \qquad \underline{\int}_B f(x)\,dx = \overline{\int}_B f(x)\,dx.$$

The common value of the lower and upper Riemann integrals is called the *integral* of f over B, written as

$$\int_B f(x)\,dx.$$ ❍

Proposition 6.2.5. *Let $f : B \to \mathbf{R}$ be a bounded function. Then the following are equivalent.*

(i) *f is Riemann integrable over B.*

(ii) *For every $\epsilon > 0$ there exists a partition $\mathcal{B}$ of B such that $\overline{S}(f, \mathcal{B}) - \underline{S}(f, \mathcal{B}) < \epsilon$.*

Proof. (i) ⇒ (ii). For every $\epsilon > 0$ there is a partition $\mathcal{B}'$ of B such that $\overline{S}(f, \mathcal{B}') < \int_B f(x)\,dx + \frac{\epsilon}{2}$, and another partition $\mathcal{B}''$ of B such that $\underline{S}(f, \mathcal{B}'') > \int_B f(x)\,dx - \frac{\epsilon}{2}$. For the common refinement $\mathcal{B}$ of $\mathcal{B}'$ and $\mathcal{B}''$ it follows immediately that

$$\overline{S}(f, \mathcal{B}) - \underline{S}(f, \mathcal{B}) \leq \overline{S}(f, \mathcal{B}') - \underline{S}(f, \mathcal{B}'') < \epsilon.$$

(ii) ⇒ (i). The existence of such a partition $\mathcal{B}$ implies $\overline{\int}_B f(x)\,dx - \underline{\int}_B f(x)\,dx < \epsilon$; and this holds for every $\epsilon > 0$. Hence follows (i). ❑

Now let $f : \mathbf{R}^n \to \mathbf{R}$ be a function satisfying

$$f \text{ is bounded on } \mathbf{R}^n \text{ and zero outside a bounded subset of } \mathbf{R}^n. \tag{6.6}$$

Then there exists a rectangle $B \subset \mathbf{R}^n$ with

$$f(x) = 0 \qquad \text{if} \qquad x \notin B. \tag{6.7}$$

If f is Riemann integrable over B, the number $\int_B f(x)\,dx$ is independent of the choice of B, provided that (6.7) is satisfied. Indeed, let B' be another such rectangle. Then $B = (B \cap B') \cup (B \setminus B')$, where $B \cap B'$ is an n-dimensional rectangle, while the closure of $B \setminus B'$ can be written as a finite union of rectangles B_i, for $i \in I$, satisfying (6.2). Consequently, $\{B \cap B'\} \cup \{B_i \mid i \in I\}$ is a partition of B. One furthermore has $\int_{B_i} f(x)\,dx = 0$ for $i \in I$, because $f(x) = 0$ for $x \in \mathrm{int}(B_i) \subset \mathbf{R}^n \setminus B'$. Here $\mathrm{int}(B)$, the interior of a rectangle B, is obtained by only allowing the inequality signs in (6.1). Therefore

$$\int_B f(x)\,dx = \int_{B \cap B'} f(x)\,dx = \int_{B'} f(x)\,dx.$$

We need the following:

Definition 6.2.6. The *support* of a function $f : \mathbf{R}^n \to \mathbf{R}$ is defined as the set, see Definition 1.2.9,

$$\mathrm{supp}(f) = \overline{\{x \in \mathbf{R}^n \mid f(x) \neq 0\}}. \qquad ❍$$

Note that $\mathrm{supp}(f)$ is compact (see the Heine–Borel Theorem 1.8.17) if $\mathrm{supp}(f)$ is bounded. If $x \notin \mathrm{supp}(f)$, then there exists a neighborhood U of x such that $f(y) = 0$, for all $y \in U$.

Definition 6.2.7. Let $\mathcal{R}(\mathbf{R}^n)$ be the collection of functions $f : \mathbf{R}^n \to \mathbf{R}$ which satisfy (6.6) and which are Riemann integrable over B, for a rectangle $B \subset \mathbf{R}^n$ as in (6.7). The elements from $\mathcal{R}(\mathbf{R}^n)$ are called *Riemann integrable functions with compact support*. Define, for every $f \in \mathcal{R}(\mathbf{R}^n)$,

$$\int_{\mathbf{R}^n} f(x)\,dx = \int f(x)\,dx = \int_B f(x)\,dx. \qquad \bigcirc$$

We recall the definition of the *characteristic function* 1_A of a subset $A \subset \mathbf{R}^n$, with

$$1_A(x) = 1 \quad \text{if} \quad x \in A, \qquad 1_A(x) = 0 \quad \text{if} \quad x \notin A.$$

It follows immediately that the characteristic function 1_B of an n-dimensional rectangle B is a Riemann integrable function with compact support. In addition

$$\int_{\mathbf{R}^n} 1_B(x)\,dx = \mathrm{vol}_n(B).$$

Theorem 6.2.8. *We have the following properties for* $\mathcal{R}(\mathbf{R}^n)$.

(i) $\mathcal{R}(\mathbf{R}^n)$ *is a linear space with pointwise addition and multiplication by a scalar.*

(ii) *The mapping* $I : \mathcal{R}(\mathbf{R}^n) \to \mathbf{R}$ *with* $f \mapsto \int_{\mathbf{R}^n} f(x)\,dx$ *is linear, and monotonic, that is, if* $f \le g$ *(that is,* $f(x) \le g(x)$*, for all* $x \in \mathbf{R}^n$*), then* $I(f) \le I(g)$.

(iii) $f \in \mathcal{R}(\mathbf{R}^n)$ *if and only if* f_+ *and* $f_- \in \mathcal{R}(\mathbf{R}^n)$*, where*

$$f_\pm = \frac{1}{2}(|f| \pm f) \ge 0, \qquad f = f_+ - f_-, \qquad |f| = f_+ + f_-;$$

and one has $I(f) = I(f_+) - I(f_-)$. *In particular we have* $|f| \in \mathcal{R}(\mathbf{R}^n)$ *if* $f \in \mathcal{R}(\mathbf{R}^n)$*, while*

$$\left| \int_{\mathbf{R}^n} f(x)\,dx \right| \le \int_{\mathbf{R}^n} |f(x)|\,dx.$$

(iv) *If* $f, g \in \mathcal{R}(\mathbf{R}^n)$*, then* $fg \in \mathcal{R}(\mathbf{R}^n)$*, with* fg *defined by pointwise multiplication.*

Proof. Parts (i) and (ii) readily follow. With respect to (iii) we note that, for a rectangle B,

$$\begin{aligned} \sup_B fg - \inf_B fg &\le \sup_B f \, \sup_B g - \inf_B f \, \inf_B g \\ &= (\sup_B f - \inf_B f)\, \sup_B g + \inf_B f\, (\sup_B g - \inf_B g). \end{aligned}$$

to see this, distinguish the cases $0 \le \inf_B f$, $\inf_B f < 0 < \sup_B f$ and $\sup_B f \le 0$. It follows that $f \in \mathcal{R}(\mathbf{R}^n)$ if and only if $f_+ \in \mathcal{R}(\mathbf{R}^n)$ and $f_- \in \mathcal{R}(\mathbf{R}^n)$; and if this is the case, then $I(f) = I(f_+) - I(f_-)$. The particular case of $|f|$ now follows from $|f| = f_+ + f_-$; and the inequality of integrals is seen to result from the estimates $f \le |f|$ and $-f \le |f|$.
(iv). Because of $fg = f_+ g_+ - f_- g_+ - f_+ g_- + f_- g_-$, it suffices to prove the assertion for the case where $f \ge 0$ and $g \ge 0$. We then have

$$\begin{aligned} \sup_B fg - \inf_B fg &\le \sup_B f \ \sup_B g - \inf_B f \ \inf_B g \\ &= (\sup_B f - \inf_B f) \ \sup_B g + \inf_B f \ (\sup_B g - \inf_B g). \end{aligned}$$

The Riemann integrability of fg now easily follows. ❑

6.3 Jordan measurability

Definition 6.3.1. Let $A \subset \mathbf{R}^n$ be bounded. Let $f : \mathbf{R}^n \to \mathbf{R}$ be a function and assume that f is bounded on A. Then f is said to be *Riemann integrable over* A if $f\,1_A$ is a function in the space $\mathcal{R}(\mathbf{R}^n)$. If such is the case we write

$$\int_A f(x)\,dx := \int_{\mathbf{R}^n} 1_A(x)\,f(x)\,dx,$$

and we speak of the *Riemann integral* or simply the *integral* of f over A.

The lower and upper Riemann integrals of 1_A are said to be the *inner* and *outer measures*, respectively, of A. If 1_A is a function in the space $\mathcal{R}(\mathbf{R}^n)$, then A is said to be *Jordan measurable*, while

$$\mathrm{vol}_n(A) := \int_A dx = \int_{\mathbf{R}^n} 1_A(x)\,dx$$

is said to be the *n-dimensional volume* or the *n-dimensional Jordan measure* of A. In particular, therefore, an n-dimensional rectangle B is a Jordan measurable set.

The set A is said to be *negligible in* $\mathbf{R}^n$ if $\mathrm{vol}_n(A) = 0$. ❍

Let $A \subset \mathbf{R}^n$ be a bounded subset. Let $B \subset \mathbf{R}^n$ be a rectangle with $A \subset B$ and let $\mathcal{B} = \{B_i \mid i \in I\}$ be a partition of B. Define

$$I_{e(xterior)} = \{\, i \in I \mid B_i \cap A \ne \emptyset \,\}, \qquad I_{i(nterior)} = \{\, i \in I \mid B_i \subset A \,\}.$$

Since $\sup_{B_i} 1_A = 1$ if $i \in I_e$, and $\inf_{B_i} 1_A = 1$ if $i \in I_i$, it follows that

$$\begin{aligned} \overline{S}(1_A,\, \mathcal{B}) - \underline{S}(1_A,\, \mathcal{B}) &= \sum_{i \in I_e} \mathrm{vol}_n(B_i) - \sum_{i \in I_i} \mathrm{vol}_n(B_i) \\ &= \sum_{\{i \in I \mid B_i \cap A \ne \emptyset \text{ and } B_i \setminus A \ne \emptyset\}} \mathrm{vol}_n(B_i). \end{aligned} \tag{6.8}$$

Theorem 6.3.2. *Let $A \subset \mathbf{R}^n$ be a bounded subset. Then the following assertions are equivalent.*

(i) *A is Jordan measurable.*

(ii) *∂A is negligible.*

Proof. **(i) $\Rightarrow$ (ii).** Let $\epsilon > 0$ be arbitrary. According to Proposition 6.2.5 there exists a partition $\mathcal{B} = \{B_i \mid i \in I\}$ of B such that (6.8) leads to

$$\sum_{i \in I_e} \mathrm{vol}_n(B_i) - \sum_{i \in I_i} \mathrm{vol}_n(B_i) < \frac{\epsilon}{2}.$$

Since $\bigcup_{i \in I_e} B_i$ is closed, being a finite union of closed sets, and since $\overline{A}$ is the smallest closed set containing A, it follows that

$$\overline{A} \subset \bigcup_{i \in I_e} B_i. \tag{6.9}$$

Replacing the rectangles B_i with $i \in I_i$ by similar rectangles B_i' whose edges are of somewhat smaller length, we can arrange that

$$A \supset \bigcup_{i \in I_i} B_i \supset \bigcup_{i \in I_i} \mathrm{int}(B_i) \supset \bigcup_{i \in I_i} B_i' \tag{6.10}$$

and

$$\sum_{i \in I_e} \mathrm{vol}_n(B_i) - \sum_{i \in I_i} \mathrm{vol}_n(B_i') < \epsilon. \tag{6.11}$$

Because $\bigcup_{i \in I_i} \mathrm{int}(B_i)$ is open, being a union of open sets, and because $\mathrm{int}(A)$ is the largest open set contained in A, (6.10) leads to

$$\mathrm{int}(A) \supset \bigcup_{i \in I_i} B_i'. \tag{6.12}$$

From (1.4), (6.9) and (6.12) we now obtain

$$\partial A = \overline{A} \setminus \mathrm{int}(A) \subset \bigcup_{i \in I_e} B_i \setminus \bigcup_{i \in I_i} B_i'. \tag{6.13}$$

But (6.13) and (6.11) together imply that the outer measure of ∂A is smaller than ϵ.

(ii) $\Rightarrow$ (i). For this proof we also apply Proposition 6.2.5. Hence let $\epsilon > 0$ be arbitrary. Then we can find a partition $\{B_i \mid i \in I\}$ of B such that

$$\partial A \subset \bigcup_{i \in I} \mathrm{int}(B_i) \qquad \text{and} \qquad \sum_{i \in I} (\sup_{B_i} 1_A - \inf_{B_i} 1_A)\, \mathrm{vol}_n(B_i) < \epsilon.$$

For every $x \in A \setminus \partial A$, we can select a rectangle B_x satisfying $x \in \mathrm{int}(B_x) \subset B_x \subset A$. Note that $\sup_{B_x} 1_A - \inf_{B_x} 1_A = 0$. Because $\overline{A}$ is compact, the open covering

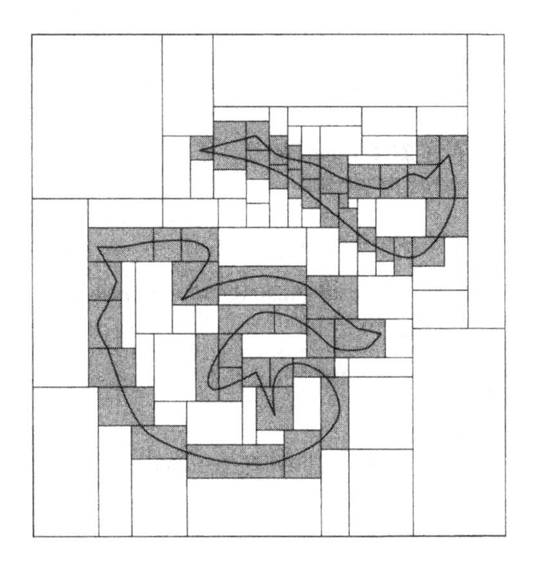

Illustration for Formula (6.13)

$\{ B_i \mid i \in I \} \cup \{ B_x \mid x \in A \setminus \partial A \}$ of $\overline{A}$ admits a finite subcovering, say $\mathcal{B}'$, on the strength of the Heine–Borel Theorem 1.8.17. Next we use the endpoints of the coordinate intervals of the rectangles in $\mathcal{B}'$ to define a partition $\mathcal{B}$ of B; then each of the rectangles in $\mathcal{B}'$ is a union of rectangles in $\mathcal{B}$. It is immediate now that $\overline{S}(f, \mathcal{B}) - \underline{S}(f, \mathcal{B}) < \epsilon$. ❑

Corollary 6.3.3. *Assume A and $B \subset \mathbf{R}^n$ are bounded and Jordan measurable. Then the following sets are also bounded and Jordan measurable:*

$$A \cap B, \qquad A \cup B, \qquad A \setminus B, \qquad \text{int}(A), \qquad \overline{A}.$$

Proof. Note that $1_{A \cap B} = 1_A 1_B$, $1_{A \cup B} = 1_A + 1_B - 1_{A \cap B}$, and $1_{A \setminus B} = 1_A - 1_{A \cap B}$. Now use Theorem 6.2.8.(i) and (iv). According to Formula (1.4) one has $\text{int}\, A = A \setminus \partial A$ and $\overline{A} = A \cup \partial A$. ❑

Remark. In what follows we shall often be dealing with compact subsets of $\mathbf{R}^n$. Such sets are not necessarily Jordan measurable (see Exercise 6.1 for an example in $\mathbf{R}$).

Definition 6.3.4. For $A \subset \mathbf{R}^n$ we denote by $\mathcal{J}(A)$ the collection of compact and Jordan measurable subsets of A. ❍

The following theorem gives a most useful criterion for the Riemann integrability of a function over a set.

Theorem 6.3.5. *Let $K \in \mathcal{J}(\mathbf{R}^n)$ and let $f : K \to \mathbf{R}$ be a continuous function. Then we have the following properties.*

(i) *f is Riemann integrable over K.*

(ii) *Next, assume that $g : K \to \mathbf{R}$ is also continuous and that $g(x) \leq f(x)$, for all $x \in K$. Then $L \in \mathcal{J}(\mathbf{R}^{n+1})$ if*

$$L = \{ (x, y) \in \mathbf{R}^{n+1} \mid x \in K,\ g(x) \leq y \leq f(x) \};$$

$$\text{and} \qquad \mathrm{vol}_{n+1}(L) = \int_K (f(x) - g(x))\, dx.$$

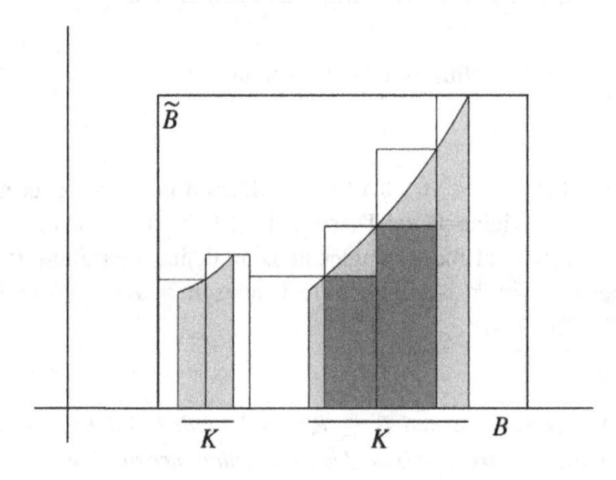

Illustration for the proof of Theorem 6.3.5.(ii)

Proof. Considering Theorem 6.2.8.(iii), it is sufficient to prove (i) for the case

$$f(x) \geq 0 \qquad (x \in K). \tag{6.14}$$

Further, let $c = \min_{x \in K} g(x)$. Since $c \leq g(x) \leq f(x)$, for $x \in K$, we have

$$\begin{aligned} L &= \{ (x, y) \in \mathbf{R}^{n+1} \mid x \in K,\ c \leq y \leq f(x) \} \\ &\quad \setminus \{ (x, y) \in \mathbf{R}^{n+1} \mid x \in K,\ c \leq y < g(x) \}. \end{aligned} \tag{6.15}$$

In view of this we first prove (ii) for the case where

$$L = \{ (x, y) \in \mathbf{R}^{n+1} \mid x \in K,\ 0 \leq y \leq f(x) \}. \tag{6.16}$$

Under the assumptions (6.14) and (6.16) we will now simultaneously prove both (i) and (ii). Let $B \subset \mathbf{R}^n$ be a rectangle with $K \subset B$ and let $\mathcal{B} = \{B_i \mid i \in I\}$ be a partition of B. Define $f(x) = 0$, if $x \in B \setminus K$. Then

$$\overline{S}(1_K\, f,\ \mathcal{B}) = \sum_{i \in I} \sup_{x \in B_i} 1_K(x)\, f(x)\, \mathrm{vol}_n(B_i) = \sum_{B_i \cap K \neq \emptyset} \sup_{B_i} f\, \mathrm{vol}_n(B_i),$$

$$\underline{S}(1_K\, f,\ \mathcal{B}) = \sum_{i \in I} \inf_{x \in B_i} 1_K(x)\, f(x)\, \mathrm{vol}_n(B_i) = \sum_{B_i \subset K} \inf_{B_i} f\, \mathrm{vol}_n(B_i).$$

Further, we note that $L \subset \widetilde{B} := B \times [\,0,\ \sup_K f\,]$, where $\widetilde{B}$ is a rectangle in $\mathbf{R}^{n+1}$. In addition

$$L \subset \bigcup_{B_i \cap K \neq \emptyset} B_i \times [\,0,\ \sup_{B_i} f\,], \tag{6.17}$$

and, if $B_i \subset K$,

$$B_i \times [\,0,\ \inf_{B_i} f\,] \subset L. \tag{6.18}$$

Let $\widetilde{\mathcal{B}}$ be a partition of $\widetilde{B}$ such that it is a common refinement of all rectangles $B_i \times [\,0,\ \sup_{B_i} f\,]$, for $B_i \cap K \neq \emptyset$, and all $B_i \times [\,0,\ \inf_{B_i} f\,]$, for $B_i \subset K$. From (6.17) and (6.18), respectively, one then finds

$$\overline{S}(1_L,\ \widetilde{\mathcal{B}}) \le \sum_{B_i \cap K \neq \emptyset} \sup_{B_i} f\, \mathrm{vol}_n(B_i) = \overline{S}(1_K f,\ \mathcal{B}),$$

$$\underline{S}(1_L,\ \widetilde{\mathcal{B}}) \ge \sum_{B_i \subset K} \inf_{B_i} f\, \mathrm{vol}_n(B_i) = \underline{S}(1_K f,\ \mathcal{B}).$$

Accordingly, (i) and (ii) result if we prove that the following difference becomes small upon a suitable choice of the partition $\mathcal{B}$:

$$\overline{S}(1_K f,\ \mathcal{B}) - \underline{S}(1_K f,\ \mathcal{B}) = \sum_{B_i \cap K \neq \emptyset} \sup_{B_i} f\, \mathrm{vol}_n(B_i) - \sum_{B_i \subset K} \inf_{B_i} f\, \mathrm{vol}_n(B_i)$$

$$= \sum_{B_i \subset K} (\sup_{B_i} f - \inf_{B_i} f)\, \mathrm{vol}_n(B_i) + \sum_{B_i \cap K \neq \emptyset,\ B_i \setminus K \neq \emptyset} \sup_{B_i} f\, \mathrm{vol}_n(B_i).$$

By Theorem 1.8.15, f is uniformly continuous on the compact set K, that is, for every $\eta > 0$ there exists a $\delta > 0$ such that $\sup_{B_i} f - \inf_{B_i} f < \eta$, if $B_i \subset K$ and B_i has edges of length $< \delta$. This may be arranged for all B_i by making them smaller where necessary. Because of Formula (6.8), the Jordan measurability of K implies

$$\sum_{B_i \cap K \neq \emptyset,\ B_i \setminus K \neq \emptyset} \mathrm{vol}_n(B_i) < \eta,$$

for suitably chosen $\mathcal{B}$. And hence, for such a $\mathcal{B}$,

$$\overline{S}(1_K f,\ \mathcal{B}) - \underline{S}(1_K f,\ \mathcal{B}) < \eta\, \mathrm{vol}_n(K) + \max_{x \in K} f(x)\, \eta.$$

We return to the general case for (ii). This is a consequence of the foregoing and of (6.15), once we know that the following sets M and M' have the same $(n+1)$-dimensional volume, where

$$M := \{ (x, y) \in \mathbf{R}^{n+1} \mid x \in K, \ c \leq y \leq g(x) \}$$

and

$$M' := \{ (x, y) \in \mathbf{R}^{n+1} \mid x \in K, \ c \leq y < g(x) \}.$$

But this is the case, because $\mathrm{vol}_{n+1}(M \setminus M') = 0$ on the basis of the inclusion $M \setminus M' \subset \partial M$ and the preceding theorem. ❑

Definition 6.3.6. We denote the space of *continuous functions* $\mathbf{R}^n \to \mathbf{R}$ with *compact support* by

$$C_c(\mathbf{R}^n) = C_c^0(\mathbf{R}^n) = \{ f : \mathbf{R}^n \to \mathbf{R} \mid f \text{ continuous and } \operatorname{supp}(f) \text{ compact} \}. \quad ❍$$

Corollary 6.3.7. $C_c(\mathbf{R}^n) \subset \mathcal{R}(\mathbf{R}^n)$, *that is, the linear space of continuous functions on* $\mathbf{R}^n$ *with compact support is contained in the linear space of the Riemann integrable functions on* $\mathbf{R}^n$ *with compact support.*

Proof. Let $B \subset \mathbf{R}^n$ be a rectangle with $\operatorname{supp}(f) \subset B$, then f is continuous on B. Hence, according to (i) of the preceding theorem, f is Riemann integrable over B. ❑

Corollary 6.3.8. *Let* $K \in \mathcal{J}(\mathbf{R}^d)$, *for* $d < n$, *and let* $f : K \to \mathbf{R}^{n-d}$ *be a continuous mapping. Then*

$$\operatorname{graph}(f) = \{ (x, \ f(x)) \mid x \in K \}$$

is a negligible set in $\mathbf{R}^n$. *More generally, finite unions of this kind of graph are negligible in* $\mathbf{R}^n$.

Proof. The boundedness of the continuous function $x \mapsto (f_1(x), \dots, f_{n-d-1}(x))$ on K implies there is a rectangle $B \subset \mathbf{R}^{n-d-1}$ such that $(f_1(x), \dots, f_{n-d-1}(x)) \in B$ if $x \in K$. Defining the continuous function $\widetilde{f} : K \times B \to \mathbf{R}$ by $\widetilde{f}(x, t) = f_{n-d}(x)$, we have

$$\begin{aligned} \operatorname{graph}(f) &= \{ (x, \ f(x)) \in \mathbf{R}^n \mid x \in K \} \\ &\subset \{ (x, \ t, \ \widetilde{f}(x, t)) \in \mathbf{R}^n \mid (x, t) \in K \times B \} = \operatorname{graph}(\widetilde{f}). \end{aligned}$$

Now let $c = \min_{(x,t) \in K \times B} \widetilde{f}(x, t)$. Then

$$\operatorname{graph}(f) \subset \partial \{ (x, t, y) \in \mathbf{R}^n \mid (x, t) \in K \times B, \ c \leq y \leq \widetilde{f}(x, t) \},$$

and so the assertion follows from the two preceding theorems. ❑

Example 6.3.9. (i) The interval $[-1, 1]$ belongs to $\mathcal{J}(\mathbf{R})$, because it is a rectangle in $\mathbf{R}$.
(ii) The circular disk $B^2 = \{ x \in \mathbf{R}^2 \mid \|x\| \le 1 \}$ belongs to $\mathcal{J}(\mathbf{R}^2)$. Indeed

$$B^2 = \{ x \in \mathbf{R}^2 \mid x_1 \in [-1, 1], \; -f(x_1) \le x_2 \le f(x_1) \},$$

where $f : [-1, 1] \to \mathbf{R}$ is the continuous function with $f(x_1) = \sqrt{1 - x_1^2}$. The assertion therefore follows from (ii) of Theorem 6.3.5.
(iii) The subset K of a solid cone in $\mathbf{R}^3$ defined by

$$K = \{ x \in \mathbf{R}^3 \mid 0 \le x_3 \le 1, \; x_1^2 + x_2^2 \le (1 - x_3)^2 \}$$

is compact; indeed, for $x \in K$ one has $x_1^2 + x_2^2 \le 1$, therefore $K \subset \{ x \in \mathbf{R}^3 \mid |x_i| \le 1 \; (1 \le i \le 3) \}$. Furthermore, K is a Jordan measurable set in $\mathbf{R}^3$. This is so because $(x_1, x_2, x_3) \in K$ implies that $(x_1, x_2) \in B^2$. Conversely, with $(x_1, x_2) \in B^2$ fixed, the inequalities $\sqrt{x_1^2 + x_2^2} \le 1 - x_3$ and $0 \le x_3$ imply that then x_3 may still vary as follows: $0 \le x_3 \le 1 - \sqrt{x_1^2 + x_2^2}$. In other words

$$K = \{ x \in \mathbf{R}^3 \mid (x_1, x_2) \in B^2, \; 0 \le x_3 \le f(x_1, x_2) \},$$

where $f : B^2 \to \mathbf{R}$ is the continuous function with $f(x_1, x_2) = 1 - \sqrt{x_1^2 + x_2^2}$. Thus, again by (ii) of Theorem 6.3.5, the assertion follows. ☆

6.4 Successive integration

We now formulate results that will enable us to reduce an n-dimensional integration to successive lower-dimensional integrations. This method is most effective if reduction is possible to one-dimensional integrations which then can be performed by computing antiderivatives.

Consider a function $f : \mathbf{R}^p \times \mathbf{R}^q \to \mathbf{R}$. If f is continuous, the function $z \mapsto f(y, z)$ is a continuous function on $\mathbf{R}^q$, for every $y \in \mathbf{R}^p$. Remarkably, for Riemann integrability there is **no** valid analogous result. If f is Riemann integrable, it does not necessarily follow, for every $y \in \mathbf{R}^p$, that the function $z \mapsto f(y, z)$ is a Riemann integrable function on $\mathbf{R}^q$. This explains the formulation of Theorem 6.4.2 below.

Definition 6.4.1. A bounded function $f : \mathbf{R}^n \to \mathbf{R}$ is said to be a *step function* if there exist a rectangle $B \subset \mathbf{R}^n$ and a partition $\mathcal{B} = \{B_i \mid i \in I\}$ of B such that $f(x) = 0$, for $x \notin B$ and such that $f|_{\text{int}(B_i)}$, for all $i \in I$, is a constant function, while, for every $1 \le j \le n$, the function $x_j \mapsto f(x_1, \dots, x_j, \dots, x_n)$ is left-continuous. If these conditions are met, f is said to be a *step function associated with* $\mathcal{B}$. Note that a step function is Riemann integrable over $\mathbf{R}^n$. ❍

Theorem 6.4.2. *Let f be a Riemann integrable function with compact support on $\mathbf{R}^{p+q}$. Then*

$$y \mapsto \underline{\int}_{\mathbf{R}^q} f(y,z)\,dz \qquad \text{and} \qquad y \mapsto \overline{\int}_{\mathbf{R}^q} f(y,z)\,dz \qquad (y \in \mathbf{R}^p)$$

are Riemann integrable functions with compact support on $\mathbf{R}^p$, and

$$\int_{\mathbf{R}^{p+q}} f(x)\,dx = \int_{\mathbf{R}^p} \underline{\int}_{\mathbf{R}^q} f(y,z)\,dz\,dy = \int_{\mathbf{R}^p} \overline{\int}_{\mathbf{R}^q} f(y,z)\,dz\,dy.$$

Proof. Let $B \subset \mathbf{R}^{p+q}$ be a rectangle such that f vanishes outside B, let $\mathcal{B} = \{B_i \mid i \in I\}$ be a partition of B, and let g_-, g_+ be step functions associated with $\mathcal{B}$ such that

$$g_-(x) \le f(x) \le g_+(x) \qquad (x \in B).$$

Then for every $y \in \mathbf{R}^p$ the functions $z \mapsto g_-(y,z)$ and $z \mapsto g_+(y,z)$ are step functions on $\mathbf{R}^q$ dominated by $z \mapsto f(y,z)$, and dominating $z \mapsto f(y,z)$, respectively. Accordingly one has, for every $y \in \mathbf{R}^p$,

$$\int_{\mathbf{R}^q} g_-(y,z)\,dz \le \underline{\int}_{\mathbf{R}^q} f(y,z)\,dz \le \overline{\int}_{\mathbf{R}^q} f(y,z)\,dz \le \int_{\mathbf{R}^q} g_+(y,z)\,dz.$$

But $y \mapsto \int_{\mathbf{R}^q} g_\pm(y,z)\,dz$ in turn are step functions on $\mathbf{R}^p$; and we obtain

$$\begin{aligned}
\int_{\mathbf{R}^{p+q}} g_-(x)\,dx &= \int_{\mathbf{R}^p}\int_{\mathbf{R}^q} g_-(y,z)\,dz\,dy \le \underline{\int}_{\mathbf{R}^p}\underline{\int}_{\mathbf{R}^q} f(y,z)\,dz\,dy \\
&\le \overline{\int}_{\mathbf{R}^p}\underline{\int}_{\mathbf{R}^q} f(y,z)\,dz\,dy \le \overline{\int}_{\mathbf{R}^p}\overline{\int}_{\mathbf{R}^q} f(y,z)\,dz\,dy \\
&\le \int_{\mathbf{R}^p}\int_{\mathbf{R}^q} g_+(y,z)\,dz\,dy = \int_{\mathbf{R}^{p+q}} g_+(x)\,dx.
\end{aligned}$$

Because the supremum of the left–hand side and the infimum of the right–hand side, both taken over all possible partitions $\mathcal{B}$, equal $\int_{\mathbf{R}^{p+q}} f(x)\,dx$, it follows that

$$\int_{\mathbf{R}^{p+q}} f(x)\,dx = \underline{\int}_{\mathbf{R}^p}\underline{\int}_{\mathbf{R}^q} f(y,z)\,dz\,dy = \overline{\int}_{\mathbf{R}^p}\underline{\int}_{\mathbf{R}^q} f(y,z)\,dz\,dy,$$

which proves the assertion about $y \mapsto \underline{\int}_{\mathbf{R}^q} f(y,z)\,dz$. ❑

Remark. The formulation of the preceding theorem may not be simplified, as the following example demonstrates. Let $f(x,y) = \frac{1}{q}$, if $x = \frac{p}{q}$ with $p, q \in \mathbf{N}$, where p and q are relatively prime, $p \le q$, and $y \in \mathbf{Q} \cap [0,1]$. Let $f(x,y) = 0$, in all other cases. Then f is Riemann integrable over $\mathbf{R}^2$ with vanishing integral. But for every $x = \frac{p}{q}$ as above, $y \mapsto f(x,y)$ is not Riemann integrable: $\underline{\int}_{\mathbf{R}} f(x,y)\,dy = 0$, while $\overline{\int}_{\mathbf{R}} f(x,y)\,dy = \frac{1}{q}$.

Remark. If in the preceding theorem the roles of $y \in \mathbf{R}^p$ and $z \in \mathbf{R}^q$ are interchanged, it follows that the functions $z \mapsto \underline{\int}_{\mathbf{R}^p} f(y, z)\, dy$ and $z \mapsto \overline{\int}_{\mathbf{R}^p} f(y, z)\, dy$ are also Riemann integrable over $\mathbf{R}^q$, and that their integrals over $\mathbf{R}^q$ both equal $\int_{\mathbf{R}^{p+q}} f(x)\, dx$. This constitutes another proof, different from the one of Theorem 2.10.7.(iii), that the order of integration may be interchanged. In particular, we have obtained the following:

Corollary 6.4.3 (Interchanging the order of integration). *Let $f : \mathbf{R}^{p+q} \to \mathbf{R}$ be a continuous function with compact support. Then, for every $y \in \mathbf{R}^p$, the integral $\int_{\mathbf{R}^q} f(y, z)\, dz$ is well-defined, and in addition, for every $z \in \mathbf{R}^q$, the integral $\int_{\mathbf{R}^p} f(y, z)\, dy$ and the functions thus defined are Riemann integrable with compact support on $\mathbf{R}^p$ and $\mathbf{R}^q$, respectively. Furthermore,*

$$\int_{\mathbf{R}^{p+q}} f(x)\, dx = \int_{\mathbf{R}^p} \int_{\mathbf{R}^q} f(y, z)\, dz\, dy = \int_{\mathbf{R}^q} \int_{\mathbf{R}^p} f(y, z)\, dy\, dz.$$

Example 6.4.4. The requirement of continuity (or of boundedness) of f does play a role, as becomes apparent from the following. One has

$$\int_0^1 \int_0^1 \frac{x-y}{(x+y)^3}\, dy\, dx = \frac{1}{2}, \qquad \int_0^1 \int_0^1 \frac{x-y}{(x+y)^3}\, dx\, dy = -\frac{1}{2}.$$

If the integrals

$$\int_0^1 \int_0^1 \frac{x}{(x+y)^3}\, dy\, dx \qquad \text{and} \qquad \int_0^1 \int_0^1 \frac{y}{(x+y)^3}\, dx\, dy$$

were both well-defined, one would, on symmetry grounds, expect them to be equal. The interchangeability of the order of integration would then imply that the original integrals vanish. But we have, for $x > 0$, which causes the convergence,

$$\begin{aligned}
\int_0^1 \frac{x-y}{(x+y)^3}\, dy &= \int_0^1 \frac{2x-(x+y)}{(x+y)^3}\, dy = \int_0^1 \Big(\frac{2x}{(x+y)^3} - \frac{1}{(x+y)^2} \Big)\, dy \\
&= \Big[-\frac{x}{(x+y)^2} + \frac{1}{x+y} \Big]_{y=0}^{y=1} \\
&= -\frac{x}{(x+1)^2} + \frac{1}{x+1} + \frac{x}{x^2} - \frac{1}{x} = \frac{1}{(x+1)^2}.
\end{aligned}$$

Consequently,

$$\int_0^1 \int_0^1 \frac{x-y}{(x+y)^3}\, dy\, dx = \int_0^1 \frac{1}{(x+1)^2}\, dx = \Big[\frac{-1}{x+1} \Big]_0^1 = -\frac{1}{2} + 1 = \frac{1}{2}.$$

Interchanging the roles of x and y, we obtain

$$\int_0^1 \int_0^1 \frac{y-x}{(x+y)^3}\, dx\, dy = \frac{1}{2}, \qquad \text{that is} \qquad \int_0^1 \int_0^1 \frac{x-y}{(x+y)^3}\, dx\, dy = -\frac{1}{2}. \quad ☆$$

Theorem 6.4.5. *Let a_1, $b_1 \in \mathbf{R}$ with $a_1 \le b_1$, and write $K_1 = [\,a_1, b_1\,]$. Assume that there have been defined, by induction over $2 \le j \le n$,*

$$\text{continuous functions } a_j,\ b_j : K_{j-1} \to \mathbf{R} \qquad \text{with} \qquad a_j \le b_j;$$

and sets $K_j \subset \mathbf{R}^j$ given by

$$\{\, x \in \mathbf{R}^j \mid (x_1, \dots, x_{j-1}) \in K_{j-1},\ a_j(x_1, \dots, x_{j-1}) \le x_j \le b_j(x_1, \dots, x_{j-1})\,\}.$$

Then the following assertions hold.

(i) *The sets K_j belong to $\mathcal{J}(\mathbf{R}^j)$.*

(ii) *For every continuous function $f : K_n \to \mathbf{R}$ one has*

$$\int_{K_n} f(x)\,dx$$

$$= \int_{a_1}^{b_1} \int_{a_2(x_1)}^{b_2(x_1)} \cdots \int_{a_n(x_1,\dots,x_{n-1})}^{b_n(x_1,\dots,x_{n-1})} f(x_1, \dots, x_{n-1}, x_n)\,dx_n \cdots dx_2\,dx_1.$$

Proof. Assertion (i) follows by induction over j and Theorem 6.3.5. Write $x' = (x_1, \dots, x_{n-1}) \in \mathbf{R}^{n-1}$, whence $x = (x', x_n) \in \mathbf{R}^n$. Note that $1_{K_n}(x) = 1_{K_{n-1}}(x')\,1_{[a_n(x'),\,b_n(x')]}(x_n)$. For every $x' \in K_{n-1}$ the function $x_n \mapsto (1_{K_n} f)(x', x_n)$ is continuous on the interval $[a_n(x'),\, b_n(x')]$; and so, by Theorem 6.3.5.(i), it is Riemann integrable over the said interval. That is, for every $x' \in K_{n-1}$,

$$\int_{\mathbf{R}} (1_{K_n} f)(x', x_n)\,dx_n = 1_{K_{n-1}}(x') \int_{a_n(x')}^{b_n(x')} f(x', x_n)\,dx_n.$$

Applying the preceding corollary, one finds

$$\begin{aligned}\int_{K_n} f(x)\,dx &= \int_{\mathbf{R}^n} (1_{K_n} f)(x)\,dx = \int_{\mathbf{R}^{n-1}} 1_{K_{n-1}}(x') \int_{a_n(x')}^{b_n(x')} f(x', x_n)\,dx_n\,dx' \\ &= \int_{K_{n-1}} \int_{a_n(x')}^{b_n(x')} f(x', x_n)\,dx_n\,dx'.\end{aligned}$$

The proof of assertion (ii) can now be completed by induction over n, provided we know

$$x' \mapsto \int_{a_n(x')}^{b_n(x')} f(x', x_n)\,dx_n \tag{6.19}$$

to be a continuous function on K_{n-1}. In order to verify this, define

$$I : \mathbf{R}^2 \times K_{n-1} \to \mathbf{R} \qquad \text{by} \qquad I(a, b, x') = \int_a^b f(x', x_n)\,dx_n.$$

Then

$$\begin{aligned}|I(a,b,x') - I(\widetilde{a},\widetilde{b},\widetilde{x}')| &\le |I(a,b,x') - I(\widetilde{a},b,x')| \\ &\quad +|I(\widetilde{a},b,x') - I(\widetilde{a},\widetilde{b},x')| + |I(\widetilde{a},\widetilde{b},x') - I(\widetilde{a},\widetilde{b},\widetilde{x}')| \\ &\le \left|\int_a^{\widetilde{a}} f(x',\, x_n)\,dx_n\right| + \left|\int_b^{\widetilde{b}} f(x',\, x_n)\,dx_n\right| \\ &\quad + \left|\int_{\widetilde{a}}^{\widetilde{b}} (f(x',\, x_n) - f(\widetilde{x}',\, x_n))\,dx_n\right|.\end{aligned}$$

Because f is uniformly continuous on K_n, there exists, for every $\eta > 0$, a $\delta > 0$ such that, uniformly in x_n,

$$|f(x',\, x_n) - f(\widetilde{x}',\, x_n)| < \eta \qquad (\|x' - \widetilde{x}'\| < \delta).$$

From this follows, for $|a - \widetilde{a}| < \min\{\eta,\, 1\}$, $|b - \widetilde{b}| < \min\{\eta,\, 1\}$, $\|x' - \widetilde{x}'\| < \delta$,

$$|I(a,\, b,\, x') - I(\widetilde{a},\, \widetilde{b},\, \widetilde{x}')| < \eta\,(2\max_{K_n}|f| + |a| + |b| + 2),$$

proving the continuity of I in $(a,\, b,\, x')$. But this implies the continuity of the function in (6.19). ❑

Remark. In the proof above the Continuity Theorem 2.10.2 is not directly applicable since the interval of integration $J = [\,a, b\,]$ of the variable x_n is also allowed to vary.

6.5 Examples of successive integration

Example 6.5.1. Let K be the compact set in $\mathbf{R}^3$, bounded by the planes

$$\{\,x \in \mathbf{R}^3 \mid x_1 + x_2 + x_3 = a\,\} \quad (a > 0), \qquad \{x \in \mathbf{R}^3 \mid x_i = 0\} \quad (1 \le i \le 3).$$

Then

$$\int_K \|x\|^2\,dx = \frac{a^5}{20}.$$

We have

$$K = \{\,x \in \mathbf{R}^3 \mid 0 \le x_1,\ 0 \le x_2,\ 0 \le x_3,\ x_1 + x_2 + x_3 \le a\,\}.$$

Therefore $x \in K$ means that $0 \le x_1 \le a - x_2 - x_3$, while $x_2 \ge 0$ and $x_3 \ge 0$, that is $x_1 \in [\,0, a\,]$. Once $x_1 \in [\,0, a\,]$ has been fixed, $0 \le x_2 \le a - x_1 - x_3$ and $x_3 \ge 0$ imply that x_2 may still vary within $[\,0, a - x_1\,]$; and for (x_1, x_2) thus fixed, x_3 may still vary within $[\,0, a - x_1 - x_2\,]$. Consequently

$$K = \{\,x \in \mathbf{R}^3 \mid 0 \le x_1 \le a,\ 0 \le x_2 \le a - x_1,\ 0 \le x_3 \le a - x_1 - x_2\,\}.$$

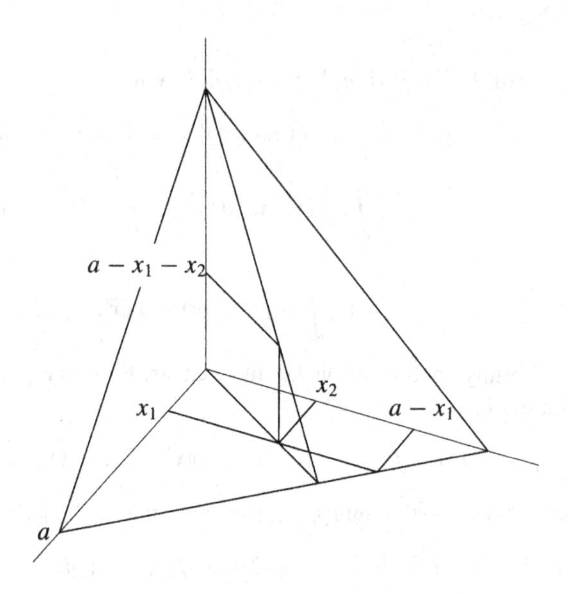

Illustration for Example 6.5.1

Hence we have, in the notation of the preceding theorem,

$$K_1 = [\,0, a\,], \quad K_2 = \{\,(x_1, x_2) \in \mathbf{R}^2 \mid x_1 \in K_1,\ 0 \le x_2 \le a - x_1\,\}, \quad K_3 = K.$$

As a consequence

$$\int_K \|x\|^2\,dx = \int_0^a \int_0^{a-x_1} \int_0^{a-x_1-x_2} (x_1^2 + x_2^2 + x_3^2)\,dx_3\,dx_2\,dx_1$$

$$= \int_0^a \int_0^{a-x_1} \Big[x_1^2 x_3 + x_2^2 x_3 + \frac{x_3^3}{3} \Big]_{x_3=0}^{x_3=a-x_1-x_2} dx_2\,dx_1$$

$$= \int_0^a \int_0^{a-x_1} (x_1^2(a - x_1 - x_2) + x_2^2(a - x_1 - x_2) + \frac{(a - x_1 - x_2)^3}{3})\,dx_2\,dx_1$$

$$= \int_0^a \Big[x_1^2(a - x_1)x_2 - \frac{x_1^2 x_2^2}{2} + \frac{(a - x_1)x_2^3}{3} - \frac{x_2^4}{4} - \frac{(a - x_1 - x_2)^4}{12} \Big]_{x_2=0}^{x_2=a-x_1} dx_1$$

$$= \int_0^a \Big(x_1^2(a - x_1)^2 - \frac{x_1^2(a - x_1)^2}{2} + \frac{(a - x_1)^4}{3} - \frac{(a - x_1)^4}{4} + \frac{(a - x_1)^4}{12} \Big)\,dx_1$$

$$= \int_0^a \Big(\frac{x_1^2(a - x_1)^2}{2} + \frac{(a - x_1)^4}{6} \Big)\,dx_1 = \frac{a^5}{20}. \qquad ☆$$

Example 6.5.2 (Intersection of two cylinders). The intersection of the two solid cylinders in $\mathbf{R}^3$ given by $\{ x \in \mathbf{R}^3 \mid x_1^2 + x_2^2 \le 1 \}$ and $\{ x \in \mathbf{R}^3 \mid x_1^2 + x_3^2 \le 1 \}$, respectively, has volume $\frac{16}{3}$.

Proof. I. From $x_1^2 + x_2^2 \le 1$ and $x_1^2 + x_3^2 \le 1$ we see that x_1 may vary in the entire interval $[-1, 1]$. We set $b(x) = \sqrt{1-x^2}$. Given x_1, we see from $x_2^2 \le 1 - x_1^2$ that x_2 may still vary in $[-b(x_1), b(x_1)]$; likewise, x_3 may vary in the same interval $[-b(x_1), b(x_1)]$. The intersection, therefore, is given by

$$\{ x \in \mathbf{R}^3 \mid -1 \le x_1 \le 1, \ -b(x_1) \le x_2 \le b(x_1), \ -b(x_1) \le x_3 \le b(x_1) \}.$$

The desired volume is

$$\int_{-1}^{1} \int_{-b(x_1)}^{b(x_1)} \int_{-b(x_1)}^{b(x_1)} dx_3\, dx_2\, dx_1 = 8 \int_0^1 \int_0^{b(x_1)} \int_0^{b(x_1)} dx_3\, dx_2\, dx_1$$

$$= 8 \int_0^1 \int_0^{b(x_1)} b(x_1)\, dx_2\, dx_1 = 8 \int_0^1 b(x_1)^2\, dx_1 = 8 \Big[x_1 - \frac{x_1^3}{3} \Big]_0^1$$

$$= 8(1 - \frac{1}{3}).$$ ❑

Proof. II. From $x_1^2 + x_2^2 \le 1$ it follows that x_2 may vary in the entire interval $[-1, 1]$. Given x_2, we see from $x_1^2 \le 1 - x_2^2$ that x_1 may vary in the interval $[-b(x_2), b(x_2)]$. Finally then, x_3 may vary in $[-b(x_1), b(x_1)]$. Thus the intersection is given by

$$\{ x \in \mathbf{R}^3 \mid -1 \le x_2 \le 1, \ -b(x_2) \le x_1 \le b(x_2), \ -b(x_1) \le x_3 \le b(x_1) \}.$$

And so the desired volume is

$$\int_{-1}^{1} \int_{-b(x_2)}^{b(x_2)} \int_{-b(x_1)}^{b(x_1)} dx_3\, dx_1\, dx_2 = 8 \int_0^1 \int_0^{b(x_2)} \int_0^{b(x_1)} dx_3\, dx_1\, dx_2$$

$$= 8 \int_0^1 \int_0^{b(x_2)} \sqrt{1 - x_1^2}\, dx_1\, dx_2.$$

Upon computation of an antiderivative of $\sqrt{1-x_1^2} = \frac{1-x_1^2}{\sqrt{1-x_1^2}}$ this becomes

$$8 \int_0^1 \Big[\frac{x_1}{2}\sqrt{1-x_1^2} + \frac{\arcsin x_1}{2} \Big]_{x_1=0}^{x_1=b(x_2)} dx_2$$

$$= 4 \int_0^1 (x_2\sqrt{1-x_2^2} + \arcsin\sqrt{1-x_2^2})\, dx_2$$

$$= 4 \int_0^1 x_2\sqrt{1-x_2^2}\, dx_2 + 4 \int_0^1 \arccos x_2\, dx_2.$$

And now

$$
\begin{aligned}
\int_0^1 x_2\sqrt{1-x_2^2}\,dx_2 &= \Big[-\frac{(1-x_2^2)^{3/2}}{3}\Big]_0^1 = \frac{1}{3},\\
\int_0^1 \arccos x_2\,dx_2 &= [\,x_2\arccos x_2\,]_0^1 + \int_0^1 \frac{x_2}{\sqrt{1-x_2^2}}\,dx_2\\
&= \arccos 1 - \Big[\sqrt{1-x_2^2}\Big]_0^1 = 1.
\end{aligned}
$$

❑

☆

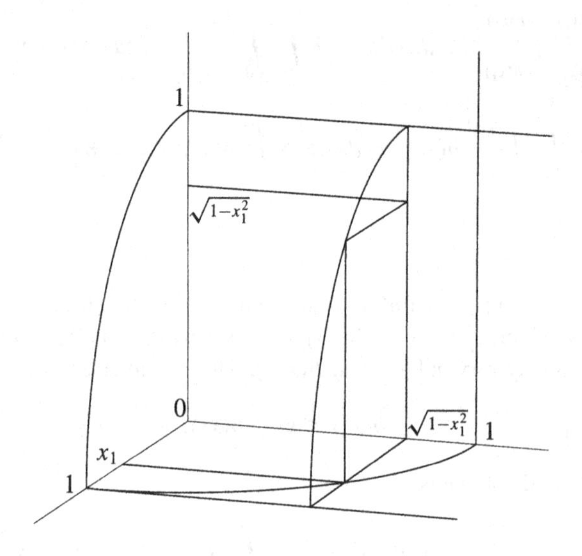

Illustration for Example 6.5.2: Intersection of two cylinders

Example 6.5.3 (Sharpening a pencil). A hexagonal pencil, not sharpened, has a thickness of 7 mm between two parallel faces. The pencil is sharpened into an exact conical point such that at the tip the angle between two generators on the conical surface is at most $\frac{\pi}{6}$ radians. The volume of the material removed equals $0.389\cdots\text{cm}^3$, if the sharpening has been done in the most economical way.

Indeed, choose the origin of the coordinate system for $\mathbf{R}^3$ at the tip of the pencil, the negative direction of the x_3-axis along the axis of the pencil, the x_1-axis perpendicular to a face of the pencil. Let $a = 3.5$. Because of the sixfold symmetry of the pencil about the x_3-axis, the planes through the x_3-axis and those edges of the pencil that lie in the plane $\{\, x \in \mathbf{R}^3 \mid x_1 = a \,\}$ form an angle of $\frac{\pi}{3}$ radians. Therefore, these planes are defined by the equations $x_2 = \pm\tan\frac{\pi}{6}\,x_1 = \pm\frac{1}{3}\sqrt{3}\,x_1$. Let $C \subset \mathbf{R}^3$ be the conical surface, part of which bounds the pencil. Note that for given (x_1, x_2) there exists a unique $x_3^0 = x_3^0(x_1, x_2)$ such that $x_3^0 \le 0$ and $(x_1, x_2, x_3^0) \in C$. Now the volume removed equals six times that of the set

$$\{x \in \mathbf{R}^3 \mid 0 \le x_1 \le a,\ -\frac{1}{3}\sqrt{3}x_1 \le x_2 \le \frac{1}{3}\sqrt{3}x_1,\ x_3^0(x_1, x_2) \le x_3 \le 0\}.$$

Next we derive an equation for C. To calculate $\tan\frac{\pi}{12}$ we use the formula for the tangent of the double angle:

$$\frac{1}{3}\sqrt{3} = \tan\frac{\pi}{6} = \frac{2\tan\frac{\pi}{12}}{1-\tan^2\frac{\pi}{12}}, \qquad \text{that is,} \qquad 1-\tan^2\frac{\pi}{12} = \frac{6}{\sqrt{3}}\tan\frac{\pi}{12}.$$

The positive root of this quadratic equation equals $\tan\frac{\pi}{12} = \frac{1}{2}(-2\sqrt{3}+\sqrt{12+4}) = 2-\sqrt{3}$, so

$$\tan\Big(\frac{\pi}{2}-\frac{\pi}{12}\Big) = \frac{1}{\tan\frac{\pi}{12}} = \frac{1}{2-\sqrt{3}} = 2+\sqrt{3}.$$

Therefore, $x \in \mathbf{R}^3$ lying on one of the two generators of the conical surface C contained in the plane $\{x \in \mathbf{R}^3 \mid x_2 = 0\}$ satisfies $x_3 = -\tan(\frac{\pi}{2}-\frac{\pi}{12})x_1 = -(2+\sqrt{3})x_1$. And C itself has the equation $x_3 = -(2+\sqrt{3})\sqrt{x_1^2+x_2^2}$, in particular

$$x_3^0 = x_3^0(x_1, x_2) = -(2+\sqrt{3})\sqrt{x_1^2+x_2^2}.$$

Consequently, the desired volume is given by

$$6\int_0^a \int_{-\frac{1}{3}\sqrt{3}x_1}^{\frac{1}{3}\sqrt{3}x_1} \int_{-(2+\sqrt{3})\sqrt{x_1^2+x_2^2}}^{0} dx_3\,dx_2\,dx_1$$

$$= 12(2+\sqrt{3})\int_0^a \int_0^{\frac{1}{3}\sqrt{3}x_1} \sqrt{x_1^2+x_2^2}\,dx_2\,dx_1.$$

Using an antiderivative of $\frac{x_1^2+x_2^2}{\sqrt{x_1^2+x_2^2}}$ with respect to x_2, we find, modulo the factor $12(2+\sqrt{3})$

$$\int_0^a \Big[\frac{x_2}{2}\sqrt{x_2^2+x_1^2} + \frac{x_1^2}{2}\log\Big(x_2+\sqrt{x_2^2+x_1^2}\Big)\Big]_{x_2=0}^{x_2=\frac{1}{3}\sqrt{3}x_1} dx_1$$

$$= \int_0^a \Big(\frac{x_1^2}{6}\sqrt{3}\sqrt{\frac{4}{3}} + \frac{x_1^2}{2}\log\Big(x_1\Big(\frac{1}{3}\sqrt{3}+\sqrt{\frac{4}{3}}\Big)\Big) - \frac{x_1^2}{2}\log x_1\Big)\,dx_1$$

$$= \int_0^a \Big(\frac{x_1^2}{3} + \frac{x_1^2}{2}\log\sqrt{3}\Big)\,dx_1 = \Big(\frac{1}{3}+\frac{\log 3}{4}\Big)\int_0^a x_1^2\,dx_1$$

$$= \frac{1}{36}\Big(4+3\log 3\Big)(3.5)^3 = 8.689\cdots.$$

☆

6.6 Change of Variables Theorem: formulation and examples

In the integral calculus on $\mathbf{R}$ an important result is the Fundamental Theorem 2.10.1,

$$\int_a^b f'(x)\,dx = f(b) - f(a),$$

which gives a relation between integration and differentiation on $\mathbf{R}$. This result has a direct analog in the calculus of integrals on $\mathbf{R}^n$ (see Theorem 7.6.1), in the form of a relation between integration and total differentiation on $\mathbf{R}^n$. On $\mathbf{R}$ the Change of Variables Theorem

$$\int_{\Psi(a)}^{\Psi(b)} f(x)\,dx = \int_a^b (f \circ \Psi)(y)\Psi'(y)\,dy, \tag{6.20}$$

for $\Psi : [\,a, b\,] \to \mathbf{R}$ continuously differentiable and $f : \Psi([\,a, b\,]) \to \mathbf{R}$ continuous, is a direct consequence of the Fundamental Theorem. Indeed, let F be an antiderivative of f on $\Psi([\,a, b\,])$, then both sides of (6.20) are equal to $F \circ \Psi(b) - F \circ \Psi(a)$. Hence, the essence is to recognize the function f as a derivative; on $\mathbf{R}^n$ with $n > 1$, however, this is not possible, because a derived function on $\mathbf{R}^n$ does not take values in $\mathbf{R}$, but in $\mathrm{Lin}(\mathbf{R}^n, \mathbf{R}) \simeq \mathbf{R}^n$. Nonetheless, a Change of Variables Theorem does exist for $\mathbf{R}^n$ with $n > 1$, but its proof is distinctly more complicated than that for $\mathbf{R}$. The proof for $\mathbf{R}^n$ is obtained by means of a reduction to $\mathbf{R}$, and requires some technical tools: the Implicit Function Theorem 3.5.1, and results which enable us to localize an integral, that is, to study an integral near a point. For this reason we prefer to begin by merely stating the theorem, and then illustrating it by some examples. Section 6.7 contains the results relating to localization, while the proof can be found in Section 6.9.

Other proofs. The appendix to this chapter, in Section 6.13, contains two more proofs of the Change of Variables Theorem which consider the problem from different perspectives.

Theorem 6.6.1 (Change of Variables Theorem). *Assume U and V to be open subsets of $\mathbf{R}^n$ and let $\Psi : V \to U$ be a C^1 diffeomorphism. Let $f : U \to \mathbf{R}$ be a bounded function with compact support. Then f is Riemann integrable over U if and only if the function*

$$y \mapsto (f \circ \Psi)(y)|\det D\Psi(y)|$$

is Riemann integrable over V. If either condition is met, one has

$$\int_{\Psi(V)} f(x)\,dx = \int_U f(x)\,dx = \int_V (f \circ \Psi)(y)|\det D\Psi(y)|\,dy.$$

An immediate consequence of this theorem and its proof is the following:

Corollary 6.6.2. *If* $L \in \mathcal{J}(V)$, *then* $\Psi(L) \in \mathcal{J}(U)$, *while*

$$\mathrm{vol}_n\,(\Psi(L)) = \int_U 1_{\Psi(L)}(x)\,dx = \int_L |\det D\Psi(y)|\,dy.$$

If, moreover, Ψ *satisfies the condition* $|\det D\Psi(y)| = 1$, *for all* $y \in L$, *we have*

$$\mathrm{vol}_n\,(\Psi(L)) = \mathrm{vol}_n(L).$$

In view of this corollary a C^1 diffeomorphism $\Psi : V \to U$ with the property $|\det D\Psi| = 1$ on V is said to be *volume-preserving*.

Remarks. Theorem 6.6.1 tells us how an integral over U transforms under a C^1 regular coordinate transformation $x = \Psi(y)$ on U. Specifically, the integral over a transformed set of a function equals the integral over that set of the product of the transformed function and the absolute value of the Jacobian of the transformation. Here we recall the terminology *Jacobi matrix* of Ψ at y to describe $D\Psi(y)$; and the *Jacobian* of Ψ at y is defined as

$$\det D\Psi(y).$$

Consider the case $n = 1$. For a C^1 diffeomorphism $\Psi : V \to U$ where V, $U \subset \mathbf{R}$, we have, on account of Example 2.4.9, $D\Psi(y) = \Psi'(y) \neq 0$, for all $y \in V$. However, the proof of the Change of Variables Theorem (6.20) on $\mathbf{R}$ did not assume $\Psi'(y) \neq 0$, for all $y \in\ ]\,a, b\,[$; thus the "old" version of the Change of Variables Theorem on $\mathbf{R}$ is stronger than the "new" one. We now consider the "new" version of the Change of Variables Theorem on $\mathbf{R}$. If $\Psi :]\,a, b\,[\ \to \mathbf{R}$ is a C^1 diffeomorphism, then by the Intermediate Value Theorem 1.9.5 the continuity of Ψ' and the fact that $\Psi'(y) \neq 0$, for all $y \in\ ]\,a, b\,[$, imply that either $\Psi'(y) > 0$, or $\Psi'(y) < 0$, for all $y \in\ ]\,a, b\,[$. But this means that Ψ is strictly monotonic, either increasing or decreasing, on $[\,a, b\,]$, and either $\Psi([\,a, b\,]) = [\Psi(a),\ \Psi(b)]$ or $\Psi([\,a, b\,]) = [\Psi(b),\ \Psi(a)]$, respectively. In the latter case $|\det D\Psi(y)| = -\Psi'(y)$, and so the Change of Variables Theorem 6.6.1 then yields

$$\int_{\Psi(b)}^{\Psi(a)} f(x)\,dx = -\int_a^b (f \circ \Psi)(y)\Psi'(y)\,dy = \int_b^a (f \circ \Psi)(y)\Psi'(y)\,dy.$$

The formulation of Theorem 6.6.1 as given above is more or less dictated by the fact that under these conditions the structure of the proof remains reasonably transparent. Thus, the present formulation of the theorem is for open sets in $\mathbf{R}^n$, because these are the natural domains of diffeomorphisms. Sometimes, the mapping $\Psi : V \to U$ which would seem the most natural, cannot be extended to a diffeomorphism $\overline{V} \to \overline{U}$, for example because $\det D\Psi$ vanishes at points of ∂V.

Furthermore, the condition on f ensures that no convergence problems will crop up.

Nevertheless, in many cases Corollary 6.6.2 in the form given above cannot be used to calculate the volume

$$\mathrm{vol}_n(K) = \int_{\mathbf{R}^n} 1_K(x)\,dx$$

of a compact and Jordan measurable set $K \subset \mathbf{R}^n$, for the following reason. The calculation requires finding a C^1 diffeomorphism $\Psi : V \to U$ and a set $L \in \mathcal{J}(V)$ with $\Psi(L) = K$. The problem often involved in this is that the obvious change of variables Ψ becomes singular on ∂L; consequently, Ψ is not a C^1 diffeomorphism on an open neighborhood of L in such cases.

Therefore the proof of the Change of Variables Theorem is followed, in Section 6.10, by limit arguments which justify calculating $\mathrm{vol}_n(K)$ by means of the theorem after all: on account of Theorem 6.3.2 one has $\mathrm{vol}_n(K) = \mathrm{vol}_n(\mathrm{int}\,K)$ for Jordan measurable sets K, and the characteristic function of $\mathrm{int}(K)$ may be approximated from within by admissible functions. Accordingly, in the examples we will use the Change of Variables Theorem for calculating the volumes of "reasonable" sets, meaning sets in $\mathbf{R}^n$ bounded by a finite number of lower-dimensional C^1 manifolds.

Example 6.6.3 (Volume of parallelepiped). Let $B \subset \mathbf{R}^n$ be a rectangle and assume $\Psi : \mathbf{R}^n \to \mathbf{R}^n$ is a linear mapping. Then $\Psi(B)$ is a *parallelepiped* in $\mathbf{R}^n$, that is, a set of the form

$$\{\, a + \sum_{1\le j\le n} t_j b^{(j)} \mid 0 \le t_j \le 1 \ (1 \le j \le n) \,\};$$

here a is one of the vertices of $\Psi(B)$ and the $b^{(j)}$, for $1 \le j \le n$, are the direction vectors of the edges that originate from a. By the Change of Variables Theorem (see Proposition 6.13.4 for a direct proof) it follows that

$$\mathrm{vol}_n\,(\Psi(B)) = |\det \Psi|\ \mathrm{vol}_n(B),$$

that is, $\mathrm{vol}_n\,(\Psi(B))/\,\mathrm{vol}_n(B)$ is independent of the set B, whereas it does depend on the linear mapping Ψ.

Note that in linear algebra this transformation property usually is one of the defining properties of volume in an n-dimensional vector space; in our development of analysis in $\mathbf{R}^n$ the Change of Variables Theorem is required to obtain this result. This comes about because in Section 6.3 we have chosen a definition of n-dimensional volume which is more elementary, one in which rectangles play a fundamental role. ☆

Example 6.6.4 (Polar coordinates). Let $U = \mathbf{R}^2 \setminus (\,]-\infty,\,0\,] \times \{0\})$ be the plane excluding the nonpositive part of the x_1-axis. Then define V and Ψ by

$$V = \mathbf{R}_+ \times \,]-\pi,\,\pi\,[\, \subset \mathbf{R}^2 \qquad \text{and} \qquad \Psi(r, \alpha) = r(\cos\alpha,\, \sin\alpha).$$

On account of Example 3.1.1, Ψ is a C^1 diffeomorphism, while

$$\det D\Psi(r, \alpha) = \det \begin{pmatrix} \cos\alpha & -r\sin\alpha \\ \sin\alpha & r\cos\alpha \end{pmatrix} = r > 0 \qquad ((r, \alpha) \in V).$$

Using the Change of Variables Theorem, together with Corollary 6.4.3, we obtain, for continuous functions f that vanish outside a compact subset contained in U,

$$\int_{\mathbf{R}^2} f(x)\,dx = \int_V (f \circ \Psi)(y)|\det D\Psi(y)|\,dy$$

$$= \int_{\mathbf{R}_+} r \int_{-\pi}^{\pi} f(r\cos\alpha,\, r\sin\alpha)\,d\alpha\,dr = \int_{-\pi}^{\pi} \int_{\mathbf{R}_+} rf(r\cos\alpha,\, r\sin\alpha)\,dr\,d\alpha.$$

Consider $-\pi \le \alpha_1 < \alpha_2 \le \pi$ and $\phi \in C(\,]\alpha_1, \alpha_2[\,)$, and define

$$V' = \{\,(r, \alpha) \in V \mid \alpha_1 < \alpha < \alpha_2,\ 0 < r < \phi(\alpha)\,\}.$$

By way of application we compute the area of the bounded open set $U' = \Psi(V')$, that is, U' is the bounded open set in $\mathbf{R}^2$ that in polar coordinates is bounded by the lines with equation $\alpha = \alpha_1$ and $\alpha = \alpha_2$ and the curve with equation $r = \phi(\alpha)$. It follows that

$$\text{area}(U') = \int_{\alpha_1}^{\alpha_2} \int_0^{\phi(\alpha)} r\,dr\,d\alpha = \frac{1}{2} \int_{\alpha_1}^{\alpha_2} \phi(\alpha)^2\,d\alpha.$$

In particular, the area of the bounded open set in $\mathbf{R}^2$ that in polar coordinates is bounded by the *lemniscate* $r^2 = a^2 \cos 2\alpha$ with $a > 0$ (see Exercise 5.18), equals

$$\text{area}(U') = 2\frac{a^2}{2} \int_{-\frac{\pi}{4}}^{\frac{\pi}{4}} \cos 2\alpha\,d\alpha = a^2. \qquad ☆$$

Example 6.6.5 (Volume of truncated cone). Let $G \subset \mathbf{R}^2$ be an open set. The *truncated cone* with apex $(0, 0, t)$ where $0 < t$, of height h where $0 < h \le t$, and with Jordan measurable basis $G \subset \mathbf{R}^2$, is defined as the open set $U \subset \mathbf{R}^3$ with

$$U = \{\,((1-s)y,\ s\,t) \in \mathbf{R}^3 \mid y \in G,\ 0 \le s \le \frac{h}{t}\,\}.$$

We now define $V \subset \mathbf{R}^3$ and $\Psi : V \to U$ by

$$V = G \times \left]0, \frac{h}{t}\right[\qquad \text{and} \qquad \Psi(y, s) = ((1-s)y,\ s\,t).$$

Then it is easy to see that $\Psi : V \to U$ is a bijective C^1 mapping, and that

$$\det D\Psi(y,s) = \det \begin{pmatrix} 1-s & 0 & -y_1 \\ 0 & 1-s & -y_2 \\ 0 & 0 & t \end{pmatrix} = (1-s)^2 t > 0 \qquad ((y,s) \in V).$$

Using the Global Inverse Function Theorem 3.2.8 we see that $\Psi : V \to U$ is a C^1 diffeomorphism, and by the Change of Variables Theorem and Theorem 6.4.5 we find

$$\begin{aligned} \mathrm{vol}_3(U) &= \int_U dx = \int_V t\,(1-s)^2\,dy\,ds = t\int_G dy \int_0^{\frac{h}{t}} (1-s)^2\,ds \\ &= t\ \mathrm{area}(G)\left[-\frac{(1-s)^3}{3}\right]_0^{\frac{h}{t}} = \frac{t}{3}\ \mathrm{area}(G)\Big(1-\Big(1-\frac{h}{t}\Big)^3\Big). \end{aligned}$$

In particular, if G is a square with edges of length a, the upper face of U is also a square, with edges of length $b = a(1-\frac{h}{t})$. Furthermore, in that case $\frac{t}{a} = \frac{h}{a-b}$. Hence the volume of a truncated pyramid of height h, having lower and upper faces with edges of length a and b, respectively, equals

$$\frac{1}{3}\frac{t}{a}(a^3 - b^3) = \frac{h}{3}(a^2 + ab + b^2). \qquad ☆$$

Example 6.6.6. It is sometimes easier to give the mapping $\Phi := \Psi^{-1} : U \to V$, and there may be no need for an explicit description of $\Psi : V \to U$. Assume we have, as in Application 3.3.A,

$$f(x) = \|x\|^2, \qquad U = \{\, x \in \mathbf{R}^2 \mid 1 < x_1^2 - x_2^2 < 9,\ 1 < 2x_1x_2 < 4 \,\},$$
$$V = \{\, y \in \mathbf{R}^2 \mid 1 < y_1 < 9,\ 1 < y_2 < 4 \,\}, \qquad \Phi(x) = (x_1^2 - x_2^2,\ 2x_1x_2) =: y.$$

According to Application 3.3.A, $\Phi : U \to V$ is a C^1 diffeomorphism, with inverse $\Psi := \Phi^{-1}$ and $\det D\Psi(y) = \frac{1}{4\|y\|}$, and we have $f(x) = \|y\|$. Therefore

$$\int_U f(x)\,dx = \int_V \|y\|\,\frac{1}{4\|y\|}\,dy = \frac{1}{4}\int_V dy = \frac{1}{4}\int_1^9 dy_1 \int_1^4 dy_2 = 6. \qquad ☆$$

Example 6.6.7 (Newton's potential of a ball). The *Newton potential* of set $A \in \mathcal{J}(\mathbf{R}^3)$ is the function $\phi_A : \mathbf{R}^3 \setminus A \to \mathbf{R}$, defined by

$$\phi_A(x) = -\frac{1}{4\pi}\int_A \frac{1}{\|x-y\|}\,dy.$$

Now assume A to be the closed ball about the origin, of radius R. In view of the rotation symmetry of A we may then assume $x = (0, 0, a)$, with $a = \|x\| > R$. Therefore

$$\phi_A(x) = -\frac{1}{4\pi}\int_{\|y\|\le R} \frac{1}{\sqrt{y_1^2 + y_2^2 + (y_3 - a)^2}}\, dy.$$

Now introduce cylindrical coordinates $y = \Psi(r, \alpha, y_3) = (r\cos\alpha,\ r\sin\alpha,\ y_3)$ (see, for example, Exercise 3.6). Then $\det D\Psi(r, \alpha, y_3) = r$. Because of this determinant the resulting integrand in cylindrical coordinates is given by

$$\frac{r}{\sqrt{r^2 + (y_3 - a)^2}}.$$

The antiderivative with respect to r of this function can then be readily obtained; hence the following. Because $\|y\|^2 = r^2 + y_3^2 \le R^2$, for $y \in A$, it follows that $U := \{\, y \in \mathbf{R}^3 \mid \|y\| < R \,\} = \Psi(V)$, with V the solid half cylinder

$$V := \{\, (r, \alpha, y_3) \in \mathbf{R}^3 \mid -\pi < \alpha < \pi,\ -R < y_3 < R,\ 0 < r < \sqrt{R^2 - y_3^2}\,\}.$$

On application of Theorem 6.4.5 this gives

$$\begin{aligned}
-\phi_A(x) &= \frac{1}{4\pi}\int_{-\pi}^{\pi}\int_{-R}^{R}\int_0^{\sqrt{R^2 - y_3^2}} \frac{r}{\sqrt{r^2 + (y_3 - a)^2}}\, dr\, dy_3\, d\alpha \\
&= \frac{1}{2}\int_{-R}^{R}\Big[\sqrt{r^2 + (y_3 - a)^2}\,\Big]_{r=0}^{r=\sqrt{R^2 - y_3^2}}\, dy_3 \\
&= \frac{1}{2}\int_{-R}^{R}(\sqrt{R^2 + a^2 - 2ay_3} - |y_3 - a|)\, dy_3 \\
&= \frac{1}{2}\Big[-\frac{1}{3a}\sqrt{(R^2 + a^2 - 2ay_3)^3} - ay_3 + \frac{y_3^2}{2}\Big]_{y_3=-R}^{y_3=R} \\
&= \frac{4}{3}\frac{\pi R^3}{4\pi}\frac{1}{a} = \frac{\mathrm{vol}(A)}{4\pi}\frac{1}{\|x\|}.
\end{aligned}$$

See Exercise 6.50.(viii) for the computation of the volume of the ball. Hence

$$\phi_A(x) = -\frac{\mathrm{vol}(A)}{4\pi}\frac{1}{\|x\|} \qquad (\|x\| > R).$$

Newton's potential therefore behaves as if the total "mass" of the ball were concentrated at the origin. The fundamental ideas for this calculation are first found in the literature,[1] in geometrical form, in Proposition LXXIV in Newton's *Principia*

[1] See also: *Littlewood's Miscellany*, Cambridge University Press, 1986, p. 169.

Mathematica, London, 1687. In Exercise 7.68 one finds two other proofs of this result.

In the Newtonian theory of gravitation the potential ϕ_A plays the following role. Consider a mass of unit density occupying the space of the solid A, that is, of total mass $M = \mathrm{vol}(A)$. Then the force F exerted by this mass M on a point mass m placed at a point $x \in \mathbf{R}^3$ is given by

$$F(x) = -\operatorname{grad} m\,\phi_A(x) = -\frac{mM}{4\pi}\frac{x}{\|x\|^3}.$$

Thus we obtain *Newton's law of gravitation*: the mass M exerts on the point mass m an **attractive** force of magnitude inversely proportional to the square of the distance $\|x\|$ between m and the center of gravity of M. (See Exercise 3.47.(i) for the relation between conservation of energy and the description of force as minus the gradient of potential energy.) Because of the minus sign in the formula for F, we have to consider negative potentials in order to get an attractive force law. Conversely, the potential at x equals the work done by the force F when we displace a unit mass from the point x to infinity; in particular, the potential equals 0 at infinity. In fact, in gravitation this work is independent of the path γ taken. Writing $\gamma : I = [\,0, 1\,] \to \mathbf{R}^3$ with $\gamma(0) = x$ and $\gamma(1) = \infty$, we obtain for this work the following line integral (see Section 8.1):

$$\begin{aligned}\int_\gamma \langle F(s), d_1 s\rangle &:= -\int_I \langle \operatorname{grad}(\phi_A \circ \gamma), D\gamma\,\rangle(t)\,dt = -\int_0^1 \frac{d}{dt}(\phi_A \circ \gamma)(t)\,dt \\ &= \phi_A(x).\end{aligned}$$

In the theory of electrostatics the forces between like charges are **repulsive**, as a consequence, in that discipline, potentials are not defined with a minus sign as in gravitation. ☆

Example 6.6.8 (Kepler's second law). The notation is that of Application 3.3.B. In particular $t \mapsto x(t)$ is a C^k curve in $\mathbf{R}^2$, $J \subset \mathbf{R}$ an open interval, $V := \,]\,0, 1\,[\, \times J \subset \mathbf{R}^2$,

$$\Psi : V \to P_J := \Psi(V) \subset \mathbf{R}^2 \qquad \text{defined by} \qquad \Psi(r, t) = r x(t).$$

From Application 3.3.B we know that $\Psi : V \to P_J$ is a C^k diffeomorphism for J suitably chosen, and that $\det D\Psi(r, t) = r \det(x(t)\ x'(t))$. By means of the Change of Variables Theorem one finds

$$\begin{aligned}\mathrm{area}(P_J) &= \int_{\Psi(V)} dx = \int_V |\det D\Psi(r, t)|\,dr\,dt \\ &= \int_0^1 r \int_J |\det(x(t)\ x'(t))|\,dt\,dr = \frac{1}{2}\int_J |\det(x(t)\ x'(t))|\,dt.\end{aligned}$$

Now let $k \geq 2$. Without loss of generality it may be assumed that $J = \,]0, s[$. Then the following holds:

$$\text{the area of } P_J \text{ is proportional to the length of } J, \tag{6.21}$$

if and only if there exists a constant $l \in \mathbf{R}$ such that, for all admissible s,

$$\int_0^s |\det(x(t)\ x'(t))|\, dt = l\, s.$$

The, permitted, calculation of the second derivative with respect to s of this identity gives, on account of the chain rule and Proposition 2.7.6,

$$\det(x'(t)\ x'(t)) + \det(x(t)\ x''(t)) = \det(x(t)\ x''(t)) = 0.$$

It follows that (6.21) is true if and only if

$$x(s) \text{ and the acceleration } x''(s) \text{ are linearly dependent vectors.}$$

In physical terms: the acceleration of a point particle orbiting in a plane is radial if and only if the particle obeys *Kepler's second law*: its radius vector sweeps out equal areas in equal times. ☆

Example 6.6.9 (Transport equation). The notation is that of Section 5.9. Let $(\Psi^t)_{t\in\mathbf{R}}$ be a one-parameter group of C^2 diffeomorphisms of $\mathbf{R}^n$ with associated tangent vector field $\psi : \mathbf{R}^n \to \mathbf{R}^n$, let $L \in \mathcal{J}(\mathbf{R}^n)$, and let $f \in C^1(\mathbf{R}^n)$. Then the following *transport equation* holds:

$$\frac{d}{dt}\int_{\Psi^t(L)} f(x)\, dx = \int_{\Psi^t(L)} (Df(x)\psi(x) + f(x)\operatorname{div}\psi(x))\, dx. \tag{6.22}$$

Indeed, by continuity $\det D\Psi^t(y) > 0$, for $t \in \mathbf{R}$ and $y \in L$. According to the Change of Variables Theorem 6.6.1, the Differentiation Theorem 2.10.4 and Formulae (5.22) and (5.31) we get

$$\frac{d}{dt}\int_{\Psi^t(L)} f(x)\, dx = \frac{d}{dt}\int_L (f\circ\Psi^t)(y)\det D\Psi^t(y)\, dy$$

$$= \int_L \Big(Df(\Psi^t(y))\psi(\Psi^t(y)) + (f\circ\Psi^t)(y)\ \operatorname{div}\psi(\Psi^t(y))\Big)\det D\Psi^t(y)\, dy.$$

And using the Change of Variables Theorem once again we obtain Formula (6.22).

Finally, choose $f = 1$, and conclude that the following assertions are equivalent.

(i) $\operatorname{vol}_n(\Psi^t(L)) = \operatorname{vol}_n(L)$, for all $t \in \mathbf{R}$ and all sets $L \in \mathcal{J}(\mathbf{R}^n)$.

(ii) $\operatorname{div}\psi = 0$ on $\mathbf{R}^n$. ☆

6.7 Partitions of unity

We introduce an important technical tool. Recall Definition 1.8.16 of an open covering.

Definition 6.7.1. Let $K \subset \mathbf{R}^n$ be a subset and let $\mathcal{O} = \{O_i \mid i \in I\}$ be an open covering of K. A C^k *partition of unity on* K *subordinate to* $\mathcal{O}$, where $k \in \mathbf{N}_0$, is a finite collection of functions $\chi_j : \mathbf{R}^n \to \mathbf{R}$ with $1 \le j \le l$ such that

(i) χ_j is a C^k function, for $1 \le j \le l$;

(ii) $0 \le \chi_j(x) \le 1$, for $x \in \mathbf{R}^n$;

(iii) for every j there exists $O \in \mathcal{O}$ such that $\operatorname{supp}(\chi_j) \subset O$;

(iv) $\sum_{1 \le j \le l} \chi_j(x) = 1$, for $x \in K$. ❍

Example 6.7.2. Let the notation be like the one above. Assume in addition K is compact and Jordan measurable, and $f : \mathbf{R}^n \to \mathbf{R}$ is a given function. Further assume that each of the functions $f\chi_j : \mathbf{R}^n \to \mathbf{R}$, with $1 \le j \le l$, is Riemann integrable over K (this may be regarded as local information). Then

$$(1_K f)(x) = 1_K(x) f(x) \sum_{1 \le j \le l} \chi_j(x) = \sum_{1 \le j \le l} 1_K(x)(f\chi_j)(x).$$

But this yields (the global information) that f is Riemann integrable over K, while

$$\int_K f(x)\,dx = \sum_{1 \le j \le l} \int_K f(x)\chi_j(x)\,dx.$$ ☆

Theorem 6.7.3. *For every compact set* $K \subset \mathbf{R}^n$ *and every open covering* $\mathcal{O}$ *of* K *there exists a continuous partition of unity over* K *subordinate to* $\mathcal{O}$.

Before proving the theorem we introduce auxiliary functions that play a role in the proof. Assume $a' < a < b < b'$ and define the piecewise affine function $f = f_{a',b',a,b} : \mathbf{R} \to \mathbf{R}$ by

$$f(x) = \begin{cases} 0 & (x \le a'); \\ \dfrac{x-a'}{a-a'} & (a' \le x \le a); \\ 1 & (a \le x \le b); \\ \dfrac{x-b'}{b-b'} & (b \le x \le b'); \\ 0 & (b' \le x). \end{cases} \tag{6.23}$$

Then f is continuous on $\mathbf{R}$. Let B and B' be rectangles in $\mathbf{R}^n$, as in (6.1), such that

$$B \subset B' \qquad \text{and} \qquad a'_j < a_j < b_j < b'_j \qquad (1 \le j \le n). \tag{6.24}$$

Then

$$f = f_{B',B} : \mathbf{R}^n \to \mathbf{R} \qquad \text{with} \qquad f(x) = \prod_{1 \le j \le n} f_{a'_j, b'_j, a_j, b_j}(x_j) \qquad (x \in \mathbf{R}^n) \tag{6.25}$$

is a continuous function such that

$$0 \le f(x) \le 1 \qquad (x \in \mathbf{R}^n); \qquad \operatorname{supp}(f) = B'; \qquad f(x) = 1 \qquad (x \in B). \tag{6.26}$$

Proof. Because $\mathcal{O}$ is a covering of K, there exists, for every $x \in K$, a set $O \in \mathcal{O}$ such that $x \in O$; denote this set O by O_x. Since O_x is open in $\mathbf{R}^n$, there exist rectangles B_x and B'_x that satisfy both (6.24) and

$$x \in I_x := \operatorname{int}(B_x) \subset B'_x \subset O_x. \tag{6.27}$$

The collection $\{ I_x \mid x \in K \}$ forms an open covering of the compact set K. By the Heine–Borel Theorem 1.8.17 there exist finitely many points $x_1, \dots, x_l \in K$ such that

$$K \subset \bigcup_{1 \le j \le l} I_{x_j}. \tag{6.28}$$

Now define, in the notation of (6.25),

$$\psi_j = f_{B'_j, B_j}, \qquad \text{where} \qquad B'_j = B'_{x_j}, \quad B_j = B_{x_j} \qquad (1 \le j \le l).$$

Then the $\psi_j : \mathbf{R}^n \to \mathbf{R}$ are continuous functions satisfying (see (6.26) and (6.27)), for $x \in B_j$,

$$0 \le \psi_j(x) \le 1 \quad (x \in \mathbf{R}^n); \qquad \operatorname{supp}(\psi_j) \subset B'_j \subset O_{x_j}; \qquad \psi_j(x) = 1. \tag{6.29}$$

Further, let

$$\chi_1 = \psi_1; \qquad \chi_{j+1} = (1 - \psi_1)(1 - \psi_2) \cdots (1 - \psi_j)\psi_{j+1} \qquad (1 \le j \le l - 1). \tag{6.30}$$

It follows from (6.29) that the $\chi_1, \dots, \chi_l$ satisfy the requirements (i)–(iii) for a partition of unity subordinate to $\mathcal{O}$. The relation

$$\sum_{1 \le i \le j} \chi_i = 1 - \prod_{1 \le i \le j} (1 - \psi_i) \tag{6.31}$$

is trivial for $j = 1$. If (6.31) is true for $j < l$, then summing (6.30) and (6.31) yields (6.31) for $j + 1$. Consequently (6.31) is valid for $j = l$. If $x \in K$, then by (6.28), (6.27) and (6.29) there exists an i such that $\psi_i(x) = 1$; thus follows $\sum_{1 \le j \le l} \chi_j(x) = 1$. ❑

Remark. After Formula (6.29) one might be tempted to finish the proof by setting $\chi_j = \frac{\psi_j}{\sum_{1\le j\le l}\psi_j}$. However, the zeros of the denominator then make it necessary to take additional measures.

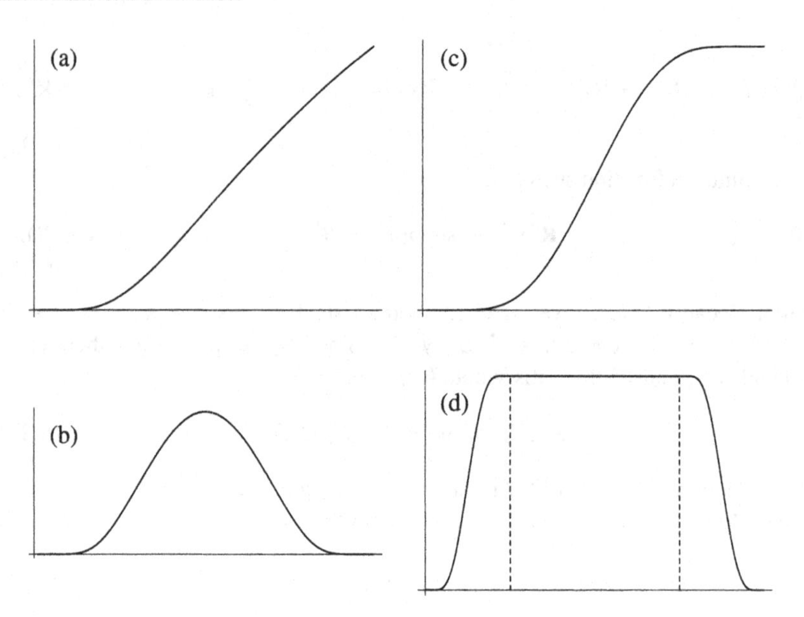

Illustration for the proof of Theorem 6.7.4
(a) g on $[0, 1]$; (b) $g_{-2,-1}$ on $[-2, -1]$;
(c) $h_{-2,-1}$ on $[-2, -1]$; (d) $h_{-2,2,-1,1}$ on $[-2, 2]$.
For clarity of display the scale in various graphs has been adjusted

With a view to applications in a subsequent section we further formulate:

Theorem 6.7.4. *For every compact set $K \subset \mathbf{R}^n$ and every open covering $\mathcal{A}$ of K there is a C^∞ partition of unity on K subordinate to $\mathcal{A}$.*

Proof. The proof proceeds in a way analogous to that of Theorem 6.7.3, provided we may replace the function f in (6.23) by a C^∞ function. To verify this we add the following remarks.

(a) The function $g : \mathbf{R} \to \mathbf{R}$ defined by

$$g(x) = 0 \qquad (x \le 0); \qquad g(x) = e^{-1/x} \qquad (x > 0)$$

is a C^∞ function; this is also true at $x = 0$, where all derivatives vanish. In fact, there exist polynomials p_k such that $g^{(k)}(x) = p_k(\frac{1}{x})g(x)$, for $x > 0$. In particular, $\lim_{x\downarrow 0} g^{(k)}(x) = \lim_{y\to\infty} p_k(y)e^{-y} = 0$. In turn, this implies $g^{(k)}(0) = 0$, for all $k \in \mathbf{N}_0$.

(b) Let $a < b$. The function $g_{a,b} : \mathbf{R} \to \mathbf{R}$ is a C^∞ function if

$$g_{a,b}(x) = g(x-a)\, g(b-x) \qquad (x \in \mathbf{R}).$$

(c) The function $h_{a,b} : \mathbf{R} \to \mathbf{R}$ is a C^∞ function if

$$h_{a,b}(x) = \int_a^x g_{a,b}(y)\, dy \Big/ \int_a^b g_{a,b}(y)\, dy,$$

while

$$h_{a,b}(x) = 0 \quad (x \le a); \qquad 0 < h_{a,b}(x) < 1 \quad (a < x < b);$$
$$h_{a,b}(x) = 1 \quad (b \le x).$$

(d) Let $a' < a < b < b'$ and define $h = h_{a',b',a,b} : \mathbf{R} \to \mathbf{R}$ by $h(x) = h_{a',a}(x)\, h_{-b',-b}(-x)$, for $x \in \mathbf{R}$. Then h is a C^∞ function, and one has

$$\begin{array}{llll} h(x) = 0 & (x \le a'); & 0 < h(x) < 1 & (a' < x < a); \\ h(x) = 1 & (a \le x \le b); & 0 < h(x) < 1 & (b < x < b'); \\ h(x) = 0 & (b' \le x). & & \end{array}$$

❑

6.8 Approximation of Riemann integrable functions

The techniques from the preceding section enable us to prove that Riemann integrable functions can be approximated, to arbitrary precision, by continuous functions.

Lemma 6.8.1. *Let $K \subset \mathbf{R}^n$ be a compact subset. For $\delta > 0$, let*

$$K_\delta = \{\, y \in \mathbf{R}^n \mid \text{there exists } x \in K \text{ with } \|y - x\| \le \delta \,\}.$$

Then K_δ is compact. If $U \subset \mathbf{R}^n$ is an open set with $K \subset U$, then $\delta > 0$ exists such that $K_\delta \subset U$.

Proof. For every $x \in K$ there exists $\delta(x) > 0$ such that $B(x;\, 2\delta(x)) \subset U$, see Definition1.2.1. Furthermore $K \subset \bigcup_{x \in K} B(x;\, \delta(x))$. On account of Definition 1.8.16.(ii) there exist $x_1, \dots, x_l \in K$ with

$$K \subset \bigcup_{1 \le j \le l} B(x_j;\, \delta(x_j)).$$

Let $\delta = \min\{\,\delta(x_j) \mid 1 \le j \le l\,\} > 0$, and let $y \in K_\delta$. Then there is an $x \in K$ with $\|y - x\| \le \delta$, and also $1 \le j \le l$ with $x \in B(x_j;\, \delta(x_j))$. Consequently

$$\|y - x_j\| \le \|y - x\| + \|x - x_j\| < \delta(x_j) + \delta(x_j) = 2\delta(x_j).$$

This gives $y \in B(x_j;\ 2\delta(x_j)) \subset U$; and hence $K_\delta \subset U$. ❑

Theorem 6.8.2. *Let $f : \mathbf{R}^n \to \mathbf{R}$ be a bounded function with compact support.*

(i) *Then f is Riemann integrable if and only if for every $\epsilon > 0$ there exist functions $g_-,\ g_+ \in C_c(\mathbf{R}^n)$, the space of continuous functions with compact support, such that*

$$g_- \le f \le g_+ \qquad \text{and} \qquad \int (g_+(x) - g_-(x))\,dx < \epsilon.$$

And in this case one also has

$$\left| \int f(x)\,dx - \int g_\pm(x)\,dx \right| < \epsilon.$$

(ii) *Let $U \subset \mathbf{R}^n$ be an open subset and assume* $\operatorname{supp}(f) \subset U$. *If f is Riemann integrable, the functions g_- and g_+ may then be chosen such that* $\operatorname{supp}(g_-)$ *and* $\operatorname{supp}(g_+) \subset U$.

Proof. (i). Consider first an arbitrary rectangle $B \subset \mathbf{R}^n$ and let $\epsilon > 0$ be arbitrary. By choosing a somewhat larger rectangle B' with $B \subset B'$ and using the function $f_{B',\,B}$ from (6.25), we can see that there exists a $g_+ \in C_c(\mathbf{R}^n)$ with

$$1_B \le g_+, \qquad B \subset \operatorname{supp}(g_+) = B', \qquad \int g_+(x)\,dx - \operatorname{vol}_n(B) < \frac{\epsilon}{2}.$$

Interchanging the roles of B and B' we obtain a $g_- \in C_c(\mathbf{R}^n)$ with similar properties. One finds

$$g_- \le 1_B \le g_+, \qquad \operatorname{supp}(g_-) \subset B \subset \operatorname{supp}(g_+) \subset B',$$

$$\int (g_+(x) - g_-(x))\,dx < \epsilon.$$

Now assume f to be Riemann integrable and let $B \subset \mathbf{R}^n$ be a rectangle with $\operatorname{supp}(f) \subset B$. According to Proposition 6.2.5 there exists a partition $\mathcal{B}$ of B with

$$\sum_{i \in I} \Big(\sup_{B_i} f\ \operatorname{vol}_n(B_i) - \inf_{B_i} f\ \operatorname{vol}_n(B_i) \Big) < \epsilon. \tag{6.32}$$

In consideration of the equality $f = f_+ - f_-$ from Theorem 6.2.8.(iii) we may then assume that $f \geq 0$. With ϵ suitably chosen, apply the argument above to the functions 1_{B_i}, and multiply the $g_-^{(i)}$ and $g_+^{(i)}$ thus found by $\inf_{B_i} f$ and $\sup_{B_i} f$, respectively. Assertion (i) follows with

$$\text{supp}(g_-) \subset \text{supp}(g_+) \subset \bigcup_{i \in I} B_i'.$$

(ii). According to the preceding lemma, there exists $\delta > 0$ such that $\text{supp}(f)_\delta \subset U$. We may assume that $\text{diameter}(B_i') < \delta$, for $i \in I$, for which possibly $\mathcal{B}$ in (6.32) must be refined. But then $B_i' \cap \text{supp}(f) \neq \emptyset$ implies $B_i' \subset U$; this proves the assertion. ❑

6.9 Proof of Change of Variables Theorem

The proof proceeds in five steps. In Step II we show that for every $y^0 \in V$ the restriction of the diffeomorphism Ψ to a suitably chosen open neighborhood $V(y^0)$ of y^0 in V can be written as a composition of "simpler" diffeomorphisms that essentially behave like diffeomorphisms in one variable. The proof relies on the Implicit Function Theorem 3.5.1. Indeed, this is an important reason for developing the theory of this theorem prior to the theory of integration in $\mathbf{R}^n$. In Step III we prove that the theorem follows if it is known to hold, for all $y^0 \in V$, for functions f with supports contained in the open sets $\Psi(V(y^0))$ in U. An essential element in this proof is the theory from Section 6.7 concerning compact sets and partitions of unity. In Step IV we show that the theorem is true for the composition of two diffeomorphisms if we already have it for the individual diffeomorphisms. In Step V the theorem is proved for the "simpler" diffeomorphisms by means of the Change of Variables Theorem for $\mathbf{R}$. Because the Change of Variables Theorem for $\mathbf{R}$ usually is proved under the assumption that f is continuous, we prove in Step I that the theorem is generally valid if it holds for continuous functions f with $\text{supp}(f) \subset U$.

Other proofs. The Change of Variables Theorem 6.6.1 can also be proved without recourse to the Implicit Function Theorem 3.5.1 as in Step II. Since it is of independent interest we give such a proof, called the second, in Appendix 6.13 to this chapter. As a consequence Chapter 6 can be studied independently from Chapters 3 through 5, which might appeal to readers who prefer to make an early start with the theory of integration. Which proof one prefers is mainly a matter of taste or prerequisites. By way of justification of the procedure followed in this section it may be remarked that the second proof uses the linear version of the decomposition from Step II, as well as Steps I and IV. Furthermore, localization as in Step III is an extremely useful technique in analysis.

The appendix contains one more proof, called the third; it might be slightly surprising yet it is quite efficient.

Step I (Reduction to continuous f). Let f be a Riemann integrable function with $\text{supp}(f) \subset U$. By Theorem 6.8.2.(ii) there exist, for every $\epsilon > 0$, continuous functions g_-, g_+ on $\mathbf{R}^n$ such that

$$\text{supp}(g_-),\ \text{supp}(g_+) \subset U; \qquad g_- \le f \le g_+; \qquad \int_U (g_+(x) - g_-(x))\,dx < \epsilon.$$

Therefore $(g_- \circ \Psi)\,|\det D\Psi| \le (f \circ \Psi)\,|\det D\Psi| \le (g_+ \circ \Psi)\,|\det D\Psi|$ on V; and so

$$\int_V (g_- \circ \Psi)(y)|\det D\Psi(y)|\,dy \le \underline{\int}_V (f \circ \Psi)(y)|\det D\Psi(y)|\,dy$$

$$\le \overline{\int}_V (f \circ \Psi)(y)|\det D\Psi(y)|\,dy \le \int_V (g_+ \circ \Psi)(y)|\det D\Psi(y)|\,dy.$$

Under the assumption that the theorem is true for f replaced by the continuous functions g_-, g_+ with supports in U, the first and last terms equal $\int_U g_-(x)\,dx$ and $\int_U g_+(x)\,dx$, respectively; but these numbers differ by less than ϵ, and moreover their difference from $\int_U f(x)\,dx$ is smaller than ϵ. Ergo, the upper and lower Riemann integrals of $(f \circ \Psi)\,|\det D\Psi|$ over V differ by less than ϵ, and therefore their difference from $\int_U f(x)\,dx$ is smaller than ϵ. Since this is true for every $\epsilon > 0$, it follows that $(f \circ \Psi)\,|\det D\Psi|$ is Riemann integrable over V, with Riemann integral $\int_U f(x)\,dx$.

Next, assume that conversely $g := (f \circ \Psi)\,|\det D\Psi|$ is Riemann integrable over V. One has

$$(f \circ \Psi)(y) = g(y)\,|\det D\Psi(y)|^{-1} \qquad (y \in V);$$

that is, for $x \in U$,

$$f(x) = (g \circ \Psi^{-1})(x)\,|\det D\Psi(\Psi^{-1}(x))|^{-1} = (g \circ \Psi^{-1})(x)\,|\det D\Psi^{-1}(x)|,$$

because the chain rule gives $\det D\Psi(\Psi^{-1}(x)) \det D\Psi^{-1}(x) = 1$. Applying the preceding argument, with f replaced by g and Ψ by Ψ^{-1}, we conclude that f is Riemann integrable over U.

Step II (Reduction to the case of dimension one). Let us write $\Psi(y) = (\Psi_1(y), \dots, \Psi_n(y))$ in $\mathbf{R}^n$. For the moment let $y^0 \in V$ be fixed. Because $D\Psi(y^0) \in \text{Aut}(\mathbf{R}^n)$, there exists an index $1 \le j \le n$ such that $D_j\Psi_n(y^0) \ne 0$. One may assume $j = n$, that is, $D_n\Psi_n(y^0) \ne 0$, which may require prior permutation of the coordinates of y. By virtue of the Implicit Function Theorem 3.5.1 this means that the equation for y_n,

$$\Psi_n(y_1, \dots, y_n) = x_n, \tag{6.33}$$

with the $y_1, \dots, y_{n-1}, x_n$ as parameters (near $(y_1^0, \dots, y_{n-1}^0, \Psi_n(y_1^0, \dots, y_n^0))$), can be solved near y_n^0. Denote the solution by

$$y_n = \phi_n(y_1, \dots, y_{n-1}, x_n); \tag{6.34}$$

it is C^1-dependent on $y_1, \dots, y_{n-1}, x_n$. But this implies that the C^1 mapping Ξ_n defined in an open neighborhood of y^0 in V by

$$\Xi_n(y) = (y_1, \dots, y_{n-1}, \Psi_n(y)) = (y_1, \dots, y_{n-1}, x_n), \tag{6.35}$$

is invertible on its image, with C^1 inverse

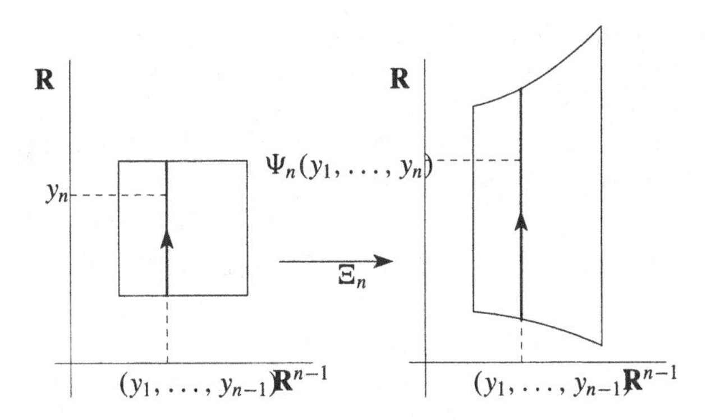

Illustration for Formula (6.35)

$$\Xi_n^{-1}(y_1, \dots, y_{n-1}, x_n) = (y_1, \dots, y_{n-1}, \phi_n(y_1, \dots, y_{n-1}, x_n)). \tag{6.36}$$

Consequently, Ξ_n is a (locally defined) C^1 diffeomorphism leaving the first $n-1$ coordinates invariant. In addition we have, on account of (6.36), (6.33) and (6.34),

$$((\Psi \circ \Xi_n^{-1})(y_1, \dots, y_{n-1}, x_n))_n = \Psi_n(y_1, \dots, y_{n-1}, \phi_n(y_1, \dots, y_{n-1}, x_n)) = x_n,$$

while, for $1 \le j \le n-1$,

$$((\Psi \circ \Xi_n^{-1})(y_1, \dots, y_{n-1}, x_n))_j = \Psi_j(y_1, \dots, y_{n-1}, \phi_n(y_1, \dots, y_{n-1}, x_n)).$$

That is, $\Psi \circ \Xi_n^{-1}$ leaves the n-th coordinate invariant and acts like a C^1 diffeomorphism (C^1-dependent on x_n) on the first $n-1$ coordinates. Finally, in an open neighborhood of y^0 in V,

$$\Psi = (\Psi \circ \Xi_n^{-1}) \circ \Xi_n.$$

Once more applying, mutatis mutandis, the foregoing procedure to $\Psi \circ \Xi_n^{-1}$, we finally obtain that the restriction of the diffeomorphism Ψ to a suitably chosen open neighborhood $V(y^0)$ of y^0 in V can be written as a composition of finitely many diffeomorphisms of a form similar to that in (6.35), save permutation of the coordinates.

Step III (Localization). Let $K = \mathrm{supp}(f) \subset U$. Then K is a compact set. We now demonstrate that the Change of Variables Theorem holds if for every $x \in K$ it is possible to find a bounded open neighborhood O_x of x in U such that the theorem holds for continuous functions f additionally satisfying

$$\mathrm{supp}(f) \subset O_x. \tag{6.37}$$

Indeed, the collection $\mathcal{O} = \{\, O_x \mid x \in K \,\}$ is an open covering of K. On account of Theorem 6.7.3 there exists a continuous partition $\chi_1, \ldots, \chi_l$ of unity on K subordinate to $\mathcal{O}$. Consequently, for every $1 \le j \le l$ there is an $x \in K$ such that $\mathrm{supp}(\chi_j f) \subset \mathrm{supp}(\chi_j) \subset O_x$; that is, $\chi_j f$ satisfies (6.37). Because χ_j is continuous, $\chi_j f$ is continuous. It follows that

$$\int_U (\chi_j f)(x)\,dx = \int_V (\chi_j f)(\Psi(y))\,|\det D\Psi(y)|\,dy.$$

Hence

$$\int_U \sum_j \chi_j(x)\, f(x)\,dx = \int_V \sum_j \chi_j(\Psi(y))\, f(\Psi(y))\,|\det D\Psi(y)|\,dy.$$

But this leads to the desired conclusion, because $\sum_j \chi_j(x) = 1$, for $x \in K = \mathrm{supp}(f)$.

Step IV (Composition). If the Change of Variables Theorem holds for a C^1 diffeomorphism $\Psi : V \to U$ and for a C^1 diffeomorphism $\Xi : W \to V$, then it also holds for the composition $\Psi \circ \Xi : W \to U$. Indeed, the chain rule gives

$$\det D(\Psi \circ \Xi)(z) = \det\left(D\Psi(\Xi(z)) \circ D\Xi(z)\right) = \det D\Psi(\Xi(z))\,\det D\Xi(z).$$

Therefore

$$\begin{aligned}
&\int_W f(\Psi \circ \Xi(z))\,|\det D(\Psi \circ \Xi)(z)|\,dz \\
&\quad = \int_W (f \circ \Psi)(\Xi(z))\,|\det D\Psi(\Xi(z))|\,|\det D\Xi(z)|\,dz \\
&\quad = \int_V (f \circ \Psi)(y)\,|\det D\Psi(y)|\,dy = \int_U f(x)\,dx.
\end{aligned}$$

Step V (Case of dimension one). The Change of Variables Theorem holds if Ψ has the special form (compare (6.35))

$$\Psi(y) = (y_1, \ldots, y_{n-1}, \psi(y)),$$

where ψ is a C^1 mapping. Indeed,

$$D\Psi(y) = \begin{pmatrix} 1 & 0 & \cdots & \cdots & 0 \\ 0 & \ddots & \ddots & & \vdots \\ \vdots & \ddots & \ddots & \ddots & \vdots \\ 0 & \cdots & 0 & 1 & 0 \\ D_1\psi(y) & \cdots & \cdots & \cdots & D_n\psi(y) \end{pmatrix}$$

implies $\det D\Psi(y) = D_n\psi(y) \neq 0$. Therefore $y_n \mapsto \psi(y_1, \dots, y_n)$ may be assumed strictly monotonically decreasing or increasing. Using Corollary 6.4.3 one then finds

$$\int_V (f \circ \Psi)(y) |\det D\Psi(y)|\, dy = \int_V f(y_1, \dots, y_{n-1}, \psi(y))\, |D_n\psi(y)|\, dy$$

$$= \int \cdots \Big(\int f(y_1, \dots, y_{n-1}, \psi(y_1, \dots, y_n))\, |D_n\psi(y_1, \dots, y_n)|\, dy_n \Big) \cdots dy_1.$$

In view of the monotony we may remove the absolute signs in the integrand and account for them in the integration limits for the integral with respect to y_n. To the last term we may subsequently apply the Change of Variables Theorem on $\mathbf{R}$, which leads to

$$\begin{aligned} \int_V (f \circ \Psi)(y) |\det D\Psi(y)|\, dy &= \int \cdots \Big(\int f(y_1, \dots, x_n)\, dx_n \Big) \cdots dy_1 \\ &= \int_U f(x)\, dx. \end{aligned}$$

6.10 Absolute Riemann integrability

The proof of the Change of Variables Theorem as it stands assumes that the integrand f is bounded and vanishes outside a compact set. In applications one is often interested in functions f like $f(x) = \log \|x\|$, for $0 < \|x\| < 1$, or $f(x) = e^{-\|x\|^2}$, for $x \in \mathbf{R}^n$; that is, an f which either is itself unbounded or has an unbounded domain. For such functions Corollary 6.10.7 below is of importance.

Definition 6.10.1. Let $U \subset \mathbf{R}^n$ be open and let $f : U \to \mathbf{R}$. The function f is said to be *locally Riemann integrable in U* if for every $x \in U$ there exists a rectangle $B \subset U$ such that $x \in \operatorname{int}(B)$ and f is Riemann integrable over B. ❍

Lemma 6.10.2. *Let $U \subset \mathbf{R}^n$ be open and let $f : U \to \mathbf{R}$. The following assertions are equivalent.*

(i) *The function f is locally Riemann integrable in U.*

(ii) *For every function $\chi \in C_c(\mathbf{R}^n)$ with $\mathrm{supp}(\chi) \subset U$ the function χf is Riemann integrable over U.*

Proof. **(ii) $\Rightarrow$ (i)** follows by choosing a suitable rectangle $B \subset U$ and subsequently approximating 1_B by functions $\chi \in C_c(\mathbf{R}^n)$, using Theorem 6.8.2.(ii).
(i) $\Rightarrow$ (ii). Choose $\chi \in C_c(\mathbf{R}^n)$ with $\mathrm{supp}(\chi) \subset U$. For every $x \in \mathrm{supp}(\chi)$ there exists a rectangle $B_x \subset U$ with $x \in \mathrm{int}(B_x)$ and f Riemann integrable over B_x. Applying Theorem 6.7.3 to the compact set $\mathrm{supp}(\chi)$ and the open covering $\{\, \mathrm{int}(B_x) \mid x \in \mathrm{supp}(\chi) \,\}$, we find $\chi_j \in C_c(\mathbf{R}^n)$ where $1 \le j \le l$, with the following properties. On $\mathrm{supp}(\chi)$ one has $\sum \chi_j = 1$, and for every index j there exists an $x \in \mathrm{supp}(\chi)$ with $\mathrm{supp}(\chi_j) \subset \mathrm{int}(B_x)$. Hence follows $\chi_j \chi f = (\chi_j \chi)(1_{B_x} f)$. By means of Corollary 6.3.7 and Theorem 6.2.8.(iv) one concludes that $\chi_j \chi f$ is Riemann integrable over U. But then $\chi f = \sum_j \chi_j \chi f$ is also Riemann integrable over U. ❑

Lemma 6.10.2 immediately has the following consequence.

Lemma 6.10.3. *Let $U \subset \mathbf{R}^n$ be open and let $f : U \to \mathbf{R}$. Then, with $f = f_+ - f_-$ as in Theorem 6.2.8.(iii), we have the following properties.*

(i) *f is locally Riemann integrable if and only if f_+ and f_- are locally Riemann integrable.*

(ii) *If f is continuous then f is locally Riemann integrable.*

(iii) *If f is locally Riemann integrable then f is integrable over every $K \in \mathcal{J}(U)$.*

(iv) *The collection of functions that are locally Riemann integrable over U forms a linear space.*

In preparation for a new definition we formulate the following:

Proposition 6.10.4. *Let $A \subset \mathbf{R}^n$ be bounded and let $f : A \to \mathbf{R}$ be bounded and Riemann integrable over A. Then, with $f = f_+ - f_-$ as in Theorem 6.2.8.(iii),*

$$\int_A f(x)\,dx = \sup_{K \in \mathcal{J}(A)} \int_K f_+(x)\,dx - \sup_{K \in \mathcal{J}(A)} \int_K f_-(x)\,dx.$$

Proof. In view of Theorem 6.2.8.(iii) it is sufficient to consider f_+ and f_- separately, hence we may assume $f \ge 0$. Suppose that B is an n-dimensional rectangle with $A \subset B$ and that $\epsilon > 0$. Consider a partition $\mathcal{B} = \{\, B_i \mid i \in I \,\}$ of B with

$\overline{S}(f, \mathcal{B}) - \underline{S}(f, \mathcal{B}) < \epsilon$. Next let $I' = \{ i \in I \mid B_i \subset A \}$, then $K := \bigcup_{i \in I'} B_i \in \mathcal{J}(A)$. Since $B_i \not\subset A$ implies $\inf_{B_i} f = 0$, we find

$$\underline{S}(f, \mathcal{B}) = \underline{S}(f, \mathcal{B}') \leq \int_K f(x)\, dx \leq \overline{S}(f, \mathcal{B}') \leq \overline{S}(f, \mathcal{B}).$$

On the other hand, we have the same inequalities with $\int_K f(x)\, dx$ replaced by $\int_A f(x)\, dx$. Accordingly

$$\int_A f(x)\, dx - \epsilon \leq \int_K f(x)\, dx \leq \int_A f(x)\, dx. \qquad \square$$

Definition 6.10.5. Let $U \subset \mathbf{R}^n$ be open and let $f : U \to \mathbf{R}$. The function f is said to be *absolutely Riemann integrable* over U if f is locally Riemann integrable in U and if

$$\sup_{K \in \mathcal{J}(U)} \int_K |f(x)|\, dx < \infty.$$

Since $0 \leq f_\pm \leq |f| = f_+ + f_-$, we have that f_+ and f_- are absolutely Riemann integrable if and only if f is so. For f absolutely Riemann integrable over U, we define

$$\int_U f(x)\, dx = \sup_{K \in \mathcal{J}(U)} \int_K f_+(x)\, dx - \sup_{K \in \mathcal{J}(U)} \int_K f_-(x)\, dx. \qquad \bigcirc$$

Proposition 6.10.4 shows that Definition 6.10.5 and Definition 6.3.1 agree for functions for which both definitions apply. The collection of functions that are absolutely Riemann integrable over U forms a linear space, but, in contrast with Theorem 6.2.8.(iv), it is not closed under pointwise multiplication of functions. For example, the function $x \mapsto \frac{1}{\sqrt{x}}$ is absolutely Riemann integrable over $]\,0, 1\,[$, whereas $x \mapsto (\frac{1}{\sqrt{x}})^2 = \frac{1}{x}$ is not.

Let $U \subset \mathbf{R}^n$ be open. In the following theorem we will consider sequences

$$(K_k)_{k \in \mathbf{N}} \quad \text{in} \quad \mathcal{J}(U) \qquad \text{with} \qquad \bigcup_{k \in \mathbf{N}} K_k = U, \qquad K_k \subset \operatorname{int}(K_{k+1}) \quad (k \in \mathbf{N}). \tag{6.38}$$

First we show how to construct such sequences. Define

$$C_k = \{ x \in U \mid \|x\| \leq k, \ \|x - y\| \geq \frac{1}{k} \text{ as } y \notin U \} \qquad (k \in \mathbf{N}).$$

Then C_k is a compact subset of U on account of the Heine–Borel Theorem 1.8.17, while $C_k \subset C_{k+1}$. To show that $\{ C_k \mid k \in \mathbf{N} \}$ is a covering of U, let $x \in U$ be arbitrary. Since U is open, we have $\inf\{ \|x - y\| \mid y \notin U \} > 0$; hence there exists $k \in \mathbf{N}$ with $x \in C_k$. Next, we note that the set

$$V_{k+1} = \{ x \in U \mid \|x\| < k + 1, \ \|x - y\| > \frac{1}{k+1} \text{ as } y \notin U \} \qquad (k \in \mathbf{N})$$

is open. Since $C_k \subset V_{k+1} \subset C_{k+1}$, it follows that $C_k \subset \mathrm{int}(C_{k+1})$. The sets C_k are not quite the sets we want, since they may not be Jordan measurable. We construct the sets K_k as follows. For each $x \in C_k$ choose a rectangle that is centered at x and is contained in $\mathrm{int}(C_{k+1})$. The interiors of these rectangles cover the compact set C_k; choose finitely many of them whose interiors cover C_k and let their union be K_k. Since K_k is a finite union of rectangles, it belongs to $\mathcal{J}(U)$. Then $C_k \subset \mathrm{int}(K_k) \subset K_k \subset \mathrm{int}(C_{k+1})$. Then $(K_k)_{k\in\mathbf{N}}$ satisfies the conditions in (6.38).

Next we give a useful criterion for the absolute Riemann integrability of a continuous function over an open set.

Theorem 6.10.6. *Let $U \subset \mathbf{R}^n$ be open and let $f : U \to \mathbf{R}$ be continuous. Suppose $(K_k)_{k\in\mathbf{N}}$ is as in (6.38). Then the following are equivalent.*

(i) *f is absolutely Riemann integrable over U.*

(ii) *$(\int_{K_k} |f(x)|\,dx)_{k\in\mathbf{N}}$ is a bounded sequence in $\mathbf{R}$.*

If one of these conditions is satisfied, we have that $(\int_{K_k} |f(x)|\,dx)_{k\in\mathbf{N}}$ is monotonically nondecreasing and

$$\lim_{k\to\infty} \int_{K_k} f(x)\,dx = \int_U f(x)\,dx. \tag{6.39}$$

Proof. **(i) $\Rightarrow$ (ii).** Since f is absolutely Riemann integrable over U if and only f_- and f_+ are so, we may assume that $f \geq 0$. Obviously (ii) follows, since for every $k \in \mathbf{N}$,

$$\int_{K_k} f(x)\,dx \leq \sup_{K\in\mathcal{J}(U)} \int_K f(x)\,dx = \int_U f(x)\,dx.$$

(ii) $\Rightarrow$ (i). Let $K \in \mathcal{J}(U)$ be arbitrary, then K is covered by the increasing collection of open sets $\{\,\mathrm{int}(K_k) \mid k \in \mathbf{N}\,\}$; hence, being compact, by finitely many of them, and therefore by one of them, say $\mathrm{int}(K_N)$. Accordingly

$$\int_K f(x)\,dx \leq \int_{K_N} f(x)\,dx \leq \lim_{k\to\infty} \int_{K_k} f(x)\,dx.$$

It follows that f is absolutely Riemann integrable over U and that

$$\int_U f(x)\,dx = \sup_{K\in\mathcal{J}(U)} \int_K f(x)\,dx \leq \lim_{k\to\infty} \int_{K_k} f(x)\,dx.$$

Finally, we prove Formula (6.39). We have

$$\lim_{k\to\infty} \int_{K_k} f(x)\,dx \leq \sup_{K\in\mathcal{J}(U)} \int_K f(x)\,dx = \int_U f(x)\,dx,$$

and the conjunction of the last two formulae proves the identity. ❑

Corollary 6.10.7. *Assume U and V to be open subsets of $\mathbf{R}^n$ and let $\Psi : V \to U$ be a C^1 diffeomorphism. Let $f : U \to \mathbf{R}$ be a function. Then f is absolutely Riemann integrable over U if and only if the function $(f \circ \Psi)\,|\det D\Psi|$ is absolutely Riemann integrable over V. If either of these conditions is met, then*

$$\int_U f(x)\,dx = \int_V (f \circ \Psi)(y)|\det D\Psi(y)|\,dy.$$

Proof. As follows from the preceding theorem and construction, we may approximate f by Riemann integrable functions with compact supports. Then apply the Change of Variables Theorem 6.6.1 to these functions. ❑

Example 6.10.8 ($\int_{\mathbf{R}} e^{-x^2}\,dx = \sqrt{\pi} = 1.772\,453\,850\,905\,516\cdots$). (See also Exercises 2.73, 6.15, 6.41 and 6.50.(i)). The function $f(x) = e^{-\|x\|^2}$ is continuous on $\mathbf{R}^2$, and therefore f is locally Riemann integrable in $\mathbf{R}^2$. Define $B(R) = \{x \in \mathbf{R}^2 \mid \|x\| \le R\}$, for $R > 0$. Using polar coordinates $x = r(\cos\alpha,\ \sin\alpha)$ one finds, for all $R > 0$,

$$\int_{B(R)} f(x)\,dx = \int_{-\pi}^{\pi}\int_0^R re^{-r^2}\,dr\,d\alpha = 2\pi\left[-\frac{e^{-r^2}}{2}\right]_0^R = \pi(1-e^{-R^2}) \le \pi. \tag{6.40}$$

Let $K \in \mathcal{J}(\mathbf{R}^2)$, then there exists a number $R > 0$ such that $K \subset B(R)$. Because f is a positive function it follows that

$$\int_K f(x)\,dx \le \int_{B(R)} f(x)\,dx \le \pi;$$

that is, f is absolutely Riemann integrable over $\mathbf{R}^2$. On account of Corollary 6.4.3 one also has, for every $R > 0$,

$$\Big(\int_{-R}^{R} e^{-x^2}\,dx\Big)^2 = \Big(\int_{-R}^{R} e^{-x_1^2}\,dx_1\Big)\Big(\int_{-R}^{R} e^{-x_2^2}\,dx_2\Big) = \int_{C(R)} e^{-(x_1^2+x_2^2)}\,dx,$$

where $C(R) = [-R,\ R] \times [-R,\ R]$. We have $B(R) \subset C(R) \subset B(R\sqrt{2})$, and by (6.40) therefore

$$\pi(1-e^{-R^2}) \le \int_{C(R)} f(x)\,dx \le \pi(1-e^{-2R^2});$$

from which

$$\Big(\int_{\mathbf{R}} e^{-x^2}\,dx\Big)^2 = \lim_{R\to+\infty}\int_{C(R)} f(x)\,dx = \pi. \qquad ☆$$

Remark. For a function f on $\mathbf{R}$ we have the notion of improper Riemann integrability. This concept is of particular importance if f is Riemann integrable over $\mathbf{R}$, while $|f|$ is not. An example of this was given in Example 2.10.11. Absolute Riemann integrability is a more stringent condition on a function f on $\mathbf{R}^n$, in that integrability of $|f|$ is required. Accordingly, cancellations due to oscillatory behavior are not taken into account. For the Riemann integral in $\mathbf{R}^n$, with $n > 1$, there is no useful analog of the concept of improper Riemann integrability. This is related to the fact that an unbounded set in $\mathbf{R}^n$, with $n > 1$, can be approximated from within by compact sets in many different ways.

6.11 Application of integration: Fourier transformation

We will show that there exists an ample class of functions that can be written as a continuous superpositions of periodic functions of a simple type, see Formula (6.43).

We recall the notation from Formula (2.24). In particular we have, for a vector $x = (x_1, \dots, x_n) \in \mathbf{R}^n$ and a multi-index $\alpha = (\alpha_1, \dots, \alpha_n) \in \mathbf{N}_0^n$,

$$x^\alpha = x_1^{\alpha_1} \cdots x_n^{\alpha_n}, \qquad |\alpha| = \alpha_1 + \cdots + \alpha_n, \qquad D^\alpha = D_1^{\alpha_1} \cdots D_n^{\alpha_n}. \tag{6.41}$$

Definition 6.11.1. We define the space $\mathcal{S}(\mathbf{R}^n)$ of *Schwartz functions* on $\mathbf{R}^n$ as the linear space of all C^∞ functions $f : \mathbf{R}^n \to \mathbf{C}$ such that, for all multi-indices α, $\beta \in \mathbf{N}_0^n$,

$$\sup\{\, |x^\beta (D^\alpha f)(x)| \mid x \in \mathbf{R}^n \,\} < \infty. \qquad \bigcirc$$

Note that if $f \in \mathcal{S}(\mathbf{R}^n)$, then $\cdot^\beta f : x \mapsto x^\beta f(x)$ and $D^\alpha f$ are also functions in $\mathcal{S}(\mathbf{R}^n)$. Furthermore, for every $f \in \mathcal{S}(\mathbf{R}^n)$ there exists a constant $c > 0$ such that, for all $x \in \mathbf{R}^n$,

$$|f(x)| \le c(1 + \|x\|)^{-(n+1)};$$

and therefore also, with $\xi \in \mathbf{R}^n$ arbitrary,

$$|e^{-i\langle x, \xi \rangle} f(x)| \le c(1 + \|x\|)^{-(n+1)}.$$

Consequently, the integrand in the following definition is absolutely Riemann integrable over $\mathbf{R}^n$.

Definition 6.11.2. For every $f \in \mathcal{S}(\mathbf{R}^n)$ we define the *Fourier transform* $\widehat{f}$ of f as the function $\widehat{f} : \mathbf{R}^n \to \mathbf{C}$ with

$$\widehat{f}(\xi) = \int_{\mathbf{R}^n} e^{-i\langle x, \xi \rangle} f(x)\, dx. \qquad \bigcirc$$

Theorem 6.11.3. (i) *The* Fourier transformation $f \mapsto \widehat{f}$ *is an endomorphism of* $\mathcal{S}(\mathbf{R}^n)$. *For every multi-index* $\alpha \in \mathbf{N}_0^n$ *and* $\xi \in \mathbf{R}^n$

$$\text{(ii)}\ \widehat{(D^\alpha f)}(\xi) = (i\xi)^\alpha \widehat{f}(\xi), \qquad \text{(iii)}\ \widehat{(\cdot^\alpha f)}(\xi) = (iD)^\alpha \widehat{f}(\xi).$$

Proof. (iii). Note that

$$|(iD_\xi)^\alpha(e^{-i\langle x,\xi\rangle} f(x))| = |x^\alpha e^{-i\langle x,\xi\rangle} f(x)| \le |x^\alpha f(x)| \qquad (x \in \mathbf{R}^n),$$

while $\cdot^\alpha f$ again belongs to $\mathcal{S}(\mathbf{R}^n)$. Therefore assertion (iii) follows from the Differentiation Theorem 2.10.13, or rather, from its direct extension to the case of integration over $\mathbf{R}^n$. (See Exercise 6.82 for a direct proof of assertion (iii).)
(ii). Since for $\beta \in \mathbf{N}_0^n$

$$(i\xi)^\beta e^{-i\langle x,\xi\rangle} = (-D_x)^\beta(e^{-i\langle x,\xi\rangle}),$$

we find from assertion (iii), by integration by parts,

$$\begin{aligned}(i\xi)^\beta((iD)^\alpha \widehat{f})(\xi) &= \int_{\mathbf{R}^n} (-D_x)^\beta(e^{-i\langle x,\xi\rangle})\, x^\alpha f(x)\, dx \\ &= \int_{\mathbf{R}^n} e^{-i\langle x,\xi\rangle} D^\beta(x^\alpha f(x))\, dx.\end{aligned} \tag{6.42}$$

Note that this is allowed, and that the boundary terms have vanishing limits, because $f \in \mathcal{S}(\mathbf{R}^n)$. In particular, assertion (ii) follows from (6.42), by taking $\alpha = 0 \in \mathbf{N}_0^n$. Finally we have, for all $\xi \in \mathbf{R}^n$,

$$|\xi^\beta(D^\alpha \widehat{f})(\xi)| \le \int_{\mathbf{R}^n} |D^\beta(x^\alpha f(x))|\, dx < \infty;$$

which proves $\widehat{f} \in \mathcal{S}(\mathbf{R}^n)$. ❑

Example 6.11.4. If $g(x) = e^{-\frac{1}{2}\|x\|^2}$, then $\widehat{g} = (2\pi)^{n/2} g$ (see Exercise 6.83 for a different proof). Indeed, we have $(D_j g)(x) = -x_j g(x)$, for $1 \le j \le n$. Therefore, by (ii) and (iii) from the preceding theorem $-\xi_j \widehat{g}(\xi) = (D_j \widehat{g})(\xi)$, for $1 \le j \le n$. Solving this system of differential equations we obtain $\widehat{g}(\xi) = c\, e^{-\frac{1}{2}\|\xi\|^2}$, with constant $c \ne 0$. Then, by Example 6.10.8,

$$c = \widehat{g}(0) = \int_{\mathbf{R}^n} g(x)\, dx = \int_{\mathbf{R}^n} e^{-\frac{1}{2}\|x\|^2}\, dx = (2\pi)^{\frac{n}{2}}. \qquad ☆$$

Example 6.11.5 (Convolution). For $f,\ g\in\mathcal{S}(\mathbf{R}^n)$ we define the *convolution* $f*g:\mathbf{R}^n\to\mathbf{C}$ by

$$f*g(x)=\int_{\mathbf{R}^n}f(x-y)g(y)\,dy.$$

Note that, for all $x,y\in\mathbf{R}^n$ one has $|f(x-y)g(y)|\le|g(y)|\ \sup|f|$; and this implies that the integrand is absolutely Riemann integrable over $\mathbf{R}^n$. In fact, $f*g\in\mathcal{S}(\mathbf{R}^n)$; for the proof see Exercise 6.85. Moreover

$$\widehat{f*g}=\widehat{f}\widehat{g}.$$

This can be proved by noting that in the following integrals the order of integration may be interchanged.

$$\begin{aligned}(\widehat{f*g})(\xi)&=\int_{\mathbf{R}^n}e^{-i\langle x,\xi\rangle}\int_{\mathbf{R}^n}f(x-y)g(y)\,dy\,dx\\&=\int_{\mathbf{R}^n}g(y)\int_{\mathbf{R}^n}e^{-i\langle x,\xi\rangle}f(x-y)\,dx\,dy\\&=\int_{\mathbf{R}^n}g(y)\int_{\mathbf{R}^n}e^{-i\langle x+y,\xi\rangle}f(x)\,dx\,dy\\&=\int_{\mathbf{R}^n}e^{-i\langle y,\xi\rangle}g(y)\,dy\int_{\mathbf{R}^n}e^{-i\langle x,\xi\rangle}f(x)\,dx=\widehat{f}(\xi)\widehat{g}(\xi).\end{aligned}$$ ☆

Theorem 6.11.6 (Fourier Inversion Theorem). *The Fourier transformation is an automorphism of $\mathcal{S}(\mathbf{R}^n)$. The inverse of Fourier transformation is given by*

$$f(x)=(2\pi)^{-n}\int_{\mathbf{R}^n}e^{i\langle x,\xi\rangle}\widehat{f}(\xi)\,d\xi\qquad(f\in\mathcal{S}(\mathbf{R}^n),\ x\in\mathbf{R}^n).\tag{6.43}$$

Proof. First we verify Formula (6.43) for $x=0$, that is, we prove

$$f(0)=(2\pi)^{-n}\int_{\mathbf{R}^n}\widehat{f}(\xi)\,d\xi.\tag{6.44}$$

As a starting assumption about $f\in\mathcal{S}(\mathbf{R}^n)$ let $f(0)=0$, then

$$f(x)=\int_0^1\frac{df}{dt}(tx)\,dt=\sum_{1\le j\le n}x_j\int_0^1D_jf(tx)\,dt=:\sum_{1\le j\le n}x_j\widetilde{g}_j(x),$$

where $\widetilde{g}_j\in C^\infty(\mathbf{R}^n)$ for $1\le j\le n$. Now use (6.26) combined with (d) in the proof of Theorem 6.7.4 to find a C^∞ function χ with compact support such that $\chi=1$ in a neighborhood of 0. Then we have $g_j\in\mathcal{S}(\mathbf{R}^n)$ for $1\le j\le n$, if we take

$$g_j(x)=(\widetilde{g}_j\chi)(x)+\frac{x_j}{\|x\|^2}(f(1-\chi))(x)\qquad(x\in\mathbf{R}^n);$$

and furthermore

$$f(x) = \sum_{1 \le j \le n} x_j g_j(x).$$

Using (iii) from the preceding theorem one obtains $\widehat{f} = i \sum_{1 \le j \le n} D_j \widehat{g}_j$. On the strength of Corollary 6.4.3, the Fundamental Theorem of Integral Calculus 2.10.1, applied to $x_j \mapsto D_j \widehat{g}_j(x)$, and the fact that $g_j \in \mathcal{S}(\mathbf{R}^n)$, it then follows that

$$\int_{\mathbf{R}^n} \widehat{f}(\xi)\, d\xi = 0. \tag{6.45}$$

This proves (6.44) if $f(0) = 0$. Now let $f \in \mathcal{S}(\mathbf{R}^n)$ be arbitrary; set $h := f - f(0)g$, with g as in Example 6.11.4. One then has $h \in \mathcal{S}(\mathbf{R}^n)$, $h(0) = 0$ and $\widehat{h} = \widehat{f} - f(0)\widehat{g}$. Thus, according to (6.45),

$$0 = \int_{\mathbf{R}^n} \widehat{h}(\xi)\, d\xi = \int_{\mathbf{R}^n} \widehat{f}(\xi)\, d\xi - f(0) \int_{\mathbf{R}^n} \widehat{g}(\xi)\, d\xi.$$

Furthermore,

$$\int_{\mathbf{R}^n} \widehat{g}(\xi)\, d\xi = (2\pi)^{\frac{n}{2}} \int_{\mathbf{R}^n} e^{-\frac{1}{2}\|x\|^2}\, dx = (2\pi)^n.$$

Therefore (6.43) holds for $x = 0$. The formula follows for arbitrary $x^0 \in \mathbf{R}^n$ by replacing the function f by $f^0 : x \mapsto f(x + x^0)$ in (6.44). Indeed

$$\widehat{f^0}(\xi) = \int_{\mathbf{R}^n} e^{-i\langle x, \xi \rangle} f(x + x^0)\, dx = \int_{\mathbf{R}^n} e^{-i\langle x - x^0, \xi \rangle} f(x)\, dx = e^{i\langle x^0, \xi \rangle} \widehat{f}(\xi). \quad \square$$

Remark. The Inversion Formula (6.43) tells us that a Schwartz function f on $\mathbf{R}^n$ can be written as a *superposition* over the different *frequencies* $\xi \in \mathbf{R}^n$ of the plane waves $x \mapsto e^{i\langle x, \xi \rangle}$ in $\mathbf{R}^n$, where $\widehat{f}(\xi)$ determines the *amplitude* of the wave of frequency ξ.

The plane waves are characterized by the fact that they are exactly the bounded eigenfunctions for the differential operator D, that is, if $f \in C^1(\mathbf{R}^n, \mathbf{C})$ satisfies the eigenvalue equation $Df(x) = \lambda\, f(x)$ with $\lambda \in \mathbf{C}^n$, and if f is a bounded function, then $\lambda = i\,\xi$ with $\xi \in \mathbf{R}^n$, and $f(x) = f(0)\, e^{i\langle x, \xi \rangle}$. The arbitrary function in $\mathcal{S}(\mathbf{R}^n)$ can therefore be written as a superposition of bounded eigenfunctions for the differential operator D acting on $C^1(\mathbf{R}^n, \mathbf{C})$. The formula $(\mathcal{F} \circ D \circ \mathcal{F}^{-1}) f(\xi) = (i\,\xi) f(\xi)$ shows that the differential operator D acting on $\mathcal{S}(\mathbf{R}^n)$ can be diagonalized by conjugation with the Fourier transformation $\mathcal{F}$, and that the action in $\mathcal{S}(\mathbf{R}^n)$ of the operator obtained by conjugation is that of multiplication by i and by the coordinate ξ.

Example 6.11.7. The *heat equation* (see Example 7.9.5 for more details) for a function $u : \mathbf{R}^n \times \mathbf{R} \to \mathbf{R}$, assumed to be differentiable sufficiently many times, reads

$$D_t u(x, t) = k\, \Delta_x u(x, t) \qquad (k > 0,\ x \in \mathbf{R}^n,\ t \in \mathbf{R}), \tag{6.46}$$

where Δ_x is the *Laplace operator* or *Laplacian*,

$$\Delta_x = \Delta = \sum_{1\le j\le n} D_j^2.$$

This is an example of a *partial differential equation*, relating different partial derivatives of u. We want to solve the *initial value problem* for this equation, that is, we look for solutions u of (6.46) satisfying the following additional condition, for $t = 0$:

$$u(x, 0) = f(x) \qquad (x \in \mathbf{R}^n), \tag{6.47}$$

where f is a given function.

We are going to apply a Fourier transformation to the function $x \mapsto u(x, t)$; therefore we assume that $u(., t) \in \mathcal{S}(\mathbf{R}^n)$, for all $t \in \mathbf{R}$, and that $f \in \mathcal{S}(\mathbf{R}^n)$. Define

$$\widehat{u}(\xi, t) = \int_{\mathbf{R}^n} e^{-i\langle x,\xi\rangle} u(x, t)\, dx.$$

Assuming that on the right–hand side differentiation under the integral sign is allowed, we obtain from (6.46) and (6.47), using Theorem 6.11.3.(ii),

$$D_t\widehat{u}(\xi, t) = -k\|\xi\|^2\widehat{u}(\xi, t), \qquad \widehat{u}(\xi, 0) = \widehat{f}(\xi) \qquad (\xi \in \mathbf{R}^n,\ t \in \mathbf{R}).$$

The function $t \mapsto \widehat{u}(\xi, t)$ therefore satisfies a first-order ordinary differential equation, with the solution

$$\widehat{u}(\xi, t) = \widehat{f}(\xi)e^{-tk\|\xi\|^2}.$$

Application of Example 6.11.4 yields

$$e^{-tk\|\xi\|^2} = \widehat{g_t}(\xi), \qquad \text{with} \qquad g_t(x) = (4\pi kt)^{-\frac{n}{2}}e^{-\frac{\|x\|^2}{4kt}}. \tag{6.48}$$

On account of Example 6.11.5 we therefore have

$$\widehat{u}(\xi, t) = \widehat{f}(\xi)\widehat{g_t}(\xi) = \widehat{(f * g_t)}(\xi).$$

It then follows from the Inversion Theorem that

$$\begin{aligned} u(x, t) = (f * g_t)(x) &= \int_{\mathbf{R}^n} f(x-y)g_t(y)\, dy \\ &= (4\pi kt)^{-\frac{n}{2}} \int_{\mathbf{R}^n} f(x-y)e^{-\frac{\|y\|^2}{4kt}}\, dy = \pi^{-\frac{n}{2}} \int_{\mathbf{R}^n} f(x - 2\sqrt{kt}\,y)e^{-\|y\|^2}\, dy. \end{aligned} \tag{6.49}$$

Because the calculation above makes several assumptions about the function u, it has to be checked that the u from Formula (6.49) does indeed satisfy (6.46) and (6.47). For this the reader is referred to Exercise 6.92.

Finally, assume $K \subset \mathbf{R}^n$ to be a bounded set with the property $\operatorname{supp}(f) \subset K$, and f to be nontrivial with $f \ge 0$ on $\mathbf{R}^n$. But for every $x \in \mathbf{R}^n$ and all $t \in \mathbf{R}_+$ there exists $y \in \mathbf{R}^n$ such that $x - 2\sqrt{kt}\,y \in \operatorname{supp}(f)$; and this implies $f(x - 2\sqrt{kt}\,y) \ge 0$. It follows, for all $x \in \mathbf{R}^n$ and $t \in \mathbf{R}_+$, that $u(x, t) > 0$; that is, the solution u has **infinite** speed of propagation. ☆

6.12 Dominated convergence

From analysis on $\mathbf{R}$ we recall the result that integration and taking a limit may be interchanged for a sequence of functions that converges **uniformly** on a closed interval in $\mathbf{R}$. We now prove that this interchange is also permitted if the sequence of functions is *boundedly convergent*, that is, if it is pointwise convergent and uniformly bounded.

The construction in the proof of Theorem 6.12.2 below leads to functions whose Riemann integrability is not guaranteed. This explains why we encounter $\underline{\int}$, the lower Riemann integral from Definition 6.2.3, in the next proposition.

Proposition 6.12.1. *Let $K \in \mathcal{J}(\mathbf{R}^n)$ and let $(f_k)_{k\in\mathbf{N}}$ be a sequence of functions on K. Assume that the following properties hold.*

(i) *f_1 is bounded, and $\lim_{k\to\infty} f_k(x) = 0$, for every $x \in K$.*

(ii) *$(f_k)_{k\in\mathbf{N}}$ is monotonically decreasing, that is, $f_{k+1}(x) \le f_k(x)$, for all $k \in \mathbf{N}$ and $x \in K$.*

Then one has

$$\lim_{k\to\infty} \underline{\int}_K f_k(x)\,dx = 0.$$

Proof. Select $\epsilon > 0$ arbitrarily. By applying Theorem 6.8.2.(i) to the step function corresponding to a suitable lower sum of f_k, we can find a sequence $(g_k)_{k\in\mathbf{N}}$ of continuous functions on K such that $g_k \le f_k$ and

$$\underline{\int}_K f_k(x)\,dx \le \int_K g_k(x)\,dx + \frac{\epsilon}{2^{k+1}} \qquad (k \in \mathbf{N}).$$

Note that the inequality above remains valid if g_k is replaced by $(g_k)_+$, hence we may assume that $0 \le g_k$. Now introduce the sequence $(h_k)_{k\in\mathbf{N}}$ of functions on K by $h_1 = g_1$ and $h_k = \min(g_k, h_{k-1})$. Then every h_k is a continuous function on K (in fact, $\min(f, g) = \frac{1}{2}(f+g-|f-g|)$, for any two functions f and g). The sequence is monotonically decreasing and satisfies $\lim_{k\to\infty} h_k(x) = 0$ for all $x \in K$, since $0 \le h_k \le g_k \le f_k$. The identity $-\min(f, g) = \max(-f, -g)$, which is valid for any two functions f and g, implies

$$\begin{aligned} f_k - h_k &= f_k - \min(g_k, h_{k-1}) = \max(f_k - g_k, f_k - h_{k-1}) \\ &\le f_k - g_k + f_k - h_{k-1} \le f_k - g_k + f_{k-1} - h_{k-1}. \end{aligned}$$

Accordingly, one proves by mathematical induction over $k \in \mathbf{N}$

$$f_k - h_k \le \sum_{1\le j\le k} (f_j - g_j); \qquad \text{so} \qquad \underline{\int}_K f_k(x)\,dx \le \int_K h_k(x)\,dx + \sum_{1\le j\le k} \frac{\epsilon}{2^{j+1}}.$$

As the sequence $(h_k)_{k\in\mathbf{N}}$ satisfies the conditions of Dini's Theorem 1.8.19 it converges uniformly on K to the function 0. Therefore we can find $k_0 \in \mathbf{N}$ such that for all $k \geq k_0$ we have

$$\int_K h_k(x)\,dx \leq \frac{\epsilon}{2}; \qquad \text{hence} \qquad 0 \leq \underline{\int}_K f_k(x)\,dx \leq \frac{\epsilon}{2} + \sum_{1\leq j\leq k} \frac{\epsilon}{2^{j+1}} < \epsilon. \qquad \square$$

Theorem 6.12.2. *Let $K \in \mathcal{J}(\mathbf{R}^n)$, and let the functions f and f_k, for $k \in \mathbf{N}$, be Riemann integrable over K. Assume that the following properties hold.*

(i) *For every $x \in K$ one has $\lim_{k\to\infty} f_k(x) = f(x)$.*

(ii) *There exists a number $m > 0$ such that $|f_k(x)| \leq m$, for every $x \in K$ and $k \in \mathbf{N}$.*

Then

$$\lim_{k\to\infty} \int_K f_k(x)\,dx = \int_K f(x)\,dx.$$

Proof. The functions $f_k - f$ are Riemann integrable over K and have pointwise limit 0; consequently there is no loss of generality in assuming $f = 0$. In addition, by Theorem 6.2.8.(iii) we may assume that $f_k \geq 0$. For each $k \in \mathbf{N}$ set $g_k = \sup\{ f_{k+j} \mid j \in \mathbf{N}_0 \}$. (Note that the g_k are not automatically Riemann integrable.) Obviously, the sequence $(g_k)_{k\in\mathbf{N}}$ satisfies the conditions in Proposition 6.12.1, and therefore

$$0 \leq \lim_{k\to\infty} \int_K f_k(x)\,dx \leq \lim_{k\to\infty} \underline{\int}_K g_k(x)\,dx = 0. \qquad \square$$

The generalization of the last theorem to the case of absolutely Riemann integrable functions is the following:

Theorem 6.12.3 (Arzelà's Dominated Convergence Theorem). *Let U be an open set in $\mathbf{R}^n$, and assume f and $f_k : U \to \mathbf{R}$, for $k \in \mathbf{N}$, to be absolutely Riemann integrable functions over U. Suppose that we have the following.*

(i) *$\lim_{k\to\infty} f_k(x) = f(x)$, for every $x \in U$.*

(ii) *There exists a function $g : U \to \mathbf{R}$ that is bounded on U and absolutely Riemann integrable over U such that $|f_k(x)| \leq g(x)$, for all $k \in \mathbf{N}$ and $x \in U$.*

Then

$$\lim_{k\to\infty} \int_U f_k(x)\,dx = \int_U f(x)\,dx.$$

Proof. Let $\epsilon > 0$ be arbitrary. Apply Theorem 6.10.6 with $|f|$ and g, to find a set $K \in \mathcal{J}(U)$ such that

$$\int_{U\setminus K} |f(x)|\,dx < \frac{\epsilon}{3}, \qquad \int_{U\setminus K} |f_k(x)|\,dx \le \int_{U\setminus K} g(x)\,dx < \frac{\epsilon}{3} \qquad (k \in \mathbf{N}),$$

respectively. According to the preceding theorem there exists $k_0 \in \mathbf{N}$ such that for $k \ge k_0$,

$$\begin{aligned}
&\left|\int_U f(x)\,dx - \int_U f_k(x)\,dx\right| \\
&\quad = \left|\int_K (f(x) - f_k(x))\,dx + \int_{U\setminus K} f(x)\,dx - \int_{U\setminus K} f_k(x)\,dx\right| \\
&\quad \le \left|\int_K (f(x) - f_k(x))\,dx\right| + \int_{U\setminus K} |f(x)|\,dx + \int_{U\setminus K} |f_k(x)|\,dx \\
&\quad < \frac{\epsilon}{3} + \frac{\epsilon}{3} + \frac{\epsilon}{3} = \epsilon.
\end{aligned}$$

❑

By way of application of this result we now give a version of the Differentiation Theorem 2.10.13 in the case of integration over (unbounded) sets in $\mathbf{R}^p$.

Theorem 6.12.4 (Differentiation Theorem). *Let $U \subset \mathbf{R}^n$ and $V \subset \mathbf{R}^p$ be open subsets, and let $f : U \times V \to \mathbf{R}$ be a function with the following properties.*

(i) *For every $x \in U$, the function $t \mapsto f(x, t)$ is absolutely Riemann integrable over V.*

(ii) *The total derivative $D_1 f : U \times V \to \mathrm{Lin}(\mathbf{R}^n, \mathbf{R})$ with respect to the variable in U exists and, for every $x \in U$, the mapping $t \mapsto D_1 f(x, t)$ is absolutely Riemann integrable over V (here the integration is by components).*

(iii) *There exists a function $g : V \to [\,0, \infty\,[$ which is bounded on V and absolutely Riemann integrable over V, such that $\|D_1 f(x, t)\|_{\mathrm{Eucl}} \le g(t)$, for all $(x, t) \in U \times V$.*

Then $F : U \to \mathbf{R}$, given by $F(x) = \int_V f(x, t)\,dt$, is a differentiable mapping satisfying

$$D_1 F(x) = \int_V D_1 f(x, t)\,dt \qquad (x \in U).$$

Proof. Let $a \in U$ and suppose $(h_k)_{k\in\mathbf{N}}$ is a sequence of arbitrary vectors in $\mathbf{R}^n$ converging to 0. The derivative of the mapping: $\mathbf{R} \to \mathbf{R}$ with $s \mapsto f(a + sh_k, t)$

is given by $s \mapsto D_1 f(a + sh_k, t)\, h_k$; hence we have, according to the Fundamental Theorem 2.10.1,

$$f(a + h_k, t) - f(a, t) = \int_0^1 D_1 f(a + sh_k, t)\, ds\; h_k.$$

Furthermore, as a function of t the right–hand side is absolutely Riemann integrable over V. Therefore

$$\begin{aligned} F(a + h_k) - F(a) &= \int_V (f(a + h_k, t) - f(a, t))\, dt \\ &= \int_V \int_0^1 D_1 f(a + sh_k, t)\, ds\, dt\; h_k =: \phi(a + h_k) h_k, \end{aligned}$$

where $\phi : U \to \mathrm{Lin}(\mathbf{R}^n, \mathbf{R})$. Applying the Dominated Convergence Theorem twice with the mappings $t \mapsto D_1 f(a + sh_k, t)$, which converge pointwise to $t \mapsto D_1 f(a, t)$ for $k \to \infty$, we obtain

$$\lim_{k\to\infty} \phi(a + h_k) = \int_V \int_0^1 \lim_{k\to\infty} D_1 f(a + sh_k, t)\, ds\, dt = \int_V D_1 f(a, t)\, dt = \phi(a).$$

On the strength of Lemma 1.3.3 and Hadamard's Lemma 2.2.7 this implies the differentiability of F at a, with derivative $\int_V D_1 f(a, t)\, dt$. ❑

We will extend Corollary 6.4.3 on changing the order of integration to continuous functions $f : U \to \mathbf{R}$ that are absolutely Riemann integrable over the open set $U \subset \mathbf{R}^{p+q}$.

Example 6.12.5. Consider $f : \mathbf{R}^2 \to \mathbf{R}$ given by $f(y, z) = e^{-y^4 z^2 - z^2}$. In view of Example 6.10.8

$$\int_{\mathbf{R}} f(y, z)\, dz = \frac{\sqrt{\pi}}{\sqrt{1 + y^4}}, \qquad \int_{\mathbf{R}} \int_{\mathbf{R}} f(y, z)\, dz\, dy = \sqrt{\pi} \int_{\mathbf{R}} \frac{1}{\sqrt{1 + y^4}}\, dy < \infty,$$

yet one has the divergent integral $\int_{\mathbf{R}} f(y, 0)\, dy = \int_{\mathbf{R}} 1\, dy$. On the other hand, $\int_{\mathbf{R}} f(y, z)\, dy < \infty$, for all $z \neq 0$. For every $K \in \mathcal{J}(\mathbf{R}^2)$ there is $R > 0$ with $K \subset [-R, R] \times [-R, R]$. Using this fact and the substitution of variables $(y, z) = (y, \frac{\tilde{z}}{\sqrt{y^4+1}})$ in $\mathbf{R}^2$, one can prove that Theorem 6.10.6.(ii) is satisfied. It follows that f is absolutely Riemann integrable over $\mathbf{R}^2$. Examples like this one explain why the conditions in the following two propositions are not automatically satisfied if f is continuous and absolutely Riemann integrable. ☆

Proposition 6.12.6. *Let $U \subset \mathbf{R}^{p+q}$ be an open set and let $f : U \to \mathbf{R}$ be a continuous function that is absolutely Riemann integrable over U. Further suppose that*

$$y \mapsto \int_{\mathbf{R}^q} f(y, z)\, dz = \int_{U''(y)} f(y, z)\, dz,$$

$$z \mapsto \int_{\mathbf{R}^p} f(y, z)\, dy = \int_{U'(z)} f(y, z)\, dy$$

both are continuous functions (in particular, they assume finite values everywhere). Here $U''(y) = \{ z \in \mathbf{R}^q \mid (y, z) \in U \}$ and $U'(z) = \{ y \in \mathbf{R}^p \mid (y, z) \in U \}$. Then

$$\int_U f(x)\, dx = \int_{\mathbf{R}^p} \int_{\mathbf{R}^q} f(y, z)\, dz\, dy = \int_{\mathbf{R}^q} \int_{\mathbf{R}^p} f(y, z)\, dy\, dz.$$

Proof. Since f is absolutely Riemann integrable over U if and only if f_+ and f_- are so, we may suppose $f \geq 0$. Let the sequence $(K_k)_{k\in\mathbf{N}}$ of sets in $\mathcal{J}(U)$ be as in (6.38) and the subsequent construction, and define $K_k(y) = \{ z \in \mathbf{R}^q \mid (y, z) \in K_k \}$, for $k \in \mathbf{N}$ and $y \in \mathbf{R}^p$. The sets $K_k(y)$ are Jordan measurable in $\mathbf{R}^q$. Therefore the following functions g_k and $g : \mathbf{R}^p \to \mathbf{R}$ are well-defined.

$$g_k(y) = \int_{K_k(y)} f(y, z)\, dz = \int_{\mathbf{R}^q} 1_{K_k(y)}(z) f(y, z)\, dz,$$

$$g(y) = \int_{\mathbf{R}^q} 1_{U''(y)}(z) f(y, z)\, dz.$$

On the strength of Theorem 6.4.2,

$$\int_{K_k} f(x)\, dx = \int_{\mathbf{R}^p} g_k(y)\, dy \qquad (k \in \mathbf{N}).$$

$(K_k(y))_{k\in\mathbf{N}}$ is a nondecreasing sequence of sets in $\mathcal{J}(\mathbf{R}^q)$ with union equal to $U''(y)$. Applying Arzelà's Dominated Convergence Theorem 6.12.3 to the sequence of functions $(f_k)_{k\in\mathbf{N}}$ satisfying $f_k = 1_{K_k(y)} f(y, \cdot) : \mathbf{R}^q \to \mathbf{R}$ and $\lim_{k\to\infty} f_k = f(y, \cdot)$, and using $f(y, \cdot)$ as the majorizing function, we obtain that $\lim_{k\to\infty} g_k(y) = g(y)$, for all $y \in \mathbf{R}^p$. Since $f \geq 0$ implies that $(g_k)_{k\in\mathbf{N}}$ is a nondecreasing sequence, with g as a limit and majorizing function, application once more of Arzelà's Dominated Convergence Theorem gives

$$\int_U f(x)\, dx = \lim_{k\to\infty} \int_{K_k} f(x)\, dx = \lim_{k\to\infty} \int_{\mathbf{R}^p} g_k(y)\, dy = \int_{\mathbf{R}^p} g(y)\, dy.$$

Interchanging the roles of y and z finally gives the equality of both iterated integrals. ❑

In practice it might be difficult to establish absolute Riemann convergence of f over U, whereas computation of an iterated integral might be feasible. Therefore results of the following type are extremely important in analysis.

Proposition 6.12.7. *Let $f : U \to \mathbf{R}$ be continuous on the open set $U \subset \mathbf{R}^{p+q}$, and assume that $y \mapsto \int_{\mathbf{R}^q} |f(y,z)|\,dz$ and $z \mapsto \int_{\mathbf{R}^p} |f(y,z)|\,dy$ both are continuous functions. Then $y \mapsto \int_{\mathbf{R}^q} f(y,z)\,dz$ and $z \mapsto \int_{\mathbf{R}^p} f(y,z)\,dy$ are continuous. Further suppose that one of the iterated integrals for $|f|$ converges; say $\int_{\mathbf{R}^p}\int_{\mathbf{R}^q} |f(y,z)|\,dz\,dy < \infty$. Then f is absolutely Riemann integrable over U and both iterated integrals for f are equal to the integral of f over U, in other words*

$$\int_U f(x)\,dx = \int_{\mathbf{R}^p}\int_{\mathbf{R}^q} f(y,z)\,dz\,dy = \int_{\mathbf{R}^q}\int_{\mathbf{R}^p} f(y,z)\,dy\,dz.$$

Proof. In order to prove the absolute integrability of f we verify that Theorem 6.10.6.(ii) is satisfied. To this end we may suppose that every K_k, for $k \in \mathbf{N}$, is a finite union of rectangles $\{ B_{ki} \mid i \in I \}$. For every $1 \le j \le n$, collect the endpoints of the j-th coordinate interval of all rectangles B_{ki}, and arrange these points in increasing order, as in the proof of Proposition 6.1.2. Thus one obtains a finite number of nonoverlapping rectangles $B'_{kl} \subset \mathbf{R}^p$, with $l \in L_k$, and, for every l, a finite number of nonoverlapping rectangles $B''_{klm} \subset \mathbf{R}^q$, with $m \in M_{kl}$, such that

$$K_k = \bigcup \{ B'_{kl} \times B''_{klm} \mid l \in L_k,\ m \in M_{kl} \} \qquad (k \in \mathbf{N}).$$

Because $|f|$ is continuous on K_k we obtain from Theorem 6.4.2, for $k \in \mathbf{N}$,

$$\begin{aligned}\int_{K_k} |f(x)|\,dx &= \sum_{l\in L_k}\sum_{m\in M_{kl}} \int_{B'_{kl}\times B''_{klm}} |f(x)|\,dx \\ &= \sum_l \sum_m \int_{B'_{kl}}\int_{B''_{klm}} |f(y,z)|\,dz\,dy = \sum_l \int_{B'_{kl}} \sum_m \int_{B''_{klm}} |f(y,z)|\,dz\,dy \\ &\le \sum_l \int_{B'_{kl}}\int_{\mathbf{R}^q} |f(y,z)|\,dz\,dy \le \int_{\mathbf{R}^p}\int_{\mathbf{R}^q} |f(y,z)|\,dz\,dy.\end{aligned}$$

Hence f is absolutely Riemann integrable over U. Therefore application of Proposition 6.12.6 implies that the iterated integrals of f both equal the integral of f over U. ❑

In the formulation above it is essential to work with the absolute value of f. Indeed, the existence of an iterated integral, if it involves cancellation, need not imply the existence of the integral of f over U. Observe that we have to deal with only one of the two iterated integrals, which is fortunate since it is often the case that one is easier to estimate.

Proposition 6.12.7 is a very special case of *Fubini's Theorem*, which is part of the theory of *Lebesgue integration*. In this theory one is able to weaken the notion of Riemann integrability to that of Lebesgue integrability and yet to integrate functions belonging to this wider class. The property of being a Lebesgue integrable (or measurable) function is preserved under the formation of partial functions and integrals thereof, and as a consequence Fubini's theorem has a more natural formulation than Proposition 6.12.7. Results like this make Lebesgue integration superior to Riemann integration.

6.13 Appendix: two other proofs of Change of Variables Theorem

In this appendix we treat two other proofs of the Change of Variables Theorem 6.6.1, which we will call the second and the third proof. Each of them highlights different aspects of the problem: the second is intuitive, geometrical, but rather technical; although less intuitive, the third proof is quite efficient.

Second proof. Contrary to the demonstration in Section 6.9 this proof does not require the Implicit function Theorem 3.5.1; on the other hand, some more detailed information from linear algebra is needed. As a consequence Chapter 6 can be studied independently from Chapters 3 through 5, which might appeal to readers who prefer to make an early start with the theory of integration. In this setup the proof of the Change of Variables Theorem in full generality requires the validity of the theorem in the special case of a diffeomorphism $\Psi : \mathbf{R}^n \to \mathbf{R}^n$ that belongs to $\mathrm{Aut}(\mathbf{R}^n)$. The treatment of the latter case needs some linear algebra, in particular Lemma 6.13.2 below.

We begin with a definition. We denote the standard basis vectors in $\mathbf{R}^n$ by e_j, for $1 \le j \le n$.

Definition 6.13.1. A transformation in $\mathrm{End}(\mathbf{R}^n)$ is said to be *basic* if it is given by one of the following formulae, for $1 \le k, l \le n$, $k \ne l$, $x \in \mathbf{R}^n$ and $\lambda \in \mathbf{R} \setminus \{0\}$:

$$F_{kl}^{\pm}(x) = x \pm x_l e_k, \qquad M_l(\lambda)(x) = x + (\lambda - 1)x_l e_l,$$
$$S_{kl}(x) = x + (x_l - x_k)(e_k - e_l). \qquad ❍$$

We note some properties of basic linear transformations. In view of

$$(F_{kl}^{\pm})^{-1} = F_{kl}^{\mp}, \qquad M_l(\lambda)^{-1} = M_l(\lambda^{-1}), \qquad S_{kl}^{-1} = S_{lk},$$

every basic linear transformation is invertible, with an inverse that is basic too.

Next we study the compositions AB and BA, where A is an arbitrary linear and B is a basic linear transformation of $\mathbf{R}^n$. In doing so we denote a linear transformation and its matrix by the same symbol. From

$$F_{kl}^{\pm}(e_j) = e_j \pm \delta_{jl} e_k = \begin{cases} e_j, & j \ne l; \\ e_l \pm e_k, & j = l, \end{cases}$$

we see that the matrix $AF_{kl}^{\pm}$ is obtained from the matrix A by the column operation of adding/subtracting the k-th column vector of A to/from the l-th column vector of A. From $(F_{kl}^{\pm})^t = F_{lk}^{\pm}$ we obtain $(F_{kl}^{\pm}A)^t = A^t(F_{kl}^{\pm})^t = A^t F_{lk}^{\pm}$, writing A^t for the transpose matrix of A. Combining this with the preceding result we immediately see that $F_{kl}^{\pm}A$ arises from A by the row operation of addition/subtraction of the l-th row vector of A to/from the k-th row vector of A. Furthermore,

$$M_l(\lambda)(e_j) = e_j + (\lambda - 1)\delta_{jl}e_l = \begin{cases} e_j, & j \neq l; \\ \lambda e_l, & j = l, \end{cases}$$

implies that $AM_l(\lambda)$ arises from A when the l-th column vector of A is multiplied by λ, and that $M_l(\lambda)A$ is the result of the analogous row operation. Also, from

$$S_{kl}(e_j) = e_j + (\delta_{jl} - \delta_{jk})(e_k - e_l) = \begin{cases} e_j, & j \neq k, l; \\ e_l, & j = k; \\ e_k, & j = l, \end{cases}$$

we see that AS_{kl} and $S_{kl}A$ are obtained from A by interchanging the k-th and the l-th column vectors and row vectors, respectively, of A. Thus right and left multiplication of a matrix A by the matrix of a basic transformation amount to performing on A one of the column and row operations, respectively, which are well-known from linear algebra.

These results enable us to prove the following:

Lemma 6.13.2. *Every element in* $\mathrm{Aut}(\mathbf{R}^n)$ *can be written as a product of basic linear transformations.*

Proof. Let $A \in \mathrm{Aut}(\mathbf{R}^n)$ be arbitrary. Since the top row vector of A is different from 0 there exist B_1 and $B_1' \in \mathrm{Aut}(\mathbf{R}^n)$, both being products of basic linear transformations, with $B_1AB_1' = \begin{pmatrix} 1 & * \\ * & * \end{pmatrix}$. But then we can find B_1 and B_1' as above such that

$$B_1AB_1' = \begin{pmatrix} 1 & 0 \\ 0 & A_1 \end{pmatrix} \qquad \text{with} \qquad A_1 \in \mathrm{End}(\mathbf{R}^{n-1}).$$

Because $\det(B_1AB_1')$ is a nonzero multiple of $\det A \neq 0$ we have $\det A_1 \neq 0$, that is, $A_1 \in \mathrm{Aut}(\mathbf{R}^{n-1})$. By induction over the dimension n we can therefore show the existence of B and $B' \in \mathrm{Aut}(\mathbf{R}^n)$ such that both are products of basic linear transformations and that $BAB' = I$; this implies $A = B^{-1}B'^{-1}$, which proves the lemma. ❑

Remark. The decomposition into basic linear transformations is not unique. For instance, $F_{kl}^{-} = M_l(-1)F_{kl}^{+}M_l(-1)$.

Now we are sufficiently prepared to establish the Change of Variables Theorem 6.6.1 in the special case of a mapping $\Psi : \mathbf{R}^n \to \mathbf{R}^n$ that belongs to $\mathrm{Aut}(\mathbf{R}^n)$. In fact, we shall need the proposition for a slightly more general class of mappings, which is given in the following:

Definition 6.13.3. A *bijective affine transformation* Ψ of $\mathbf{R}^n$ is a mapping $\mathbf{R}^n \to \mathbf{R}^n$ that can be written in the form $\Psi(y) = x^0 + Ay$, where $x^0 \in \mathbf{R}^n$ and $A \in \mathrm{Aut}(\mathbf{R}^n)$.○

Note that the bijective affine transformation Ψ above is a C^1 mapping and that its inverse $\Psi^{-1} : \mathbf{R}^n \to \mathbf{R}^n$ is given by

$$\Psi^{-1}(x) = -A^{-1}x^0 + A^{-1}x. \tag{6.50}$$

Furthermore, x^0 and A are uniquely determined by Ψ since $x^0 = \Psi(0)$ and $A = \Psi - \Psi(0)$.

Proposition 6.13.4. *Let $x^0 \in \mathbf{R}^n$ and $A \in \mathrm{Aut}(\mathbf{R}^n)$, and denote by Ψ the corresponding bijective affine transformation of $\mathbf{R}^n$. Suppose $f : \mathbf{R}^n \to \mathbf{R}$ to be a continuous function with compact support. Then*

$$\int_{\mathbf{R}^n} f(x)\,dx = \int_{\mathbf{R}^n} (f \circ \Psi)(y)|\det D\Psi(y)|\,dy = |\det A| \int_{\mathbf{R}^n} f(x^0 + Ay)\,dy.$$

Proof. Using $D\Psi(y) = A$, for all $y \in \mathbf{R}^n$, and Corollary 6.4.3 we reduce the problem to the case of dimension one,

$$\int_{\mathbf{R}^n} (f \circ \Psi)(y)|\det D\Psi(y)|\,dy = |\det A| \int_{\mathbf{R}^n} f(x^0 + Ay)\,dy$$

$$= |\det A| \int_{\mathbf{R}} \cdots \left(\int_{\mathbf{R}} f(x^0 + A(y_1, \dots, y_n))\,dy_1 \right) \cdots dy_n$$

$$= |\det A| \int_{\mathbf{R}} \cdots \left(\int_{\mathbf{R}} f(A(y_1, \dots, y_n))\,dy_1 \right) \cdots dy_n.$$

Here we used the invariance under translation of the one-dimensional integration. In view of the preceding Lemma 6.13.2 and Step IV on composition in Section 6.9 we may assume A to be a basic linear transformation. Now in the case of $A = F_{kl}^{\pm}$ we have $\det A = 1$ and we write the last integral as the iteration of an $(n-2)$-dimensional integral over $\mathbf{R}^{n-2}$ and of

$$\int_{\mathbf{R}} \int_{\mathbf{R}} f(y_1, \dots, y_l \pm y_k, \dots, y_n)\,dy_l\,dy_k = \int_{\mathbf{R}} \int_{\mathbf{R}} f(y_1, \dots, y_l, \dots, y_n)\,dy_l\,dy_k,$$

employing again the invariance under translation of the one-dimensional integration. Thus we obtain

$$\int_{\mathbf{R}} \cdots \left(\int_{\mathbf{R}} f(x)\,dx_1 \right) \cdots dx_n = \int_{\mathbf{R}^n} f(x)\,dx.$$

If $A = M_l(\lambda)$ we use $\det A = \lambda$ and

$$\int_{\mathbf{R}} |\lambda| f(y_1, \dots, \lambda y_l, \dots, y_n)\,dy_l = \int_{\mathbf{R}} f(y_1, \dots, y_l, \dots, y_n)\,dy_l.$$

Finally, for $A = S_{kl}$ the result follows from Corollary 6.4.3. ❑

Proof of Change of Variables Theorem. On the basis of Step I on reduction to continuous f in Section 6.9 we may assume that f is continuous with $\operatorname{supp}(f) \subset U$. The proof then proceeds in three steps. In Step I the volume of a small cube C is compared with that of its image $\Psi(C)$. To this end Ψ is replaced by its first-order Taylor polynomial at the point y^0 at which C is centered, that is, by the affine mapping $T = T_{y^0}\Psi : \mathbf{R}^n \to \mathbf{R}^n$ satisfying

$$T(y) = \Psi(y^0) + D\Psi(y^0)(y - y^0) = \Psi(y^0) - D\Psi(y^0)y^0 + D\Psi(y^0)y.$$

Thus we compare $\Psi(C)$ with the parallelepiped $T(C)$, and the error estimate is a direct consequence of the definition of differentiability of Ψ. As T is bijective because $D\Psi(y^0)$ is, we may as well estimate the difference between $(T^{-1} \circ \Psi)(C)$ and C itself. It turns out that for every $\epsilon > 0$ there exists $\delta > 0$ such that

$$C \text{ cube of diameter less than } \delta \qquad \Longrightarrow \qquad (T^{-1} \circ \Psi)(C) \subset C^{\epsilon},$$

where C^{ϵ} is the cube concentric with C whose sides are multiplied by $1 + \epsilon$. As Proposition 6.13.4 is applicable with Ψ replaced by T we find the upper bound (6.56) below. Technically it is convenient to compare $(T^{-1} \circ \Psi)(C)$ and C, instead of $\Psi(C)$ and $T(C)$, because estimating the volume of the set of points at a distance less than ϵ from a given set is easier if the latter set is a cube rather than a parallelepiped. A complication in the argument is that we do not know right away whether $\Psi(C)$ is Jordan measurable.

In Step II the upper bound (6.56) is used for majorizing the integral $\int_U f(x)\,dx$ by $(1 + \epsilon)^{n+1}$ times an upper sum of $f \circ \Psi|\det D\Psi|$, and thus by

$$(1 + \epsilon)^{n+1} \int_V f \circ \Psi(y)|\det D\Psi(y)|\,dy.$$

Sending ϵ to 0 leads to (6.58) below.

Obtaining a lower bound for $\operatorname{vol}_n(\Psi(C))$ is a delicate matter. For example, suppose $n = 2$ and let $\Psi(C)$ be a long rectangle $[\,0, \delta^{-1}\,] \times [\,0, \delta\,]$, with volume 1. Then by moving each point in $\Psi(C)$ over a distance of only δ, by sending $x \in \mathbf{R}^2$ to $(x_1, 0)$, the rectangle collapses to the interval $[\,0, \delta^{-1}\,] \times \{0\}$ along the x_1-axis, which has volume 0. In Step III this difficulty is avoided by applying the preceding arguments to the inverse of Ψ.

Step I (Local inequality). We begin with the preparations needed to establish the estimate (6.56) below. In view of Theorem 1.8.3,

$$K := \operatorname{supp}(f \circ \Psi) = \Psi^{-1}(\operatorname{supp} f) \subset V$$

is a compact subset of V. Below we shall work with cubes covering K; these are not necessarily contained in K and therefore we enlarge K in a suitable way. According to Lemma 6.8.1 we can find $\delta > 0$ such that

$$K_{\delta} = \{\, y \in \mathbf{R}^n \mid \text{ there exists } y' \in K \text{ with } \|y - y'\| \le \delta \,\} \subset V. \tag{6.51}$$

In particular, every cube $C \subset \mathbf{R}^n$ centered at a point of K and with diameter less than δ is contained in the compact set K_δ.

From the definition of differentiability of Ψ at $y^0 \in V$, there exists for every $\eta > 0$ a $\delta = \delta(y^0, \eta) > 0$ such that for every $y \in V$ with $\|y - y^0\| < \delta$,

$$\|\Psi(y) - T_{y^0}\Psi(y)\| = \|\Psi(y) - \Psi(y^0) - D\Psi(y^0)(y - y^0)\| < \eta \, \|y - y^0\|.$$

Actually this estimate is valid uniformly for $y^0 \in K_\delta$; more precisely, for every $\eta > 0$ there exists $\delta > 0$ such that every y and $y^0 \in K_\delta$ for which the line segment between y and y^0 lies entirely in K_δ satisfy

$$\|y - y^0\| < \delta \qquad \Longrightarrow \qquad \|\Psi(y) - T_{y^0}\Psi(y)\| \leq \eta\|y - y^0\|. \tag{6.52}$$

Indeed, use Formula (2.25) and the fact that $D\Psi|_{K_\delta} : K_\delta \to \mathrm{Aut}(\mathbf{R}^n)$ is continuous, and therefore uniformly continuous, on the compact set K_δ.

From (6.50) and (2.5) we obtain

$$\|(T_{y^0}\Psi)^{-1}(x) - (T_{y^0}\Psi)^{-1}(x')\| \leq \|D\Psi(y^0)^{-1}\|_{\mathrm{Eucl}} \|x - x'\| \qquad (x, x' \in \mathbf{R}^n). \tag{6.53}$$

Once more using that $D\Psi$ is continuous on K and that K is compact, we find

$$0 < m = \max_{y \in K} \|D\Psi(y)^{-1}\|_{\mathrm{Eucl}} < \infty. \tag{6.54}$$

We now claim that for arbitrary $\eta > 0$ we can find $\delta > 0$ with the following properties. Condition (6.51) is satisfied and for all $y^0 \in K$ and $y \in K_\delta$ with $\|y - y^0\| < \delta$ we have, in view of (6.53), (6.54) and (6.52),

$$\begin{aligned}\|(T_{y^0}\Psi)^{-1}(\Psi(y)) - y\| &= \|(T_{y^0}\Psi)^{-1}(\Psi(y)) - (T_{y^0}\Psi)^{-1}(T_{y^0}\Psi(y))\| \\ &\leq \|D\Psi(y^0)^{-1}\|_{\mathrm{Eucl}} \|\Psi(y) - T_{y^0}\Psi(y)\| \leq m\eta\|y - y^0\|.\end{aligned}$$

Now let $\epsilon > 0$ be arbitrary, select $\eta = \dfrac{\epsilon}{m}$ and a $\delta > 0$ corresponding to this η, and deduce

$$\|(T_{y^0}\Psi)^{-1}(\Psi(y)) - y\| < \epsilon\,\delta \qquad \text{for} \qquad y \in C,$$

where

$$C \text{ is a cube centered at an arbitrary } y^0 \in K \text{ with diameter less than } \delta. \tag{6.55}$$

Hence we get, writing C^ϵ for the cube concentric with C whose sides are multiplied by $1 + \epsilon$,

$$(T_{y^0}\Psi)^{-1}(\Psi(y)) \in C_{\epsilon\,\delta} \subset C^\epsilon, \qquad \text{and so} \qquad \Psi(C) \subset T_{y^0}\Psi(C^\epsilon).$$

It follows that the following upper Riemann integral satisfies the inequality

$$\overline{\mathrm{vol}_n}(\Psi(C)) := \overline{\int_{\mathbf{R}^n}} 1_{\Psi(C)}(x)\, dx \leq \overline{\int_{\mathbf{R}^n}} 1_{T_{y^0}\Psi(C^\epsilon)}(x)\, dx = \mathrm{vol}_n\,(T_{y^0}\Psi(C^\epsilon)).$$

Applying Proposition 6.13.4 with Ψ replaced by the bijective affine mapping $T_{y^0}\Psi$, we find

$$\mathrm{vol}_n\,(T_{y^0}\Psi(C^\epsilon)) = |\det D\Psi(y^0)|\,\mathrm{vol}_n(C^\epsilon) = (1+\epsilon)^n|\det D\Psi(y^0)|\,\mathrm{vol}_n(C).$$

Thus, for any cube C as in (6.55),

$$\overline{\mathrm{vol}_n}(\Psi(C)) \le (1+\epsilon)^n|\det D\Psi(y^0)|\,\mathrm{vol}_n(C). \tag{6.56}$$

Step II (Global inequality). Let $\mathcal{B} = \{\, C_i \mid i \in I\,\}$ be a finite set of nonoverlapping cubes C_i centered at $y_i^0 \in K$ with diameters less than δ as above, satisfying

$$K \subset \bigcup_{i\in I} C_i \subset K_\delta \subset V.$$

Since $f \circ \Psi|_{C_i} : C_i \to \mathbf{R}$ is continuous, there exist $y_i \in C_i$ such that we have $\max_{y\in C_i} f \circ \Psi(y) = f \circ \Psi(y_i)$ for $i \in I$. Define $j : V \to \mathbf{R}$ by $j(y) = |\det D\Psi(y)|$. Then $j > 0$ and, for $y,\, y^0 \in V$,

$$j(y)^{-1}j(y^0) = |j(y)^{-1}(j(y^0) - j(y)) + 1| \le j(y)^{-1}|j(y^0) - j(y)| + 1.$$

Furthermore, j is uniformly continuous on K_δ. Therefore we have, by shrinking $\delta > 0$ if necessary,

$$|\det D\Psi(y_i^0)| = j(y_i)j(y_i)^{-1}j(y_i^0) \le (1+\epsilon)|\det D\Psi(y_i)| \qquad (i \in I).$$

We now combine this estimate with (6.56) for $C = C_i$ in order to get

$$\overline{\mathrm{vol}_n}(\Psi(C_i)) \le (1+\epsilon)^{n+1}|\det D\Psi(y_i)|\,\mathrm{vol}_n(C_i) \qquad (i \in I). \tag{6.57}$$

After these preparations we are able to treat the global problem. We have $f = f_+ - f_-$ with $f_\pm \ge 0$ as in Theorem 6.2.8.(iii) and $f_\pm$ continuous. By going over to the $f_\pm$ and using the linearity of integration we may assume $0 \le f$. In view of

$$f = f1_{\Psi(K)} \le \sum_{i\in I} f1_{\Psi(C_i)} \le \sum_{i\in I} f\circ\Psi(y_i)1_{\Psi(C_i)}$$

we obtain from (6.57)

$$\begin{aligned}
\int_U f(x)\,dx &\le \sum_{i\in I} f\circ\Psi(y_i)\overline{\int}_{\mathbf{R}^n} 1_{\Psi(C_i)}(x)\,dx = \sum_{i\in I} f\circ\Psi(y_i)\overline{\mathrm{vol}_n}(\Psi(C_i))\\
&\le (1+\epsilon)^{n+1}\sum_{i\in I} f\circ\Psi(y_i)|\det D\Psi(y_i)|\,\mathrm{vol}_n(C_i)\\
&\le (1+\epsilon)^{n+1}\sum_{i\in I} \max_{y\in C_i}(f\circ\Psi|\det D\Psi|)(y)\,\mathrm{vol}_n(C_i).
\end{aligned}$$

The sum at the right–hand side is an upper sum of $f \circ \Psi |\det D\Psi|$ determined by the partition $\mathcal{B}$ that covers $K_\cdot = \mathrm{supp}(f \circ \Psi)$; thus

$$\int_U f(x)\,dx \le (1+\epsilon)^{n+1} \int_V f \circ \Psi(y) |\det D\Psi(y)|\,dy.$$

Since the estimate is valid for every $\epsilon > 0$ it implies

$$\int_U f(x)\,dx \le \int_V f \circ \Psi(y) |\det D\Psi(y)|\,dy. \tag{6.58}$$

Step III (Reverse inequality). Now apply this inequality with $\Psi : V \to U$ replaced by $\Psi^{-1} : U \to V$, and $f : U \to \mathbf{R}$ by $f \circ \Psi |\det D\Psi| : V \to \mathbf{R}$, respectively; the conditions are satisfied since $\mathrm{supp}(f \circ \Psi |\det D\Psi|) = K$ is compact in V. As $|\det D\Psi(\Psi^{-1}(x))|\,|\det D\Psi^{-1}(x)| = 1$ we find

$$\int_V f \circ \Psi(y) |\det D\Psi(y)|\,dy \le \int_U f(x)\,dx.$$

The desired equality from the Change of Variables Theorem follows by combining these two inequalities.

Remark. Note that the proof makes repeated use of the fact that Ψ is a C^1 diffeomorphism, fully exploiting all implications thereof.

Third proof. In the remainder of this section we give a third proof of the Change of Variables Theorem 6.6.1. Again the main ingredient is reduction to the case of a bijective affine transformation as treated in Proposition 6.13.4. In Step IV below this reduction is effectuated by showing that

$$\frac{d}{dt} \int_V (f \circ \Psi^t)(y) \det D\Psi^t(y)\,dy = 0,$$

if $(\Psi^t)_{t\in[0,1]}$ is a one-parameter C^1 family of C^1 diffeomorphisms with the property that $\mathrm{supp}(f \circ \Psi^t) \cap \partial V = \emptyset$. In particular, we construct such a family (Ψ^t) that transforms Ψ into a local best affine approximation to Ψ. The other technical tool is the Global Inverse Function Theorem 3.2.8. This approach might be the least intuitive of the three. The circle of ideas involved here originates from *homotopy theory*, a subject in *algebraic topology*.

Step I (Localization). This is the same as Step III on localization in Section 6.9. The precise nature of the bounded open neighborhoods O_x of $x \in U$ that are used to cover the compact set $K = \mathrm{supp}(f) \subset U$ is specified at the end of the following step.

Step II (Deformation). We consider a C^1 diffeomorphism $\Psi : V \to U$. According to Step I it is sufficient to study the diffeomorphism in suitable neighborhoods of an arbitrary but fixed point $\Psi^{-1}(x) = y \in V$. We will show that locally, near y, the diffeomorphism Ψ can be deformed through a one-parameter C^1 family of C^1 diffeomorphisms into its best affine approximation at y.

From the definition of differentiability of Ψ at y (compare with (2.10)) we see, for $y + h \in V$,

$$\Psi(y+h) = \Psi(y) + D\Psi(y)h + \epsilon_y(h) =: Ah + \epsilon_y(h),$$
$$\|\epsilon_y(h)\| = o(\|h\|), \qquad h \to 0.$$

Here A is the best (bijective) affine approximation to Ψ at y. Next define, for $t \in [\,0, 1\,]$,

$$\Psi(t, y+h) = \Psi(y+h) - t\epsilon_y(h) = Ah + (1-t)\epsilon_y(h). \tag{6.59}$$

Then $\Psi(0, y+h) = \Psi(y+h)$ and $\Psi(1, y+h) = Ah$. Further introduce

$$\widetilde{\Psi} : [\,0, 1\,] \times V \to [\,0, 1\,] \times \mathbf{R}^n \qquad \text{given by} \qquad \widetilde{\Psi}(t, y) = (t, \Psi(t, y)).$$

In order to prove that $\widetilde{\Psi}$ is a C^1 diffeomorphism onto its image we now verify that the conditions of the Global Inverse Function Theorem 3.2.8 are satisfied. In fact, $D_y\Psi(t, y) = D\Psi(y) \in \mathrm{Aut}(\mathbf{R}^n)$ immediately gives $D\widetilde{\Psi}(t, y) \in \mathrm{Aut}(\mathbf{R}^{n+1})$, for $(t, y) \in [\,0, 1\,] \times V$. Moreover, $\widetilde{\Psi}(t, y+h) = \widetilde{\Psi}(t', y+h')$ implies $t = t'$ and also

$$Ah + (1-t)\epsilon_y(h) = Ah' + (1-t)\epsilon_y(h'),$$

so

$$h - h' = (1-t)D\Psi(y)^{-1}(\epsilon_y(h') - \epsilon_y(h)).$$

Next we apply the result of Example 2.5.4 with the mapping ϵ_y and ϵ equal to $(2\|D\Psi(y)^{-1}\|_{\mathrm{Eucl}})^{-1}$ in order to find $\delta > 0$ as in the example. Thus we obtain, for $h, h' \in B(0; \delta)$ and $t \in [\,0, 1\,]$,

$$\|h - h'\| \le (1-t)\|D\Psi(y)^{-1}\|_{\mathrm{Eucl}}\|\epsilon_y(h') - \epsilon_y(h)\| \le \frac{1}{2}\|h - h'\|.$$

Consequently $h = h'$, which proves the injectivity of $\widetilde{\Psi}$. On account of the Global Inverse Function Theorem, $\widetilde{\Psi}$ restricted to $[\,0, 1\,] \times B(y; \delta)$ is a C^1 diffeomorphism onto its image. Finally, take O_x equal to the open neighborhood $\Psi(B(y; \delta))$ of $x = \Psi(y)$; note that δ depends on the choice of $y \in V$.

Step III (Supports). Write $\Psi^t = \Psi(t, \cdot)$. We need control over $\mathrm{supp}(f \circ \Psi^t) \subset V$. Recall that K is compact in U. According to Lemma 6.8.1 we can find $\epsilon > 0$ such that $K \subset K_\epsilon \subset U$, with K_ϵ compact too. And as a consequence of Theorem 1.8.3 the image set $L := \Psi^{-1}(K_\epsilon)$ is compact in V. In view of Steps I and II it is sufficient to consider the subsets $K \cap O_x$. According to (6.59),

$$\Psi^t(y+h) \in K \cap O_x \qquad \Longrightarrow \qquad y + h \in \Psi^{-1}(K + t\epsilon_y(h)).$$

By shrinking $\delta > 0$ as in Step II if necessary, we may arrange that $K + t\epsilon_y(h) \subset K_\epsilon$, for all $(t, h) \in [\,0, 1\,] \times B(0; \delta)$. Consequently we obtain

$$\operatorname{supp}(f \circ \Psi^t) \subset L \subset V \qquad (t \in [\,0, 1\,]). \tag{6.60}$$

Step IV (Differentiation). Suppose $\widetilde{\Psi}$ restricted to $[\,0, 1\,] \times B(y^0; \delta)$ is a C^1 diffeomorphism as in Step II. Then $(\Psi^t)_{t\in[\,0,1\,]}$ is a C^1 family of C^1 diffeomorphisms defined on a subset of V. In the formulae below we write D for the total derivative with respect to the variable $y \in V$, in particular, $D\Psi^t(y)$ for $D_y\Psi(t, y)$. Further assume that $f \in C^1(O_{x^0})$ with $\Psi(y^0) = x^0$. Then we introduce

$$I(t) = \int_V (f \circ \Psi^t)(y) \det D\Psi^t(y)\, dy \qquad (t \in [\,0, 1\,]).$$

Under these assumptions we will prove that $I(t)$ actually is independent of t, which implies

$$\int_V (f \circ \Psi)(y) \det D\Psi(y)\, dy = \det D\Psi(y^0) \int_V f(\Psi(y^0) + D\Psi(y^0)y)\, dy.$$

Given t and $t + u \in [\,0, 1\,]$ we define

$$\Xi^u \in C^1(V, \mathbf{R}^n) \qquad \text{by} \qquad \Xi^u = (\Psi^t)^{-1}\Psi^{t+u}.$$

Observe that $\Xi^0 = I$. The chain rule now implies the following identity of functions on V:

$$f \circ \Psi^{t+u} \,\det D\Psi^{t+u} = (f \circ \Psi^t) \circ \Xi^u \,(\det D\Psi^t) \circ \Xi^u \,\det D\Xi^u = g \circ \Xi^u \,\det D\Xi^u, \tag{6.61}$$

where $g = f \circ \Psi^t \,\det D\Psi^t \in C^1(V)$. Next define the C^1 vector field

$$\xi : V \to \mathbf{R}^n \qquad \text{by} \qquad \xi(y) = \frac{d}{du}\Big|_{u=0} \Xi^u(y).$$

Note that the mapping $t \mapsto D\Psi^t(y)$ is continuously differentiable, while

$$y \mapsto \frac{d}{dt}\Psi^t(y) = -\epsilon_{y^0}(y - y^0) = \Psi(y^0) + D\Psi(y^0)(y - y^0) - \Psi(y)$$

is continuously differentiable near y^0 too. On account of Theorem 2.7.2 on the equality of mixed partial derivatives and Exercise 2.44.(i) we therefore obtain

$$\frac{d}{du}\Big|_{u=0} \det D\Xi^u(y) = \operatorname{tr} D\frac{d}{du}\Big|_{u=0} \Xi^u(y) = \operatorname{tr} D\xi(y) = \operatorname{div}\xi(y),$$

where we recall the definition of the divergence of ξ from Formula (5.30). Then we have, by the Differentiation Theorem 2.10.13 or 6.12.4 and Formula (6.61),

$$
\begin{aligned}
I'(t) &= \frac{d}{du}\Big|_{u=0} I(t+u) = \int_V \frac{d}{du}\Big|_{u=0} (g \circ \Xi^u)(y) \det D\Xi^u(y)\,dy \\
&= \int_V (Dg(y)\xi(y) + g(y)\operatorname{div}\xi(y))\,dy = \int_V \operatorname{div}(g\xi)(y)\,dy \\
&= \int_V \sum_{1\le j\le n} D_j(g\xi_j)(y)\,dy = 0.
\end{aligned}
$$

The last equality is obtained by the Fundamental Theorem of Integral Calculus on $\mathbf{R}$, see Case I in the proof of Theorem 7.6.1 on integration of a total derivative, and by the inclusion (6.60). In Example 8.9.3 one can find an alternative computation.

Step V (Case of affine transformation). The previous steps enable a reduction of the proof of the theorem to the case treated in Proposition 6.13.4. Two details still have to be settled. Note that $\det D\Psi(y) \neq 0$, for all $y \in V$. Hence it is a consequence of the Intermediate Value Theorem 1.9.5 that $\det D\Psi(y)$ has constant sign on the connected components of V, see Definition 1.9.7. Considering V as the union of its connected components and splitting the integral accordingly, we can take the sign of $\det D\Psi(y)$ outside the integrals, which justifies the omission of the absolute value signs in Step IV. Furthermore, we may use approximation arguments from Section 7.7 to replace the condition of $f \in C^1(U)$ by that of $f \in C(U)$.

Remark. Actually, in the notation of the Steps II and IV above one can find a C^1 family $(\Psi^t)_{t\in[0,1]}$ of C^1 diffeomorphisms satisfying $\Psi^0 = \Psi$ and $\Psi^1 = \operatorname{sgn}(\det D\Psi(y))I$, where I is the identity mapping. This is a consequence of the connectedness of the subset $\operatorname{Aut}^\circ(n, \mathbf{R}) = \{A \in \operatorname{Aut}(\mathbf{R}^n) \mid \det A > 0\}$ of $\operatorname{End}(\mathbf{R}^n)$, which can be proved using Lemma 6.13.2. Arguing in this way one may bypass Proposition 6.13.4.

Chapter 7

Integration over Submanifolds

The knowledge acquired about manifolds and integration will now be used to develop the theory of integration over a submanifold in $\mathbf{R}^n$ of dimension $d < n$. In particular the d-dimensional volume (length, (hyper)area, etc.) of bounded submanifolds will be defined. By way of application we study the generalization to $\mathbf{R}^n$ of the Fundamental Theorem of Integral Calculus on $\mathbf{R}$. This is a problem with two aspects: finding correct formulae on the one hand, and antidifferentiation of a function of several variables on the other. The first aspect culminates in the theorem which asserts equality between the integral of the total derivative of a function over an open set, and an integral of the function itself over the boundary of that open set. Gauss' Divergence Theorem then is a direct corollary.

7.1 Densities and integration with respect to density

In Chapter 6 we introduced in particular the integral of (absolutely) Riemann integrable functions defined on open subsets of $\mathbf{R}^n$, which we shall regard here as C^k submanifolds in $\mathbf{R}^n$ of dimension n. We now wish to develop a theory of integration over C^k submanifolds, for $k \geq 1$, in $\mathbf{R}^n$ of dimension $d < n$. Note that according to Corollary 6.3.8 such submanifolds are negligible in $\mathbf{R}^n$, if they are compact. We shall therefore have to be somewhat careful to take due account with respect to integration of the manifolds V in $\mathbf{R}^n$ of dimension $d < n$.

First we consider this problem locally, that is, in a neighborhood U in $\mathbf{R}^n$ of a point $x \in V$. According to Theorem 4.7.1 there exist an open subset $D \subset \mathbf{R}^d$ and a C^k embedding $\phi : D \to \mathbf{R}^n$ such that $\phi(D) = V \cap U$. Let $f : V \to \mathbf{R}$ be a bounded function for which

$$\operatorname{supp}(f) \subset \phi(D) \text{ is a compact set.} \tag{7.1}$$

Then $f \circ \phi : D \to \mathbf{R}$ has a compact support in $\mathbf{R}^d$; and it is tempting to call f Riemann integrable over V if $f \circ \phi$ is Riemann integrable over $D \subset \mathbf{R}^d$. And further, in that case, to define the integral of f over V, with the notation $I_\phi(f)$, as the integral of $f \circ \phi$ over D,

$$I_\phi(f) = \int_D (f \circ \phi)(y)\, dy. \tag{7.2}$$

Now let $\widetilde{D} \subset \mathbf{R}^d$ be open, let $\widetilde{\phi} : \widetilde{D} \to V$ be another C^k embedding, and assume

$$\operatorname{supp}(f) \subset \phi(D) \cap \widetilde{\phi}(\widetilde{D}). \tag{7.3}$$

Then

$$\operatorname{supp}(f \circ \phi) \subset D_{\widetilde{\phi},\phi} := \phi^{-1}(\widetilde{\phi}(\widetilde{D})) \subset D.$$

It follows from Lemma 4.3.3.(iii) that the mapping $\phi^{-1} \circ \widetilde{\phi} : D_{\phi,\widetilde{\phi}} \to D_{\widetilde{\phi},\phi}$ is a C^k diffeomorphism of open subsets in $\mathbf{R}^d$. Furthermore,

$$(f \circ \phi) \circ (\phi^{-1} \circ \widetilde{\phi}) = f \circ \widetilde{\phi} \qquad \text{on} \qquad D_{\phi,\widetilde{\phi}}.$$

On account of the Change of Variables Theorem 6.6.1, therefore, $f \circ \phi$ is Riemann integrable over D if and only if

$$(f \circ \phi) \circ (\phi^{-1} \circ \widetilde{\phi})\, |\det D(\phi^{-1} \circ \widetilde{\phi})| = (f \circ \widetilde{\phi})\, |\det D(\phi^{-1} \circ \widetilde{\phi})|$$

is Riemann integrable over $\widetilde{D}$; the latter applies if and only if $f \circ \widetilde{\phi}$ is Riemann integrable over $\widetilde{D}$. Subject to this assumption we then have

$$\int_D (f \circ \phi)(y)\, dy = \int_{\widetilde{D}} (f \circ \widetilde{\phi})(\widetilde{y})\, |\det D(\phi^{-1} \circ \widetilde{\phi})(\widetilde{y})|\, d\widetilde{y}.$$

In general, therefore,

$$I_\phi(f) = \int_D (f \circ \phi)(y)\, dy \neq \int_{\widetilde{D}} (f \circ \widetilde{\phi})(\widetilde{y})\, d\widetilde{y} = I_{\widetilde{\phi}}(f).$$

This result shows that the Riemann integrability of f over V is independent of the choice of the parametrization ϕ of V, but that the value $I_\phi(f)$ of the integral of f over V depends on ϕ, unless by chance $|\det D(\phi^{-1} \circ \widetilde{\phi})| \equiv 1$, that is, unless $\phi^{-1} \circ \widetilde{\phi}$ is a volume-preserving coordinate transformation in $\mathbf{R}^d$.

We are thus confronted with the presence of extra factors $|\det D(\phi^{-1} \circ \widetilde{\phi})|$ in the integrand. It is natural, therefore, to incorporate these from the start in the definition of the integral $I_\phi(f)$ of f over V, in such a way that this integral is independent of the chosen parametrization ϕ. Thus, assume there is a continuous function $\rho_\phi : D \to \mathbf{R}$ associated with the embedding $\phi : D \to \mathbf{R}^n$ and let, unlike (7.2), the integral $I_\phi(f)$ of f over V be defined by

$$I_\phi(f) = \int_D (f \circ \phi)(y)\, \rho_\phi(y)\, dy.$$

We then require

$$
\begin{aligned}
I_\phi(f) &= \int_D (f \circ \phi)(y)\, \rho_\phi(y)\, dy \\
&= \int_{\widetilde{D}} (f \circ \phi) \circ (\phi^{-1} \circ \widetilde{\phi})(\widetilde{y})\, \rho_\phi(\phi^{-1} \circ \widetilde{\phi}(\widetilde{y}))\, |\det D(\phi^{-1} \circ \widetilde{\phi})(\widetilde{y})|\, d\widetilde{y} \\
&= \int_{\widetilde{D}} (f \circ \widetilde{\phi})(\widetilde{y})\, \rho_{\widetilde{\phi}}(\widetilde{y})\, d\widetilde{y} = I_{\widetilde{\phi}}(f).
\end{aligned}
$$

Hence the following:

Definition 7.1.1. A *continuous d-dimensional density* ρ on a C^k submanifold V in $\mathbf{R}^n$ of dimension d is a mapping which assigns to every C^k embedding $\phi : D_\phi \to V$, where $D_\phi \subset \mathbf{R}^d$ is open, a continuous function $\rho_\phi : D_\phi \to \mathbf{R}$ in such a way that

$$
\rho_{\widetilde{\phi}}(\widetilde{y}) = \rho_\phi(\phi^{-1} \circ \widetilde{\phi}(\widetilde{y}))\, |\det D(\phi^{-1} \circ \widetilde{\phi})(\widetilde{y})|, \tag{7.4}
$$

for any C^k embedding $\widetilde{\phi} : D_{\widetilde{\phi}} \to V$ and every $\widetilde{y} \in \widetilde{\phi}^{-1}(\phi(D_\phi))$. ❍

Remark. A continuous density ρ on V is uniquely determined by a collection of continuous functions $\rho_\phi : D_\phi \to \mathbf{R}$ satisfying (7.4), where $\phi \in \Phi$, if Φ is a collection of C^k embeddings such that

$$
V \subset \bigcup_{\phi \in \Phi} \phi(D_\phi). \tag{7.5}
$$

To see this, consider an arbitrary C^k embedding $\widetilde{\phi} : \widetilde{D} \to V$. For every $\widetilde{y} \in \widetilde{D}$ we can find a $\phi \in \Phi$ with $\widetilde{\phi}(\widetilde{y}) \in \phi(D_\phi)$. We now define

$$
\rho_{\widetilde{\phi}}(\widetilde{y}) = \rho_\phi(\phi^{-1} \circ \widetilde{\phi}(\widetilde{y}))\, |\det D(\phi^{-1} \circ \widetilde{\phi})(\widetilde{y})|.
$$

It is easy to verify that this definition does not depend on the choice of $\phi \in \Phi$ for which $\widetilde{\phi}(\widetilde{y}) \in \phi(D_\phi)$, and that the collection $\{\, \rho_{\widetilde{\phi}} \mid \widetilde{\phi} \text{ arbitrary embedding} \,\}$ defined in this way satisfies requirement (7.4). Hence all that is required for the introduction of a continuous density on V is a minimal collection Φ of embeddings that satisfy (7.5); thus for a sphere in $\mathbf{R}^3$ two embeddings suffice.

Remark. It is also possible to formulate the theory above in terms of coordinatizations or charts (see Definition 4.2.4),

$$
\kappa := \phi^{-1} : \phi(D) \to D =: U_\kappa; \qquad \text{hence} \qquad \kappa : \kappa^{-1}(U_\kappa) \to U_\kappa,
$$

instead of the embeddings ϕ. A density ρ then assigns to every chart κ a continuous function $\rho_\kappa : U_\kappa \to \mathbf{R}$ such that

$$\rho_{\widetilde{\kappa}}(\widetilde{y}) = \rho_\kappa(\kappa \circ \widetilde{\kappa}^{-1}(\widetilde{y}))\,|\det D(\kappa \circ \widetilde{\kappa}^{-1})(\widetilde{y})| \qquad (\widetilde{y} \in U_{\widetilde{\kappa}} \cap (\widetilde{\kappa} \circ \kappa^{-1})(U_\kappa)).$$

Note that $\kappa \circ \widetilde{\kappa}^{-1}$ is the transition mapping from the last Remark in Section 4.3.

We now want to free ourselves from the requirements in (7.1) and (7.3) that the support of f be contained in $\phi(D)$, or even in the intersection of several such image sets. Therefore we have:

Definition 7.1.2. – Theorem. Let V be a C^k submanifold, with $k \geq 1$, in $\mathbf{R}^n$ of dimension d. Let $f : V \to \mathbf{R}$ be a bounded function with compact support $\operatorname{supp}(f) =: K$. Let Φ' be a collection of C^k embeddings $\phi : D_\phi \to V$ with $D_\phi \subset \mathbf{R}^d$ open and with $K \subset \bigcup\{\phi(D_\phi) \mid \phi \in \Phi'\}$. Let $\{\chi_\phi \mid \phi \in \Phi\}$ be a continuous partition of unity on K subordinate to the open covering $\{U_\phi \mid \phi \in \Phi'\}$ of K (see Theorem 6.7.3); here $\phi(D_\phi) = V \cap U_\phi$, with U_ϕ open in $\mathbf{R}^n$. Then f is said to be *Riemann integrable* over V if for every $\phi \in \Phi$ the function

$$(\chi_\phi f) \circ \phi : D_\phi \to \mathbf{R}$$

is Riemann integrable over D_ϕ. If this is the case, the *integral of f over V with respect to the density ρ*, notation $\int_V f(x)\rho(x)\,dx$, is defined by

$$\int_V f(x)\rho(x)\,dx = \sum_{\phi\in\Phi} \int_{D_\phi} (\chi_\phi f) \circ \phi(y)\,\rho_\phi(y)\,dy. \tag{7.6}$$

Here it is of course essential that the left–hand side in (7.6) is in fact independent of the choice of the collection Φ and of the partition $\{\chi_\phi \mid \phi \in \Phi\}$. Indeed, let $\widetilde{\Phi}$ and $\{\chi_{\widetilde{\phi}} \mid \widetilde{\phi} \in \widetilde{\Phi}\}$, respectively, be another such choice. Then

$$\begin{aligned}
&\sum_{\phi\in\Phi} \int_{D_\phi} (\chi_\phi f) \circ \phi(y)\,\rho_\phi(y)dy \\
&\quad= \sum_{\phi\in\Phi} \int_{D_\phi} \sum_{\widetilde{\phi}\in\widetilde{\Phi}} \chi_{\widetilde{\phi}} \circ \phi(y)\,\chi_\phi \circ \phi(y)\,f \circ \phi(y)\,\rho_\phi(y)\,dy \\
&\quad= \sum_{\widetilde{\phi}\in\widetilde{\Phi},\,\phi\in\Phi} \int_{D_{\widetilde{\phi}}} \chi_{\widetilde{\phi}} \circ \widetilde{\phi}(\widetilde{y})\,\chi_\phi \circ \widetilde{\phi}(\widetilde{y})\,f \circ \widetilde{\phi}(\widetilde{y})\,\rho_\phi(\phi^{-1} \circ \widetilde{\phi}(\widetilde{y})) \\
&\qquad\qquad\qquad\qquad\qquad\cdot|\det D(\phi^{-1} \circ \widetilde{\phi})(\widetilde{y})|\,d\widetilde{y} \\
&\quad= \sum_{\widetilde{\phi}\in\widetilde{\Phi}} \int_{D_{\widetilde{\phi}}} \sum_{\phi\in\Phi} \chi_\phi \circ \widetilde{\phi}(\widetilde{y})\,\chi_{\widetilde{\phi}} \circ \widetilde{\phi}(\widetilde{y})\,f \circ \widetilde{\phi}(\widetilde{y})\,\rho_{\widetilde{\phi}}(\widetilde{y})\,d\widetilde{y} \\
&\quad= \sum_{\widetilde{\phi}\in\widetilde{\Phi}} \int_{D_{\widetilde{\phi}}} (\chi_{\widetilde{\phi}} f) \circ \widetilde{\phi}(\widetilde{y})\,\rho_{\widetilde{\phi}}(\widetilde{y})\,d\widetilde{y}.
\end{aligned}$$

Here we have successively used $\sum_{\widetilde{\phi}\in\widetilde{\Phi}} \chi_{\widetilde{\phi}} \equiv 1$ on K; the substitution $y = \phi^{-1} \circ \widetilde{\phi}(\widetilde{y})$ and the Change of Variables Theorem 6.6.1; Formula (7.4); and, finally, $\sum_{\phi\in\Phi} \chi_\phi \equiv 1$ on K. □

The continuous density ρ is said to be *positive* if $\rho_\phi(y) > 0$, for every embedding ϕ and all $y \in D_\phi$. In this case ρ may be regarded as a continuous "ubiquitous" *mass density* on V. In cases where a ρ_ϕ also takes values ≤ 0, the physical analog is a continuous *charge density*. ❍

The following lemma gives a description of all continuous d-dimensional densities on V in terms of one positive density.

Lemma 7.1.3. *Let ρ be a fixed positive continuous d-dimensional density on V.*

(i) *For every continuous function f on V the mapping $f\,\rho$ with $f\,\rho : \phi \mapsto f \circ \phi\, \rho_\phi$ defines a continuous d-dimensional density on V.*

(ii) *The mapping $f \mapsto f\,\rho$ is a bijection from the space of all continuous functions on V to the collection of all continuous d-dimensional densities on V.*

Proof. (i). The function $f \circ \phi\, \rho_\phi$ satisfies (7.4), because

$$(f \circ \widetilde{\phi}\, \rho_{\widetilde{\phi}})(\widetilde{y}) = f \circ \phi(\phi^{-1} \circ \widetilde{\phi}(\widetilde{y}))\, \rho_\phi(\phi^{-1} \circ \widetilde{\phi}(\widetilde{y}))\, |\det D(\phi^{-1} \circ \widetilde{\phi}(\widetilde{y}))|.$$

(ii). This assertion follows if we can prove that $f \mapsto f\,\rho$ is a surjection. Thus, given a continuous density $\widetilde{\rho}$ we have to determine a function f such that $\widetilde{\rho} = f\,\rho$. We define, for every embedding ϕ, the continuous function $f_\phi : \phi(D_\phi) \to \mathbf{R}$ by

$$f_\phi(x) = \frac{\widetilde{\rho}_\phi(\phi^{-1}(x))}{\rho_\phi(\phi^{-1}(x))} \qquad (x \in \phi(D_\phi)). \tag{7.7}$$

Note that positivity of ρ is essential here. If $x \in \phi(D_\phi) \cap \widetilde{\phi}(D_{\widetilde{\phi}})$, one has, by (7.4),

$$f_{\widetilde{\phi}}(x) = \frac{\widetilde{\rho}_{\widetilde{\phi}}(\widetilde{\phi}^{-1}(x))}{\rho_{\widetilde{\phi}}(\widetilde{\phi}^{-1}(x))} = \frac{\widetilde{\rho}_\phi(\phi^{-1} \circ \widetilde{\phi}(\widetilde{\phi}^{-1}(x)))}{\rho_\phi(\phi^{-1} \circ \widetilde{\phi}(\widetilde{\phi}^{-1}(x)))} = f_\phi(x).$$

Consequently there exists a unique function f on V such that $f(x) = f_\phi(x)$, for all x in $\phi(D_\phi)$. Because the function $f|_{\phi(D_\phi)} = f_\phi$ is always continuous, it follows that f is continuous on V. Furthermore, (7.7) implies that $\widetilde{\rho} = f\,\rho$. ❑

7.2 Absolute Riemann integrability with respect to density

In Definition 7.1.2 – Theorem a restriction was made to functions $f : V \to \mathbf{R}$ with compact support. As in Chapter 6, this is not without its drawbacks. Therefore we imitate Lemma 6.10.2 and Definition 6.10.5 in the following:

Definition 7.2.1. Let the notation be as in Section 7.1. Let ρ be a positive continuous density on V, and let $f : V \to \mathbf{R}$. The function f is said to be *absolutely Riemann integrable over V with respect to ρ* if f is locally Riemann integrable in V, and if

$$\sup_{K \in \mathcal{J}(V)} \int_K |f(x)|\rho(x)\,dx < \infty.$$

For such an f one sees, as in Definition 6.10.5, the existence of the *integral* $\int_V f(x)\rho(x)\,dx$ *of f over V with respect to ρ*, with the following property:

$$\int_V f(x)\rho(x)\,dx = \sup_{K \in \mathcal{J}(V)} \int_K f_+(x)\rho(x)\,dx - \sup_{K \in \mathcal{J}(V)} \int_K f_-(x)\rho(x)\,dx.$$

A subset A of V is said to be *Jordan measurable with respect to ρ* if the characteristic function 1_A is absolutely Riemann integrable over V with respect to ρ; in that case

$$\text{vol}_d(A) := \int_A \rho(x)\,dx := \int_V 1_A(x)\rho(x)\,dx$$

is said to be the *d-dimensional volume* or the *d-dimensional Jordan measure of A* with respect to ρ.

The set A is said to be *negligible in V* if $\text{vol}_d(A) = 0$. Note that the question of A being negligible or not is independent of the choice of the positive continuous density ρ on V; this follows from Lemma 7.1.3. ❍

By means of the techniques from Sections 6.8 and 6.7 one easily proves the following:

Lemma 7.2.2. *Let $K \subset V$ be a compact set which is negligible in V. Then there exist, for every $\epsilon > 0$, a continuous function $\chi_K : V \to \mathbf{R}$ with compact support, and an open set U_K in V such that*

$$0 \leq \chi_K \leq 1, \qquad K \subset U_K, \qquad \chi_K = 1 \text{ on } U_K, \qquad \int_V \chi_K(x)\rho(x)\,dx < \epsilon.$$

Proposition 7.2.3. *Let $U \subset V$ be an open subset in V and assume that $\partial_V U$ is negligible in V. Let $f : V \to \mathbf{R}$ be a locally bounded function. Then $1_U f$ is absolutely Riemann integrable over V with respect to ρ if and only if $f|_U$ is absolutely Riemann integrable over U with respect to $\rho|_U$. If this is the case, then*

$$\int_V (1_U f)(x)\rho(x)\,dx = \int_U (f|_U)(y)(\rho|_U)(y)\,dy.$$

Proof. In conformity with the definition of absolute Riemann integrability over V, consider arbitrary $K \in \mathcal{K}(V)$. One has that (see Definition 1.2.16)

$$L := K \cap \partial_V U$$

is compact in V; to see this, use the fact that the intersection of two closed sets is also closed. Furthermore, L is negligible in V, because L is a subset of the negligible $\partial_V U$. Let $\epsilon > 0$ and let χ_L and U_L be as in the preceding lemma. Set $\chi = 1_K(1 - \chi_L)$. Because $1 - \chi_L$ vanishes in the neighborhood U_L of $\partial_V U \cap K$, it follows that χ vanishes in a neighborhood of $\partial_V U$. Because $1_K - \chi = 1_K\, \chi_L$, one has, as f is locally bounded,

$$\left| \int_V (1_K\, 1_U\, f)(x)\rho(x)\, dx - \int_V (\chi\, 1_U\, f)(x)\rho(x)\, dx \right|$$

$$\leq \sup_{x \in K} |f(x)| \int_V \chi_L(x)\rho(x)\, dx < \epsilon \sup_K |f|.$$

A similar estimate holds if f is replaced by $|f|$. But this shows that, in examining the absolute Riemann integrability of $1_K 1_U\, f$ over V with respect to ρ, it is immaterial whether or not one imposes on the set $K \in \mathcal{K}(V)$ the extra condition that $K \subset U$. And this last point corresponds to the examination of the absolute Riemann integrability of $f|_U$ over U with respect to $\rho|_U$. ❑

Definition 7.1.2 – Theorem suffers from the complication that $\int_V f(x)\rho(x)\, dx$ is defined in terms of bump functions χ_ϕ, if the manifold V has to be described by more than one parametrization ϕ. Indeed, the functions χ_ϕ ensure that overlapping subsets $\phi(D_\phi)$ of V give proper contributions to the integral. The following theorem demonstrates that when the case arises, the integral can be calculated without these bump functions.

Theorem 7.2.4. *Let ρ be a positive continuous density on V and let Φ be a finite collection of C^1 embeddings $\phi : D_\phi \to V$ such that*

(i) $\phi(D_\phi) \cap \widetilde{\phi}(D_{\widetilde{\phi}}) = \emptyset$, *if* $\phi \neq \widetilde{\phi}$;

(ii) $N := V \setminus \bigcup_{\phi \in \Phi} \phi(D_\phi)$ *is a negligible set in* V.

Let $f : V \to \mathbf{R}$ be a locally bounded function. Then f is absolutely Riemann integrable over V with respect to ρ if and only if $f \circ \phi\, \rho_\phi$ is absolutely Riemann integrable over D_ϕ, for every $\phi \in \Phi$. If this is the case, it follows that

$$\int_V f(x)\rho(x)\, dx = \sum_{\phi \in \Phi} \int_{D_\phi} (f \circ \phi)(y)\rho_\phi(y)\, dy. \tag{7.8}$$

Proof. One has

$$f = \sum_{\phi \in \Phi} 1_{\phi(D_\phi)} f + 1_N f.$$

Because N is negligible in V, it follows that f is absolutely Riemann integrable over V with respect to ρ if and only if this holds for

$$\widetilde{f} := \sum_{\phi \in \Phi} 1_{\phi(D_\phi)} f.$$

When this holds, f and $\widetilde{f}$ have the same integral over V with respect to ρ. For $\widetilde{f}$ absolutely Riemann integrable over V with respect to ρ we then see, multiplying $\widetilde{f}$ only by continuous functions $\chi : V \to \mathbf{R}$ with compact support $\operatorname{supp}(\chi) \subset \phi(D_\phi)$, that $f|_{\phi(D_\phi)}$ is absolutely Riemann integrable over $\phi(D_\phi)$ with respect to ρ. But that is equivalent to the absolute Riemann integrability of $f \circ \phi \, \rho_\phi$ over D_ϕ.

Now assume, conversely,

$$f \circ \phi \, \rho_\phi \text{ absolutely Riemann integrable over } D_\phi, \quad \text{for all } \phi \in \Phi. \tag{7.9}$$

Then

$$\phi(D_\phi) \subset W := V \setminus \bigcup_{\widetilde{\phi} \neq \phi} \widetilde{\phi}(D_{\widetilde{\phi}});$$

and so W, being a complement of open sets in V, is closed in V. Because $\overline{\phi(D_\phi)}^V$ is the smallest closed set in V containing $\phi(D_\phi)$, it follows that $\overline{\phi(D_\phi)}^V \subset W$. Because $\phi(D_\phi)$ is open in V, we have the disjoint union

$$\partial_V(\phi(D_\phi)) \cup \phi(D_\phi) = \overline{\phi(D_\phi)}^V.$$

Consequently

$$\partial_V(\phi(D_\phi)) \subset W \setminus \phi(D_\phi) = V \setminus \bigcup_{\phi \in \Phi} \phi(D_\phi) = N.$$

Hence $\partial_V(\phi(D_\phi))$ is negligible in V. On account of the preceding proposition, it follows from (7.9) that $1_{\phi(D_\phi)} f$ is absolutely Riemann integrable over V with respect to ρ, and that

$$\int_V (1_{\phi(D_\phi)} f)(x)\rho(x)\,dx = \int_{\phi(D_\phi)} f(x)\rho(x)\,dx = \int_{D_\phi} (f \circ \phi)(y)\rho_\phi(y)\,dy.$$

Summation over $\phi \in \Phi$ now leads to (7.8). ❑

Remark. In practice, Φ often consists of a single element ϕ, that is, in such a case there exists a C^1 embedding $\phi : D \to V$ such that $V \setminus \phi(D)$ is negligible in V. Then f is absolutely Riemann integrable over V with respect to ρ if and only if $f \circ \phi \, \rho_\phi$ is absolutely Riemann integrable over D. If this is the case, then

$$\int_V f(x)\rho(x)\,dx = \int_D (f \circ \phi)(y)\rho_\phi(y)\,dy.$$

7.3 Euclidean d-dimensional density

Let V be a C^k submanifold in $\mathbf{R}^n$ of dimension d. Then V, by virtue of its being a submanifold of $\mathbf{R}^n$, possesses a "natural" positive d-dimensional density ω; this ω is said to be the *Euclidean d-dimensional density* on V, to be introduced below. It will turn out that, if $d = n$, integration with respect to this ω is identical with the n-dimensional integration from Chapter 6.

From Example 2.9.6 it follows, for a mapping $A \in \mathrm{Lin}(\mathbf{R}^d, \mathbf{R}^n)$, that $A^tA \in \mathrm{End}(\mathbf{R}^d)$ satisfies $\det(A^tA) \geq 0$. Accordingly we can now define

$$\omega : \mathrm{Lin}(\mathbf{R}^d, \mathbf{R}^n) \to \mathbf{R} \qquad \text{by} \qquad \omega(A) = \sqrt{\det(A^tA)} = \sqrt{\det(\langle a_i, a_j \rangle)}.$$

We then have the following properties for ω:

$$\omega(AB) = |\det B|\,\omega(A) \qquad (B \in \mathrm{End}(\mathbf{R}^d)); \tag{7.10}$$

$$\omega(CA) = \omega(A) \qquad (C \in \mathbf{O}(\mathbf{R}^n)). \tag{7.11}$$

Indeed, (7.10) follows from

$$\det(AB)^t\,(AB) = \det B^t(A^tA)B = \det B^t\,\det(A^tA)\,\det B = (\det B)^2\,\omega(A)^2.$$

According to Definition 2.9.4 any $C \in \mathbf{O}(\mathbf{R}^n)$ satisfies $C^tC = I$, and (7.11) is found from

$$\det(CA)^t\,(CA) = \det A^t(C^tC)A = \omega(A)^2.$$

Definition 7.3.1. – Theorem. Define, for every C^k embedding $\phi : D_\phi \to V$, with $k \geq 1$, the function $\omega_\phi : D_\phi \to \mathbf{R}$ by

$$\omega_\phi(y) = \omega(D\phi(y)) = \sqrt{\det\left(D\phi(y)^t \circ D\phi(y)\right)} \qquad (y \in D_\phi).$$

Then $\omega : \phi \mapsto \omega_\phi$ is a positive d-dimensional C^k density on V, the *Euclidean d-dimensional density* on V. Indeed, let $\widetilde{\phi}$ be another embedding; one then has, by the chain rule,

$$D\widetilde{\phi}(\widetilde{y}) = D(\phi \circ (\phi^{-1} \circ \widetilde{\phi}))(\widetilde{y}) = D\phi((\phi^{-1} \circ \widetilde{\phi})(\widetilde{y})) \circ D(\phi^{-1} \circ \widetilde{\phi})(\widetilde{y}),$$

where $D(\phi^{-1} \circ \widetilde{\phi})(\widetilde{y}) \in \mathrm{End}(\mathbf{R}^d)$. Using (7.10), one then finds

$$\begin{aligned} \omega_{\widetilde{\phi}}(\widetilde{y}) &= \omega(D\widetilde{\phi}(\widetilde{y})) = |\det D(\phi^{-1} \circ \widetilde{\phi})(\widetilde{y})|\,\omega\big(D\phi((\phi^{-1} \circ \widetilde{\phi})(\widetilde{y}))\big) \\ &= \omega_\phi(\phi^{-1} \circ \widetilde{\phi}(\widetilde{y}))\,|\det D(\phi^{-1} \circ \widetilde{\phi})(\widetilde{y})|; \end{aligned}$$

that is, requirement (7.4) for a density is met.

If f is Riemann integrable over V, then by Definition 7.1.2 we have the *integral of f over V with respect to the Euclidean density*, with notation

$$\int_V f(x)\,d_dx.$$

The subscript d in d_dx emphasizes that we are dealing with integration with respect to the Euclidean density on a d-dimensional manifold. ❍

Motivation for the preceding definition. In the special case of $d = n$ one has, on account of the Global Inverse Function Theorem 3.2.8, that $\phi : D \to \phi(D)$ is a C^k diffeomorphism of open sets in $\mathbf{R}^n$. Additionally, because the matrix of $D\phi(y) : \mathbf{R}^n \to \mathbf{R}^n$ is square, one has in this case

$$\omega_\phi(y) = \sqrt{\det(D\phi(y)^t \circ D\phi(y))} = \sqrt{\det D\phi(y)^t \, \det D\phi(y)} = |\det D\phi(y)|.$$

In view of the Change of Variables Theorem 6.6.1 it now follows for $\int_{\phi(D)} f(x)\, d_n x$, the integral of f over $\phi(D)$ with respect to the Euclidean density ω on $\phi(D)$, that

$$\int_{\phi(D)} f(x)\, d_n x = \int_D (f \circ \phi)(y)\, |\det D\phi(y)|\, dy = \int_{\phi(D)} f(x)\, dx.$$

In other words, integration with respect to the Euclidean density on $\phi(D)$ on the one hand and n-dimensional integration over $\phi(D)$ on the other coincide.

(The case $d < n$). If $\phi : D \to V$ with $D \subset \mathbf{R}^d$ open and $d < n$, we should like to imitate the foregoing and define $\int_{\phi(D)} f(x)\, d_d x$, in accordance with the Change of Variables Theorem 6.6.1, as

$$\int_{\phi(D)} f(x)\, d_d x = \int_D (f \circ \phi)(y)\, |\det D\phi(y)|\, dy. \tag{7.12}$$

However, there is a problem in that $D\phi(y) : \mathbf{R}^d \to \mathbf{R}^n$, and that as a result $\det D\phi(y)$ is undefined. But we do know, from Theorem 5.1.2, if $\phi(y) = x$, that

$$D\phi(y) : \mathbf{R}^d \to T_x V \subset \mathbf{R}^n$$

is a linear isomorphism onto the tangent space $T_x V$ to V at x. But there exists $C \in \mathbf{O}(\mathbf{R}^n)$ by which $T_x V$ is "laid flat", that is, for which

$$C(T_x V) = \mathbf{R}^d \times \{0_{\mathbf{R}^{n-d}}\}.$$

Therefore $C \circ D\phi(y)$ is a bijective linear transformation from $\mathbf{R}^d$ to $\mathbf{R}^d \times \{0_{\mathbf{R}^{n-d}}\}$. On its image, the projection $P_d \in \mathrm{Lin}(\mathbf{R}^n, \mathbf{R}^d)$ onto the first d coordinates is a linear isomorphism, that is

$$D(y) := P_d \circ C \circ D\phi(y) \in \mathrm{Aut}(\mathbf{R}^d).$$

The correct version of (7.12) therefore is

$$\int_{\phi(D)} f(x)\, d_d x = \int_D (f \circ \phi)(y)\, |\det D(y)|\, dy. \tag{7.13}$$

Since $P_d^t \circ P_d \in \mathrm{End}(\mathbf{R}^n)$ equals the identity on $\mathbf{R}^d \times \{0_{\mathbf{R}^{n-d}}\} = \mathrm{im}(C)$, we find

$$\begin{aligned}(\det D(y))^2 &= \det D(y)^t \, \det D(y) = \det(D(y)^t \circ D(y)) \\ &= \det(D\phi(y)^t \circ C^{-1} \circ P_d^t \circ P_d \circ C \circ D\phi(y)) \\ &= \det(D\phi(y)^t \circ D\phi(y)) = \omega_\phi(y)^2.\end{aligned}$$

And this in fact yields, by (7.13),

$$\int_{\phi(D)} f(x)\,d_d x = \int_D (f \circ \phi)(y)\,\omega_\phi(y)\,dy.$$

The Euclidean d-dimensional density ω on V can therefore be characterized as follows: ω equals the standard volume factor in $\mathbf{R}^d \simeq \mathbf{R}^d \times \{0_{\mathbf{R}^{n-d}}\} \simeq T_x V$, if $T_x V$ "lies flat", and in addition ω is invariant under elements in $\mathbf{O}(\mathbf{R}^n)$.

(Area equals volume divided by length). It is possible to formulate the foregoing in a somewhat different manner. The column vectors $D_1\phi(y), \dots, D_d\phi(y)$ in $\mathbf{R}^n$ of the matrix $D\phi(y)$ span $T_x V$. Now $\dim(T_x V)^\perp = n - d$, and we can therefore choose vectors $\nu_{d+1}(y)$ through $\nu_n(y)$ in $\mathbf{R}^n$ such that together they form an orthonormal basis for $(T_x V)^\perp$. Define $\overline{D}\phi(y) \in \mathbf{GL}(n, \mathbf{R})$ as the matrix having in the first d columns the column vectors of $D\phi(y)$, and in the last $n - d$ columns the vectors $\nu_{d+1}(y), \dots, \nu_n(y)$,

$$\overline{D}\phi(y) = {}_{n}\overset{\qquad d \qquad\qquad\qquad n-d}{\left(D\phi(y) \;\middle|\; \nu_{d+1}(y) \;\; \cdots \;\; \nu_n(y) \right)}. \tag{7.14}$$

One then has

$$\begin{aligned}
(\det \overline{D}\phi(y))^2 &= \det(\overline{D}\phi(y)^t \circ \overline{D}\phi(y)) \\
&= \begin{array}{c} d \\ \\ n-d \end{array}
\left| \begin{array}{c|c}
\overset{d}{\langle D_j\phi(y),\, D_1\phi(y) \rangle} & \overset{n-d}{\langle \nu_{d+j}(y),\, D_1\phi(y) \rangle} \\
\vdots & \vdots \\
\langle D_j\phi(y),\, D_d\phi(y) \rangle & \langle \nu_{d+j}(y),\, D_d\phi(y) \rangle \\
\hline
\langle D_j\phi(y),\, \nu_{d+1}(y) \rangle & \langle \nu_{d+j}(y),\, \nu_{d+1}(y) \rangle \\
\vdots & \vdots \\
\underset{1 \le j \le d}{\langle D_j\phi(y),\, \nu_n(y) \rangle} & \underset{1 \le j \le n-d}{\langle \nu_{d+j}(y),\, \nu_n(y) \rangle}
\end{array} \right| \\
&= \begin{array}{c} d \\ n-d \end{array}
\left| \begin{array}{c|c}
\overset{d}{\langle D_j\phi(y),\, D_i\phi(y) \rangle} & \overset{n-d}{0} \\
\hline
0 & I_{n-d}
\end{array} \right| \\
&= \det(D\phi(y)^t \circ D\phi(y)).
\end{aligned}$$

Therefore

$$\omega_\phi(y) = |\det \overline{D}\phi(y)|, \tag{7.15}$$

in other words, $\omega_\phi(y)$ is the n-dimensional volume of the parallelepiped in $\mathbf{R}^n$ (see Example 6.6.3) spanned by the vectors $D_1\phi(y), \dots, D_d\phi(y), \nu_{d+1}(y), \dots, \nu_n(y)$ in $\mathbf{R}^n$. This volume is independent of the choice of the vectors $\nu_{d+1}(y), \dots, \nu_n(y)$, provided only that they form an orthonormal basis for $(T_x V)^\perp$.

7.4 Examples of Euclidean densities

I. Arc length. Let $d = 1$. If $D \subset \mathbf{R}$ is open and $\phi : D \to \mathbf{R}^n$ a C^k embedding, for $k \geq 1$, then $V = \mathrm{im}(\phi)$ is a C^k curve in $\mathbf{R}^n$, while, if we identify linear operators and matrices,

$$D\phi(y) = \phi'(y) = \begin{pmatrix} \phi_1'(y) \\ \vdots \\ \phi_n'(y) \end{pmatrix} \in \mathrm{Lin}(\mathbf{R}, \mathbf{R}^n) \qquad (y \in D).$$

We have $D\phi(y)^t \circ D\phi(y) = (\langle D\phi(y), D\phi(y) \rangle)$, and therefore

$$\omega_\phi(y) = \omega(D\phi(y)) = \|D\phi(y)\| \qquad (y \in D). \tag{7.16}$$

The Euclidean density ω in this case is said to be the *arc length*, and accordingly the integration with respect to ω is said to be the *integration with respect to arc length*. Thus we have, for f Riemann integrable over V,

$$\int_V f(x)\, d_1x = \int_D (f \circ \phi)(y)\|D\phi(y)\|\, dy.$$

In particular, the *arc length* of V is defined as

$$\int_V d_1x,$$

if the integral converges. Note that the arc length of V has now been defined independently of the parametrization of V.

Example 7.4.1 (Circle, ellipse and cycloid). The segment of the unit circle $\{ x \in \mathbf{R}^2 \mid \|x\| = 1 \}$ lying between $(1, 0)$ and $(\cos x, \sin x)$, for $0 < x < 2\pi$, is parametrized by

$$\phi(t) = (\cos t, \sin t) \qquad (0 < t < x).$$

Therefore the arc length of this segment is

$$\int_0^x \sqrt{(-\sin t)^2 + \cos^2 t}\, dt = x.$$

We now define the *angle* between two intersecting lines in $\mathbf{R}^2$ as the length of the shortest segment, lying between those lines, of the unit circle about the point of intersection. The result above implies that the angle between the positive direction of the x_1-axis and the line through the points $(0, 0)$ and $(\cos x, \sin x)$, (indeed) equals x.

k^2	0	1	2	3	4	5	6	7	8	9
0.0	5708	5669	5629	5589	5550	5510	5470	5429	5389	5348
0.1	5308	5267	5226	5184	5143	5101	5059	5017	4975	4933
0.2	4890	4848	4805	4762	4718	4675	4631	4587	4543	4498
0.3	4454	4409	4364	4318	4273	4227	4181	4134	4088	4041
0.4	3994	3947	3899	3851	3803	3754	3705	3656	3606	3557
0.5	3506	3456	3405	3354	3302	3250	3198	3145	3092	3038
0.6	2984	2930	2875	2819	2763	2707	2650	2593	2534	2476
0.7	2417	2357	2296	2235	2173	2111	2047	1983	1918	1852
0.8	1785	1717	1648	1578	1507	1434	1360	1285	1207	1129
0.9	1048	0965	0879	0791	0700	0605	0505	0399	0286	0160

Illustration for Example 7.4.1
$E(k) = 1.a$, table giving first four decimals of a (rounded off) with k^2 increasing in steps of 0.01

The length of the ellipse $\{ x \in \mathbf{R}^2 \mid \frac{x_1^2}{a^2} + \frac{x_2^2}{b^2} = 1 \}$, with $a > b > 0$, is given by $4aE(e)$. Here

$$e = \frac{\sqrt{a^2 - b^2}}{a}$$

is said to be the *eccentricity* of the ellipse, and

$$E(k) = \int_0^{\frac{\pi}{2}} \sqrt{1 - k^2 \sin^2 t}\, dt \qquad (0 < k < 1)$$

is *Legendre's form of the complete elliptic integral of the second kind with modulus* k. Indeed, the ellipse is the image under the embedding $t \mapsto (a \cos t,\ b \sin t)$.

The arc length of the segment of the cycloid $\phi : t \mapsto (t - \sin t,\ 1 - \cos t)$ from Example 5.3.6 lying between $\phi(0)$ and $\phi(2\pi)$ is given by

$$\int_0^{2\pi} \sqrt{(1 - \cos t)^2 + \sin^2 t}\, dt = \sqrt{2} \int_0^{2\pi} \sqrt{1 - \cos t}\, dt$$

$$= 2 \int_0^{2\pi} \sin \frac{t}{2}\, dt = \Big[-4 \cos \frac{t}{2} \Big]_0^{2\pi} = 8. \quad ☆$$

Special case. Let $n = 2$ and assume $\phi : \mathbf{R} \to \mathbf{R}^2$ has the form $\phi(t) = (t, f(t))$ of the graph of a C^k function $f : \mathbf{R} \to \mathbf{R}$, for $k \geq 1$. Then the arc length of the segment of the graph of f lying between $(a, f(a))$ and $(b, f(b))$ is given by

$$\int_a^b \|(1, f'(t))\|\, dt = \int_a^b \sqrt{1 + f'(t)^2}\, dt. \qquad (7.17)$$

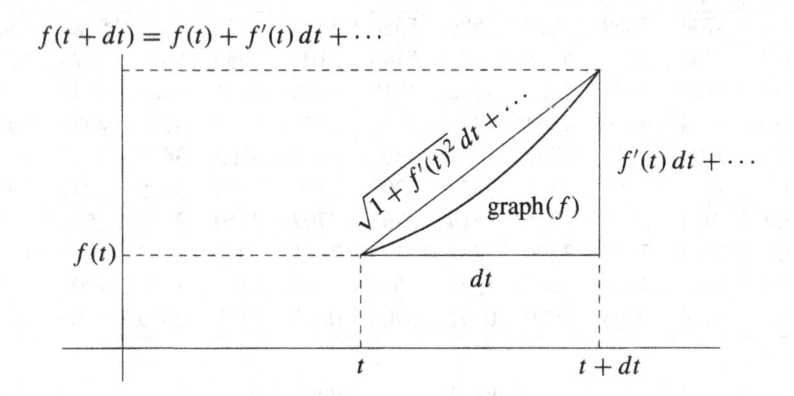

Illustration for Example 7.4.1: Special case

Example 7.4.2. The length l of a circle in $\mathbf{R}^2$ of radius R is $2\pi R$. Indeed, let $f(t) = \sqrt{R^2 - t^2}$. Then

$$1 + f'(t)^2 = 1 + \left(\frac{-t}{\sqrt{R^2 - t^2}}\right)^2 = 1 + \frac{t^2}{R^2 - t^2} = \frac{R^2}{R^2 - t^2}.$$

And therefore

$$l = 2\int_{-R}^{R} \frac{R}{\sqrt{R^2 - t^2}}\,dt = 2R\int_{-1}^{1} \frac{1}{\sqrt{1 - t^2}}\,dt = 2R\Big[\arcsin t\Big]_{-1}^{1} = 2\pi R. \quad ☆$$

Special case. Let $n = 2$. Many curves in $\mathbf{R}^2$ are described in polar coordinates (r, α) for $\mathbf{R}^2$ by an equation of the form $r = f(\alpha)$, for example

$$r = R \qquad (\textit{circle}); \qquad r = 2(1 + \cos\alpha) \qquad (\textit{cardioid});$$
$$r = \cos 2\alpha \qquad (\textit{rose with four petals}).$$

Such curves therefore occur as images under mappings of the form $\phi : \alpha \mapsto f(\alpha)(\cos\alpha,\ \sin\alpha)$. If this is a C^k mapping, for $k \geq 1$, then

$$D\phi(\alpha) = \begin{pmatrix} f'(\alpha)\cos\alpha - f(\alpha)\sin\alpha \\ f'(\alpha)\sin\alpha + f(\alpha)\cos\alpha \end{pmatrix}, \qquad \|D\phi(\alpha)\| = \sqrt{f(\alpha)^2 + f'(\alpha)^2}.$$

Example 7.4.3. The length of a circle in $\mathbf{R}^2$ of radius R is $\int_{-\pi}^{\pi} R\,d\alpha = 2\pi R$; and of the cardioid

$$2\int_{-\pi}^{\pi}\sqrt{(1+\cos\alpha)^2+\sin^2\alpha}\,d\alpha = 2\sqrt{2}\int_{-\pi}^{\pi}\sqrt{1+\cos\alpha}\,d\alpha = 4\int_{-\pi}^{\pi}\cos\frac{\alpha}{2}\,d\alpha = 16. \quad ☆$$

Example 7.4.4 (Parametrization by arc length). Assume the C^1 embedding $\phi : \,]a, b[\, \to \mathbf{R}^n$ has an arc length of l. We then have the corresponding *arc-length function* (compare (7.16))

$$\lambda : \,]a, b[\, \to \,]0, l[\qquad \text{with} \qquad \lambda(t) = \int_a^t \|\phi'(y)\|\,dy.$$

Because $\lambda'(t) = \|\phi'(t)\| > 0$, the function λ is strictly monotonically increasing on $]a, b[$; therefore λ has a differentiable inverse function $\lambda^{-1} : \,]0, l[\, \to \,]a, b[$. If $\lambda(t) = s$, then by the chain rule

$$(\lambda^{-1})'(\lambda(t))\,\lambda'(t) = 1, \qquad \text{that is,} \qquad (\lambda^{-1})'(s) = \|\phi'(t)\|^{-1}.$$

Consequently, we find for the curve

$$\psi : \,]0, l[\, \to \mathbf{R}^n \qquad \text{with} \qquad \psi(s) = \phi(\lambda^{-1}(s)) = \phi(t),$$

by means of the chain rule

$$\|\psi'(s)\| = \|\phi'(\lambda^{-1}(s))\|\,|(\lambda^{-1})'(s)| = \|\phi'(t)\|\,\|\phi'(t)\|^{-1} = 1.$$

That is, the curve $\psi : \,]0, l[\, \to \mathbf{R}^n$ is a C^1 parametrization of $V = \phi(\,]a, b[\,)$ with the arc length as a parameter, and the tangent vector $\psi'(s)$ to V at $\psi(s)$ is always of unit length. ☆

II. Area. If $d = 2$, then V is a C^k surface in $\mathbf{R}^n$, while, if we identify linear operators and matrices, for $y \in D$,

$$D\phi(y) = \begin{pmatrix} D_1\phi(y) & D_2\phi(y) \end{pmatrix} = \begin{pmatrix} \dfrac{\partial\phi_1}{\partial y_1}(y) & \dfrac{\partial\phi_1}{\partial y_2}(y) \\ \vdots & \vdots \\ \dfrac{\partial\phi_n}{\partial y_1}(y) & \dfrac{\partial\phi_n}{\partial y_2}(y) \end{pmatrix} \in \mathrm{Lin}(\mathbf{R}^2, \mathbf{R}^n).$$

We obtain

$$D\phi(y)^t \circ D\phi(y) = \begin{pmatrix} \|D_1\phi\|^2(y) & \langle D_1\phi,\, D_2\phi\rangle(y) \\ \langle D_2\phi,\, D_1\phi\rangle(y) & \|D_2\phi\|^2(y) \end{pmatrix},$$

and consequently

$$\omega_\phi(y) = \sqrt{\|D_1\phi\|^2\,\|D_2\phi\|^2 - \langle\, D_1\phi,\ D_2\phi\,\rangle^2}\,(y) \qquad (y \in D). \tag{7.18}$$

If α is equal to the angle between the vectors $D_1\phi(y)$ and $D_2\phi(y)$, we find

$$\frac{\langle\, D_1\phi(y),\ D_2\phi(y)\,\rangle}{\|D_1\phi(y)\|\,\|D_2\phi(y)\|} = \cos\alpha; \qquad \text{so} \qquad \omega_\phi(y) = \|D_1\phi(y)\|\,\|D_2\phi(y)\|\,\sin\alpha. \tag{7.19}$$

In this case the Euclidean density ω is said to be the *Euclidean area*; accordingly, the integration with respect to ω is said to be the *integration with respect to Euclidean area*. Thus, for f Riemann integrable over V,

$$\int_V f(x)\,d_2x = \int_D (f\circ\phi)(y)\sqrt{\|D_1\phi\|^2\,\|D_2\phi\|^2 - \langle\, D_1\phi,\ D_2\phi\,\rangle^2}\,(y)\,dy.$$

In particular, the *Euclidean area* of V is defined as

$$\int_V d_2x,$$

if the integral converges. Note that the area of V is now defined independently of the parametrization of V.

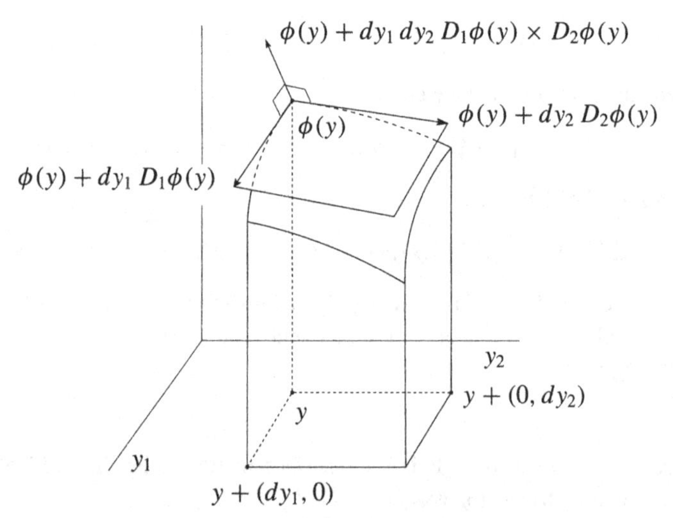

Illustration for 7.4.II. Area
Area of parallelogram spanned by $dy_1\,D_1\phi(y)$ and $dy_2\,D_2\phi(y)$ equals
$\|(D_1\phi \times D_2\phi)(y)\|\,dy_1\,dy_2 = \|(D_1\phi \times D_2\phi)(y)\|\,dy$

Special case. In the case where $n = 3$, Formula (7.19) takes the form, see the remark on linear algebra in Section 5.3,

$$\omega_\phi(y) = \|(D_1\phi \times D_2\phi)(y)\|. \tag{7.20}$$

Example 7.4.5 (Torus). According to Example 5.7.2, the toroidal surface $T \subset \mathbf{R}^3$ has the parametrization $\phi :]-\pi,\ \pi\,[\times]-\pi,\ \pi\,[\to T$, with

$$\phi : (\alpha,\ \theta) \mapsto ((2+\cos\theta)\cos\alpha,\ (2+\cos\theta)\sin\alpha,\ \sin\theta),$$

while

$$\frac{\partial\phi}{\partial\alpha}(\alpha,\ \theta) \times \frac{\partial\phi}{\partial\theta}(\alpha,\ \theta) = (2+\cos\theta)\begin{pmatrix} \cos\alpha\cos\theta \\ \sin\alpha\cos\theta \\ \sin\theta \end{pmatrix}.$$

Therefore, from (7.20),

$$\omega_\phi(\alpha,\ \theta) = 2+\cos\theta.$$

Consequently the Euclidean area of the torus equals

$$\int_{[-\pi,\,\pi]\times[-\pi,\,\pi]} (2+\cos\theta)\,d\alpha\,d\theta = \int_{-\pi}^{\pi} d\alpha \int_{-\pi}^{\pi} 2\,d\theta = 2\pi\cdot 4\pi = 8\pi^2.$$

This result leads to the following remark. Intersect the toroidal surface with a plane through the x_3-axis. We slit the toroidal surface open along one of the circles thus created. Next, we straighten the toroidal surface, in such a way that the central circle retains its length, and we obtain the form of a right cylinder. This cylinder then has a ruling of length 4π, and a perimeter of length 2π; therefore the Euclidean area of the cylinder equals $8\pi^2$ (check this). Evidently, the total area is not altered by this deformation from torus to straight cylinder.

Furthermore, the average of the Gaussian curvature K from Example 5.7.2 over the torus V equals

$$\frac{1}{\text{area } V}\int_V K(x)\,dx = \frac{1}{8\pi^2}\int_{[-\pi,\,\pi]\times[-\pi,\,\pi]} \frac{\cos\theta}{2+\cos\theta}(2+\cos\theta)\,d\alpha\,d\theta = 0. \quad ☆$$

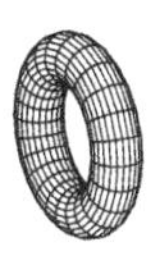

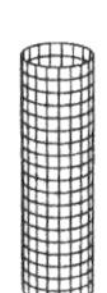
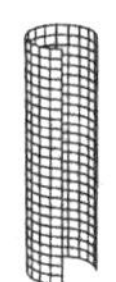
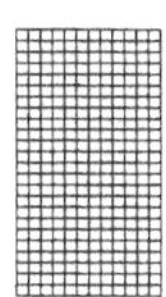

Illustration for Example 7.4.5: Torus

Example 7.4.6 (Cap of sphere and kissing number). Let V be that part of the sphere $\{\, x \in \mathbf{R}^3 \mid \|x\| = R\,\}$ lying inside the lateral surface of the cone given by the equation $x_3 = \sqrt{x_1^2 + x_2^2}\,\tan\psi$, for $0 < \psi < \frac{\pi}{2}$. In spherical coordinates for $\mathbf{R}^3$

$$x = r(\cos\alpha\cos\theta,\ \sin\alpha\cos\theta,\ \sin\theta) \qquad (0 < r,\ -\pi \le \alpha \le \pi,\ -\frac{\pi}{2} \le \theta \le \frac{\pi}{2}),$$

the set V is described as the image under the mapping

$$\phi : (\alpha,\ \theta) \mapsto R(\cos\alpha\cos\theta,\ \sin\alpha\cos\theta,\ \sin\theta),$$

where $-\pi \le \alpha \le \pi$ and $\psi \le \theta \le \frac{\pi}{2}$. Now

$$\begin{aligned}
\frac{\partial\phi}{\partial\alpha}(\alpha,\ \theta) &= R(-\sin\alpha\cos\theta,\ \cos\alpha\cos\theta,\ 0),\\
\frac{\partial\phi}{\partial\theta}(\alpha,\ \theta) &= R(-\cos\alpha\sin\theta,\ -\sin\alpha\sin\theta,\ \cos\theta),\\
\frac{\partial\phi}{\partial\alpha}\times\frac{\partial\phi}{\partial\theta}(\alpha,\ \theta) &= R^2(\cos\alpha\cos^2\theta,\ \sin\alpha\cos^2\theta,\ \cos\theta\sin\theta),\\
\left\|\frac{\partial\phi}{\partial\alpha}\times\frac{\partial\phi}{\partial\theta}(\alpha,\ \theta)\right\| &= R^2\cos\theta.
\end{aligned}$$

And so

$$\begin{aligned}
\text{area}(V) &= \int_{[-\pi,\,\pi\,]\times[\,\psi,\,\frac{\pi}{2}\,]} R^2\cos\theta\, d\alpha\, d\theta = R^2\int_{-\pi}^{\pi} d\alpha \int_{\psi}^{\frac{\pi}{2}}\cos\theta\, d\theta\\
&= 2\pi R^2(1-\sin\psi).
\end{aligned}$$

In particular, for $\psi = 0$, we find that the Euclidean area of a hemisphere equals $2\pi R^2$; therefore that of the entire sphere equals $4\pi R^2$.

We now give an application of the preceding calculation. Consider two spheres, say S_1 and S_2, in $\mathbf{R}^3$ of radius 1 that have one point in common. The lines through the center of S_1 that are tangent to S_2 form a conical surface, and the maximum angle between two of these tangent lines equals $\frac{\pi}{3}$. The solid angle subtended by S_2 from the center of S_1 therefore equals $2\pi(1-\frac{1}{2}\sqrt{3})$. This implies that in $\mathbf{R}^3$ a maximum of

$$14 < \frac{4\pi}{\pi(2-\sqrt{3})} = 8 + 4\sqrt{3} < 15$$

unit spheres, pairwise having at most one point in common, can be tangent to a unit sphere.

The actual maximum number, the so-called *kissing number*, equals 12; a proof of this is not altogether simple to give.[1] A configuration in which this number is realized is obtained by taking the 12 vertices of a regular *icosahedron* (εἴκοσιν = twenty) of diameter 4 as the centers of spheres of radius 1. Then all of these 12 spheres are tangent to the sphere of radius 1 about the center of the icosahedron. Note that the ($\frac{20\cdot 3}{2} = 30$) edges of this icosahedron are of length $\sqrt{8-\frac{8}{\sqrt{5}}} = 2.1029\cdots$. In this configuration, therefore, the 12 spheres are not tangent to each other. ☆

[1] See Chapter 12 in Aigner, M., Ziegler, G. M.: *Proofs from THE BOOK*. Springer-Verlag, Berlin and Heidelberg 1998.

Example 7.4.7. Let V be the surface in $\mathbf{R}^4$ that occurs as the intersection of the

$$\text{ellipsoid} \qquad \{\, x \in \mathbf{R}^4 \mid x_1^2 + x_2^2 + x_3^2 + 3x_4^2 = 1 \,\},$$

with the

$$\text{cone} \qquad \{\, x \in \mathbf{R}^4 \mid x_1^2 + x_2^2 + x_3^2 - x_4^2 = 0 \,\}.$$

Subtracting the equations we see that $x \in V$ satisfies

$$x_4 = \pm\frac{1}{2}, \qquad \text{hence} \qquad x_1^2 + x_2^2 + x_3^2 = \frac{1}{4}.$$

It thus emerges that V is the disjoint union of two spheres (of dimension two). We now have parametrizations

$$\phi_\pm : \,]-\pi,\, \pi\,[\times \Big]-\frac{\pi}{2},\, \frac{\pi}{2}\Big[\to V,$$

$$\phi_\pm(\alpha,\, \theta) = \frac{1}{2}(\cos\alpha\cos\theta,\ \sin\alpha\cos\theta,\ \sin\theta,\ \pm 1).$$

Then, for $\phi = \phi_\pm$,

$$\begin{aligned}
\frac{\partial\phi}{\partial\alpha}(\alpha,\, \theta) &= \frac{1}{2}(-\sin\alpha\cos\theta,\ \cos\alpha\cos\theta,\ 0,\ 0),\\
\frac{\partial\phi}{\partial\theta}(\alpha,\, \theta) &= \frac{1}{2}(-\cos\alpha\sin\theta,\ -\sin\alpha\sin\theta,\ \cos\theta,\ 0),\\
\Big\|\frac{\partial\phi}{\partial\alpha}(\alpha,\, \theta)\Big\|^2 &= \frac{\cos^2\theta}{4}, \qquad \Big\|\frac{\partial\phi}{\partial\theta}(\alpha,\, \theta)\Big\|^2 = \frac{1}{4},\\
\Big\langle \frac{\partial\phi}{\partial\alpha}(\alpha,\, \theta),\, \frac{\partial\phi}{\partial\theta}(\alpha,\, \theta) \Big\rangle &= 0.
\end{aligned}$$

Consequently, one finds from (7.18)

$$\omega_{\phi_\pm}(\alpha,\, \theta) = \Big\|\frac{\partial\phi}{\partial\alpha}(\alpha,\, \theta)\Big\|\,\Big\|\frac{\partial\phi}{\partial\theta}(\alpha,\, \theta)\Big\| = \frac{\cos\theta}{4}.$$

Therefore it follows that V has Euclidean area

$$2\int_{[\,-\pi,\,\pi\,]\times[\,-\frac{\pi}{2},\,\frac{\pi}{2}\,]} \frac{\cos\theta}{4}\, d\alpha\, d\theta = \frac{1}{2}\int_{-\pi}^{\pi} d\alpha \int_{-\frac{\pi}{2}}^{\frac{\pi}{2}} \cos\theta\, d\theta = 2\pi. \qquad ☆$$

Example 7.4.8 (Newton's potential of a sphere). We employ the notation of the preceding Example 7.4.6 and that of Example 6.6.7. The Newton potential ϕ_A of the sphere $A = \{\, x \in \mathbf{R}^3 \mid \|x\| = R \,\}$ is the zero function, because A is a negligible set in $\mathbf{R}^3$. However, if we now define the Newton potential ϕ_A as an integral with respect to the Euclidean density ω on A, then, for $x \notin A$,

$$\phi_A(x) = -\frac{1}{4\pi}\int_A \frac{1}{\|x - y\|}\, d_2 y.$$

In view of the rotational symmetry of A we may assume that $x = (0, 0, a)$, with $a = \|x\| \neq R$. Upon introducing spherical coordinates we obtain, for $y \in A$,

$$y = R(\cos\alpha\cos\theta,\ \sin\alpha\cos\theta,\ \sin\theta).$$

Hence

$$\|y - x\| = \sqrt{R^2\cos^2\theta + (R\sin\theta - a)^2} = \sqrt{R^2 + a^2 - 2aR\sin\theta}.$$

Therefore

$$\begin{aligned}
-\phi_A(x) &= \frac{1}{4\pi}\int_{-\pi}^{\pi}\left(\int_{-\frac{\pi}{2}}^{\frac{\pi}{2}} \frac{R^2\cos\theta}{\sqrt{R^2 + a^2 - 2aR\sin\theta}}\,d\theta\right)d\alpha \\
&= \frac{R^2}{2aR}\Big[-\sqrt{R^2 + a^2 - 2aR\sin\theta}\,\Big]_{-\frac{\pi}{2}}^{\frac{\pi}{2}} = \frac{R}{2a}(-|R-a| + R + a) \\
&= \begin{cases} \dfrac{R}{2a}(-(R-a) + R + a) = R & (a < R); \\ \dfrac{R}{2a}(-(a-R) + R + a) = \dfrac{R^2}{a} & (a > R); \end{cases}
\end{aligned}$$

with the result that

$$\phi_A(x) = \begin{cases} -R, & \text{if } x \text{ inside } A; \\ -\dfrac{R^2}{\|x\|}, & \text{if } x \text{ outside } A. \end{cases}$$

Note that ϕ_A can be continuously extended over A. ✩

Example 7.4.9 (Geometrical interpretation of Gauss curvature). The notation employed is that of Section 5.7. Let V be a C^2 surface in $\mathbf{R}^3$. Let $x \in V$ be fixed and let B_r be the closed ball in $\mathbf{R}^3$ about x and of radius r. Then, with $K(x)$ the Gauss curvature of V at x and $n : V \to S^2$ the Gauss mapping, we have

$$|K(x)| = \lim_{r\downarrow 0}\frac{\operatorname{area} n(V \cap B_r)}{\operatorname{area}(V \cap B_r)}. \tag{7.21}$$

Indeed, let $\phi : D_r \to V \cap B_r$ be a parametrization of $V \cap B_r$. Note that, by restriction, this gives a parametrization of $V \cap B_{r'}$, for $r' < r$. Then

$$n \circ \phi : D_r \to n(V \cap B_r)$$

is a parametrization of the image set $n(V \cap B_r)$. The tangent space to $V \cap B_r$ at $x = \phi(y)$ is spanned by the tangent vectors $D_1\phi(y)$ and $D_2\phi(y)$, while the tangent space to $n(V \cap B_r)$ at $n(x) = n \circ \phi(y)$ is spanned by (see Formula (5.12))

$$D_j(n \circ \phi)(y) = Dn(x)\,D_j\phi(y) \qquad (1 \leq j \leq 2). \tag{7.22}$$

If $A \in \mathrm{Aut}(\mathbf{R}^3)$, one has for $u,\ v,\ w \in \mathbf{R}^3$,

$$\begin{aligned}\langle\, Au \times Av,\ Aw \,\rangle &= \det(Au\ Av\ Aw) = \det A\ \det(u\ v\ w) = \det A\ \langle\, u \times v,\ w \,\rangle \\ &= \det A\ \langle\, u \times v,\ A^{-1}Aw \,\rangle = \det A\ \langle\, (A^{-1})^t(u \times v),\ Aw \,\rangle.\end{aligned}$$

That is

$$Au \times Av = \det A\ (A^{-1})^t(u \times v). \tag{7.23}$$

Because we regard $Dn(x)$ as element of $\mathrm{End}(T_x V)$ and because $K(x) = \det Dn(x)$, it follows from (7.22) and (7.23), with $A : \mathbf{R}^3 \to \mathbf{R}^3$ defined by $A|_{T_x V} = Dn(x)$ and $A|_{(T_x V)^\perp} = I$, that

$$D_1(n \circ \phi)(y) \times D_2(n \circ \phi)(y) = K(x)\,(D_1\phi \times D_2\phi)(y).$$

We obtain

$$\operatorname{area} n(V \cap B_r) = \int_{D_r} |K \circ \phi(y)|\ \|(D_1\phi \times D_2\phi)(y)\|\, dy, \tag{7.24}$$

$$\operatorname{area}(V \cap B_r) = \int_{D_r} \|(D_1\phi \times D_2\phi)(y)\|\, dy.$$

The assertion (7.21) now follows by application of an argument analogous to that of the Mean Value Theorem of integral calculus to the integral in (7.24). ✩

III. Hyperarea. (See also Example 5.3.11.) Let $d = n - 1$, then $V = \mathrm{im}(\phi)$ is a C^k *hypersurface* in $\mathbf{R}^n$. According to Formula (5.3) one has

$$\det\,(D\phi(y)^t \circ D\phi(y)) = \|(D_1\phi \times \cdots \times D_{n-1}\phi)(y)\|^2.$$

Therefore

$$\omega_\phi(y) = \|(D_1\phi \times \cdots \times D_{n-1}\phi)(y)\| \qquad (y \in D). \tag{7.25}$$

In the case where $d = 2$ (and so $n = 3$), note the agreement between Formulae (7.20) and (7.25).

In the case $d = n - 1$ the Euclidean density ω is said to be the *Euclidean hyperarea*, and accordingly the integration with respect to ω is said to be the *integration with respect to Euclidean hyperarea.* Therefore, for f Riemann integrable over V,

$$\int_V f(x)\, d_{n-1}x = \int_D (f \circ \phi)(y)\|(D_1\phi \times \cdots \times D_{n-1}\phi)(y)\|\, dy.$$

In particular, the *Euclidean hyperarea* of V is defined as

$$\operatorname{hyperarea}(V) = \int_V d_{n-1}x,$$

if the integral converges. Note that the hyperarea of V is now defined independently of the parametrization of V.

Special case. Let $D \subset \mathbf{R}^{n-1}$ be an open set and let $\phi(y) = (y,\ h(y))$, for a C^k function $h : D \to \mathbf{R}$. One then has, on account of Formula (5.7), for every $y \in D$,

$$\omega_\phi(y) = \sqrt{1 + \|Dh(y)\|^2}.$$

Note that the special case in (7.17) follows from this equation.

We have $\cos\psi = \dfrac{1}{(1 + \|Dh(y)\|^2)^{1/2}}$.

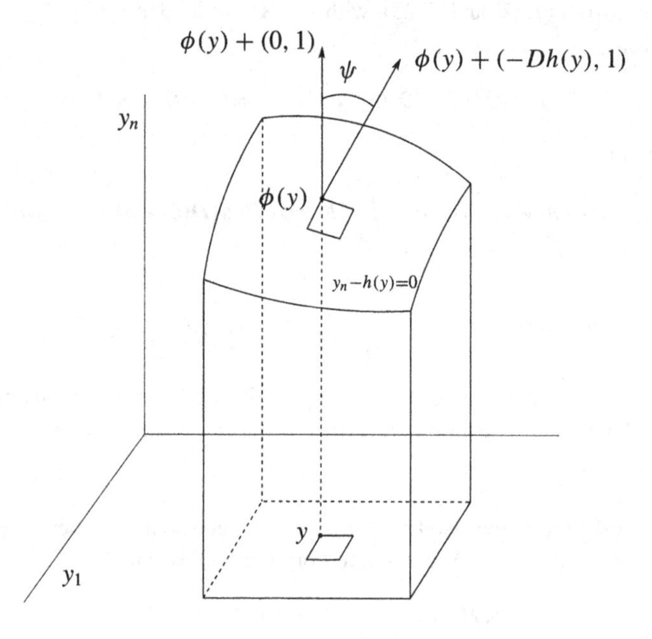

For the hyperarea $\omega_\phi(y)\, dy_1 \cdots dy_{n-1}$ of an element of the hypersurface $y_n - h(y) = 0$ over a rectangle in $\mathbf{R}^{n-1}$ with vertex y and edges of lengths $dy_1, \ldots, dy_{n-1}$, one has approximately

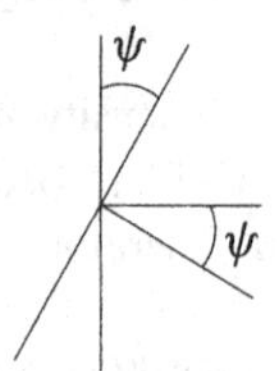

$$\frac{dy_1 \cdots dy_{n-1}}{\omega_\phi(y)\, dy_1 \cdots dy_{n-1}} = \cos\psi; \text{ hence } \omega_\phi(y) = \frac{1}{\cos\psi} = (1 + \|Dh(y)\|^2)^{1/2}$$

Illustration for 7.4.III. Hyperarea

Example 7.4.10 (Area of two-sphere). The unit sphere S^2 in $\mathbf{R}^3$ is the union of two surfaces, determined as indicated above by the functions $h_\pm : B^2 \mapsto \mathbf{R}$, with B^2 the closed unit disk in $\mathbf{R}^2$ and

$$h_\pm(y) = \pm\sqrt{1 - \|y\|^2}.$$

We have

$$Dh_\pm(y) = -\frac{1}{h_\pm(y)}y, \qquad \sqrt{1 + \|Dh_\pm(y)\|^2} = \frac{1}{\sqrt{1 - \|y\|^2}}.$$

The two surfaces have the one-dimensional equator in common; accordingly, for the two-dimensional calculation of the area this is negligible. Consequently, by Example 6.6.4,

$$\mathrm{area}(S^2) = 2\int_{B^2} \frac{1}{\sqrt{1 - \|y\|^2}}\,dy = 2\int_{-\pi}^{\pi} d\phi \int_0^1 \frac{r}{\sqrt{1 - r^2}}\,dr = 4\pi. \qquad ☆$$

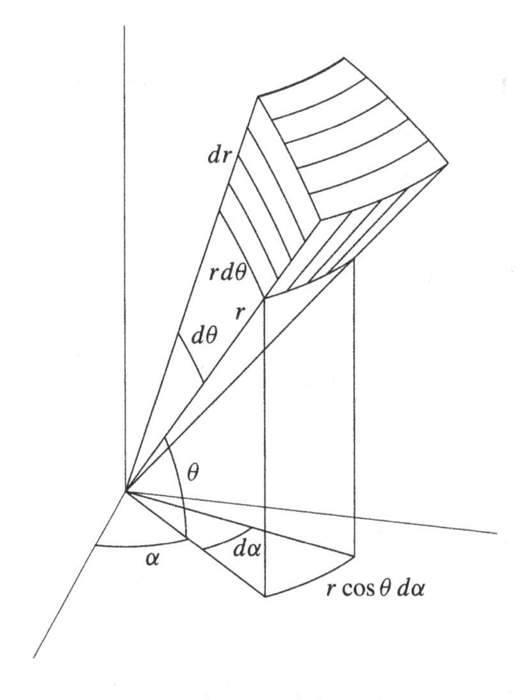

Illustration for Example 7.4.11
The shaded volume approximately equals $r\cos\theta\,d\alpha\,r\,d\theta\,dr = r^2\cos\theta\,dr\,d\alpha\,d\theta$

Example 7.4.11 (Hyperarea of three-sphere). The unit sphere S^3 in $\mathbf{R}^4$ is the union of two hypersurfaces, determined as indicated above by the functions $h_\pm : B^3 \to \mathbf{R}$, with B^3 the closed unit ball in $\mathbf{R}^3$ and

$$h_\pm(y) = \pm\sqrt{1 - \|y\|^2}.$$

The two surfaces have the two-dimensional equator in common; for the three-dimensional calculation of the hyperarea this is negligible. Therefore we now have, as in Example 7.4.10,

$$\text{hyperarea}(S^3) = 2\int_{B^3} \frac{1}{\sqrt{1 - \|y\|^2}}\, dy.$$

Parametrize B^3 with

$$\Psi : (r, \alpha, \theta) \mapsto r(\cos\alpha\cos\theta,\ \sin\alpha\cos\theta,\ \sin\theta),$$

$$(0 < r < 1,\ -\pi < \alpha < \pi,\ -\frac{\pi}{2} < \theta < \frac{\pi}{2}).$$

Then

$$D\Psi(r, \alpha, \theta) = \begin{pmatrix} \cos\alpha\cos\theta & -\sin\alpha & -\cos\alpha\sin\theta \\ \sin\alpha\cos\theta & \cos\alpha & -\sin\alpha\sin\theta \\ \sin\theta & 0 & \cos\theta \end{pmatrix} \begin{pmatrix} 1 & 0 & 0 \\ 0 & r\cos\theta & 0 \\ 0 & 0 & r \end{pmatrix}.$$

Hence $\det D\Psi(r, \alpha, \theta) = r^2\cos\theta$. By the Change of Variables Theorem 6.6.1, therefore

$$\begin{aligned} \text{hyperarea}(S^3) &= 2\int_{-\pi}^{\pi} d\alpha \int_{-\frac{\pi}{2}}^{\frac{\pi}{2}} \cos\theta\, d\theta \int_0^1 \frac{r^2}{\sqrt{1-r^2}}\, dr \\ &= 4\pi\, [\,\sin\theta\,]_{-\frac{\pi}{2}}^{\frac{\pi}{2}} \int_0^1 \left(\frac{1}{\sqrt{1-r^2}} - \sqrt{1-r^2} \right) dr \\ &= 8\pi \left[\frac{\arcsin r}{2} - \frac{r}{2}\sqrt{1-r^2} \right]_0^1 = 4\pi \arcsin 1 = 2\pi^2, \end{aligned}$$

where use has been made of the computation of the antiderivative as in Example 6.5.2. ☆

Example 7.4.12 (Spherical coordinates in $\mathbf{R}^n$). Let $S^{n-1} = \{\, x \in \mathbf{R}^n \mid \|x\| = 1 \,\}$ be the $(n-1)$-dimensional unit sphere in $\mathbf{R}^n$. Assume that $\phi : D \to S^{n-1}$ with $D \subset \mathbf{R}^{n-1}$ open is a C^1 parametrization of an open part of S^{n-1} having negligible complement (see Exercise 7.21.(vii) for explicit formulae). Then the mapping

$$\Psi : \mathbf{R}_+ \times D \to \mathbf{R}^n \qquad \text{given by} \qquad \Psi(r, y) = r\,\phi(y)$$

is a C^1 diffeomorphism onto an open dense subset in $\mathbf{R}^n$, with

$$D\Psi(r, y) = (\phi(y) \mid r\,D\phi(y)).$$

Therefore we find

$$|\det D\Psi(r, y)| = r^{n-1}|\det(\phi(y) \mid D\phi(y))| = r^{n-1}|\det \overline{D}\phi(y)|,$$

in the notation of Formula (7.14), because $\phi(y)$ is a unit vector perpendicular to $T_{\phi(y)}S^{n-1}$. But then, by Formula (7.15)

$$|\det D\Psi(r, y)| = r^{n-1}\omega_\phi(y).$$

Hence it follows, from the Change of Variables Theorem 6.6.1 and Corollary 6.4.3, that for every $f \in C_c(\mathbf{R}^n)$ (compare with Example 6.6.4),

$$\begin{aligned}\int_{\mathbf{R}^n} f(x)\,dx &= \int_{\mathbf{R}_+} r^{n-1} \int_{S^{n-1}} f(ry)\,d_{n-1}y\,dr \\ &= \int_{S^{n-1}} \int_{\mathbf{R}_+} r^{n-1} f(ry)\,dr\,d_{n-1}y.\end{aligned} \tag{7.26}$$

☆

7.5 Open sets at one side of their boundary

The Fundamental Theorem of Integral Calculus 2.10.1 on $\mathbf{R}$ asserts that, for C^1 functions f on $[\,a,\ b\,] \subset \mathbf{R}$,

$$\int_a^b \frac{df}{dx}(x)\,dx = f(b) - f(a).$$

This establishes a relation between the integral over the open set $]\,a, b\,[$ of the derivative of a function f on the one hand, and the values of that f at the boundary points b and a on the other. It is a characteristic feature of this relation that $]\,a, b\,[$ is bounded by its boundary points a and b, and that the boundary points b and a are counted as positive and negative, respectively. In Section 7.6 we formulate an analog for $\mathbf{R}^n$ of the Fundamental Theorem, applicable for suitable open sets Ω in $\mathbf{R}^n$. Admissible sets in that respect are bounded open sets Ω which are bounded by their boundary $\partial\Omega$ (see Definition 7.5.2), and whose boundary $\partial\Omega$ is a C^1 manifold. At this point, recall that according to Formula (1.4) one has $\partial\Omega = \overline{\Omega} \setminus \Omega$, if Ω is open. Undesirable situations are outlined below. Note that in situation IV a point $x^0 \in \partial\Omega$ possesses an open neighborhood U in $\mathbf{R}^n$ with

$$U \cap \Omega = U \setminus \partial\Omega. \tag{7.27}$$

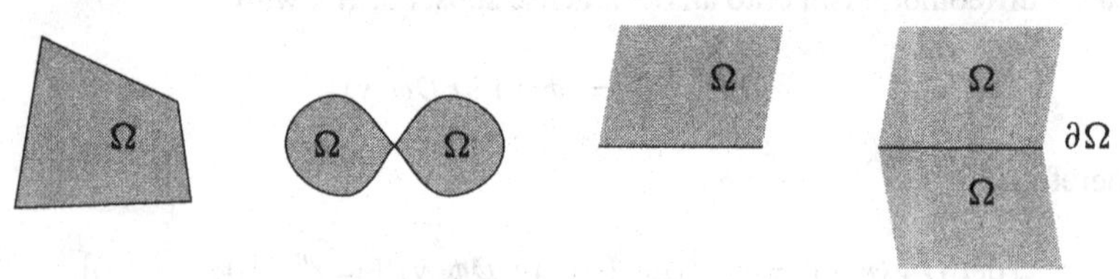

I: $\partial\Omega$ not C^1 II: $\partial\Omega$ image under immersion, not embedding III: Ω unbounded IV: Ω not bounded by $\partial\Omega$

Notation. Let $\Omega \subset \mathbf{R}^n$ be an open set that satisfies, in the terminology of Example 5.3.11, the requirement

$$\partial\Omega \quad \text{is a } C^k \text{ hypersurface in } \mathbf{R}^n \qquad (k \geq 1).$$

According to the theory in that example, $\partial\Omega$ can locally be described as follows. For every $x^0 \in \partial\Omega$ there exist an open neighborhood U of x^0 in $\mathbf{R}^n$, an open subset $D \subset \mathbf{R}^{n-1}$, and a C^k embedding $\phi : D \to \mathbf{R}^n$ with

$$\partial\Omega \cap U = \mathrm{im}(\phi) = \{\, \phi(y) \mid y \in D \,\}. \tag{7.28}$$

For $x = \phi(y) \in \partial\Omega \cap U$ the column vectors $D_j\phi(y)$ in the matrix $D\phi(y)$, for $1 \leq j \leq n-1$, form a basis for $T_x(\partial\Omega)$, the tangent space to $\partial\Omega$ at x. Let $x^0 = \phi(y^0)$, and let $v \in \mathbf{R}^n$ with $v \notin T_{x^0}(\partial\Omega)$ be chosen arbitrarily, then $\det(v \mid D\phi(y^0)) \neq 0$ because of the linear independence of the occurring vectors. Now define

$$\Psi : \mathbf{R} \times D \to \mathbf{R}^n \qquad \text{by} \qquad \Psi(t, y) = tv + \phi(y). \tag{7.29}$$

One then has

$$D\Psi(t, y) = (v \mid D\phi(y)) \qquad ((t, y) \in \mathbf{R} \times D), \tag{7.30}$$

and so

$$\det D\Psi(t, y^0) \neq 0 \qquad (t \in \mathbf{R}). \tag{7.31}$$

The Local Inverse Function Theorem 3.2.4 asserts that we can find $\delta > 0$ and that we can shrink D and U if necessary, in such a way that

$$\Psi : \,]-\delta, \delta[\, \times D \to U \quad \text{is a } C^k \text{ diffeomorphism of open sets.} \tag{7.32}$$

Note that $\partial\Omega \cap U \subset \Psi(\{0\} \times D)$; on the other hand $tv + \phi(y) = \phi(y') \in \partial\Omega \cap U$, with $|t| < \delta$ and $y, y' \in D$, implies $t = 0$ and $y = y'$. As a result we have

$$\partial\Omega \cap U = \Psi(\{0\} \times D). \tag{7.33}$$

That is, $\partial\Omega$ has now been flattened locally. Indeed, (7.33) asserts that in (t, y)-coordinates $\partial\Omega \cap U$ is given by the condition $t = 0$.

Also, according to Example 5.3.11, there exists a C^k function $g : U \to \mathbf{R}$ with $Dg \neq 0$ on U and

$$\partial\Omega \cap U = N(g) = \{\, x \in U \mid g(x) = 0 \,\}. \tag{7.34}$$

Theorem 7.5.1. *For an open subset $\Omega \subset \mathbf{R}^n$ the following three assertions are equivalent.*

(i) *$\partial\Omega$ is a C^k hypersurface and for every $x^0 \in \partial\Omega$ there does* **not** *exist an open neighborhood U of x^0 in $\mathbf{R}^n$ such that $U \cap \Omega = U \setminus \partial\Omega$ (compare with (7.27)).*

(ii) *For every $x^0 \in \partial\Omega$ and for every $v \in \mathbf{R}^n$ with $v \notin T_{x^0}(\partial\Omega)$, there exist U, Ψ, δ and D as in notations (7.28), (7.29) and (7.32) with*

$$\Omega \cap U = \Psi(\,]-\delta,\ 0\,[\times D) = \{\, tv + \phi(y) \mid -\delta < t < 0,\ y \in D \,\} \qquad \textit{or}$$
$$\Omega \cap U = \Psi(\,]\,0,\ \delta\,[\times D).$$

(iii) *For every $x^0 \in \partial\Omega$ there exist U and g as in notation (7.34), for which*

$$\Omega \cap U = \{\, x \in U \mid g(x) < 0 \,\}.$$

Proof. (i) $\Rightarrow$ (ii). Let U be as in (7.32) and write

$$U_0 = \{\, \Psi(0, y) \mid y \in D \,\}, \qquad U_\pm = \{\, \Psi(t, y) \mid |t| < \delta,\ t \gtrless 0,\ y \in D \,\}.$$

From (7.33) we know already that $\partial\Omega \cap U = U_0$; and because $\Omega \cap \partial\Omega = \emptyset$, it now follows that $\Omega \cap U = (\Omega \cap U_+) \cup (\Omega \cap U_-)$. We will prove that either $\Omega \cap U = U_+$ or $\Omega \cap U = U_-$.

We may assume that D is a convex set in $\mathbf{R}^{n-1}$ (this may require U to be shrunk). Because $x^0 \in \overline{\Omega}$ one has $\Omega \cap U \neq \emptyset$; and this leads to either $\Omega \cap U_+ \neq \emptyset$ or $\Omega \cap U_- \neq \emptyset$. We now prove

$$\Omega \cap U_\pm \neq \emptyset \qquad \Longrightarrow \qquad \Omega \cap U_\pm = U_\pm. \tag{7.35}$$

For the proof, let $x = \Psi(t, y) \in \Omega \cap U_+$, and assume $x' = \Psi(t', y') \in U_+$ is chosen arbitrarily; then $t > 0$ and $t' > 0$. Next define, for $s \in [\,0, 1\,]$, the vector $x(s) \in U$ by

$$x(s) = \Psi((1 - s)t + st',\ (1 - s)y + sy').$$

In particular, then, $x(0) = x$ and $x(1) = x'$. Because $t > 0$ and $t' > 0$, it follows that

$$(1-s)t + st' > 0 \qquad (s \in [\,0, 1\,]); \tag{7.36}$$

which implies $x(s) \in U_+$, for $s \in [\,0, 1\,]$. If $x' \notin \Omega$, the point $x(\sigma) \in U_+$ with

$$\sigma = \sup\{\, s \in [\,0, 1\,] \mid x(s) \in \Omega \,\}$$

would be contained in $\partial\Omega \cap U$. But in view of (7.33) and (7.36) this forms a contradiction, and so (7.35) does indeed follow. If the two sets $\Omega \cap U_+$ and $\Omega \cap U_-$ are nonempty, we obtain

$$U \cap \Omega = U_+ \cup U_- = U \setminus \partial\Omega;$$

but this was ruled out in (i). As a result, one has either $\Omega \cap U = U_+$ or U_-; and this is precisely (ii).
(ii) $\Rightarrow$ (iii). Define $g = p_1 \circ \Psi^{-1}$, where $p_1 : x \mapsto x_1 : \mathbf{R}^n \to \mathbf{R}$ is the projection onto the first coordinate. Then $g(x) = t$ if $x = \Psi(t, y)$, and by Formula (7.31) it follows that

$$Dg(x^0) = p_1 \circ (D\Psi(t, y^0))^{-1} \neq 0.$$

Therefore g is a submersion in x^0.
(iii) $\Rightarrow$ (i). Let $U \subset \mathbf{R}^n$ be as in (7.34). Consider the curve $\gamma : t \mapsto x^0 + t \operatorname{grad} g(x^0)$ in $\mathbf{R}^n$. We then have in view of Formula (2.18)

$$g \circ \gamma(0) = 0, \qquad (g \circ \gamma)'(0) = \|\operatorname{grad} g(x^0)\|^2 > 0.$$

Hence $\gamma(t) \in U$ and $g(\gamma(t)) > 0$, for $0 < t < \delta$, where $\delta > 0$ should be sufficiently small; and consequently $\gamma(t) \notin \Omega \cup \partial\Omega$; but this implies

$$\gamma(t) \in U \setminus \partial\Omega \qquad \text{and} \qquad \gamma(t) \notin U \cap \Omega,$$

that is, $U \setminus \partial\Omega \neq U \cap \Omega$. ❑

Definition 7.5.2. We say that the open set $\Omega \subset \mathbf{R}^n$ *at the point* $x^0 \in \partial\Omega$ *has a* C^k *boundary* $\partial\Omega$ *and lies at one side of* $\partial\Omega$, if any one of the equivalent properties (i) – (iii) in the preceding theorem is satisfied. If this is the case for every $x^0 \in \partial\Omega$, we say that Ω *has a* C^k *boundary* $\partial\Omega$ *and lies at one side of* $\partial\Omega$. ❍

Now let $x \in \partial\Omega$ be such a boundary point, let $v \in \mathbf{R}^n$ with $v \notin T_x(\partial\Omega)$. In terms of the function g in Theorem 7.5.1.(iii), there are two possibilities:

$$\text{(i)} \quad \langle \operatorname{grad} g(x),\ v \rangle > 0 \qquad \text{or} \qquad \text{(ii)} \quad \langle \operatorname{grad} g(x),\ v \rangle < 0.$$

As in part (iii) $\Rightarrow$ (i) in the proof of the theorem, we see that in case (i) there exists, for every differentiable curve γ in $\mathbf{R}^n$ with $\gamma(0) = x$ and $D\gamma(0) = v$, a $\delta > 0$ such that

$$\text{(i)}' \quad \gamma(t) \in \Omega \qquad (-\delta < t < 0), \qquad \gamma(t) \notin \overline{\Omega} \qquad (0 < t < \delta);$$

in other words, γ leaves Ω via the boundary. In case (ii) we have instead of (i)$'$

$$\text{(ii)}' \quad \gamma(t) \notin \overline{\Omega} \qquad (-\delta < t < 0), \qquad \gamma(t) \in \Omega \qquad (0 < t < \delta);$$

that is, γ enters the set Ω via the boundary. This makes conditions (i) and (ii) independent of the choice of the function g in Theorem 7.5.1.(iii).

We recall the definition from Example 5.3.11 of a normal $n(x)$ to $\partial\Omega$ at x

$$n(x) \perp T_x(\partial\Omega) \qquad \text{and} \qquad \|n(x)\| = 1.$$

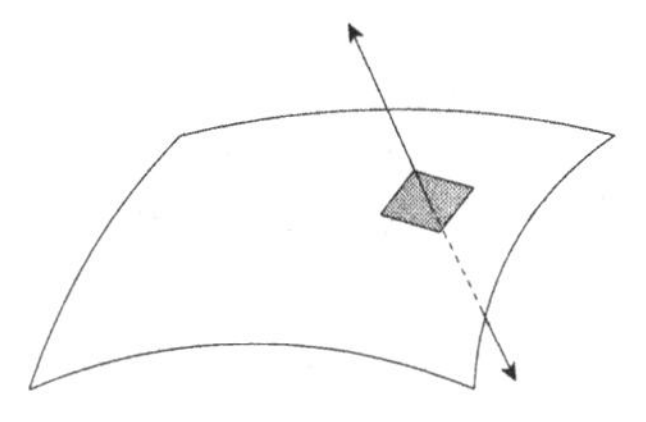
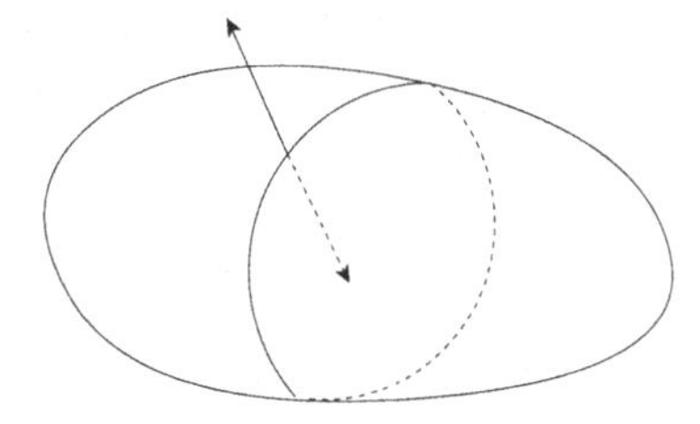

Illustration for Definition 7.5.3: Outer and inner normals

Definition 7.5.3. Assume the open set Ω has a C^k boundary $\partial\Omega$ and that Ω lies at one side of $\partial\Omega$. If $x \in \partial\Omega$, $v \in \mathbf{R}^n$, $v \notin T_x(\partial\Omega)$, we say that v *points outward from* Ω if condition (i) or (i)$'$ above is met; or that it *points inward in* Ω if (ii) or (ii)$'$ above holds. In particular, the outward-pointing normal to $\partial\Omega$ at x, notation $\nu(x)$, is said to be the *outer normal* to $\partial\Omega$ at x; the other normal, $-\nu(x)$, is said to be the *inner normal* to $\partial\Omega$ at x. Thus, these normals are defined independently of the choice of the function g in Theorem 7.5.1.(iii). ❍

Lemma 7.5.4. *Let $\Omega \subset \mathbf{R}^n$ be an open set having a C^k boundary $\partial\Omega$ and lying at one side of $\partial\Omega$. Let $x \in \partial\Omega$ and let $v \in \mathbf{R}^n$ with $v \notin T_x(\partial\Omega)$ be pointing outward from Ω. Condition (ii) in Theorem 7.5.1 then takes the following form:*

$$\Omega \cap U = \Psi(]-\delta,\, 0[\, \times D) = \{\, tv + \phi(y) \mid -\delta < t < 0,\ y \in D \,\}.$$

The outer normal $\nu(x)$ to $\partial\Omega$ at $\phi(y) = x$, with the substitution of $-y_1$ for the variable y_1 when applicable, is given by

$$\nu(\phi(y)) = \|(D_1\phi \times \cdots \times D_{n-1}\phi)(y)\|^{-1}(D_1\phi \times \cdots \times D_{n-1}\phi)(y) \in \mathbf{R}^n. \quad (7.37)$$

In what follows it will invariably be assumed that the outer normal is given by Formula (7.37). Furthermore, then

$$\det D\Psi(t, y) = \langle\, v,\ (D_1\phi \times \cdots \times D_{n-1}\phi)(y) \,\rangle > 0 \qquad ((t, y) \in \mathbf{R} \times D). \quad (7.38)$$

In the special case of $x = \phi(y) = (y, h(y))$, for a C^k function $h : D \to \mathbf{R}$, one has

$$\nu(x) = (-1)^{n-1}(1 + \|Dh(y)\|^2)^{-1/2}\,(-Dh(y),\ 1) \in \mathbf{R}^n. \quad (7.39)$$

If $\partial\Omega$ is locally described as in Theorem 7.5.1.(iii), then

$$\nu(x) = \|\operatorname{grad} g(x)\|^{-1}\ \operatorname{grad} g(x) \in \mathbf{R}^n. \quad (7.40)$$

Proof. By Formula (5.5), Equation (7.37) is correct to within its sign. When this cross product yields the inner normal, we replace the variable y_1 in the embedding ϕ by $-y_1$, as a result of which the cross product changes its sign. Formula (7.38) follows from (7.30). Further, (7.39) is derived from (5.6). And finally, (7.40) follows in a straightforward manner from Example 5.3.11, plus the fact that the direction of $\operatorname{grad} g(x)$ is the direction in which, starting at x, the function g increases the most rapidly, see Theorem 2.6.3.(i). ❑

Remark. In the considerations above, $\operatorname{grad} g(x)$, for $x \in \partial\Omega \cap U$, plays an important role. Note that Exercise 4.32 implies that, for all $x \in \partial\Omega \cap U$, the vector $\operatorname{grad} g(x)$, to within a scalar factor $f(x) \neq 0$, is uniquely determined by the set $\partial\Omega \cap U = \{\, x \in U \mid g(x) = 0 \,\}$. Since the function f is continuous on $\partial\Omega \cap U$, it further follows that the sign of f is constant on the connected components of $\partial\Omega \cap U$.

Remark on the description of a boundary. Given a point x^0 in an $(n-1)$-dimensional C^k hypersurface $\partial\Omega$ in $\mathbf{R}^n$, there are various customary ways to describe a full neighborhood of x^0 in $\mathbf{R}^n$.

(i) Describe $\partial\Omega$ near x^0 as the image under a C^k embedding ϕ, and move away from $\partial\Omega$ following a fixed direction $v \in \mathbf{R}^n$ which at x^0 is transverse to $\partial\Omega$ (this is the situation described in Lemma 7.5.4), that is

$$x = \Psi(t, y) = tv + \phi(y) \qquad \text{with} \qquad v \notin T_{x^0}(\partial\Omega).$$

(ii) Describe $\partial\Omega$ near x^0 as the image under a C^k embedding ϕ, and move away orthogonally with respect to $\partial\Omega$, that is (see Exercise 5.11)

$$x = \Psi(t, y) = t\nu(\phi(y)) + \phi(y).$$

(iii) Describe $\partial\Omega$ near x^0 as the image under a C^k embedding ϕ, and move away from $\partial\Omega$ by following a transverse C^k flow, that is (see Section 5.9)

$$x = \Psi(t, y) = \Phi^t(\phi(y)) \qquad \text{with} \qquad \frac{\partial \Phi}{\partial t}(0, x^0) \notin T_{x^0}(\partial\Omega).$$

The situation in (i) is a special case of this.

(iv) Describe $\partial\Omega$ near x^0 as zero-set of a C^k submersion g, and apply the Submersion Theorem 4.5.2.(iii); this gives

$$g(x) = t \qquad \Longleftrightarrow \qquad x = \Psi(t, y).$$

In all of these cases $\Psi : \mathbf{R} \times \mathbf{R}^{n-1} \supset\!\!\to \mathbf{R}^n$ is a C^k diffeomorphism (except in (ii), where one merely obtains C^{k-1}). In case (iii) the proof is obtained, as for (i), by using the Local Inverse Function Theorem 3.2.4.

Example 7.5.5. Assume a, b and $c > 0$, and let V be the ellipsoid in $\mathbf{R}^3$

$$V = \{\, x \in \mathbf{R}^3 \mid g(x) = \frac{x_1^2}{a^2} + \frac{x_2^2}{b^2} + \frac{x_3^2}{c^2} - 1 = 0 \,\}.$$

Let $d(x)$ be the distance in $\mathbf{R}^3$ from 0 to the geometric tangent plane $x + T_x V$ to V at x. Then

$$\frac{1}{d(x)} = \frac{1}{2}\langle\, \operatorname{grad} g(x),\ \nu(x)\,\rangle \qquad (x \in V). \tag{7.41}$$

Indeed, $Dg(x) = 2(\frac{x_1}{a^2}, \frac{x_2}{b^2}, \frac{x_3}{c^2})$, and so

$$h \in T_x V \qquad \Longleftrightarrow \qquad \frac{1}{2}\langle\, \operatorname{grad} g(x),\ h\,\rangle = \frac{x_1 h_1}{a^2} + \frac{x_2 h_2}{b^2} + \frac{x_3 h_3}{c^2} = 0.$$

Because $k \in x + T_x V$ if and only if $k - x \in T_x V$, it follows that

$$x + T_x V = \{\, k \in \mathbf{R}^3 \mid \frac{x_1 k_1}{a^2} + \frac{x_2 k_2}{b^2} + \frac{x_3 k_3}{c^2} = 1 \,\}. \tag{7.42}$$

Since $\nu(x)$ is perpendicular to $T_x V$ and has unit length, $d(x)$ is determined by the requirement

$$d(x)\,\nu(x) \in x + T_x V.$$

One has

$$\nu(x) = \frac{1}{c(x)}\Big(\frac{x_1}{a^2}, \frac{x_2}{b^2}, \frac{x_3}{c^2}\Big) \qquad \text{with} \qquad c(x) = \sqrt{\frac{x_1^2}{a^4} + \frac{x_2^2}{b^4} + \frac{x_3^2}{c^4}}.$$

Therefore, according to (7.42),

$$d(x)\,c(x) = \frac{d(x)}{c(x)}\Big(\frac{x_1^2}{a^4} + \frac{x_2^2}{b^4} + \frac{x_3^2}{c^4}\Big) = 1.$$

Consequently, (7.41) results from

$$\frac{1}{d(x)} = c(x) = c(x)^2\,\frac{1}{c(x)} = \Big\langle \Big(\frac{x_1}{a^2}, \frac{x_2}{b^2}, \frac{x_3}{c^2}\Big), \frac{1}{c(x)}\Big(\frac{x_1}{a^2}, \frac{x_2}{b^2}, \frac{x_3}{c^2}\Big)\Big\rangle. \qquad ☆$$

7.6 Integration of a total derivative

Now we are prepared enough to prove the following:

Theorem 7.6.1 (Integration of a total derivative). *Let $\Omega \subset \mathbf{R}^n$ be a bounded open subset having a C^1 boundary $\partial\Omega$ and lying at one side of $\partial\Omega$. Let $\nu(y)^t$ be the outer normal to $\partial\Omega$ at $y \in \partial\Omega$ considered as a row vector, and let $d_{n-1}y$ be integration with respect to the Euclidean $(n-1)$-dimensional density on the C^1 hypersurface $\partial\Omega$. Let $f : \Omega \to \mathbf{R}$ be a C^1 function such that f and its total derivative $Df : \Omega \to \mathrm{Lin}(\mathbf{R}^n, \mathbf{R}) \simeq \mathbf{R}^n$ can both be extended to continuous mappings on $\overline{\Omega}$. Then the following identity of row vectors holds in $\mathbf{R}^n$, where the integration is performed by components:*

$$\int_\Omega Df(x)\,dx = \int_{\partial\Omega} f(y)\,\nu(y)^t\,d_{n-1}y. \tag{7.43}$$

Remarks. According to Corollary 6.3.8, the boundary $\partial\Omega$ is an n-dimensional negligible set in $\mathbf{R}^n$, and therefore, by Theorem 6.3.2, Ω is Jordan measurable. Hence, the integral on the left–hand side in Formula (7.43) is well-defined.

By taking the j-th component in (7.43), for $1 \le j \le n$, we find

$$\int_\Omega D_j f(x)\,dx = \int_{\partial\Omega} f(y)\,\nu_j(y)\,d_{n-1}y. \tag{7.44}$$

Further, Formula (7.43) is equivalent to the assertion

$$\int_\Omega Df(x)v\,dx = \int_{\partial\Omega} f(y)\,\langle\,\nu(y), v\,\rangle\,d_{n-1}y \qquad (v \in \mathbf{R}^n), \tag{7.45}$$

and also to

$$\int_{\Omega} \operatorname{grad} f(x)\,dx = \int_{\partial\Omega} f(y)\,\nu(y)\,d_{n-1}y.$$

Let $n = 1$ and $\Omega = \,]\,a, b\,[$. Then $\partial\Omega = \{a, b\}$, $\nu(a) = -1$, $\nu(b) = +1$. Integrating a function over a point (a zero-dimensional manifold) with respect to the Euclidean zero-dimensional density is, by definition, tantamount to evaluating the function at that point. Therefore the Fundamental Theorem of Integral Calculus on $\mathbf{R}$ is a special case of (7.43),

$$\int_a^b \frac{df}{dx}(x)\,dx = \int_{\{a\}} f(y)\,(-1)\,d_0 y + \int_{\{b\}} f(y)\,(+1)\,d_0 y = f(b) - f(a).$$

Proof. We first demonstrate that, for every point $x \in \overline{\Omega} = \Omega \cup \partial\Omega$, we can find an open neighborhood U_x of x in $\mathbf{R}^n$ such that Formula (7.45) holds for functions f as specified in the theorem, if moreover these functions satisfy $\operatorname{supp}(f) \subset U_x$. Since both sides of Equation (7.45) depend linearly on $v \in \mathbf{R}^n$, it suffices to prove that formula for vectors v belonging to a basis for $\mathbf{R}^n$. Note that the choice of that basis may depend on the point x considered.

Case I. Assume $x \in \Omega$. Because Ω is an open set in $\mathbf{R}^n$, we can find vectors a and $b \in \Omega$ such that $x \in U_x := \{\, y \in \mathbf{R}^n \mid a_j < y_j < b_j \ (1 \le j \le n)\,\} \subset \Omega$; and therefore $U_x \cap \partial\Omega = \emptyset$. We now successively choose $v = e_j$, the standard j-th basis vector in $\mathbf{R}^n$, for $1 \le j \le n$. For a function f as in the theorem, also satisfying

$$\operatorname{supp}(f) \subset U_x \subset \Omega, \tag{7.46}$$

Corollary 6.4.3 gives

$$\begin{aligned} \int_{\Omega} D_j f(x)\,dx &= \int_{\mathbf{R}^n} D_j f(x)\,dx \\ &= \int \cdots \int \int D_j f(x)\,dx_j\,dx_1 \cdots dx_{j-1}\,dx_{j+1} \cdots dx_n. \end{aligned} \tag{7.47}$$

On account of (7.46) one has, for fixed $(x_1, \dots, x_{j-1}, x_{j+1}, \dots, x_n)$,

$$f(x_1, \dots, x_{j-1}, x_j, x_{j+1}, \dots, x_n) \ne 0 \qquad \Longrightarrow \qquad a_j < x_j < b_j.$$

Therefore, application of the Fundamental Theorem of Integral Calculus on $\mathbf{R}$ to the function $x_j \mapsto f(x_1, \dots, x_{j-1}, x_j, x_{j+1}, \dots, x_n)$, defined on $[a_j, b_j]$, gives

$$\begin{aligned} &\int D_j f(x_1, \dots, x_j, \dots, x_n)\,dx_j \\ &\quad = f(x_1, \dots, b_j, \dots, x_n) - f(x_1, \dots, a_j, \dots, x_n) = 0 - 0 = 0. \end{aligned}$$

Consequently, (7.47) implies

$$\int_\Omega D_j f(x)\,dx = 0.$$

On the other hand, $f \equiv 0$ on $\partial\Omega$ under the assumption (7.46), and therefore one also has

$$\int_{\partial\Omega} f(y)\,\nu_j(y)\,d_{n-1}y = 0,$$

which proves (7.44) in this case.

Case II. Assume $x \in \partial\Omega$. In this case we select each of the basis vectors v such that v is not included in $T_x(\partial\Omega)$ and points outward from Ω. This we do by choosing a basis $(h_1, \dots, h_{n-1})$ for the linear subspace $T_x(\partial\Omega)$ of dimension $n-1$, and $v_0 \notin T_x(\partial\Omega)$. The vectors $v_1 = h_1 + v_0$, ..., $v_{n-1} = h_{n-1} + v_0$, $v_n = v_0$ then form a basis of $\mathbf{R}^n$. They are not contained in $T_x(\partial\Omega)$, and after changing over to their opposites if necessary, we may assume that they point outwards from Ω. Now successively consider $v = v_j$, for $1 \le j \le n$. By Lemma 7.5.4, the pair (x, v) defines an open neighborhood $U = U(v)$ of x; let $\Psi : \,]-\delta, 0\,[\times D \to \Omega \cap U$ with $\Psi(t, y) = tv + \phi(y)$ be the corresponding C^1 substitution of variables. Application of the Change of Variables Theorem 6.6.1 to a function f as in the theorem, with f moreover satisfying $\operatorname{supp}(f) \subset \overline{\Omega} \cap U$, gives

$$\int_\Omega Df(x)v\,dx = \int_{]-\delta,0[\times D} Df(\Psi(t, y))v\,|\det D\Psi(t, y)|\,dt\,dy. \tag{7.48}$$

Now, by the chain rule and Formula (7.29),

$$Df(\Psi(t, y))v = Df(\Psi(t, y))\frac{\partial\Psi}{\partial t}(t, y) = \frac{\partial(f\circ\Psi)}{\partial t}(t, y). \tag{7.49}$$

And by means of (7.38) we find

$$|\det D\Psi(t, y)| = \langle\, v,\ (D_1\phi \times \cdots \times D_{n-1}\phi)(y)\,\rangle \qquad ((t, y) \in \mathbf{R} \times D); \tag{7.50}$$

note that the expression on the right–hand side is independent of t. Then use Corollary 6.4.3, (7.49) and (7.50), subsequently the Fundamental Theorem of Integral Calculus on $[\,-\delta, 0\,]$, plus the fact that $(f\circ\Psi)(-\delta, y) = 0$, to write the right–hand side in (7.48) as

$$\begin{aligned}
&\int_D \int_{-\delta}^0 \frac{\partial(f\circ\Psi)}{\partial t}(t, y)\,dt\,\langle\, v,\ (D_1\phi \times \cdots \times D_{n-1}\phi)(y)\,\rangle\,dy \\
&= \int_D (f\circ\Psi)(0, y)\,\langle\,(D_1\phi \times \cdots \times D_{n-1}\phi)(y),\ v\,\rangle\,dy \\
&= \int_D (f\circ\phi)(y)\langle\,(\nu\circ\phi)(y),\ v\,\rangle\,\omega_\phi(y)\,dy = \int_{\partial\Omega} f(y)\,\langle\,\nu(y),\ v\,\rangle\,d_{n-1}y.
\end{aligned} \tag{7.51}$$

The penultimate identity follows because of Formulae (7.37) and (7.25). Combination of (7.48) and (7.51) now gives (7.45). By intersecting the neighborhoods $U(v_j)$ of x, for $1 \leq j \leq n$, we find the desired open neighborhood U_x of x in $\mathbf{R}^n$.

End of Proof. $\overline{\Omega}$ is compact, according to the Heine–Borel Theorem 1.8.17. By virtue of Theorem 6.7.4 there exists a C^1 partition $\{\, \chi_i \mid i \in I \,\}$ of unity on $\overline{\Omega}$ subordinate to the open covering $\{\, U_x \mid x \in \overline{\Omega} \,\}$ of $\overline{\Omega}$. Because Formula (7.43) has been proved above, for f replaced by $\chi_i f$, for every $i \in I$, summation over $i \in I$ gives Formula (7.43) for f, since $\sum_{i \in I} \chi_i = 1$ on $\overline{\Omega}$ and therefore also $D(\sum_{i \in I} \chi_i f) = Df$ on Ω. ❑

Corollary 7.6.2 (Integration by parts in $\mathbf{R}^n$). *Assume Ω and f to be as in the theorem above, and assume g satisfies the same conditions as f. Then, for $1 \leq j \leq n$,*

$$\int_\Omega D_j f(x)\, g(x)\, dx = \int_{\partial\Omega} (f\, g)(y)\, \nu_j(y)\, d_{n-1}y - \int_\Omega f(x)\, D_j g(x)\, dx.$$

Proof. Replace f by fg in Formula (7.44) and apply Leibniz' formula $D_j(f\, g) = g\, D_j f + f\, D_j g$. ❑

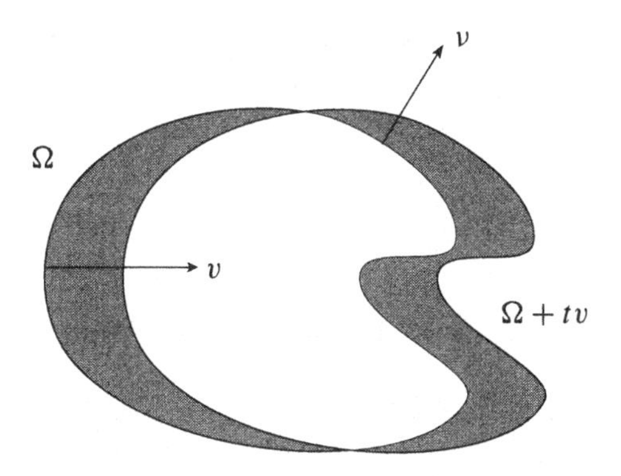

Illustration for Remark in Section 7.6

Remark. It is also possible to prove Theorem 7.6.1 starting from

$$\begin{aligned}\int_{\Omega} Df(x)v\,dx &= \int_{\Omega} \lim_{t\to 0} \frac{1}{t}(f(x+tv) - f(x))\,dx \\ &= \lim_{t\to 0}\frac{1}{t}\Big(\int_{\Omega} f(x+tv)\,dx - \int_{\Omega} f(x)\,dx\Big) \\ &= \lim_{t\to 0}\frac{1}{t}\Big(\int_{\Omega+tv} f(x)\,dx - \int_{\Omega} f(x)\,dx\Big).\end{aligned}$$

Here $v \in \mathbf{R}^n$ is a fixed vector and

$$\Omega + tv = \{\, x + tv \mid x \in \Omega \,\}.$$

Where $\Omega + tv$ and Ω overlap, the integrals cancel each other; what remains, therefore, is an integral over a strip along the boundary. To first-order approximation in t, the thickness of this strip at a point $y \in \partial\Omega$ equals $t \langle\, \nu(y),\, v \,\rangle$, where $\langle\, \nu(y),\, v \,\rangle < 0$ corresponds to a region where the strip is to be counted negative. If, finally, it can be shown that the integral over a strip along the boundary, of thickness δ, to first-order approximation (for $\delta \downarrow 0$) equals δ times the integral over the boundary with respect to the Euclidean density, the proof will be completed (see the Motivation in Section 7.3, and Exercise 7.35.(iii)). The advantage of this proof is that it is more transparent geometrically, and furthermore that it provides an interesting additional interpretation of integration over an $(n-1)$-dimensional manifold with respect to the Euclidean density. The price to be paid is, of course, that at several stages limit arguments will have to be given, in which uniform convergence may be expected to play an important role. A fully detailed proof along these lines will therefore be of considerable length.

In our development of the theory the formula below will serve as justification for this line of reasoning. Let the notation be that of Example 6.6.9. Let $v \in \mathbf{R}^n$, and choose $\Psi^t(x) = x + tv$, for $(t, x) \in \mathbf{R} \times \mathbf{R}^n$; then $\psi(x) = v$ and $\operatorname{div} \psi(x) = 0$, for $x \in \mathbf{R}^n$. By combining the transport equation (6.22) and Theorem 7.6.1 we find

$$\frac{d}{dt}\Big|_{t=0} \int_{\Omega+tv} f(x)\,dx = \int_{\Omega} Df(x)v\,dx = \int_{\partial\Omega} f(y)\langle\, \nu(y),\, v \,\rangle\, d_{n-1}y. \qquad (7.52)$$

7.7 Generalizations of the preceding theorem

In many of the applications envisaged, the conditions of Theorem 7.6.1, such as those relating to continuity of f and Df on $\overline{\Omega}$, or those relating to smoothness of the boundary $\partial\Omega$, are not fully met.

The assumption concerning the continuity of Df on $\overline{\Omega}$ may be relaxed. It is sufficient to assume that f is continuous on $\overline{\Omega}$, and that $x \mapsto Df(x)v$, with $v \in \mathbf{R}^n$, is continuous on Ω and absolutely Riemann integrable over Ω. Indeed, in

Formula (7.51)

$$\int_{-\delta}^{0} \frac{\partial (f \circ \Psi)}{\partial t}(t, y)\, dt$$

may be replaced by

$$\int_{-\delta}^{-\epsilon} \frac{\partial (f \circ \Psi)}{\partial t}(t, y)\, dt.$$

The arguments from the proof can be applied to this integral, and in the resulting identity we finally take the limit for $\epsilon \downarrow 0$. This leads to (7.43), without the continuity of Df up to and including the boundary having been needed.

In addition, the assumption that $\partial\Omega$ is a C^1 hypersurface may be relaxed; it is sufficient that the nondifferentiability of $\partial\Omega$ be localized in a "relatively small" subset S of $\partial\Omega$. Begin by assuming that S is a closed (and hence compact) subset of $\partial\Omega$ such that

$$W := \partial\Omega \setminus S$$

is in fact an $(n-1)$-dimensional C^1 manifold, with Ω at one side of W at each point of W. The idea is to approximate f by functions whose support is disjoint from S, because Theorem 7.6.1 does apply to such functions. In order to limit the number of technical complications, we assume that f and $D_j f$ can both be extended to continuous functions on $\overline{\Omega}$, for $1 \leq j \leq n$. Now assume the subset S of $\partial\Omega$ satisfies the following condition:

$$\begin{gathered} \text{for all } \epsilon > 0 \text{ and for every open neighborhood } U \text{ of } S \text{ in } \mathbf{R}^n \\ \text{there exist an open neighborhood } U' \text{ of } S \text{ in } \mathbf{R}^n \text{ and } \chi \in C^1(\mathbf{R}^n) \text{ with} \\ U' \subset U, \qquad 0 \leq \chi \leq 1, \qquad \chi = 1 \text{ on } U', \qquad \mathrm{supp}(\chi) \subset U, \\ \int_{\mathbf{R}^n} \chi(x)\, dx < \epsilon, \qquad \int_{\mathbf{R}^n} \|\,\mathrm{grad}\, \chi(x)\|\, dx < \epsilon. \end{gathered} \tag{7.53}$$

Because $\mathrm{supp}\,((1-\chi) f) \cap S = \emptyset$, one has, by Theorem 7.6.1,

$$\int_\Omega D_j((1-\chi) f)(x)\, dx = \int_W ((1-\chi) f \, \nu_j)(y)\, d_{n-1}y. \tag{7.54}$$

Let $\|\cdot\|$ be the uniform norm on the linear space of bounded functions on Ω, with

$$\|g\| = \sup\{\, |g(x)| \mid x \in \Omega \,\},$$

then, by Leibniz' rule,

$$\begin{aligned} \Big| \int_\Omega D_j(\chi f)(x)\, dx \Big| &\leq \int_\Omega |(D_j\chi)(x)|\, |f(x)|\, dx + \int_\Omega \chi(x)\, |(D_j f)(x)|\, dx \\ &\leq \|f\| \int_\Omega |(D_j\chi)(x)|\, dx + \|D_j f\| \int_\Omega \chi(x)\, dx \\ &< \epsilon\, (\|f\| + \|D_j f\|). \end{aligned}$$

As a result, the left–hand side in (7.54) converges to $\int_\Omega (D_j f)(x)\,dx$, if $\epsilon \downarrow 0$; and the right–hand side in (7.54) converges to $\int_W (f\,\nu_j)(y)\,d_{n-1}y$, if U shrinks to S.

We now wish to replace condition (7.53) on S by one that is more easily verified in practice. This problem has a new aspect in that we now also need to control the magnitude of the partial derivatives of the bump function χ, and the support of χ as well. A number of pertinent remarks follow; in fact, these are part of a theory of simultaneous uniform approximation of a function and of its derivatives (see also Exercise 6.103).

Choose a C^1 function $\psi \in C_c(\mathbf{R}^n)$ (see Definition 6.3.6) such that

$$\psi \geq 0; \qquad \psi(x) = 0 \qquad (\|x\| \geq 1); \qquad \int_{\mathbf{R}^n} \psi(x)\,dx = 1.$$

Next define, for $t > 0$, the function $\psi_t : \mathbf{R}^n \to \mathbf{R}$ by $\psi_t(x) = t^{-n}\,\psi(t^{-1}x)$. Then

$$\psi_t(x) = 0 \qquad (\|x\| \geq t); \qquad \int_{\mathbf{R}^n} \psi_t(x)\,dx = \int_{\mathbf{R}^n} \psi(t^{-1}x)\,d(t^{-1}x) = 1.$$

Subsequently define, for every $g \in C(\mathbf{R}^n)$ and $t > 0$, the function $g_t : \mathbf{R}^n \to \mathbf{R}$ by (compare with Example 6.11.5 on convolution)

$$g_t(x) = g * \psi_t(x) = \int_{\mathbf{R}^n} g(x - y)\,\psi_t(y)\,dy = \int_{\mathbf{R}^n} \psi_t(x - y)\,g(y)\,dy. \tag{7.55}$$

Note that the integration in (7.55) is in fact over a subset of $\operatorname{supp}(\psi_t)$; that is why g_t is well-defined. It follows immediately that

$$|g_t(x)| \leq \int_{\mathbf{R}^n} |g(x - y)|\,\psi_t(y)\,dy \leq \|g\| \int_{\mathbf{R}^n} \psi_t(y)\,dy = \|g\| \qquad (x \in \mathbf{R}^n);$$

that is

$$\|g_t\| \leq \|g\| \qquad (t > 0). \tag{7.56}$$

Using the Differentiation Theorem 2.10.4 or 6.12.4 and Theorem 2.3.4 one proves that g_t is a C^1 function on $\mathbf{R}^n$, with

$$(D_j g_t)(x) = \int_{\mathbf{R}^n} (D_j \psi_t)(x - y)\,g(y)\,dy.$$

Because $(D_j \psi_t)(x) = t^{-n}\,D_j(x \mapsto \psi(t^{-1}x)) = t^{-n-1}(D_j\psi)(t^{-1}x)$, this leads to

$$\begin{aligned}
|(D_j g_t)(x)| &\leq t^{-n-1} \int_{\mathbf{R}^n} |(D_j\psi)(t^{-1}(x - y))|\,|g(y)|\,dy \\
&\leq t^{-1}\,\|g\| \int_{\mathbf{R}^n} |(D_j\psi)(t^{-1}(x - y))|\,d(t^{-1}y) \\
&\leq t^{-1}\,\|g\| \int_{\mathbf{R}^n} \|\operatorname{grad}\psi(y)\|\,dy =: t^{-1}\,\|g\|\,k.
\end{aligned}$$

Therefore

$$\|D_j g_t\| \le t^{-1} \|g\| \, k \qquad (1 \le j \le n, \ t > 0). \tag{7.57}$$

We now recall the sets $S_\delta = \{ y \in \mathbf{R}^n \mid \text{ there exists } x \in S \text{ with } \|x - y\| \le \delta \}$, for $\delta > 0$, defined in Lemma 6.8.1. Again, every S_δ is compact in $\mathbf{R}^n$. Suppose U is an open neighborhood of S in $\mathbf{R}^n$, then by Lemma 6.8.1 there exists a $\delta > 0$ such that $S_{2\delta} \subset U$. Hence $S \subset S_\delta \subset \operatorname{int}(S_{2\delta}) \subset U$, where $\operatorname{int}(S_{2\delta})$ is an open covering of S_δ. Applying Theorem 6.7.3 we find a continuous function $\chi_\delta : \mathbf{R}^n \to \mathbf{R}$ with

$$0 \le \chi_\delta \le 1, \qquad \chi_\delta = 1 \text{ on } S_\delta, \qquad \operatorname{supp}(\chi_\delta) \subset S_{2\delta}. \tag{7.58}$$

Next we define, for $0 < t < \delta$ (compare with (7.55)),

$$\chi_{\delta,\,t}(x) := (\chi_\delta)_t(x) = \int_{\mathbf{R}^n} \psi_t(y)\, \chi_\delta(x - y)\, dy \qquad (x \in \mathbf{R}^n).$$

Now let $x \in \mathbf{R}^n$ with $\chi_{\delta,t}(x) \ne 0$. This can only occur if there exists $y \in \mathbf{R}^n$ with

$$\psi_t(y) > 0, \qquad \chi_\delta(x - y) > 0; \qquad \text{that is,} \qquad \|y\| \le t, \qquad x - y \in S_{2\delta}.$$

Consequently, there exists $z \in S$ with $\|x - y - z\| \le 2\delta$, and so

$$\|x - z\| = \|(x - y - z) + y\| \le \|x - y - z\| + \|y\| \le 2\delta + t.$$

That is

$$\operatorname{supp}(\chi_{\delta,t}) \subset S_{2\delta+t}; \qquad \text{in particular,} \qquad \operatorname{supp}(D_j \chi_{\delta,t}) \subset S_{2\delta+t} \qquad (1 \le j \le n). \tag{7.59}$$

From (7.59), (7.57) and (7.58) it then follows, for $1 \le j \le n$,

$$\int_{\mathbf{R}^n} |(D_j \chi_{\delta,t})(x)|\, dx \le \|D_j \chi_{\delta,t}\| \cdot \text{outer measure}(S_{2\delta+t}) \le k\, \frac{\text{outer measure}(S_{2\delta+t})}{t}.$$

Setting $t = \frac{\delta}{2}$ one obtains

$$\int_{\mathbf{R}^n} |(D_j \chi_{\delta,\,\delta/2})(x)|\, dx \le 5k\, \frac{\text{outer measure}(S_{5\delta/2})}{5\delta/2}. \tag{7.60}$$

Moreover, from (7.56), (7.58) and (7.59)

$$\int_{\mathbf{R}^n} \chi_{\delta,\,\delta/2}(x)\, dx \le \text{outer measure}(S_{5\delta/2}). \tag{7.61}$$

Definition 7.7.1. A compact subset S of $\mathbf{R}^n$ is said to be $(n-1)$*-dimensional negligible* if

$$\lim_{\delta \downarrow 0} \frac{\text{outer measure}(S_\delta)}{\delta} = 0.$$

❍

One now obtains from (7.60) and (7.61):

Lemma 7.7.2. *Let $S \subset \mathbf{R}^n$ be compact in $\mathbf{R}^n$ and $(n-1)$-dimensional negligible. Then S satisfies (7.53).*

One readily verifies that $S_1 \cup S_2$ is an $(n-1)$-dimensional negligible set if S_1 and S_2 are; and further, that a compact subset S of an $(n-2)$-dimensional manifold in $\mathbf{R}^n$ is an $(n-1)$-dimensional negligible set. Indeed, on the strength of Theorem 4.7.1 one has, locally at least,

$$S \subset \{\,(h_1(y), h_2(y), y) \in \mathbf{R}^n \mid y \in D \subset \mathbf{R}^{n-2} \text{ open}, h_i \in C(D, \mathbf{R})\ (1 \le i \le 2)\,\}.$$

Because

$$\|(x_1, x_2, y) - (x_1', x_2', y')\| < \delta \qquad \Longrightarrow \qquad |x_i - x_i'| < \delta, \qquad \|y - y'\| < \delta,$$

one certainly has

$$S_\delta \subset \{\,(x_1, x_2, y) \in \mathbf{R}^n \mid |x_i - h_i(y)| < \delta\ (1 \le i \le 2),\ y \in D_\delta\,\}.$$

This implies

$$\begin{aligned} \text{outer measure}(S_\delta) &\le \int_{D_\delta} dy \int_{h_1(y)-\delta}^{h_1(y)+\delta} dx_1 \int_{h_2(y)-\delta}^{h_2(y)+\delta} dx_2 \\ &= 4\delta^2 \operatorname{vol}_{n-2}(D_\delta) = \mathcal{O}(\delta^2), \quad \delta \downarrow 0. \end{aligned}$$

As a result

$$\frac{\text{outer measure}(S_\delta)}{\delta} = \mathcal{O}(\delta), \quad \delta \downarrow 0.$$

This also makes the finite union of compact subsets of $(n-2)$-dimensional C^1 manifolds in $\mathbf{R}^n$ an $(n-1)$-dimensional negligible set, for example, the edges of a cube in $\mathbf{R}^3$.

The following generalization of Theorem 7.6.1 has now been proved. Note that on the strength of this generalization (the interior of) all known figures from the box of bricks (rectangle, part of cylinder, part of cone, bridge etc.) qualify as Ω.

Theorem 7.7.3. *Let $\Omega \subset \mathbf{R}^n$ be a bounded open subset with boundary $\partial\Omega$. Let S be a closed subset of $\partial\Omega$ such that S is an $(n-1)$-dimensional negligible set, $\partial'\Omega = \partial\Omega \setminus S$ a C^1 manifold in $\mathbf{R}^n$ of dimension $n-1$, and Ω lies at one side of $\partial'\Omega$ at each point of $\partial'\Omega$. Let $f : \Omega \to \mathbf{R}$ be a C^1 function such that f can be extended to a continuous function on $\overline{\Omega}$, and its total derivative $Df : \Omega \to \operatorname{Lin}(\mathbf{R}^n, \mathbf{R}) \simeq \mathbf{R}^n$ is continuous and absolutely Riemann integrable over Ω. Let ν and $d_{n-1}y$ be as in Theorem 7.6.1, but now defined with respect to $\partial'\Omega$. Then*

$$\int_\Omega Df(x)\,dx = \int_{\partial'\Omega} f(y)\,\nu(y)^t\,d_{n-1}y.$$

7.8 Gauss' Divergence Theorem

In *vector analysis* one applies variants of Theorems 7.6.1 and 7.7.3 to vector-valued functions. Hence the following:

Definition 7.8.1. Let U be an open subset of $\mathbf{R}^n$ and let $f = (f_1, \dots, f_n) : U \to \mathbf{R}^n$ be a mapping. Such a mapping is often referred to as a *vector field* on U, particularly when f is regarded as a mapping which assigns to $x \in U$ a tangent vector $f(x)$ in the tangent space $T_x U$ to U at x, where, for all $x \in U$, the latter space is identified with $\mathbf{R}^n$. In other words, the image vector $f(x) \in \mathbf{R}^n$ is seen as an arrow in $\mathbf{R}^n$, originating at $x \in U$ and with its head at $x + f(x) \in \mathbf{R}^n$. ❍

Example 7.8.2. Examples of vector fields are

- A C^1 diffeomorphism $\Phi : U \to V$, with U and V open subsets in $\mathbf{R}^n$.
- The *gradient vector field* (see Definition 2.6.2)

$$\operatorname{grad} g = \sum_{1 \le j \le n} D_j g \, e_j : U \to \mathbf{R}^n,$$

  associated with a C^1 function $g : U \to \mathbf{R}$. Note that the definition of grad g seemingly depends on the choice of coordinates on $\mathbf{R}^n$. However, this is not actually the case, in view of the characterization of grad $g(x)$ as the vector in $\mathbf{R}^n$ that points from x in the direction of steepest increase of the function g, and whose length equals the rate of increase in that direction (see Theorem 2.6.3). Exercise 3.12.(iii) contains formulae for the gradient vector field with respect to arbitrary coordinates.

- The tangent vector field of a one-parameter group of diffeomorphisms $(\Phi^t)_{t \in \mathbf{R}}$ on $\mathbf{R}^n$, as defined in Formula (5.21). ☆

Let $f : U \to \mathbf{R}^n$ be a C^1 vector field; one then has the total derivative $Df(x) \in \operatorname{End}(\mathbf{R}^n)$, for every $x \in U$. Further, we recall the *trace* tr $Df(x)$ of $Df(x)$, defined as the coefficient of $-\lambda^{n-1}$ in the *characteristic polynomial*, of $Df(x)$

$$\det(\lambda I - Df(x)) = \lambda^n - \lambda^{n-1} \operatorname{tr} Df(x) + \cdots + (-1)^n \det Df(x). \qquad (7.62)$$

The definition of tr $Df(x)$ is most obviously independent of the choice of coordinates on $\mathbf{R}^n$; with respect to the standard basis in $\mathbf{R}^n$ we find

$$\operatorname{tr} Df(x) = \sum_{1 \le j \le n} D_j f_j(x) \qquad (x \in U).$$

Definition 7.8.3. Let U be an open subset of $\mathbf{R}^n$, and let $f : U \to \mathbf{R}^n$ be a C^1 vector field. Then we define the function $\operatorname{div} f : U \to \mathbf{R}$, the *divergence of the vector field* f, by

$$\operatorname{div} f = \operatorname{tr} Df = \sum_{1\le j\le n} D_j f_j.$$

The *operator* ∇ (this symbol is pronounced as *nabla* or *del*; the name nabla derives from an ancient stringed instrument in the shape of a harp) is the n-tuple of partial differentiations

$$\nabla = \sum_{1\le j\le n} D_j e_j = \begin{pmatrix} \frac{\partial}{\partial x_1} \\ \vdots \\ \frac{\partial}{\partial x_n} \end{pmatrix} = \begin{pmatrix} D_1 \\ \vdots \\ D_n \end{pmatrix}.$$

In particular we have the formal notations, with $g \in C^1(U)$, and $f \in C^1(U, \mathbf{R}^n)$ a vector field,

$$\nabla g = \operatorname{grad} g; \qquad \langle \nabla, f \rangle = \nabla \cdot f = \langle (D_1, \dots, D_n), (f_1, \dots, f_n) \rangle = \operatorname{div} f;$$

$$\Delta := \langle \nabla, \nabla \rangle = \nabla \cdot \nabla = \sum_{1\le j\le n} D_j^2.$$

Here Δ is the *Laplace operator*, or *Laplacian*, acting on $g \in C^2(U)$ via

$$\Delta g = \operatorname{div}(\operatorname{grad} g) = \sum_{1\le j\le n} D_j^2 g. \tag{7.63}$$

(See Exercises 3.14 and 7.60, and 3.16 and 7.61, for formulae giving the divergence and the Laplacian, respectively, with respect to arbitrary coordinates.) ❍

Example 7.8.4 (Newton vector field). Define the vector field $f : \mathbf{R}^n \setminus \{0\} \to \mathbf{R}^n$ by

$$f(x) = \frac{1}{\|x\|^n}\, x, \qquad \text{that is,} \qquad f_j(x) = \frac{x_j}{\|x\|^n} \qquad (1 \le j \le n).$$

Using Example 2.4.8 we get $D_j f_j(x) = \frac{1}{\|x\|^n} - \frac{n x_j^2}{\|x\|^{n+2}}$, for $1 \le j \le n$. Consequently

$$\operatorname{div} f(x) = \frac{n}{\|x\|^n} - \frac{n\|x\|^2}{\|x\|^{n+2}} = 0 \qquad (x \in \mathbf{R}^n \setminus \{0\}).$$

Note that if $n > 2$ (see Exercises 2.30.(ii) and 2.40.(iv))

$$f(x) = \operatorname{grad}\Big(\frac{1}{2-n}\,\frac{1}{\|x\|^{n-2}}\Big), \qquad \text{and so} \qquad \Delta\Big(\frac{1}{\|\cdot\|^{n-2}}\Big) = 0 \quad \text{on} \quad \mathbf{R}^n \setminus \{0\},$$

while for $n = 2$

$$f(x) = \operatorname{grad} \log \|x\|, \qquad \text{and so} \qquad \Delta(\log \|\cdot\|) = 0 \quad \text{on} \quad \mathbf{R}^2 \setminus \{0\}.$$

Indeed, combination of the condition $f(x) = \operatorname{grad} g(x)$ and the assumption $g(x) = g_0(\|x\|)$ gives, for $1 \le j \le n$,

$$\frac{x_j}{\|x\|^n} = g_0'(\|x\|)\frac{x_j}{\|x\|};$$

hence

$$g_0'(r) = r^{1-n} \qquad \text{and} \qquad g_0(r) = \begin{cases} \dfrac{1}{2-n} r^{2-n}, & n > 2; \\ \log r, & n = 2. \end{cases}$$

The vector field

$$x \mapsto \frac{1}{|S^{n-1}|\, \|x\|^n} x : \mathbf{R}^n \setminus \{0\} \to \mathbf{R}^n$$

is said to be the *Newton vector field*; in this connection, see Exercise 7.21.(ii) for $|S^{n-1}| := \text{hyperarea}_{n-1}(S^{n-1})$. ☆

We employ the notations of Section 7.6. Henceforth consider a C^1 vector field $f : \Omega \to \mathbf{R}^n$, instead of a function $f : \Omega \to \mathbf{R}$, as we did in Section 7.6. For $y \in \partial\Omega$ we introduce $(f\,\nu^t)(y) \in \operatorname{End}(\mathbf{R}^n)$, the matrix of which is equal to the matrix product of the column vector $f(y) \in \mathbf{R}^n$ and the row vector $\nu(y)^t \in \mathbf{R}^n$; that is

$$(f\,\nu^t)(y) = \begin{pmatrix} f_1(y)\,\nu_1(y) & \cdots & f_1(y)\,\nu_n(y) \\ \vdots & & \vdots \\ f_n(y)\,\nu_1(y) & \cdots & f_n(y)\,\nu_n(y) \end{pmatrix} \in \operatorname{Mat}(n, \mathbf{R}).$$

This implies $\operatorname{tr}(f\,\nu^t) = \langle f, \nu \rangle : \partial\Omega \to \mathbf{R}$. Applying Formula (7.43) we find the following identity of elements in $\operatorname{End}(\mathbf{R}^n)$, or matrices in $\operatorname{Mat}(n, \mathbf{R})$, where the integration is performed by coefficients:

$$\int_\Omega Df(x)\,dx = \int_{\partial\Omega} (f\,\nu^t)(y)\,d_{n-1}y.$$

Forming the traces on the left and on the right, we obtain the following:

Theorem 7.8.5 (Gauss' Divergence Theorem). *Let $\Omega \subset \mathbf{R}^n$ be as in Theorem 7.6.1 or 7.7.3. Let $f : \Omega \to \mathbf{R}^n$ be a vector field whose component functions f_i, for $1 \le i \le n$, satisfy the conditions from Theorem 7.6.1. Then*

$$\int_\Omega \operatorname{div} f(x)\,dx = \int \langle f, \nu \rangle(y)\,d_{n-1}y,$$

where the integration on the right–hand side is performed over $\partial\Omega$ or W, respectively.

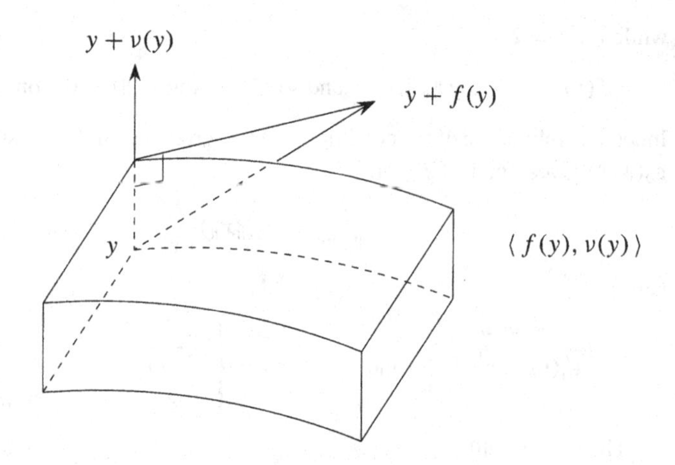

Illustration for Gauss' Divergence Theorem
$\langle f(y), \nu(y) \rangle \, d_{n-1}y$ is the volume of the cylinder whose base plane has area $d_{n-1}y$, and whose height equals the length of the normal component of the vector field $f(y)$, that is, equals the length of the component of $f(y)$ orthogonal to the base plane; this is an approximate description, of course

Definition 7.8.6. Let $V \subset \mathbf{R}^n$ be a C^1 hypersurface for which a continuous choice $y \mapsto \nu(y)$ of a normal $\nu(y)$ at the points $y \in V$ has been made. Let $f : V \to \mathbf{R}^n$ be a continuous vector field. Then the *flux* of f across V with respect to the choice of ν is defined as (see Formula (8.32) for additional details)

$$\int_V \langle f, \nu \rangle(y) \, d_{n-1}y. \qquad ❍$$

The Divergence Theorem can be formulated in words as follows. The integral of the divergence of a vector field over an open set Ω equals the flux of the vector field through the closed hypersurface $\partial\Omega$ with respect to the outer normal. The Divergence Theorem explains why the divergence of a vector field is also known as the flux of the vector field across closed surfaces per unit enclosed volume.

7.9 Applications of Gauss' Divergence Theorem

Example 7.9.1. Let $\Omega = \{ x \in \mathbf{R}^n \mid \|x\| < 1 \}$, then $\overline{\Omega} = B^n$ and $\partial\Omega = S^{n-1}$, and $\nu(y) = y$ is the outer normal to S^{n-1} at $y \in S^{n-1}$. Consider the vector field $f : \mathbf{R}^n \to \mathbf{R}^n$ given by $f(x) = x$. Then $\operatorname{div} f(x) = n$, and Gauss' Divergence Theorem yields (compare with Exercises 7.21.(iv) and 7.45.(ii))

$$n \operatorname{vol}_n(B^n) = \int_{B^n} \operatorname{div} f(x) \, dx = \int_{S^{n-1}} d_{n-1}y = \text{hyperarea}_{n-1}(S^{n-1}).$$

More generally, let $\Omega \subset \mathbf{R}^n$ be an open set that satisfies the conditions of Theorem 7.6.1 and is contained in a ball of radius r. Then placing the origin at the center of the given ball and noting that $|\langle f, \nu \rangle(y)| \le r$, for $y \in \partial\Omega$, we obtain $n \,\mathrm{vol}(\Omega) \le r \,\mathrm{hyperarea}_{n-1}(\partial\Omega)$. ☆

Example 7.9.2. We return to Example 7.5.5, that is

$$\Omega = \{\, x \in \mathbf{R}^3 \mid g(x) = \frac{x_1^2}{a^2} + \frac{x_2^2}{b^2} + \frac{x_3^2}{c^2} < 1 \,\},$$

and $d(y)$ is the distance from the origin to the geometric tangent plane $y + T_y(\partial\Omega)$. From Formula (7.41) it follows that

$$\int_{\partial\Omega} \frac{1}{d(y)}\, d_2 y = \frac{1}{2} \int_{\Omega} \mathrm{div}\, \mathrm{grad}\, g(x)\, dx = \frac{4\pi}{3}\Big(\frac{ab}{c} + \frac{bc}{a} + \frac{ca}{b}\Big).$$

Indeed,

$$\Delta g(x) = 2\Big(\frac{1}{a^2} + \frac{1}{b^2} + \frac{1}{c^2}\Big), \qquad \mathrm{vol}_3(\Omega) = \frac{4\pi abc}{3};$$

for $\mathrm{vol}_3(\Omega)$ use the substitution

$$\Psi : (r, \alpha, \theta) \mapsto r(a \cos\alpha \cos\theta,\ b \sin\alpha \cos\theta,\ c \sin\theta),$$

satisfying $\det D\Psi(r, \alpha, \theta) = abc\, r^2 \cos\theta$. In particular, if $a = b = c = 1$, then $\Omega = B^3$, $\partial\Omega = S^2$, and $d(y) = 1$, for all $y \in \partial\Omega$. Thus follows (compare with Example 7.4.10)

$$\mathrm{area}(S^2) = \int_{\partial\Omega} d_2 y = \frac{4\pi}{3}(1 + 1 + 1) = 4\pi.$$

☆

Example 7.9.3. Let $f : \mathbf{R}^3 \to \mathbf{R}^3$ be the vector field with

$$f(x) = (x_1 x_3^2,\ x_1^2 x_2 - x_3^3,\ 2x_1 x_2 + x_2^2 x_3),$$

let $a > 0$, and let V be the half sphere in $\mathbf{R}^3$ with $V = \{\, y \in \mathbf{R}^3 \mid \|y\| = a,\ y_3 > 0 \,\}$. Then the following applies to the flux of f across V with respect to the choice $\nu(y) = \frac{1}{a} y$, for $y \in V$:

$$\int_V \langle f, \nu \rangle(y)\, d_2 y = \frac{2\pi a^5}{5}. \tag{7.64}$$

Indeed, we have

$$\mathrm{div}\, f(x) = x_3^2 + x_1^2 + x_2^2 = \|x\|^2.$$

Therefore, let Ω be the open half ball in $\mathbf{R}^3$

$$\Omega = \{\, x \in \mathbf{R}^3 \mid \|x\| < a,\ x_3 > 0 \,\}.$$

Then $\partial\Omega = V_1 \cup V_2 \cup S$, with

$$V_1 = V, \qquad V_2 = \{ y \in \mathbf{R}^3 \mid y_1^2 + y_2^2 < a^2, \ y_3 = 0 \},$$
$$S = \{ y \in \mathbf{R}^3 \mid y_1^2 + y_2^2 = a^2, \ y_3 = 0 \}.$$

Then S is a one-dimensional submanifold in $\mathbf{R}^3$, and therefore a two-dimensional negligible set in $\mathbf{R}^3$. One has accordingly, with $\nu(y)$ now the outer normal to $\partial\Omega$ at y,

$$\sum_{i=1,2} \int_{V_i} \langle f, \nu\rangle(y)\, d_2 y = \int_\Omega \|x\|^2\, dx.$$

Now we may write (see Example 7.4.11)

$$\int_\Omega \|x\|^2\, dx = \int_{-\pi}^{\pi} d\alpha \int_0^{\frac{\pi}{2}} \cos\theta\, d\theta \int_0^a r^4\, dr = 2\pi \Big[\sin\theta \Big]_0^{\frac{\pi}{2}} \Big[\frac{r^5}{5} \Big]_0^a = \frac{2\pi a^5}{5}.$$

Concerning the calculation of the integral over the subset V_2 in the plane $x_3 = 0$, we note that for $y \in V_2$ one has $\nu(y) = (0, 0, -1)$, and further $d_2 y = dy_1\, dy_2$. It follows, therefore, that

$$\begin{aligned} \int_{V_2} \langle f, \nu\rangle(y)\, d_2 y &= \int_{\{ y \in \mathbf{R}^2 \mid y_1^2 + y_2^2 \le a^2 \}} -2 y_1 y_2\, dy_1\, dy_2 \\ &= -2 \int_{-a}^{a} y_1 \int_{-\sqrt{a^2 - y_1^2}}^{\sqrt{a^2 - y_1^2}} y_2\, dy_2\, dy_1 = 0, \end{aligned}$$

because the integrand in the inner integral is an odd function. This yields (7.64). Note that we have avoided calculating the integral over V by means of a parametrization of V. ☆

Example 7.9.4 (Flux of Newton vector field). Let $f : \mathbf{R}^n \setminus \{0\} \to \mathbf{R}^n$ be the Newton vector field from Example 7.8.4, with $f(x) = \frac{1}{|S^{n-1}|\,\|x\|^n}\, x$, and $\operatorname{div} f = 0$. Further, let Ω be an open set in $\mathbf{R}^n$ that satisfies the conditions of Theorem 7.6.1 or 7.7.3. Then (see Example 8.11.9 for a generalization)

$$\int_{\partial\Omega} \langle f, \nu\rangle(y)\, d_{n-1} y = \begin{cases} 1, & 0 \in \Omega; \\ 0, & 0 \notin \overline{\Omega}. \end{cases}$$

Indeed, if $0 \in \Omega$, there exists a number $r > 0$ such that the sphere $S^{n-1}(r)$ in $\mathbf{R}^n$ about 0 and of radius r is entirely contained in Ω. Let $\Omega' \subset \Omega$ be the open set bounded by $S^{n-1}(r)$ and $\partial\Omega$. Because $0 \notin \Omega'$, we may write

$$0 = \int_{\Omega'} \operatorname{div} f(x)\, dx = \int_{S^{n-1}(r)} \langle f, \nu\rangle(y)\, d_{n-1} y + \int_{\partial\Omega} \langle f, \nu\rangle(y)\, d_{n-1} y.$$

We now have, for $y \in S^{n-1}(r)$, that $\nu(y) = -\frac{1}{\|y\|} y$, and so

$$|S^{n-1}| \, \langle f, \nu \rangle(y) = \Big\langle \frac{1}{\|y\|^n} y, \; -\frac{1}{\|y\|} y \Big\rangle = -\frac{1}{\|y\|^{n-1}} = -\frac{1}{r^{n-1}} \qquad (y \in S^{n-1}(r)).$$

Therefore

$$\int_{S^{n-1}(r)} \langle f, \nu \rangle(y) \, d_{n-1}y \;=\; -\frac{1}{|S^{n-1}| \, r^{n-1}} \int_{S^{n-1}(r)} d_{n-1}y = -1. \qquad ☆$$

Example 7.9.5 (Heat equation). In physics the transfer of heat to a body $\Omega \subset \mathbf{R}^3$ of constant mass density m, whose temperature at point $x \in \Omega$ and time $t \in \mathbf{R}$ equals $T(x, t)$, is described, in first approximation, by the following laws.

(i) The amount of heat, ΔQ, required to raise the temperature of the part $x + \Delta x$ of Ω, during the interval of time from t to $t + \Delta t$, from $T(x, t)$ to $T(x, t) + \Delta T(x, t)$, is proportional to the mass $m \Delta x$ of $x + \Delta x$, and to the temperature difference $\Delta T(x, t)$. That is, there exists a constant $c_1 > 0$ such that

$$\Delta Q = c_1 m \, \Delta x \, \Delta T(x, t).$$

(ii) Let $y + \Delta y$ be a part of the boundary $\partial\Omega$ of Ω, with outer normal $\nu(y)$ at y, and of area $\Delta_2 y$. Then the amount of heat, ΔQ, moving inward during the interval of time from t to $t + \Delta t$ across the part $y + \Delta y$ of $\partial\Omega$ is proportional to the following three quantities:

(a) the interval of time Δt,

(b) the variation of T in the direction of $\nu(y)$, that is, to

$$D_{\nu(y)} T(y, t) = D_y T(y, t) \nu(y) = \langle \operatorname{grad}_y T(y, t), \; \nu(y) \rangle.$$

(This is because the flow of heat is caused by temperature differences, and then takes place from hot to cold, its magnitude being proportional to the component orthogonal to $\partial\Omega$ at y of the opposite of the gradient with respect to the spatial variable of T),

(c) the area $\Delta_2 y$.

That is, there exists a constant $c_2 > 0$ with

$$\Delta Q = c_2 \, \Delta t \langle \operatorname{grad}_y T(y, t), \; \nu(y) \rangle \, \Delta_2 y.$$

(iii) The total amount of heat, Q_1, absorbed by Ω during an interval Δt, equals the total amount of heat, Q_2, which has moved inward in the course of the same interval Δt across $\partial\Omega$.

According to (i), therefore

$$Q_1 = c_1 m \int_\Omega \Delta T(x,t)\,dx.$$

And according to (ii),

$$Q_2 = c_2\,\Delta t \int_{\partial\Omega} \langle\, \mathrm{grad}_y\, T(y,t),\ \nu(y)\,\rangle\, d_2 y.$$

Assertion (iii) tells us that $Q_1 = Q_2$. Thus we find the following equation for the temperature $T(x,t)$:

$$\int_\Omega \frac{\Delta T(x,t)}{\Delta t}\,dx = \frac{c_2}{c_1 m}\int_{\partial\Omega} \langle\, \mathrm{grad}_y\, T(y,t),\ \nu(y)\,\rangle\, d_2 y.$$

Taking the limit for $\Delta t \downarrow 0$ and applying the Divergence Theorem, we obtain, with $k := \frac{c_2}{c_1 m}$,

$$\int_\Omega \frac{\partial T}{\partial t}(x,t)\,dx = k \int_\Omega \mathrm{div}_x\, \mathrm{grad}_x\, T(x,t)\,dx.$$

Because this is valid for any body $\Omega \subset \mathbf{R}^3$, one infers the *heat equation* for the temperature T,

$$\frac{\partial T}{\partial t}(x,t) = k\,\Delta_x T(x,t) = k\Big(\frac{\partial^2 T}{\partial x_1^2} + \frac{\partial^2 T}{\partial x_2^2} + \frac{\partial^2 T}{\partial x_3^2}\Big)(x,t),$$

where Δ_x is the Laplace operator with respect to the spatial variable from (7.63). This is an example of a *partial differential equation* for T, defining a relation between different partial derivatives of T. ☆

Example 7.9.6 (Green's identities). Applying Corollary 7.6.2 concerning integration by parts in $\mathbf{R}^n$, with f replaced by $D_j f : \Omega \to \mathbf{R}$, for $1 \le j \le n$, then summing over j, we obtain *Green's first identity*,

$$\int_\Omega (g\,\Delta f)(x)\,dx = \int_{\partial\Omega}\Big(g\,\frac{\partial f}{\partial \nu}\Big)(y)\,d_{n-1}y - \int_\Omega \langle\, \mathrm{grad}\, f,\ \mathrm{grad}\, g\,\rangle(x)\,dx. \qquad (7.65)$$

Here $\frac{\partial f}{\partial \nu}$, the *derivative of f in the direction of the outer normal* ν, is defined by

$$\frac{\partial f}{\partial \nu}(y) = D_{\nu(y)} f(y) = Df(y)\nu(y) = \langle\, \mathrm{grad}\, f(y),\ \nu(y)\,\rangle \qquad (y \in \partial\Omega).$$

Subtracting a similar identity from Formula (7.65), but with the roles of f and g interchanged, we obtain *Green's second identity*,

$$\int_\Omega (f\,\Delta g - g\,\Delta f)(x)\,dx = \int_{\partial\Omega}\Big(f\,\frac{\partial g}{\partial \nu} - g\,\frac{\partial f}{\partial \nu}\Big)(y)\,d_{n-1}y.$$

In the theory of the Laplace operator this identity is an important tool. ☆

Example 7.9.7 (Dirichlet problem). Let g and h be given continuous functions on Ω and $\partial\Omega$, respectively. The *partial differential equation*

$$\Delta f = g \quad \text{on} \quad \Omega,$$

together with the *boundary condition*

$$f|_{\partial\Omega} = h,$$

is said to be a *Dirichlet problem* on Ω for a C^2 function f. We state without proof[2] that, for functions g and h that can be differentiated sufficiently many times, and for sufficiently well-behaved $\partial\Omega$, the Dirichlet problem on Ω has a solution f. See also Exercises 6.99, 7.65, 7.69.(iii), 7.70.(v) and 8.23. What will be proved here is the uniqueness of a solution f whose partial derivatives of order ≤ 2 can be continuously extended to $\overline{\Omega}$. Indeed, suppose that $\widetilde{f}$ is another such solution. Then

$$\Delta(f - \widetilde{f}) = \Delta f - \Delta \widetilde{f} = 0 \quad \text{on} \quad \Omega, \qquad f - \widetilde{f} = 0 \quad \text{on} \quad \partial\Omega.$$

Therefore consider a function f which is *harmonic* on Ω, that is, which satisfies

$$\Delta f(x) = 0 \qquad (x \in \Omega);$$

and for which, in addition

$$f|_{\partial\Omega} = 0. \tag{7.66}$$

For such f one has, by Formula (7.65),

$$\int_{\Omega} \|\operatorname{grad} f(x)\|^2 \, dx = 0.$$

Because the integrand is continuous and ≥ 0, we conclude that $\operatorname{grad} f(x) = 0$, for $x \in \Omega$. Therefore f is constant along any line segment contained in Ω; consequently, from (7.66) follows $f(x) = 0$, for every $x \in \Omega$. ☆

[2]See, for example, Chapters 27 and 12, respectively, in: Treves, F.: *Basic Linear Partial Differential Equations.* Academic Press, New York 1975; and Loomis, L. H., Sternberg, S.: *Advanced Calculus.* Addison-Wesley Publishing Company, Reading 1968.

Chapter 8

Oriented Integration

For an open set at one side of its boundary one has a natural prescription for the direction of **the** *normal. However, this is not the case for a manifold of lower dimension (consider a surface in* $\mathbf{R}^3$*, for example), and an orientation must then be chosen. This choice plays a role in the oriented integration over the manifold, introducing a dependence on the sense in which the manifold is swept out. The generalization to* $\mathbf{R}^n$ *of the second aspect of the Fundamental Theorem of Integral Calculus on* $\mathbf{R}$*, antidifferentiation of a function of several variables, leads to the concept of curl of a vector field. Oriented integration and curl together form the ingredients for the integral theorems of vector analysis and the theory of complex functions. Antidifferentiation of a function of several variables culminates in the theory of differential forms. Through systematic use of linear algebra an orgy of indices and partial derivatives is avoided.*

8.1 Line integrals and properties of vector fields

The divergence is one of the infinitesimal invariants of a vector field. There are others that are also needed in vector analysis. For an understanding of their meaning it is of advantage to be familiar with the concept of line integral, which we therefore introduce at the outset; the notion is also important for intrinsic reasons.

Definition 8.1.1. Let $I \subset \mathbf{R}$ be a closed interval, $\gamma : I \to \mathbf{R}^n$ a C^1 curve, and let $f : \operatorname{im}(\gamma) \to \mathbf{R}^n$ be a continuous vector field. Define $\int_\gamma \langle f(s), d_1 s \rangle$, the *oriented line integral* of f along γ, by

$$\int_\gamma \langle f(s), d_1 s \rangle = \int_I \langle f \circ \gamma, \ D\gamma \rangle(t)\, dt. \tag{8.1}$$

The integral on the right–hand side contains the inner product of the vectors $f(\gamma(t))$ and $\gamma'(t)$ in $\mathbf{R}^n$. If γ is a closed curve, the integral on the left is sometimes referred to as the *circulation* of f around γ. ❍

Example 8.1.2. Let $x \in \mathbf{R}^n$ and define $\gamma_x : I = [0, 1] \to \mathbf{R}^n$ by $\gamma_x(t) = tx$. Then $D\gamma_x(t) = x$, for all $t \in I$, and accordingly we have, for every $f \in C(\mathbf{R}^n, \mathbf{R}^n)$,

$$\int_{\gamma_x} \langle f(s), d_1 s \rangle = \int_0^1 \langle f(tx), x \rangle \, dt.$$ ☆

The term oriented line integral is used because the value of the line integral changes sign when the curve is traced in the opposite direction, as is shown in the following:

Lemma 8.1.3. *The right–hand side in (8.1) does not change upon a* reparametrization *of the interval I that preserves the order of the endpoints. That is, let $J = [j_-, j_+]$ and let $\psi : J \to I$ be a C^1 mapping with $I = [\psi(j_-), \psi(j_+)]$. Then*

$$\int_{\gamma \circ \psi} \langle f(s), d_1 s \rangle = \int_\gamma \langle f(s), d_1 s \rangle.$$

If, however, $I = [\psi(j_+), \psi(j_-)]$, then

$$\int_{\gamma \circ \psi} \langle f(s), d_1 s \rangle = -\int_\gamma \langle f(s), d_1 s \rangle.$$

Proof. On account of the chain rule, $D(\gamma \circ \psi)(t) = (D\gamma) \circ \psi(t) \, D\psi(t) = \psi'(t)(D\gamma) \circ \psi(t)$, and thus, by the Change of Variables Theorem 6.6.1 on $\mathbf{R}$,

$$\int_{\gamma \circ \psi} \langle f(s), d_1 s \rangle = \int_J \langle f \circ (\gamma \circ \psi), D(\gamma \circ \psi) \rangle(t) \, dt$$

$$= \int_{j_-}^{j_+} \langle (f \circ \gamma) \circ \psi, (D\gamma) \circ \psi \rangle(t) \psi'(t) \, dt = \int_{\psi(j_-)}^{\psi(j_+)} \langle f \circ \gamma, D\gamma \rangle(t) \, dt.$$ ❑

We now introduce some other invariants of a vector field, first giving motivations and definitions, and then filling in the background.

Theorem 7.6.1 is a generalization of the Fundamental Theorem of Integral Calculus on $\mathbf{R}$, but there exists another variant which will also be needed. In the integral calculus on $\mathbf{R}$, one of the possible formulations of the Fundamental Theorem (see Theorem 2.10.1), valid for continuous f on $[a, b]$, is

$$f(x) = \frac{d}{dx} \int_a^x f(t) \, dt \qquad (x \in [a, b]).$$

Consequently, on an interval every continuous function f has an antiderivative, that is, every continuous f is the derivative of a differentiable function g on that interval. The n-dimensional analog would be that a continuous vector field $f : U \to \mathbf{R}^n$ possesses an *antiderivative*, *integral*, or *scalar potential* $g : U \to \mathbf{R}$, that is, f is the total derivative of a differentiable real-valued function g; another way of saying is that f is a *gradient vector field*,

$$f = \operatorname{grad} g, \qquad \text{or, equivalently,} \qquad f_i = D_i g \quad (1 \le i \le n). \tag{8.2}$$

Assume that this is the case for a C^2 function $g : U \to \mathbf{R}$, while $n > 1$. By Theorem 2.7.2, the order of differentiation of the C^2 function g is interchangeable; hence

$$D_j f_i = D_j D_i g = D_i D_j g = D_i f_j \qquad (1 \le i,\ j \le n). \tag{8.3}$$

Thus, in contrast to the case $n = 1$, continuity of f as such is not sufficient for the existence of an integral: f also has to satisfy the *integrability conditions* in (8.3), which can be rewritten in matrix form as

$$Af(x) = 0 \qquad \text{with} \qquad Af(x)_{ij} = D_j f_i(x) - D_i f_j(x). \tag{8.4}$$

Therefore the nonvanishing on U of the infinitesimal invariant Af of the vector field f is an obstruction for finding an antiderivative for f on U; moreover, it turns out that the geometric properties of the set $U \subset \mathbf{R}^n$ also play a role in this problem. The most general result we shall formulate is Theorem 8.2.9.

Definition 8.1.4. Let $U \subset \mathbf{R}^n$ be an open subset, and let $f : U \to \mathbf{R}^n$ be a C^1 vector field. According to Lemma 2.1.4, for every $x \in U$, the derivative $Df(x) \in \operatorname{End}(\mathbf{R}^n)$ of f at x can be additively and uniquely decomposed into a self-adjoint operator $\frac{1}{2}Sf(x) \in \operatorname{End}^+(\mathbf{R}^n)$, with real eigenvalues, and an anti-adjoint operator, $\frac{1}{2}Af(x) \in \operatorname{End}^-(\mathbf{R}^n)$, with purely imaginary eigenvalues; this is known as the *Stokes decomposition*, or the *(infinitesimal) Cartan decomposition* of $Df(x)$,

$$Df(x) = \frac{1}{2}(Df(x) + Df(x)^t) + \frac{1}{2}(Df(x) - Df(x)^t) =: \frac{1}{2}Sf(x) + \frac{1}{2}Af(x). \tag{8.5}$$

Here $Df(x)^t$ is the adjoint linear operator of $Df(x)$ with respect to the standard inner product on $\mathbf{R}^n$; its matrix with respect to the standard basis is given by the transpose matrix $Df(x)^t$ of $Df(x)$. ❍

Thus the definitions of $Sf(x)$ and $Af(x)$ are independent of the choice of coordinates on $\mathbf{R}^n$, whereas they do depend on the choice of the inner product on $\mathbf{R}^n$. We note in addition

$$\operatorname{div} f = \operatorname{tr} Df = \frac{1}{2} \operatorname{tr} Sf. \tag{8.6}$$

For the cases $n = 2$ and 3 we now take a closer look at the anti-adjoint operator $Af(x)$ from the Stokes decomposition (8.5). We see immediately:

Lemma 8.1.5. *For $n = 2$ we have, in the notation used above,*

$$Af(x) = \begin{pmatrix} 0 & D_2 f_1 - D_1 f_2 \\ D_1 f_2 - D_2 f_1 & 0 \end{pmatrix}(x) = \operatorname{div}(J^t f)(x)\, J,$$

$$\textit{with} \qquad J = \begin{pmatrix} 0 & -1 \\ 1 & 0 \end{pmatrix}.$$

Here J is the matrix of the rotation in $\mathbf{R}^2$ by $\frac{\pi}{2}$ in the positive direction.

Definition 8.1.6. Let $U \subset \mathbf{R}^2$ be an open subset and let $f : U \to \mathbf{R}^2$ be a C^1 vector field. Define the function $\operatorname{curl} f : U \to \mathbf{R}$, the *curl of the vector field* f, by

$$\operatorname{curl} f = \operatorname{div}(J^t f) = D_1 f_2 - D_2 f_1 = \langle\, Af\, e_1,\, e_2 \,\rangle. \qquad \bigcirc$$

Example 8.1.7. In the definition of $\operatorname{curl} f$ one has to make a choice for the sign (this is done by introducing J). From the mapping $Af(x)$ itself one can only deduce $\det Af(x) = (D_1 f_2 - D_2 f_1)^2(x)$. Under our convention, $\operatorname{curl} J = 2$ for $J : \mathbf{R}^2 \to \mathbf{R}^2$, because $J^t J = I$. In other words, the curl of the vector field $J : x \mapsto (-x_2, x_1)$, of rotating in the positive direction, is positive. If $f : \mathbf{R}^2 \setminus \{0\} \to \mathbf{R}^2$ is given by $f(x) = \frac{1}{\|x\|^2} Jx$, then $J^t f(x) = \frac{1}{\|x\|^2} x$; hence it follows by Example 7.8.4 that $\operatorname{curl} f = 0$ on $\mathbf{R}^2 \setminus \{0\}$. ☆

We now deal with the case $n = 3$.

Lemma 8.1.8. *Let $A = (A_{ij}) \in \operatorname{Mat}(3, \mathbf{R})$ be an antisymmetric matrix. Then there exists a unique vector $a \in \mathbf{R}^3$ with*

$$Ah = a \times h \qquad (h \in \mathbf{R}^3), \qquad \textit{namely} \qquad a = (A_{32},\, A_{13},\, A_{21}).$$

Furthermore, $\langle\, Ah,\, k \,\rangle = \langle\, a,\, h \times k \,\rangle$, for $h, k \in \mathbf{R}^3$.

Proof. The uniqueness of a follows immediately. We further have

$$A = \begin{pmatrix} 0 & -A_{21} & A_{13} \\ A_{21} & 0 & -A_{32} \\ -A_{13} & A_{32} & 0 \end{pmatrix}.$$

But this is seen to be the matrix of the linear mapping $h \mapsto a \times h : \mathbf{R}^3 \to \mathbf{R}^3$ with a as above. The second equality follows from the identity $\langle\, k,\, a \times h \,\rangle = \det(k\ a\ h) =$

$\det(a\ h\ k) = \langle a,\ h \times k \rangle$, which holds for any triple of vectors a, h and $k \in \mathbf{R}^3$ (see Formula (5.2)). ❑

Application of Lemma 8.1.8 in the case where $A = Af(x)$ from the Stokes decomposition (8.5) gives a vector $a \in \mathbf{R}^3$ with j-th component given by

$$\begin{aligned} a_j &= Af(x)_{j+2,j+1} = Df(x)_{j+2,j+1} - Df(x)_{j+1,j+2} \\ &= D_{j+1} f_{j+2}(x) - D_{j+2} f_{j+1}(x), \end{aligned}$$

where the indices $1 \le j \le 3$ are taken modulo 3.

Definition 8.1.9. Let $U \subset \mathbf{R}^3$ be an open subset and let $f : U \to \mathbf{R}^3$ be a C^1 vector field. Then define the vector field $\operatorname{curl} f : U \to \mathbf{R}^3$, the *curl of the vector field* f, by

$$\operatorname{curl} f = (D_2 f_3 - D_3 f_2,\ D_3 f_1 - D_1 f_3,\ D_1 f_2 - D_2 f_1) = \nabla \times f, \qquad (8.7)$$

where ∇ is the nabla operator from Definition 7.8.3 and where the cross product is formally calculated. ❍

Corollary 8.1.10. *Let $U \subset \mathbf{R}^3$ be an open subset and let $f : U \to \mathbf{R}^3$ be a C^1 vector field. Then, in the notation of (8.5) and (8.7), for $x \in U$, h and $k \in \mathbf{R}^3$,*

$$Af(x)\,h = \operatorname{curl} f(x) \times h, \qquad \langle Af(x)\,h,\ k \rangle = \langle \operatorname{curl} f(x),\ h \times k \rangle.$$

Example 8.1.11 (Curl of infinitesimal rotation). See Example 5.9.3 for the notation. In particular, let $\phi_a = r_a : \mathbf{R}^3 \to \mathbf{R}^3$ be the tangent vector field from that example, that is

$$\phi_a(x) = (a_2 x_3 - a_3 x_2,\ a_3 x_1 - a_1 x_3,\ a_1 x_2 - a_2 x_1) \qquad (x \in \mathbf{R}^3).$$

Since r_{2a} is an anti-adjoint operator, the Stokes decomposition reads $D\phi_a(x) = r_a = \frac{1}{2} r_{2a}$, for all $x \in \mathbf{R}^3$. Therefore

$$\operatorname{curl} \phi_a(x) = 2a \qquad (x \in \mathbf{R}^3).$$ ☆

Remark. We have the following relation between the curl of a vector field in $\mathbf{R}^2$ and one in $\mathbf{R}^3$. If $f : \mathbf{R}^2 \to \mathbf{R}^2$ is a C^1 vector field, define $\widetilde{f} : \mathbf{R}^3 \to \mathbf{R}^3$ by

$$\widetilde{f}(x) = (f_1(x_1, x_2),\ f_2(x_1, x_2),\ 0).$$

Then

$$\operatorname{curl} \widetilde{f}(x) = (0, 0,\ \operatorname{curl} f(x_1, x_2)). \qquad (8.8)$$

Remarks of an analytical nature, motivating the definitions above. Let $U \subset \mathbf{R}^n$ be open and $f : U \to \mathbf{R}^n$ a C^1 vector field. Let $x \in U$ and $h, k \in \mathbf{R}^n$, then we obtain from the definition of differentiability, for $t > 0$ sufficiently small,

$$\langle f(x+th) - f(x),\ tk \rangle = t^2 \langle Df(x)h,\ k \rangle + o(t^2), \qquad t \downarrow 0,$$
$$\langle f(x+tk) - f(x),\ th \rangle = t^2 \langle Df(x)^t h,\ k \rangle + o(t^2), \qquad t \downarrow 0.$$

Subtracting these identities we find

$$\langle f(x),\ th \rangle + \langle f(x+th),\ tk \rangle - \langle f(x+tk),\ th \rangle - \langle f(x),\ tk \rangle$$
$$= t^2 \langle Af(x)\,h,\ k \rangle + o(t^2), \quad t \downarrow 0.$$

In terms of Definition 8.1.1 we recognize the expression on the left as an approximation of

$$\int_{\gamma(x,h,k,t)} \langle f(s), d_1 s \rangle,$$

the oriented line integral of the vector field f along $\gamma(x, h, k, t)$, the boundary of the parallelogram based at the point x and spanned by the vectors $t\,h$ and $t\,k$. In other words, $\gamma(x, h, k, t)$ equals the union of: the line segment in $\mathbf{R}^n$ from x to $x + t\,h$, the line segment from $x + t\,h$ to $x + t\,h + t\,k$, the line segment from $x + t\,k$ to $x + t\,k + t\,h$ traced in the opposite direction, and finally, likewise, the line segment from x to $x + t\,k$. Thus we find an interpretation of $Af(x)$, namely

$$\langle Af(x)\,h,\ k \rangle = \lim_{t\downarrow 0} \frac{1}{t^2} \int_{\gamma(x,h,k,t)} \langle f(s), d_1 s \rangle. \tag{8.9}$$

The arguments above are infinitesimal; in the following proposition, which for $n = 3$ is a special case of Stokes' Integral Theorem 8.4.4, they are made global.

Proposition 8.1.12. *Let $I = [\,0, 1\,]$, let $U \subset \mathbf{R}^n$ be open, and assume $\Gamma \in C^1(I^2, U)$ has continuous mixed second-order partial derivatives $D_2D_1\Gamma$ and $D_1D_2\Gamma : I^2 \to \mathbf{R}^n$. Then the following holds for a C^1 vector field $f : U \to \mathbf{R}^n$:*

$$\int_{\Gamma|_{\partial(I^2)}} \langle f(s),\ d_1 s \rangle = \int_{I^2} \langle (Af) \circ \Gamma \cdot D_1\Gamma,\ D_2\Gamma \rangle(x)\, dx. \tag{8.10}$$

Here $\partial(I^2)$ is the boundary of the square I^2, given by, successively, with $x_1,\ x_2 \in I$,

$$x_1 \mapsto (x_1, 0); \qquad x_2 \mapsto (1, x_2); \qquad x_1 \mapsto (1 - x_1, 1); \qquad x_2 \mapsto (0, 1 - x_2).$$

The expression on the left–hand side in (8.10) equals, by Definition 8.1.1,

$$\int_I \langle f \circ \Gamma,\ D_1\Gamma \rangle(x_1, 0)\, dx_1 + \int_I \langle f \circ \Gamma,\ D_2\Gamma \rangle(1, x_2)\, dx_2$$
$$- \int_I \langle f \circ \Gamma,\ D_1\Gamma \rangle(x_1, 1)\, dx_1 - \int_I \langle f \circ \Gamma,\ D_2\Gamma \rangle(0, x_2)\, dx_2.$$

Furthermore, the notation $\cdot$ on the right–hand side in (8.10) signifies application of a linear mapping to a vector.

Proof. On I^2 we have the following identities of functions:

$$D_1\langle f \circ \Gamma,\ D_2\Gamma \rangle = \langle (Df) \circ \Gamma \cdot D_1\Gamma,\ D_2\Gamma \rangle \ + \langle f \circ \Gamma,\ D_1 D_2\Gamma \rangle,$$

$$D_2\langle f \circ \Gamma,\ D_1\Gamma \rangle = \langle (Df)^t \circ \Gamma \cdot D_1\Gamma,\ D_2\Gamma \rangle + \langle f \circ \Gamma,\ D_2 D_1\Gamma \rangle.$$

One has, by Theorem 2.7.2, that $D_1 D_2 \Gamma = D_2 D_1 \Gamma$. Subtraction therefore yields

$$D_1\langle f \circ \Gamma,\ D_2\Gamma \rangle - D_2\langle f \circ \Gamma,\ D_1\Gamma \rangle = \langle (Af) \circ \Gamma \cdot D_1\Gamma,\ D_2\Gamma \rangle.$$

Integrating this identity over I^2 and using Corollary 6.4.3, we find

$$\int_I \int_I D_1\langle f \circ \Gamma,\ D_2\Gamma \rangle(x)\, dx_1\, dx_2 - \int_I \int_I D_2\langle f \circ \Gamma,\ D_1\Gamma \rangle(x)\, dx_2\, dx_1$$

$$= \int_{I^2} \langle (Af) \circ \Gamma \cdot D_1\Gamma,\ D_2\Gamma \rangle(x)\, dx.$$

When we apply the Fundamental Theorem of Integral Calculus 2.10.1 on $\mathbf{R}$ to the left–hand side we obtain

$$\int_I (\langle f \circ \Gamma,\ D_2\Gamma \rangle(1,\ x_2) - \langle f \circ \Gamma,\ D_2\Gamma \rangle(0,\ x_2))\, dx_2$$

$$- \int_I (\langle f \circ \Gamma,\ D_1\Gamma \rangle(x_1,\ 1) - \langle f \circ \Gamma,\ D_1\Gamma \rangle(x_1,\ 0))\, dx_1. \qquad \square$$

Remark. By means of differentiation under the integral sign and of integration by parts we find the following result, which is related to Formula (8.10):

$$\begin{aligned} D_1 \int_I \langle f \circ \Gamma,\ D_2\Gamma \rangle(x)\, dx_2 &= \int_I \langle (Af) \circ \Gamma \cdot D_1\Gamma,\ D_2\Gamma \rangle(x)\, dx_2 \\ &+ \langle f \circ \Gamma,\ D_1\Gamma \rangle(x_1,\ 1) - \langle f \circ \Gamma,\ D_1\Gamma \rangle(x_1, 0). \end{aligned} \tag{8.11}$$

This formula gives the derivative of a line integral along a curve $x_2 \mapsto \Gamma(x_1, x_2)$ that has an additional dependence on a parameter x_1, with respect to that parameter.

Remark in preparation for Section 8.6. In the preceding proposition curves $\gamma : I \to U$ were seen to play a role, and, in addition, scalar functions $y \mapsto \langle (f \circ \gamma)(y),\ D\gamma(y) \rangle$, which occur as a result of pairing the vector $f(\gamma(y)) \in \mathbf{R}^n$ and the tangent vector $D\gamma(y) \in T_{\gamma(y)}U$, for $y \in I$ (see Definition 5.1.1). Obviously, then, the vector $f(x) \in \mathbf{R}^n$ induces the mapping

$$\omega(x) = \omega_f(x) : h \mapsto \langle f(x),\ h \rangle \qquad \text{in} \qquad \mathrm{Lin}(T_x U, \mathbf{R}).$$

The mapping $\omega(x)$ then is an element of

$$\bigwedge^1 T_x^* U := T_x^* U := \mathrm{Lin}(T_x U, \mathbf{R}),$$

the linear space of all linear mappings from the tangent space T_xU to $\mathbf{R}$, which is isomorphic with $\mathbf{R}^n$. A mapping ω which assigns to every $x \in U$ an element $\omega(x) \in \bigwedge^1 T_x^*U$ is said to be a *differential* 1-*form* on U, and one writes $\omega \in \Omega^1(U)$ (see also Exercise 5.76).

Likewise, $Af(x) \in \mathrm{End}(\mathbf{R}^n)$ induces a mapping

$$\omega(x) = \omega_{Af}(x) \in \mathrm{Lin}^2(T_xU, \mathbf{R}) \qquad \text{with} \qquad \omega_{Af}(x)(h, k) = \langle\, Af(x)\,h,\ k\,\rangle.$$

This mapping $\omega(x)$ then is an element of

$$\bigwedge^2 T_x^*U = \{\, \eta \in \mathrm{Lin}^2(T_xU, \mathbf{R}) \mid \eta(h, k) = -\eta(k, h)\,\},$$

by definition the linear space of all antisymmetric bilinear mappings from the Cartesian product of T_xU with itself to $\mathbf{R}$. Indeed, $Af(x)^t = -Af(x)$ implies

$$\omega(x)(k, h) = \langle\, Af(x)\,k,\ h\,\rangle = \langle\, k,\ Af(x)^t\,h\,\rangle = -\omega(x)(h, k).$$

A mapping ω which assigns to every $x \in U$ an element $\omega(x) \in \bigwedge^2 T_x^*U$ is said to be a *differential* 2-*form* on U, and one writes $\omega \in \Omega^2(U)$.

In this context the differential 2-form ω_{Af} is said to be the *exterior derivative* of the differential 1-form ω_f. The linear mapping $\omega_f \mapsto \omega_{Af} : \Omega^1(U) \to \Omega^2(U)$ is said to be the *exterior differentiation*, and is often denoted by $d : \Omega^1(U) \to \Omega^2(U)$ instead of A (from antisymmetric) as we have done, that is, $d\,\omega_f = \omega_{Af}$. The condition $Af = 0$ for f then becomes $d\,\omega_f = 0$ for ω_f, and ω_f is said to be a *closed differential* 1-*form*. Conversely, a closed differential 1-form ω is not necessarily of the form $\omega_{\mathrm{grad}\,g}$, for a function g; but if it is, ω is said to be an *exact differential* 1-*form*.

Remarks of a geometrical nature, motivating the definitions above. The notation is that of Section 5.9. Assume that $(\Phi^t)_{t\in\mathbf{R}}$ is a one-parameter group of diffeomorphisms of $\mathbf{R}^n$ with tangent vector field $\phi : \mathbf{R}^n \to \mathbf{R}^n$. For fixed $x \in \mathbf{R}^n$, we wish to study the images $\Phi^t(x + v)$, for small values of $t \in \mathbf{R}$ and $v \in \mathbf{R}^n$. Let $w \in \mathbf{R}^n$ be arbitrary and define $f : \mathbf{R}^2 \to \mathbf{R}^n$ by $f(t, u) = \Phi^t(x + uw)$. Then we have $D_2D_1f(0, 0) = D\phi(x)w$, if we assume that $D\phi(x) \in \mathrm{End}(\mathbf{R}^n)$, the total derivative of ϕ at x, exists (note that ϕ does not depend on t). The identity $\lim_{(t,u)\to(0,0)} r(t, u) = D_2D_1f(0, 0)$ from Formula (2.19) now implies

$$\lim_{(t,u)\to(0,0)} \frac{1}{tu}(\Phi^t(x + uw) - \Phi^t(x) - uw) = D\phi(x)w.$$

Hence, setting $uw = v$ we obtain, for small values of $t \in \mathbf{R}$ and $v \in \mathbf{R}^n$,

$$\widetilde{\Phi^t_x}(v) := \Phi^t(x + v) - \Phi^t(x) = (I + tD\phi(x))v + \sigma(t\|v\|) \equiv (I + tD\phi(x))v. \tag{8.12}$$

This $\widetilde{\Phi^t_x}$ is said to be the local effect near x of the flow Φ^t.

The Stokes decomposition of $D\phi(x)$ from Formula (8.5) asserts

$$D\phi(x) = \frac{1}{2}S\phi(x)+\frac{1}{2}A\phi(x), \qquad S\phi(x) \in \mathrm{End}^+(\mathbf{R}^n), \qquad A\phi(x) \in \mathrm{End}^-(\mathbf{R}^n). \tag{8.13}$$

In view of the Spectral Theorem 2.9.3 the operator $S\phi(x)$ can be diagonalized, having eigenvalues $\lambda_1, \dots, \lambda_n \in \mathbf{R}$, and corresponding eigenspaces spanned by eigenvectors $v_1, \dots, v_n \in \mathbf{R}^n$, say. Therefore $S\phi(x)$ is an anisotropic (ἡ τροπή = change) linear dilatation in $\mathbf{R}^n$; in the direction of the eigenvector v_j its action is that of multiplication by λ_j, for $1 \le j \le n$. (In other words, $S\phi(x)$ maps a ball about 0 to an ellipsoid.) According to Example 2.4.10 we have $e^{\frac{t}{2}S\phi(x)} \in \mathrm{Aut}(\mathbf{R}^n)$, for $t \in \mathbf{R}$. It is an anisotropic dilatation in $\mathbf{R}^n$ with eigenvalues $e^{\frac{t}{2}\lambda_1}, \dots, e^{\frac{t}{2}\lambda_n}$; and under this transformation volumes change by a factor $e^{\frac{t}{2}\lambda_1} \cdots e^{\frac{t}{2}\lambda_n}$. It now follows, by Formula (8.6) (also compare with Formula (5.33)), that

$$\det e^{\frac{t}{2}S\phi(x)} = e^{\frac{t}{2}(\lambda_1+\dots+\lambda_n)} = e^{\frac{t}{2}\,\mathrm{tr}\,S\phi(x)} = e^{t\,\mathrm{div}\,\phi(x)}. \tag{8.14}$$

Since $A\phi(x)$ is anti-adjoint, Exercise 4.23.(iv) implies that $e^{\frac{t}{2}A\phi(x)} \in \mathrm{Aut}(\mathbf{R}^n)$, for $t \in \mathbf{R}$, is a rotation in $\mathbf{R}^n$. In particular, therefore, this transformation is volume-preserving. Formulae (8.12) and (8.13) now yield

$$\widetilde{\Phi^t_x} \equiv \left(I + \frac{t}{2}\,S\phi(x)\right)\left(I + \frac{t}{2}\,A\phi(x)\right) \equiv e^{\frac{t}{2}S\phi(x)}e^{\frac{t}{2}A\phi(x)}. \tag{8.15}$$

This means that the local effect $\widetilde{\Phi^t_x}$ near x of the flow Φ^t, **in the approximation of small values for** t, can **locally** be written as a composition of the rotation $e^{\frac{t}{2}A\phi(x)}$ in $\mathbf{R}^n$ and the subsequent anisotropic dilatation $e^{\frac{t}{2}S\phi(x)}$ in $\mathbf{R}^n$. The local effect $\widetilde{\Phi^t_x}$ is approximated by a linear mapping; denoting the latter also by $\widetilde{\Phi^t_x}$, we have, by Formula (8.14),

$$\det \widetilde{\Phi^t_x} = e^{t\,\mathrm{div}\,\phi(x)}.$$

Thus, this formula gives the geometrical interpretation of $\mathrm{div}\,\phi(x)$, for a tangent vector field ϕ associated with a one-parameter group $(\Phi^t)_{t\in\mathbf{R}}$ acting on $\mathbf{R}^n$: it equals the rate of change of volume at time $t = 0$, resulting from the local effect $\widetilde{\Phi^t_x}$ near x of the flow Φ^t. This conclusion also follows from Formula (5.31), that is

$$\frac{d}{dt}\Big|_{t=0} \det D\Phi^t(x) = \mathrm{div}\,\phi(x).$$

In particular, for $n = 3$ it follows from Corollary 8.1.10 and the theory of rotations in Exercise 5.58 that $t \mapsto e^{\frac{t}{2}A\phi(x)}$ is the one-parameter group of rotations $t \mapsto R_{t,\,\frac{1}{2}\mathrm{curl}\,\phi(x)}$ of the space $\mathbf{R}^3$, about the axis spanned by $\mathrm{curl}\,\phi(x) \in \mathbf{R}^3$ and with angular velocity $\frac{1}{2}\|\,\mathrm{curl}\,\phi(x)\|$. In combination with Formula (8.15) this leads to the geometrical interpretation of $\mathrm{curl}\,\phi(x)$, for a tangent vector field ϕ associated with a one-parameter group $(\Phi^t)_{t\in\mathbf{R}}$ acting on $\mathbf{R}^3$, that is, $\mathrm{curl}\,\phi(x)$ determines the "rotational component" of the local effect $\widetilde{\Phi^t_x}$ near x of the flow Φ^t.

Example 8.1.13. In the terminology of Section 5.9, the vector fields J and f from Example 8.1.7 both are tangent vector fields of one-parameter groups of diffeomorphisms of $\mathbf{R}^2$, having concentric circles about the origin as orbits. Now $\|J(x)\| = \|x\|$, while $\|f(x)\| = \frac{1}{\|x\|}$. Thus, as x moves away from the origin, the orbits under the flow associated with f are traced at decreasing rates. Evidently, $\operatorname{curl} f = 0$ implies that this phenomenon exactly compensates the rotation of x under the influence of the flow associated with J. ☆

8.2 Antidifferentiation

We return to the problem of finding an antiderivative (see Formula (8.2)).

Definition 8.2.1. Let $U \subset \mathbf{R}^n$ be an open subset and let $f : U \to \mathbf{R}^n$ be a C^1 vector field. We say that f satisfies the *integrability conditions* if (see Formula (8.4))

$$Af = 0, \qquad \text{that is,} \qquad D_j f_i = D_i f_j \qquad (1 \le i,\ j \le n).$$

The vector field f is said to be *divergence-free, source-free* or *incompressible* on U if $\operatorname{div} f = 0$. Furthermore, f is said to be *harmonic* on U if both $Af = 0$ and $\operatorname{div} f = 0$.

In particular, let $n = 3$. Then f is said to be *curl-free, vortex-free* or *irrotational* on U if $\operatorname{curl} f = 0$ (and therefore also $Af = 0$). ❍

Example 8.2.2 (curl grad and div curl). Let $U \subset \mathbf{R}^n$ be an open subset and let $g \in C^2(U)$, then (see Formula (8.3))

$$A(\operatorname{grad} g) = 0. \tag{8.16}$$

Indeed, $D(\operatorname{grad} g)(x)$ is self-adjoint due to the symmetry in the indices of the second-order partial derivatives of g (see Theorem 2.7.2). A gradient vector field $\operatorname{grad} g$ therefore satisfies the integrability conditions. The gradient vector field $\operatorname{grad} g$ of a harmonic function g, that is, a function with $\operatorname{div} \operatorname{grad} g = 0$, is harmonic. The component functions of a harmonic vector field are harmonic functions. The Newton vector field from Example 7.8.4 is harmonic.

One has in particular, for $n = 3$ and $h : U \to \mathbf{R}^3$ a C^2 vector field with $U \subset \mathbf{R}^3$,

$$\operatorname{curl} \operatorname{grad} g = \nabla \times (\nabla g) = 0 \qquad \text{and} \qquad \operatorname{div} \operatorname{curl} h = \nabla \cdot (\nabla \times h) = 0. \tag{8.17}$$

Indeed, the matrix of $D(\operatorname{curl} h)(x)$ is antisymmetric, for every $x \in U$. ☆

Definition 8.2.3. Let $U \subset \mathbf{R}^n$ be an open subset and let $f : U \to \mathbf{R}^n$ be a continuous vector field. A C^1 function $g : U \to \mathbf{R}$ is said to be an *antiderivative, integral*, or *scalar potential* for f on U if $f = \operatorname{grad} g$.

And, if $n = 3$, a C^1 vector field $h : U \to \mathbf{R}^3$ is said to be a *vector potential* for f on U if $f = \operatorname{curl} h$. ❍

From (8.16) and (8.17) it follows that $Af = 0$ (or $\operatorname{div} f = 0$ if $n = 3$) on U is a necessary condition for the existence of a scalar potential (or a vector potential, respectively) for f on U. And, under an additional condition on U, these conditions on f are also sufficient, as shown in Lemma 8.2.6 below. The necessity of additional conditions on U is apparent from the following. Assume that the continuous vector field $f : U \to \mathbf{R}^n$ possesses a scalar potential $g : U \to \mathbf{R}$. Further, let x and $y \in U$ be fixed. Then, for every C^1 curve $t \mapsto \gamma(t) : I \to U$ from x to y, the integral $\int_\gamma \langle f(s), d_1 s \rangle$ is independent of the choice of the curve γ, as long as the latter runs from the fixed point x to the fixed point y. Indeed, the value of the integral is given by the *potential difference*

$$\int_I \langle (\operatorname{grad} g) \circ \gamma,\ D\gamma \rangle(t)\, dt = \int_I \frac{d(g \circ \gamma)}{dt}(t)\, dt = g(y) - g(x). \qquad (8.18)$$

For a closed C^1 curve γ one has in particular $\int_\gamma \langle f(s), d_1 s \rangle = 0$.

Example 8.2.4. Let $U = \mathbf{R}^2 \setminus \{0\}$, and consider the vector field $f : U \to \mathbf{R}^2$ from Example 8.1.7, given by $f(x) = \frac{1}{\|x\|^2} Jx$; then we know that $\operatorname{curl} f = 0$ on U. Define $\gamma : \,]-\pi, \pi[\, \to U$ by $\gamma(t) = (\cos t, \sin t)$. Then $D\gamma(t) = (-\sin t, \cos t) = J\gamma(t)$ and therefore

$$\langle f \circ \gamma,\ D\gamma \rangle(t) = \frac{\|J\gamma(t)\|^2}{\|\gamma(t)\|^2} = 1; \qquad \text{hence} \qquad \int_\gamma \langle f(s), d_1 s \rangle = 2\pi.$$

This result can also be derived from Example 7.9.4, because $J^t f : x \mapsto \frac{1}{\|x\|^2} x$ is the Newton vector field on U (neglecting the constant 2π).

As a consequence, f can not have an antiderivative on U. On the other hand, f does have an antiderivative on $U' = \mathbf{R}^2 \setminus (\,]-\infty, 0] \times \{0\})$, because $f = \operatorname{grad} \arg$ on U', where $\arg : U' \to \mathbf{R}$ is the argument function, defined by (see Examples 3.1.1 and 2.4.8)

$$\arg(x) = 2 \arctan\left(\frac{x_2}{x_1 + \|x\|}\right).$$

Note that U' is the maximal domain of definition for the function arg.

According to Example 7.8.4 the vector field $J^t f$ has an antiderivative on U, and so

$$\int_\gamma \langle J^t f(s), d_1 s \rangle = 0. \qquad ☆$$

Definition 8.2.5. A set $U \subset \mathbf{R}^n$ is said to be *star-shaped* if there exists a point $x^0 \in U$ such that for all $x \in U$ the line segment from x^0 to x lies in U, that is, $\mathrm{im}(\gamma_x) \subset U$ with $\gamma_x : [\,0, 1\,] \to U$ defined by $\gamma_x(t) = x^0 + t(x - x^0)$. ❍

Lemma 8.2.6 (Poincaré). *Let $U \subset \mathbf{R}^n$ be star-shaped and open, and let $f : U \to \mathbf{R}^n$ be a C^1 vector field. Then the following two assertions are equivalent.*

(i) $Af = 0$ *on* U.

(ii) *There exists $g \in C^2(U)$ which is a scalar potential for f; for example the following, with γ_x as in Definition 8.2.5:*

$$g(x) = \int_{\gamma_x} \langle\, f(s),\, d_1 s\,\rangle \qquad (x \in U). \tag{8.19}$$

For $n = 3$, other equivalent assertions are as follows.

(iii) $\operatorname{div} f = 0$ *on* U.

(iv) *There exists a C^1 vector field $h : U \to \mathbf{R}^3$ which is a vector potential for f; for example*

$$h(x) = \int_0^1 ((f \circ \gamma_x) \times \gamma_x)(v)\, dv \qquad (x \in U).$$

Here the integration is carried out by components.

Proof. In order to simplify the formulae we assume that $x^0 = 0$.
(i) ⇒ (ii). According to Example 8.1.2 one has $g(x) = \int_0^1 k(v, x)\, dv$, where

$$k : [\,0, 1\,] \times U \to \mathbf{R} \qquad \text{with} \qquad k(v, x) = \langle\, f(vx),\, x\,\rangle = f(vx)^t x.$$

From $Af = 0$ and Formula (8.5) it follows that $Df(vx)^t = Df(vx)$, and therefore, if grad_x is the gradient with respect to the variable $x \in U$,

$$\begin{aligned} \mathrm{grad}_x\, k(v, x) &= v\, Df(vx)^t x + f(vx) = v\, Df(vx) x + f(vx) \\ &= v \frac{d}{dv} f(vx) + f(vx) = \frac{d}{dv}(v\, f)(vx). \end{aligned}$$

By means of differentiation under the integral sign we find

$$\mathrm{grad}\, g(x) = \int_0^1 \frac{d}{dv}(v\, f)(vx)\, dv = f(x).$$

(iii) ⇒ (iv) is proved in an analogous manner. Begin by $h(x) = \int_0^1 m(v, x)\, dv$, where

$$m : [\,0, 1\,] \times U \to \mathbf{R}^3 \qquad \text{with} \qquad m(v, x) = f(vx) \times vx = -v(r_x \circ f)(vx).$$

Here $r_x \in \mathrm{End}(\mathbf{R}^3)$ is given by $r_x(y) = x \times y$. This yields

$$D_x m(v, x) = v\, r_{f(vx)} - v^2 r_x \circ Df(vx).$$

Because $r_{f(vx)}$ and r_x are anti-adjoint operators, it follows that

$$\begin{aligned} A_x m(v, x) &:= D_x m(v, x) - D_x m(v, x)^t \\ &= 2v\, r_{f(vx)} - v^2(r_x \circ Df(vx) + Df(vx)^t \circ r_x). \end{aligned}$$

Application of the remark below, with $L = Df(vx)$, gives

$$A_x m(v, x) = 2v\, r_{f(vx)} + v^2 r_{Df(xv)x} = r_{2vf(vx)+v^2 Df(vx)x}.$$

Consequently, if curl_x is the curl with respect to the variable $x \in U$,

$$\mathrm{curl}_x\, m(v, x) = 2vf(vx) + v^2 Df(vx)x = \frac{d}{dv}(v^2 f)(vx).$$

By differentiation under the integral sign we find

$$\mathrm{curl}\, h(x) = \int_0^1 \frac{d}{dv}(v^2 f)(vx)\, dv = f(x). \qquad \square$$

Remark. Let $L \in \mathrm{End}(\mathbf{R}^3)$, with $\mathrm{tr}\, L = 0$. Then, for all $x \in \mathbf{R}^3$,

$$r_x \circ L + L^t \circ r_x = -r_{Lx}.$$

Indeed, the fact that $\mathrm{tr}\, L = 0$ implies, for all $h, k \in \mathbf{R}^3$,

$$\det(Lx\; h\; k) + \det(x\; Lh\; k) + \det(x\; h\; Lk) = 0.$$

This can be shown by calculating the coefficient of λ^2 in (see Formula (7.62))

$$\det(Lx - \lambda x\;\; Lh - \lambda h\;\; Lk - \lambda k) = \det(L - \lambda I)\, \det(x\; h\; k).$$

In view of $\det(p\, q\, r) = \langle\, p \times q, r\,\rangle$, for p, q and $r \in \mathbf{R}^3$ (see Formula (5.2)), this becomes

$$\langle\, Lx \times h, k\,\rangle + \langle\, x \times Lh, k\,\rangle + \langle\, x \times h, Lk\,\rangle = 0.$$

Therefore

$$\langle\, (r_{Lx} + r_x \circ L + L^t \circ r_x)\, h, k\,\rangle = 0.$$

Remark. From Lemma 8.2.6 it follows immediately that a harmonic vector field on $\mathbf{R}^3$ possesses a scalar potential that is itself a harmonic function.

We now formulate the property of U being simply connected, which is weaker than that of being star-shaped (see Definition 8.2.5), but which will still guarantee that a vector field satisfying the integrability conditions on U possesses a scalar potential on U (see Theorem 8.2.9). This concept originates from *homotopy theory*, a subject in *algebraic topology*.

Definition 8.2.7. An open set $U \subset \mathbf{R}^n$ is said to be *simply connected* if, for every point $x \in U$ and every C^1 curve $\gamma : I \to U$ beginning and ending at $x \in U$, there exists a C^1 *homotopy* between γ and the constant curve $\gamma_x : I \to \{x\}$, that is, if there exists a C^1 mapping (with $I = [\,0, 1\,]$, for simplicity)

$$\Gamma : I^2 \to U \qquad \text{with} \qquad \Gamma(0, t) = \gamma(t), \qquad \Gamma(1, t) = \Gamma(s, 0) = \Gamma(s, 1) = x,$$

for which the mixed second-order partial derivatives $D_2 D_1 \Gamma$ and $D_1 D_2 \Gamma : I^2 \to \mathbf{R}^n$ exist and are continuous. ❍

For example, $\mathbf{R}^n$, as well as every star-shaped open set $U \subset \mathbf{R}^n$, are each simply connected, because $\Gamma(s, t) = s\,x + (1 - s)\gamma(t) \in U$, in view of the definition of being star-shaped. Note that $D_1 D_2 \Gamma(s, t) = D_2 D_1 \Gamma(s, t) = -\gamma'(t)$.

Lemma 8.2.8. *Let $U \subset \mathbf{R}^n$ be simply connected and open, and let $f : U \to \mathbf{R}^n$ be a C^1 vector field with $Af = 0$. Then, for every closed C^1 curve $\gamma : I \to U$,*

$$\int_\gamma \langle\, f(s), d_1 s \,\rangle = 0.$$

Proof. Assume that γ begins and ends in $x \in U$, and let Γ be the corresponding C^1 homotopy. On the strength of Definition 8.1.1, the conclusion follows from Proposition 8.1.12. Indeed, $D_2\Gamma(0, t) = D\gamma(t)$, while $D_2\Gamma(1, t) = D_1\Gamma(s, 0) = D_1\Gamma(s, 1) = 0$ because $\Gamma(1, t) = \Gamma(s, 0) = \Gamma(s, 1) = x$. ❑

Theorem 8.2.9. *Let $U \subset \mathbf{R}^n$ be simply connected and open, and let $f : U \to \mathbf{R}^n$ be a C^1 vector field with $Af = 0$. Then there exists a C^2 function which is a scalar potential on U for f.*

Proof. Let $x^0 \in U$ be fixed. By analogy with Formula (8.19), we define the function $g : U \to \mathbf{R}$ by

$$g(x) = \int_{\gamma_x} \langle\, f(s), d_1 s \,\rangle;$$

here $\gamma_x : I \to U$ is an arbitrary C^1 curve in U from x^0 to $x \in U$. Then g is well-defined on U. Indeed, two different curves γ_x and $\widetilde{\gamma}_x$, meeting in C^1 fashion at x, together form a closed C^1 curve γ which begins and ends at x^0. Application of Lemma 8.2.8 to γ then yields the result that the oriented line integral of f along γ_x equals the oriented line integral of f along $\widetilde{\gamma}_x$. One then notes that, for all y, $x \in U$,

$$g(y) - g(x) = \int_{\gamma_{xy}} \langle\, f(s), d_1 s \,\rangle,$$

where γ_{xy} is a curve in U from x to y. But now one immediately concludes that $\operatorname{grad} g(x) = f(x)$, for all $x \in U$. ❑

8.3 Green's and Cauchy's Integral Theorems

Gauss' Divergence Theorem 7.8.5 can be used to prove other integral theorems: that of Green for open sets in the plane $\mathbf{R}^2$, that of Cauchy for open sets in the complex plane $\mathbf{C}$, and that of Stokes for surfaces in $\mathbf{R}^3$.

We now introduce the concept of a positive parametrization of the boundary of an admissible open set in $\mathbf{R}^2$. First we consider an example. The unit circle S^1 equals $\partial\Omega$, with Ω the bounded open subset $\{\, x \in \mathbf{R}^2 \mid \|x\| < 1 \,\}$. Consider the C^∞ parametrization $]-\pi,\ \pi\,[\;\to S^1 \setminus \{(-1, 0)\}$ of S^1 given by $t \mapsto y(t) = (\cos t,\ \sin t)$. One then has $\nu(y(t)) = y(t) = (\cos t,\ \sin t)$, for the outer normal to S^1 at $y(t)$, and $Dy(t) = (-\sin t,\ \cos t)$ (see Theorem 5.1.2), for the tangent vector to S^1 at $y(t)$. Accordingly, the direction of the tangent vector is obtained from that of the outer normal by application of J, the rotation in $\mathbf{R}^2$ by $\frac{\pi}{2}$ in the positive direction, whose matrix with respect to the standard basis in $\mathbf{R}^2$ is

$$J = \begin{pmatrix} 0 & -1 \\ 1 & 0 \end{pmatrix}. \tag{8.20}$$

In addition we find

$$\det(\nu \circ y \mid Dy)(t) = \begin{vmatrix} \cos t & -\sin t \\ \sin t & \cos t \end{vmatrix} = 1 > 0.$$

In linear algebra this is precisely the condition for $\nu \circ y(t)$ and $Dy(t)$ to form a pair of positively oriented vectors in $\mathbf{R}^2$. The geometrical equivalent of the positivity of the determinant is that the point $y(t) \in \partial\Omega$ moves counterclockwise in the plane $\mathbf{R}^2$ when t in $\mathbf{R}$ ranges from $-\pi$ to π. As this goes on, Ω constantly lies "to the left", that is, in the direction of the inner normal, and the complement of Ω lies "to the right", that is, in that of the outer normal. This suggests the following:

Definition 8.3.1. Let Ω be a bounded open subset of $\mathbf{R}^2$ having a C^1 boundary $\partial\Omega$ and lying at one side of $\partial\Omega$. Assume that $I \to \partial\Omega$ with $t \mapsto y(t)$ is a C^1 parametrization of $\partial\Omega$ by the disjoint union I of intervals in $\mathbf{R}$, with

$$\det(\nu \circ y \mid Dy)(t) > 0 \qquad (t \in I).$$

We then say that $t \mapsto y(t)$ is a *positive parametrization* of $\partial\Omega$, or, alternatively, that the parametrization $t \mapsto y(t)$ endows $\partial\Omega$ with a *positive orientation*. ❍

Remark. If $t \mapsto y(t)$ is a positive parametrization of $\partial\Omega$, the direction of the tangent vector $Dy(t)$ to $\partial\Omega$ at $y(t)$ is uniquely determined by the outer normal $\nu \circ y(t)$ to $\partial\Omega$ at $y(t)$, by means of

$$\|Dy(t)\|^{-1} Dy(t) = J\,\nu(y(t)) \qquad (t \in I). \tag{8.21}$$

Here J is the matrix from (8.20). In particular, there is no longer a freedom of choice as regards sign. In other words, every positive parametrization of $\partial\Omega$ induces the same unit tangent vector field τ on $\partial\Omega$, namely

$$\tau = J \circ \nu : \partial\Omega \to \mathbf{R}^2. \tag{8.22}$$

Example 8.3.2 (Annular domain). Let $0 < r < R$ and consider the annular bounded open set $\Omega = \{ x \in \mathbf{R}^2 \mid r < \|x\| < R \}$. Then $\partial\Omega$ is the (disconnected) set $\{ x \in \mathbf{R}^2 \mid \|x\| = r \} \cup \{ x \in \mathbf{R}^2 \mid \|x\| = R \}$. Furthermore, $t \mapsto y(t)$ is a positive C^∞ parametrization of $\partial\Omega$ if

$$y(t) = \begin{cases} R(\cos t,\ \sin t), & (0 \le t < 2\pi); \\ r(\cos t,\ -\sin t), & (4\pi \le t < 6\pi). \end{cases}$$

Omitting from Ω the points on a fixed line through the origin, one obtains two sets Ω_1 and Ω_2. Check that both $\partial\Omega_1$ and $\partial\Omega_2$ are connected sets. Now assume that positive parametrizations are chosen for both of these. Then the line segments in the intersection $\partial\Omega_1 \cap \partial\Omega_2$, considered as subsets of $\partial\Omega_1$, are traced in a direction opposite to that in which they are traced as subsets of $\partial\Omega_2$. ☆

We repeat the definition of line integral (compare with Definition 8.1.1) in the present context.

Definition 8.3.3. Let $\Omega \subset \mathbf{R}^2$ be as in Definition 8.3.1, and let $t \mapsto y(t)$ be a positive parametrization of $\partial\Omega$. Let $f : \partial\Omega \to \mathbf{R}^2$ be a continuous vector field. Then define $\int_{\partial\Omega} \langle f(y), d_1 y \rangle$, the *oriented line integral of the vector field f along* $\partial\Omega$, by

$$\int_{\partial\Omega} \langle f(y), d_1 y \rangle = \int_I \langle f \circ y, Dy \rangle(t)\, dt. \tag{8.23}$$

Here the right–hand side contains the inner product of vectors in $\mathbf{R}^2$. The expression on the left is also known as the *circulation of f around $\partial\Omega$ with respect to the choice of τ* (see (8.22)). ❍

Lemma 8.3.4. *Let $\Omega \subset \mathbf{R}^2$ be as in Definition 8.3.1, and let $t \mapsto y(t)$ be a positive parametrization of $\partial\Omega$.*

(i) *A parametrization $\widetilde{t} \mapsto \widetilde{y}(\widetilde{t}) : \widetilde{I} \to \partial\Omega$ is a positive parametrization of $\partial\Omega$ if and only if*

$$\det D(y^{-1} \circ \widetilde{y})(\widetilde{t}) > 0 \qquad (\widetilde{t} \in \widetilde{I}).$$

(ii) *The integral on the right in (8.23) does not change when a different choice is made for the positive parametrization $t \mapsto y(t)$ of $\partial\Omega$; therefore the integral on the left is defined independently of the choice of a positive parametrization.*

Proof. Let $y(t) = \widetilde{y}(\widetilde{t})$, with $t \in I$ and $\widetilde{t} \in \widetilde{I}$; then $t = (y^{-1} \circ \widetilde{y})(\widetilde{t}) =: \Psi(\widetilde{t})$. Application of the chain rule to $\widetilde{y} = y \circ \Psi$ on $\widetilde{I}$ therefore gives $D\widetilde{y}(\widetilde{t}) = \det D\Psi(\widetilde{t})\, Dy(t)$ for $\widetilde{t} \in \widetilde{I}$ and $t = \Psi(\widetilde{t}) \in I$. Assertion (i) now follows from

$$\det(\nu \circ \widetilde{y} \mid D\widetilde{y})(\widetilde{t}) = \det D\Psi(\widetilde{t})\, \det(\nu \circ y \mid Dy)(t).$$

To prove (ii) we apply the Change of Variables Theorem 6.6.1 with $\Psi : \widetilde{I} \to I$; then we obtain

$$\int_I \langle f \circ y, Dy \rangle(t)\, dt = \int_{\widetilde{I}} \langle f \circ \widetilde{y}, D\widetilde{y} \rangle(\widetilde{t})\, \frac{|\det D\Psi(\widetilde{t})|}{\det D\Psi(\widetilde{t})}\, d\widetilde{t}. \qquad ❑$$

Remark. See Lemma 8.1.3 for another formulation of Lemma 8.3.4. Furthermore, Lemma 8.3.4.(ii) can also be proved in a different way. Formulae (8.21) and (8.22) yield

$$\begin{aligned} \int_{\partial\Omega} \langle f(y), d_1 y \rangle &= \int_I \langle f \circ y(t),\ \|Dy(t)\|^{-1} Dy(t) \rangle \|Dy(t)\|\, dt \\ &= \int_{\partial\Omega} \langle f,\ J\nu \rangle(y)\, d_1 y = \int_{\partial\Omega} \langle f,\ \tau \rangle(y)\, d_1 y. \end{aligned} \tag{8.24}$$

Thus the oriented line integral of the vector field f along $\partial\Omega$ is seen to equal the integral over $\partial\Omega$ with respect to the arc length on $\partial\Omega$ of the function $y \mapsto \langle f,\ \tau \rangle(y)$; and according to Definition 7.1.2 – Theorem the latter integral is independent of the choice of the parametrization.

Theorem 8.3.5 (Green's Integral Theorem). *Let $\Omega \subset \mathbf{R}^2$ be a bounded open subset having a C^1 boundary $\partial\Omega$ and lying at one side of $\partial\Omega$. Assume that $t \mapsto y(t)$ is a positive parametrization of $\partial\Omega$. Let $f : \Omega \to \mathbf{R}^2$ be a C^1 vector field such that f and its derivative $Df : \Omega \to \mathrm{End}(\mathbf{R}^2)$ can both be extended to continuous mappings on $\overline{\Omega}$. Then*

$$\int_{\partial\Omega} \langle f(y), d_1 y \rangle = \int_{\Omega} \operatorname{curl} f(x)\, dx. \tag{8.25}$$

Proof. Note that it follows by Formula (8.24) that

$$\int_{\partial\Omega} \langle f(y), d_1 y \rangle = \int_{\partial\Omega} \langle J^t f, \nu \rangle (y)\, d_1 y.$$

Here J^t is the transpose of the matrix J from (8.20). Formula (8.25) now follows by means of the Divergence Theorem 7.8.5, because $\operatorname{div}(J^t f) = \operatorname{curl} f$, according to Definition 8.1.6. ❑

Example 8.3.6 (Descartes' folium). From Example 8.1.7 we know that $\operatorname{curl} J = 2$. We therefore immediately conclude

$$\begin{aligned} \operatorname{area}(\Omega) &= \int_{\Omega} dx = \frac{1}{2}\int_{\partial\Omega} \langle Jy, d_1 y \rangle = \frac{1}{2}\int_I (y_1\, y_2' - y_2\, y_1')(t)\, dt \\ &= \int_I \frac{1}{2} \det(y \mid Dy)(t)\, dt. \end{aligned} \tag{8.26}$$

Note that $\frac{1}{2}\det(y \mid Dy)(t)$ is the area of the triangle with vertices 0, $y(t)$ and $y + Dy(t)$.

Descartes' folium (compare with Example 5.3.7) is the curve in $\mathbf{R}^2$ defined by

$$y(t) = \frac{3at}{t^3+1}(1,\, t) \quad (-\infty \le t \le \infty,\ t \ne -1), \qquad \text{with} \quad y_i(\pm\infty) = 0.$$

We know that its points satisfy the equation $y_1^3 + y_2^3 = 3a y_1 y_2$. One has

$$(y_1\, y_2' - y_2\, y_1')(t) = y_1(t)^2 \left(\frac{y_2(t)}{y_1(t)}\right)' = \frac{9a^2t^2}{(t^3+1)^2}.$$

Let $\Omega \subset \mathbf{R}^2$ be the bounded subset which is bounded by the part of the folium parametrized by $]\,0, \infty\,[$. We have

$$\operatorname{area}(\Omega) = \frac{3a^2}{2}\int_0^\infty \frac{3t^2}{(t^3+1)^2}\, dt = \frac{3a^2}{2}\int_1^\infty \frac{1}{u^2}\, du = \frac{3a^2}{2}\left[-\frac{1}{u}\right]_1^\infty = \frac{3a^2}{2}.$$

This result can also be obtained by means of the following C^1 diffeomorphism $\Psi : V \to \Omega$. Here V is the triangle

$$V = \{\, y \in \mathbf{R}^2_+ \mid y_1 + y_2 < 3a \,\}, \qquad \Omega = \{\, x \in \mathbf{R}^2_+ \mid x_1^3 + x_2^3 < 3a x_1 x_2 \,\},$$
$$\Psi(y) = (y_1^{2/3} y_2^{1/3},\ y_1^{1/3} y_2^{2/3}).$$

Then $\det D\Psi(y) = \frac{1}{3}$, and hence $\text{area}(\Omega) = \int_V \frac{1}{3}\, dy = \frac{1}{3}\, 3a\, \frac{3a}{2} = \frac{3a^2}{2}$. ☆

As preparation for Cauchy's Integral Theorem we make the following remarks. We identify $\mathbf{C}$ with $\mathbf{R}^2$ as in (1.2). In this context, the matrix J from (8.20) is also known as the *complex structure* on $\mathbf{R}^2$, because this linear mapping yields the action of multiplication by $i = \sqrt{-1}$ on $\mathbf{C}$ identified with $\mathbf{R}^2$. As in Formula (1.2) we identify a function $f : \mathbf{C} \to \mathbf{C}$ with the vector field $f = (f_1, f_2) : \mathbf{R}^2 \to \mathbf{R}^2$ by means of

$$f(x_1{+}ix_2) = \text{Re}\, f(x_1{+}ix_2){+}i\, \text{Im}\, f(x_1{+}ix_2) \quad \longleftrightarrow \quad (f_1(x_1, x_2),\ f_2(x_1, x_2)).$$

Definition 8.3.7. Let $\Omega \subset \mathbf{C} \simeq \mathbf{R}^2$ be as in Definition 8.3.1, and let $t \mapsto z(t)$ be a positive parametrization of $\partial\Omega$. Let $f : \partial\Omega \to \mathbf{C}$ be a continuous function. In view of the equality

$$(f \circ z)(t) Dz(t) = (f_1 + i f_2)(z(t))(Dz_1 + i Dz_2)(t),$$

we define $\int_{\partial\Omega} f(z)\, dz$, the *complex line integral* of the function f *along* $\partial\Omega$, by

$$\int_{\partial\Omega} f(z)\, dz = \int_{\partial\Omega} \langle\, \overline{f}(z), d_1 z \,\rangle + i \int_{\partial\Omega} \langle\, (J\overline{f})(z), d_1 z \,\rangle, \tag{8.27}$$

where $\overline{f} = (f_1, -f_2)$ is the complex conjugate function of f; and therefore $J\overline{f} = (f_2, f_1)$. ❍

As in Lemma 8.3.4.(ii), one proves that the expression on the left does not change upon choosing another positive parametrization of $\partial\Omega$.

Example 8.3.8. If $z = x_1 + i x_2$, then $g(z) := \frac{1}{z} = \frac{\overline{z}}{|z|^2} \leftrightarrow \left(\frac{x_1}{x_1^2+x_2^2}, \frac{-x_2}{x_1^2+x_2^2}\right)$. Thus, with f as in Example 8.2.4,

$$\overline{g} = \frac{1}{\|x\|^2} x = J^t f, \qquad \text{and so} \qquad J\overline{g} = f;$$

and from the integrals in that example we obtain at once, where γ denotes the unit circle with positive orientation,

$$\int_\gamma \frac{1}{z}\, dz = 2\pi i. \qquad ☆$$

As in complex analysis we have the following:

Definition 8.3.9. Let $\Omega \subset \mathbf{C}$ be open and let $f : \Omega \to \mathbf{C}$. Then f is said to be *holomorphic* or *complex-differentiable* on Ω if f is continuously differentiable over the field $\mathbf{R}$ and the following limit exists, for every $z \in \Omega$:

$$f'(z) = \lim_{w \to 0} \frac{f(z+w) - f(z)}{w}. \qquad ○$$

The next lemma essentially follows from the definitions.

Lemma 8.3.10 (Cauchy–Riemann equation). *For $f = f_1 + i f_2 : \Omega \to \mathbf{C}$ and $(f_1, f_2) : \Omega \to \mathbf{R}^2$, respectively, one has the equivalent assertions.*

(i) *f is holomorphic.*

(ii) *The C^1 vector field (f_1, f_2) satisfies the* Cauchy–Riemann equation

$$(Df) \circ J = J \circ Df \in \mathrm{End}(\mathbf{R}^2),$$

$$\textit{that is,} \qquad D_1 f_1 = D_2 f_2 \quad \textit{and} \quad D_1 f_2 = -D_2 f_1.$$

(iii) *The C^1 vector field $\overline{f} = (f_1, -f_2)$ is harmonic (see Definition 8.2.1), that is*

$$\mathrm{curl}\, \overline{f} = 0, \qquad \mathrm{div}\, \overline{f} = \mathrm{curl}(J\overline{f}) = 0.$$

In other words, $A\overline{f} = A(J\overline{f}) = 0$.

(iv) *The real-differentiable C^1 function f satisfies*

$$\frac{\partial f}{\partial \overline{z}} = 0, \qquad \textit{where} \qquad \frac{\partial}{\partial \overline{z}} = \frac{1}{2}(D_1 + i D_2); \qquad \textit{that is,} \qquad i D_1 f = D_2 f.$$

Proof. (i) ⇔ (ii). The identity, valid for $h \in \mathbf{R}$,

$$\lim_{h \to 0} \frac{f(z+h) - f(z)}{h} = \lim_{h \to 0} \frac{f(z+ih) - f(z)}{ih}$$

corresponds to $D_1 f_1 + i D_1 f_2 = D_2 f_2 - i D_2 f_1$ (since $\frac{1}{i}(f_1 + i f_2) = f_2 - i f_1$); in turn, this is equivalent to $(Df) \circ J = J \circ Df$. ❑

The notation in part (iv) of the lemma above is explained as follows. For $z = x_1 + i x_2 \in \mathbf{C}$ we have $x_1 = \frac{1}{2}(z + \overline{z})$ and $x_2 = -\frac{i}{2}(z - \overline{z}) \in \mathbf{R}$. Regarding z and $\overline{z}$ as though they were independent variables, we find

$$\frac{\partial x_1}{\partial z} = \frac{\partial x_1}{\partial \overline{z}} = \frac{1}{2}, \qquad -\frac{\partial x_2}{\partial z} = \frac{\partial x_2}{\partial \overline{z}} = \frac{i}{2}.$$

Then the chain rule implies

$$\frac{\partial f}{\partial z} = D_1 f \frac{\partial x_1}{\partial z} + D_2 f \frac{\partial x_2}{\partial z} = \frac{1}{2}(D_1 - i D_2) f,$$

$$\frac{\partial f}{\partial \bar{z}} = D_1 f \frac{\partial x_1}{\partial \bar{z}} + D_2 f \frac{\partial x_2}{\partial \bar{z}} = \frac{1}{2}(D_1 + i D_2) f.$$

Theorem 8.3.11 (Cauchy's Integral Theorem). *Let $\Omega \subset \mathbf{C} \simeq \mathbf{R}^2$ be a bounded open subset having a C^1 boundary $\partial\Omega$ and lying at one side of $\partial\Omega$. Let $f : \Omega \to \mathbf{C}$ be a holomorphic function such that the vector field f and its derivative $Df : \Omega \to$ $\mathrm{End}(\mathbf{R}^2)$ can be extended to continuous functions on $\overline{\Omega}$. Then*

$$\int_{\partial\Omega} f(z)\, dz = 0.$$

Proof. The conclusion follows immediately from Lemma 8.3.10.(iii) and Green's Integral Theorem. ❑

By means of Definition 8.1.1, a version of (8.27), Lemma 8.2.8, and finally Lemma 8.3.10.(iii) we find the following variant (see Exercise 8.11 for another proof):

Theorem 8.3.12 (Cauchy's Integral Theorem). *Let $U \subset \mathbf{C} \simeq \mathbf{R}^2$ be a simply connected and open subset. Let $f : U \to \mathbf{C}$ be a holomorphic function. Then, for every closed C^1 curve $\gamma : I \to U$,*

$$\int_{\gamma} f(z)\, dz = 0.$$

In fact, the theorem is valid for every closed piecewise C^1 curve $\gamma : I \to U$.

8.4 Stokes' Integral Theorem

We now prepare the formulation of Stokes' Integral Theorem. Let $\Omega \subset \mathbf{R}^2$ be as in Green's Integral Theorem, and let $t \mapsto y(t) : I \to \partial\Omega$ be a positive C^1 parametrization of $\partial\Omega$. Further, let

$$\phi : \Omega \to \mathbf{R}^3 \qquad \text{with} \qquad y \mapsto \phi(y) = x,$$

be a C^2 embedding whose image is the C^2 surface in $\mathbf{R}^3$

$$\Xi = \phi(\Omega).$$

According to Formula (5.5),

$$\nu \circ \phi(y) = \|(D_1\phi \times D_2\phi)(y)\|^{-1}\,(D_1\phi \times D_2\phi)(y) \in \mathbf{R}^3$$

is a normal to Ξ at $\phi(y)$. We assume that ϕ and the total derivative $D\phi : \Omega \to \mathrm{Lin}(\mathbf{R}^2, \mathbf{R}^3)$ can be extended to continuous mappings on $\overline{\Omega}$. In particular, we may then define

$$\partial\,\Xi = \phi(\partial\Omega).$$

The extension of ϕ to $\partial\Omega$ is not necessarily injective on $\partial\Omega$, that is, a subset $S \subset \partial\,\Xi$ may exist with $S = \phi(V_1) = \phi(V_2)$ for V_1 and $V_2 \subset \partial\Omega$, while $V_1 \neq V_2$. A part S such as this does not contribute to $\partial\,\Xi$, if under the positive parametrization of $\partial\Omega$ the sets $\phi(V_1)$ and $\phi(V_2)$ are traced in opposite directions (these parts of $\partial\,\Xi$ then "cancel each other out"). One has that $\partial\,\Xi$, less the above-mentioned sets, is a piecewise C^1 submanifold in $\mathbf{R}^3$ of dimension 1 that does indeed play the role of the boundary of Ξ. The mapping

$$\gamma : t \mapsto (\phi \circ y)(t) : I \to \partial\,\Xi,$$

is a parametrization of $\partial\,\Xi$. Considering Theorem 5.1.2, we see that $D\gamma(t)$ is a tangent vector to $\partial\,\Xi$ at $\gamma(t)$.

Remark. Let the notation be as above. Let $f : \partial\,\Xi \to \mathbf{R}^3$ be a continuous vector field. In Definitions 8.1.1 and 8.3.3 we have introduced $\int_{\partial\Xi}\langle f(s), d_1 s\rangle$, the oriented line integral of the vector field f along $\partial\,\Xi$, by

$$\int_{\partial\Xi}\langle f(s), d_1 s\rangle = \int_I \langle f \circ \gamma, D\gamma\rangle(t)\,dt. \tag{8.28}$$

The integral on the left is sometimes referred to as the circulation of f around $\partial\,\Xi$ with respect to the choice of τ, the unit tangent vector field along $\partial\,\Xi$. The right–hand side in (8.28) does not change when a different positive C^1 parametrization of $\partial\Omega$ is chosen, nor does it change when another choice is made of a C^2 parametrization $\widetilde{\phi} : \widetilde{\Omega} \to \Xi$, provided $\det D(\phi^{-1} \circ \widetilde{\phi})(\widetilde{y}) > 0$, for all $\widetilde{y} \in \widetilde{\Omega}$. Consequently, the integral on the left is defined independently of the choice of the parametrization γ of $\partial\,\Xi$, provided the positivity conditions are met.

Proposition 8.4.1. *In the notation used above we may write*

$$\int_{\partial\Xi}\langle f(s), d_1 s\rangle = \int_{\partial\Omega}\langle ((D\phi)^t \cdot (f \circ \phi))(y),\, d_1 y\rangle, \tag{8.29}$$

where $D\phi(y)^t \in \mathrm{Lin}(\mathbf{R}^3, \mathbf{R}^2)$, *for* $y \in \partial\Omega$, *while* $\cdot$ *means application of a linear mapping to a vector. Thus we have here the vector field*

$$(D\phi)^t \cdot (f \circ \phi) : y \mapsto (f \circ \phi)(y) \mapsto D\phi(y)^t (f \circ \phi)(y) : \partial\Omega \to \mathbf{R}^3 \to \mathbf{R}^2. \tag{8.30}$$

Proof. By virtue of Formula (8.28) we find

$$\begin{aligned}\int_{\partial\Xi} \langle f(s), d_1 s \rangle &= \int_I \langle (f \circ \phi) \circ y,\ (D\phi) \circ y \cdot Dy \rangle(t)\, dt \\ &= \int_I \langle ((D\phi)^t \cdot (f \circ \phi)) \circ y,\ Dy \rangle(t)\, dt,\end{aligned}$$

where we have taken the adjoint. According to Formula (8.23), the expression on the right equals that in Formula (8.29). ❑

Formula (8.29) tells us that the line integral along $\partial\Xi$ can be "pulled back" to a line integral along $\partial\Omega$. We then wish to apply Green's Integral Theorem to the latter, to convert it into a surface integral over Ω. Accordingly, we calculate the curl of the vector field in Formula (8.30) by means of the following:

Lemma 8.4.2. *Assume that $f : \Xi \to \mathbf{R}^3$ is a C^1 vector field. In the notation above we have the following identity of functions on Ω:*

$$\operatorname{curl}((D\phi)^t \cdot (f \circ \phi)) = \langle (\operatorname{curl} f) \circ \phi,\ D_1\phi \times D_2\phi \rangle.$$

Proof. For the vector field in (8.30) we have

$$g := (D\phi)^t \cdot (f \circ \phi) = \begin{pmatrix} \langle D_1\phi,\ f \circ \phi \rangle \\ \langle D_2\phi,\ f \circ \phi \rangle \end{pmatrix} : \Omega \to \mathbf{R}^2.$$

Therefore

$$Dg = S + (D\phi)^t \cdot (Df) \circ \phi \cdot D\phi \qquad \text{with} \qquad S = \big(\langle D_j D_i\phi,\ f \circ \phi \rangle\big).$$

Theorem 2.7.2 implies that $S(y) \in \operatorname{End}(\mathbf{R}^2)$ is self-adjoint, for $y \in \Omega$. Furthermore, the adjoint of $(D\phi)^t \cdot (Df) \circ \phi \cdot D\phi$ equals $(D\phi)^t \cdot (Df)^t \circ \phi \cdot D\phi$; and so

$$Ag = Dg - (Dg)^t = (D\phi)^t \cdot (Df - (Df)^t) \circ \phi \cdot D\phi = (D\phi)^t \cdot (Af) \circ \phi \cdot D\phi.$$

Using Definition 8.1.6 and Corollary 8.1.10 we then find

$$\operatorname{curl} g = \langle Ag\, e_1,\ e_2 \rangle = \langle (Af) \circ \phi \cdot D\phi\, e_1,\ D\phi\, e_2 \rangle = \langle (\operatorname{curl} f) \circ \phi,\ D_1\phi \times D_2\phi \rangle. \text{❑}$$

With the help of Lemma 8.4.2 we now apply Green's Integral Theorem to the right–hand side of (8.29); this gives

$$\int_{\partial\Xi} \langle f(s), d_1 s \rangle = \int_\Omega \langle (\operatorname{curl} f) \circ \phi,\ D_1\phi \times D_2\phi \rangle(y)\, dy. \tag{8.31}$$

Finally, because the left–hand side of this formula is expressed in terms of Ξ, we want to recognize the surface integral over Ω on the right as a surface integral over Ξ which has been "pulled back"; hence the following:

Definition 8.4.3. Let $g : \overline{\Xi} \to \mathbf{R}^3$ be a continuous vector field. Then define $\int_\Xi \langle g(x), d_2x \rangle$, the *oriented surface integral of the vector field* g over Ξ, by

$$\int_\Xi \langle g(x), d_2x \rangle = \int_\Omega \langle g \circ \phi,\ D_1\phi \times D_2\phi \rangle(y)\, dy = \int_\Omega \det(g \circ \phi \mid D\phi)(y)\, dy. \bigcirc$$

The right–hand side does not change when we make a different choice $\widetilde{\phi} : \widetilde{\Omega} \to \Xi$ for the parametrization of Ξ, provided that $\det D(\phi^{-1} \circ \widetilde{\phi})(\widetilde{y}) > 0$, for $\widetilde{y} \in \widetilde{\Omega}$. Indeed, the second integral in Definition 8.4.3 is recognized by means of Formulae (7.37) and (7.20); this yields

$$\int_\Xi \langle g(x), d_2x \rangle = \int_\Omega \langle g \circ \phi,\ \nu \circ \phi \rangle(y)\, \omega_\phi(y)\, dy = \int_\Xi \langle g,\ \nu \rangle(x)\, d_2x, \qquad (8.32)$$

where the last integral is the integral with respect to the Euclidean area on Ξ of the function $x \mapsto \langle g,\ \nu \rangle(x)$ on Ξ. Formula (8.32) also shows that $\int_\Xi \langle g(x), d_2x \rangle$ equals the *flux* of g *across* Ξ *with respect to the choice of* ν (see Definition 7.8.6).

Formula (8.31) and Definition 8.4.3 imply the following:

Theorem 8.4.4 (Stokes' integral theorem). *Let $\Omega \subset \mathbf{R}^2$ be as in Green's Integral Theorem, let $t \mapsto y(t)$ be a positive C^1 parametrization of $\partial\Omega$, and let $\phi : \Omega \to \phi(\Omega) = \Xi$ be a C^2 embedding in $\mathbf{R}^3$ such that ϕ and the total derivative $D\phi$ can be extended to continuous mappings on $\overline{\Omega}$. Let $f : \Xi \to \mathbf{R}^3$ be a C^1 vector field such that f and the total derivative Df can be extended to continuous mappings on $\overline{\Xi}$. Then*

$$\int_{\partial\Xi} \langle f(s), d_1s \rangle = \int_\Xi \langle \operatorname{curl} f(x), d_2x \rangle.$$

Stokes' Integral Theorem can be worded as follows. The circulation of a vector field around the closed oriented curve $\partial\Xi \subset \mathbf{R}^3$ equals the flux of the curl of that vector field across the oriented surface $\Xi \subset \mathbf{R}^3$. The theorem explains why the curl of a vector field is also known as the circulation of the vector field around closed curves per unit enclosed area.

Further note that Formulae (8.9) and (8.10) for $n = 3$ immediately follow from Stokes' Integral Theorem.

Remark on compatible orientations. A phrase commonly encountered is the requirement that the orientation of the surface Ξ and the orientation of the boundary $\partial\Xi$ be compatible with each other. Here we give an exact formulation. According to Lemma 7.5.4 we may assume

$$\Omega \cap U = \{\, y \in \mathbf{R}^2 \mid y_1 < 0 \,\} \cap U.$$

Then $\nu(y) = e_1$, for $y \in \partial\Omega \cap U$, and $y : t \mapsto te_2$ is a positive parametrization of $\partial\Omega \cap U$. Consequently, $D_1\phi(y(t))$ is a tangent vector to Ξ at $x = (\phi \circ y)(t) \in \partial\Xi$

"pointing away from" Ξ, that is, a differentiable curve δ in $\mathbf{R}^n$ with $\delta(0) = x$ and $D\delta(0) = D_1\phi(y(t))$, leaves Ξ via the boundary. Furthermore, $D_2\phi(y(t))$ is a tangent vector to Ξ, tangent to $\partial\,\Xi$ at x, while Ξ "lies at the left of" $D_2\phi(y(t))$. Now the vectors

$$D_1\phi(y(t)), \qquad D_2\phi(y(t)), \qquad (D_1\phi \times D_2\phi)(y(t))$$

form a positively oriented triple in $\mathbf{R}^3$. Thus it is evident that, in the formulation of Stokes' Integral Theorem given above, the orientations of Ξ and of $\partial\,\Xi$ are compatible in the following sense:

- a tangent vector to Ξ, pointing outward from Ξ,
- the "positive" tangent vector to $\partial\,\Xi$ (determined by the positive orientation of $\partial\,\Xi$),
- and the normal to Ξ,

form a positively oriented triple of vectors at every point of $\partial\,\Xi$.

8.5 Applications of Stokes' Integral Theorem

Example 8.5.1. Define the vector field

$$f : \mathbf{R}^3 \to \mathbf{R}^3 \qquad f(x) = (x_3, x_1, x_2), \qquad \text{then} \qquad \operatorname{curl} f(x) = (1, 1, 1).$$

If $\Omega = \,]-\pi, \pi\,[\, \times \,]\,\psi, \frac{\pi}{2}\,[\, \subset \mathbf{R}^2$, one has, less four points, that $\partial\Omega$ equals

$$(]-\pi, \pi\,[\, \times \{\psi\}) \cup (\{\pi\} \times \,]\,\psi, \frac{\pi}{2}\,[) \cup (]-\pi, \pi\,[\, \times \{\frac{\pi}{2}\}) \cup (\{-\pi\} \times \,]\,\psi, \frac{\pi}{2}\,[),$$

where under positive orientation this set is traced counterclockwise. Further, let $\Xi = V = \phi(\Omega) \subset \mathbf{R}^3$ be the spherical surface from Example 7.4.6, described by

$$\phi : (\alpha, \theta) \mapsto R(\cos\alpha\cos\theta,\ \sin\alpha\cos\theta,\ \sin\theta) \qquad ((\alpha, \theta) \in \Omega).$$

It follows that

$$\begin{aligned} \frac{\partial\phi}{\partial\alpha} \times \frac{\partial\phi}{\partial\theta}(\alpha, \theta) &= R^2(\cos\alpha\cos^2\theta,\ \sin\alpha\cos^2\theta,\ \sin\theta\cos\theta), \\ \Big\langle (\operatorname{curl} f) \circ \phi,\ \frac{\partial\phi}{\partial\alpha} \times \frac{\partial\phi}{\partial\theta} \Big\rangle(\alpha, \theta) &= R^2((\cos\alpha + \sin\alpha)\cos^2\theta + \sin\theta\cos\theta). \end{aligned}$$

Therefore

$$\int_{\Xi} \langle \operatorname{curl} f(x),\ d_2x \rangle$$

$$= R^2 \int_{[-\pi,\,\pi]\times[\psi,\,\frac{\pi}{2}]} ((\cos\alpha + \sin\alpha)\cos^2\theta + \sin\theta\cos\theta)\, d\alpha\, d\theta$$

$$= R^2 \int_{-\pi}^{\pi} (\cos\alpha + \sin\alpha)\, d\alpha \int_{\psi}^{\frac{\pi}{2}} \cos^2\theta\, d\theta + 2\pi R^2 \int_{\psi}^{\frac{\pi}{2}} \sin\theta\cos\theta\, d\theta$$

$$= \pi R^2 \left[\sin^2\theta\right]_{\psi}^{\frac{\pi}{2}} = \pi R^2 (1 - \sin^2\psi) = \pi R^2 \cos^2\psi.$$

On the other hand, $\phi\,(\,]-\pi, \pi\,[\, \times \{\psi\}) \subset \partial\,\Xi$ has the parametrization

$$\gamma : \alpha \mapsto R(\cos\alpha\cos\psi,\ \sin\alpha\cos\psi,\ \sin\psi) \qquad (-\pi < \alpha < \pi).$$

And thus

$$\begin{aligned} f\circ\gamma(\alpha) &= R(\sin\psi,\ \cos\alpha\cos\psi,\ \sin\alpha\cos\psi), \\ D\gamma(\alpha) &= R(-\sin\alpha\cos\psi,\ \cos\alpha\cos\psi,\ 0), \\ \langle f\circ\gamma,\ D\gamma\rangle(\alpha) &= R^2(-\sin\alpha\cos\psi\sin\psi + \cos^2\alpha\cos^2\psi). \end{aligned}$$

Furthermore, we have $\phi\,(\,]-\pi, \pi\,[\, \times \{\frac{\pi}{2}\}) = \{(0, 0, 1)\}$, and because this is a zero-dimensional manifold, this part of $\partial\,\Xi$ does not contribute to the line integral along $\partial\,\Xi$. Finally,

$$\begin{aligned} \phi(\{\pi\} \times\,]\,\psi, \frac{\pi}{2}\,[\,) &= \{\, x \in \mathbf{R}^3 \mid x_1 < 0,\ x_2 = 0,\ x_3 > \sin\psi,\ \|x\| = 1\,\} \\ &= \phi(\{-\pi\} \times\,]\,\psi, \frac{\pi}{2}\,[\,), \end{aligned}$$

but the points $x = \phi(y)$ of these two image sets under ϕ are traced in opposite directions when y ranges counterclockwise over the boundary $\partial\Omega$. Therefore, these parts of $\partial\,\Xi$ do not contribute to the line integral along $\partial\,\Xi$. As a result

$$\begin{aligned} \int_{\partial\,\Xi} \langle f(s),\ d_1 s\rangle &= R^2 \int_{-\pi}^{\pi} (-\sin\alpha\cos\psi\sin\psi + \cos^2\alpha\cos^2\psi)\, d\alpha \\ &= R^2\cos^2\psi \int_{-\pi}^{\pi} \frac{1}{2}(1 + \cos 2\alpha)\, d\alpha = \pi R^2 \cos^2\psi, \end{aligned}$$

as expected on the basis of Stokes' Integral Theorem. ☆

Remark. In certain applications of Stokes' Integral Theorem, or of Gauss' Divergence Theorem, one is only given a closed C^1 curve C, or a closed C^1 surface D, in an open subset $U \subset \mathbf{R}^3$; it then remains to determine a suitable C^1 surface Ξ, or an open set Ω, **in** U such that $C = \partial\Xi$, or $D = \partial\Omega$, and such that the theorem can be applied to the pair (Ξ, C), or the pair (Ω, D), respectively. Whether or not this can be achieved is a subject of study in *homology theory*, which is a branch of *algebraic topology*. In addition, one has to be careful in defining the integrals, especially with regard to the orientations of curves and surfaces.

Formula (8.28) recalls the definition of the oriented line integral of a continuous vector field f along a C^1 curve in $\mathbf{R}^3$ forming the boundary of an embedded C^2 surface in $\mathbf{R}^3$. In that case the data define a preferential direction for a unit tangent vector field along the curve. But the same definition is also meaningful for an arbitrary embedded C^1 curve $\gamma : I \to \mathbf{R}^3$, if we fix a continuous choice for a tangent vector to that curve; in other words, γ is such that $D\gamma(t)$ points in the prescribed direction, for all $t \in I$.

This also applies, mutatis mutandis, to Definition 8.4.3 of the oriented surface integral of a continuous vector field g over an embedded C^1 surface in $\mathbf{R}^3$. There the data also define a preferential direction for a normal to the surface. And again this definition is meaningful for an arbitrary embedded C^1 surface $\phi : D \to \mathbf{R}^3$, if we fix a continuous choice for a normal to that surface; in other words, ϕ is such that $(D_1\phi \times D_2\phi)(y)$ points in the prescribed direction, for all $y \in D$.

In both cases we say that we have *oriented* the curve, or the surface, by choosing the direction of the tangent vector field, or the normal vector field, respectively.

We therefore need, in addition, the following:

Definition 8.5.2. Let $D \subset \mathbf{R}^{n-1}$ be open, let $\phi : D \to \mathbf{R}^n$ be an oriented C^1 embedding, and assume $f : \mathrm{im}(\phi) \to \mathbf{R}^n$ to be a continuous vector field. Define $\int_\phi \langle f(x), d_{n-1}x \rangle$, the *oriented hypersurface integral* of f over ϕ, by

$$\int_\phi \langle f(x), d_{n-1}x \rangle = \int_D \det(f \circ \phi \mid D\phi)(y)\, dy. \qquad ❍$$

The definition is independent of the choice of the embedding, provided the total derivative of the reparametrization has a positive determinant. Indeed, if $\phi(y) = \widetilde{\phi}(\widetilde{y})$, with $y \in D$ and $\widetilde{y} \in \widetilde{D}$, then $y = (\phi^{-1} \circ \widetilde{\phi})(\widetilde{y}) =: \Psi(\widetilde{y})$, see Lemma 4.3.3. Application of the chain rule to the identity $\widetilde{\phi} = \phi \circ \Psi$ on $\widetilde{D}$ now yields $D\widetilde{\phi}(\widetilde{y}) = D\phi(y)\, D\Psi(\widetilde{y})$. Given $\widetilde{y} \in \widetilde{D}$, the function $\mathrm{End}(\mathbf{R}^{n-1}) \to \mathbf{R}$ given by

$$D\Psi(\widetilde{y}) \mapsto \det(f \circ \widetilde{\phi} \mid D\widetilde{\phi})(\widetilde{y}) = \det(f \circ \phi(y) \mid D\phi(y)\, D\Psi(\widetilde{y}))$$

is antisymmetric and $(n-1)$-linear in the $n-1$ column vectors in $\mathbf{R}^{n-1}$ of $D\Psi(\widetilde{y})$. It is therefore given by $D\Psi(\widetilde{y}) \mapsto c(\widetilde{y}) \det D\Psi(\widetilde{y})$, where $c(\widetilde{y}) = \det(f \circ \phi(y) \mid D\phi(y))$. This yields

$$\det(f \circ \widetilde{\phi} \mid D\widetilde{\phi})(\widetilde{y}) = \det D\Psi(\widetilde{y})\, \det(f \circ \phi \mid D\phi)(y).$$

Applying the Change of Variables Theorem 6.6.1, with $\Psi : \widetilde{D} \to D$, we now find

$$\int_D \det(f \circ \phi \mid D\phi)(y)\, dy = \int_{\widetilde{D}} \det(f \circ \widetilde{\phi} \mid D\widetilde{\phi})(\widetilde{y})\, \frac{|\det D\Psi(\widetilde{y})|}{\det D\Psi(\widetilde{y})}\, d\widetilde{y}.$$

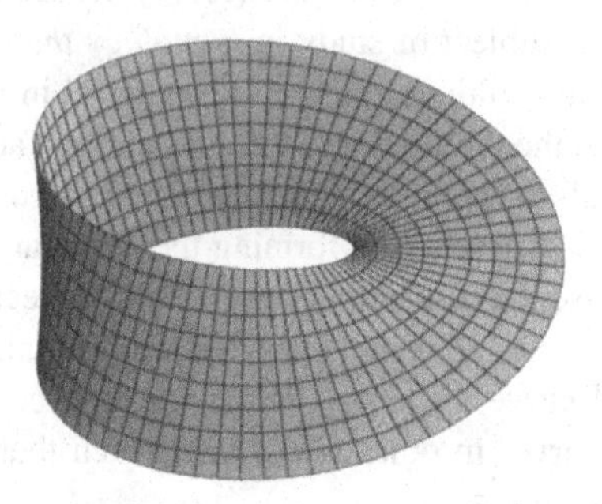

Illustration for Example 8.5.3: Möbius strip

Example 8.5.3 (Möbius strip). A C^1 surface in $\mathbf{R}^3$ cannot always be oriented; as a case in point, consider the *Möbius strip*. This can be obtained as $\overline{\Xi}$ with $\Xi = \mathrm{im}(\phi)$, where $\phi : \Omega \to \mathbf{R}^3$ is the following C^∞ embedding:

$$\Omega = \left]-\frac{1}{2}, \frac{1}{2}\right[\times \left]-\pi, \pi\right[,$$

$$\phi(t, \alpha) = \Big(\cos\alpha\big(1 + t\cos\frac{\alpha}{2}\big),\ \sin\alpha\big(1 + t\cos\frac{\alpha}{2}\big),\ t\sin\frac{\alpha}{2}\Big).$$

For a positive parametrization of Ω, the boundary $\partial\,\Xi$ is successively parametrized by

$$\begin{aligned}
&\phi(t, -\pi) &&= (-1, 0, -t) &&(-\tfrac{1}{2} < t < \tfrac{1}{2}),\\
&\phi(\tfrac{1}{2}, \alpha) &&=: \gamma_+(\alpha) &&(-\pi < \alpha < \pi),\\
&\phi(t, \pi) &&= (-1, 0, t) &&(\tfrac{1}{2} > t > -\tfrac{1}{2}),\\
&\phi(-\tfrac{1}{2}, \alpha) &&=: \gamma_-(\alpha) &&(\pi > \alpha > -\pi).
\end{aligned}$$

The surface Ξ itself is orientable. The nonorientability of $\overline{\Xi}$ is a consequence of the fact that different parts of $\partial\Omega$ (that is, $\left]-\frac{1}{2}, \frac{1}{2}\right[\times \{-\pi\}$ and $\left]-\frac{1}{2}, \frac{1}{2}\right[\times \{\pi\}$) have the same image under ϕ in $\partial\,\Xi$ (that is, the "vertical part" of $\partial\,\Xi$), whereas the orientations on these images are equal (both are traced from the top down). As a result, these parts "do not cancel each other out" (in contrast to what happens when $\partial\Omega$ is glued in such a way as to form a cylinder).

Stokes' Integral Theorem can be applied to Ξ, provided one does not equate the overlapping parts of $\partial\,\Xi$ with each other, but considers each with its own orientation. For example, consider the vector field

$$f : \{ x \in \mathbf{R}^3 \mid x_1^2 + x_2^2 \neq 0 \} \to \mathbf{R}^3 \qquad \text{with} \qquad f(x) = \frac{1}{x_1^2 + x_2^2}(-x_2, x_1, 0).$$

By means of Formula (8.8) and Example 8.1.7 we see that curl $f = 0$. By virtue of Stokes' Integral Theorem,

$$\int_{\gamma_+} \langle\, f(s), d_1 s \,\rangle + \int_{\gamma_-} \langle\, f(s), d_1 s \,\rangle = 0,$$

because the integral over (twice) the vertical part of $\partial\,\Xi$ vanishes.

On the other hand, consider the C^∞ embedding $\widetilde{\phi} : \widetilde{\Omega} \to \mathbf{R}^3$ with the property $\overline{\widetilde{\Xi}} = \overline{\Xi}$, if $\widetilde{\Xi} = \operatorname{im}(\widetilde{\phi}(\widetilde{\Omega}))$, given by

$$\widetilde{\Omega} = \left]0, \frac{1}{2}\right[\times \left]-2\pi, 2\pi\right[, \qquad \widetilde{\phi}(r, \alpha) = \phi(r, \alpha); \qquad \text{then}$$

$$\begin{aligned}
\widetilde{\phi}(r, -2\pi) &= (1 - r, 0, 0) && (0 < r < \tfrac{1}{2}),\\
\widetilde{\phi}(\tfrac{1}{2}, \alpha) &=: \gamma(\alpha) && (-2\pi < \alpha < 2\pi),\\
\widetilde{\phi}(r, 2\pi) &= (1 - r, 0, 0) && (\tfrac{1}{2} > r > 0),\\
\widetilde{\phi}(0, \alpha) &=: \gamma_0(\alpha) && (2\pi > \alpha > -2\pi).
\end{aligned}$$

Here $\gamma_0(\alpha) = (\cos\alpha, \sin\alpha, 0)$. In this case the nonorientability of $\overline{\widetilde{\Xi}}$ is evident from the fact that the "central circle", the image under γ_0, is traced twice in the same direction. Because the integrals over the "horizontal part" of $\partial\,\widetilde{\Xi}$ cancel each other out, we find

$$\int_{\gamma} \langle\, f(s), d_1 s \,\rangle = 4\pi, \qquad \text{since} \qquad \int_{\gamma_0} \langle\, f(s), d_1 s \,\rangle = -4\pi.$$

Now $\operatorname{im}(\gamma)$ is the disjoint union of $\operatorname{im}(\gamma_+)$ and $\operatorname{im}(\gamma_-)$; and, furthermore, γ and γ_+ have the same orientation on their intersection, whereas γ and γ_- have opposite orientations on their intersection. Consequently

$$\int_{\gamma} \langle\, f(s), d_1 s \,\rangle = \int_{\gamma_+} \langle\, f(s), d_1 s \,\rangle - \int_{\gamma_-} \langle\, f(s), d_1 s \,\rangle,$$

$$\text{and so} \qquad \int_{\gamma_\pm} \langle\, f(s), d_1 s \,\rangle = \pm 2\pi.$$

This example shows, moreover, that a closed C^1 curve in $\mathbf{R}^3$ that is not "knotted", to wit γ, can be the boundary of a nonorientable C^1 surface, to wit $\overline{\Xi}$. ☆

Remark. We now study the global properties, as opposed to the local ones from (8.16) and (8.17), possessed by a vector field in the presence of a potential.

Assume that the continuous vector field $f : U \to \mathbf{R}^n$ has a scalar potential $g : U \to \mathbf{R}$. Further, let x and $y \in U$ be fixed. Then we know that for every C^1 curve $t \mapsto \gamma(t) : I \to U$ from x to y, the integral $\int_\gamma \langle f(s), d_1 s \rangle$ is independent of the choice of the curve γ. According to Formula (8.18), its value is given by the *potential difference*

$$\int_\gamma \langle f(s), d_1 s \rangle = g(y) - g(x).$$

Now assume that $n = 3$ and that f possesses a vector potential $h : U \to \mathbf{R}^3$; further, let the closed oriented C^1 curve γ in U be fixed. Let $\Xi \subset U$ be an oriented C^2 surface with $\partial \Xi = \gamma$, while at every point of γ a tangent vector pointing outward from Ξ, the tangent vector to γ in the positive direction (determined by the orientation of γ), and the normal to Ξ (determined by the orientation of Ξ), form a positively oriented triple of vectors. Then, by Stokes' Integral Theorem

$$\int_\Xi \langle f(x), d_2 x \rangle = \int_\gamma \langle h(s), d_1 s \rangle.$$

That is, the oriented surface integral of f over Ξ is independent of the choice of the surface Ξ, provided the latter has the fixed curve γ as its boundary, and the orientations on Ξ and γ are compatible.

Definition 8.5.4. Let $U \subset \mathbf{R}^n$ be open and let $f : U \to \mathbf{R}^n$ be a continuous vector field.

Then f is said to be *conservative* on U if, for any two points x and $y \in U$, the oriented line integral of f along a C^1 curve γ in U from x to y is independent of the choice of γ.

Furthermore, f is said to be *solenoidal* (ὁ σωλήν = tube) on U if the oriented hypersurface integral of f over every compact oriented C^1 submanifold in U of dimension $n - 1$ equals 0. ❍

The terminology solenoidal is related to the fact that for every f having that property the oriented hypersurface integral over hypersurfaces is preserved if those hypersurfaces can be "connected" by a hypersurface $\Gamma \subset U$ "consisting of streamlines for f", that is, for which the normal to Γ at every point of Γ is perpendicular to f. In other words, the integral of f over a hypersurface is preserved if that hypersurface has transverse intersection with the same "stream tube". More precisely, f is solenoidal on U if the oriented hypersurface integral of f over Ξ_1 is equal to the opposite of that over Ξ_2, for all C^1 hypersurfaces Ξ_1 and Ξ_2 in U with the following properties: there exists a C^1 homotopy $\Gamma : I \times I^{n-1} \to U$ such that $\operatorname{im} \Gamma(0, \cdot) = \Xi_1$, $\operatorname{im} \Gamma(1, \cdot) = \Xi_2$, moreover $\partial(\operatorname{im} \Gamma) \setminus (\Xi_1 \cup \Xi_2)$ is a C^1 submanifold in U of dimension $n-1$, and $f(x) \in T_x(\partial(\operatorname{im} \Gamma))$ if x is in $\partial(\operatorname{im} \Gamma) \setminus (\Xi_1 \cup \Xi_2)$, while the orientations on Ξ_1 and Ξ_2 coincide with the outer normal on $\partial(\operatorname{im} \Gamma)$.

Theorem 8.2.9 and Lemma 8.2.6 now yield:

Proposition 8.5.5. *Let $U \subset \mathbf{R}^n$ be open and let $f : U \to \mathbf{R}^n$ be a C^1 vector field.*

(i) *Then f is conservative on U if U is simply connected and $Af = 0$, in particular, for $n = 3$, if f is curl-free.*

(ii) *Let $n = 3$. Then f is solenoidal on U if U is star-shaped and f is source-free.*

For arbitrary $n \in \mathbf{N}$ it is nontrivial to prove that a connected compact oriented C^k submanifold Ξ in $\mathbf{R}^n$ with $k \geq 1$ of dimension $n-1$ is the boundary of a bounded open set Ω that lies at one side of its boundary (see Example 8.11.10 below on the Jordan–Brouwer Separation Theorem). Moreover, it depends on the properties of U whether Ω can be found lying inside of U. If all of this is the case for every Ξ as above, Gauss' Divergence Theorem 7.8.5 implies that a divergence-free vector field on U is solenoidal on U. Here the relation between these two notions will not be discussed systematically.

Example 8.5.6. It is possible for a vector field to be curl-free, or source-free, on an open set $U \subset \mathbf{R}^3$, while the oriented integral along a closed oriented curve, or over a closed oriented surface in U, respectively, differs from 0. We limit the discussion to a few characteristic examples only.

The set $U_1 = \mathbf{R}^2 \setminus \{0\}$ is not simply connected. Let f_1 be the vector field f from Example 8.2.4; that example then tells us that $\int_{S^1} \langle f_1(s), d_1 s \rangle = 2\pi \neq 0$, if $S^1 \subset U_1$ is the unit circle with positive orientation.

Likewise, $U_2 = \mathbf{R}^3 \setminus \{0\}$ is not simply connected either. Let f_2 be the Newton vector field on U_2. According to Example 7.8.4, the vector field f_2 is source-free on U_2, but, by Formula (8.32) and Example 7.9.4, one has $\int_{S^2} \langle f_2(x), d_2 x \rangle = 1 \neq 0$, if $S^2 \subset U_2$ is the unit sphere oriented according to the outer normal.

On the other hand, it can be shown that every closed oriented C^2 curve γ in U_2 is the boundary of an oriented C^1 surface Ξ in U_2 with $\partial \Xi = \gamma$ and with compatible orientations on Ξ and γ; therefore, Stokes' Integral Theorem then yields $\int_\gamma \langle f_2(s), d_1 s \rangle = 0$. ☆

8.6 Apotheosis: differential forms and Stokes' Theorem

(ἡ ἀποθέωσις = deification). The theory of differential forms is an extension of the foregoing, and provides a natural framework for the theory of integration over oriented manifolds. Moreover, this generalization leads to unification, because Gauss' Divergence Theorem 7.8.5 and Stokes' Integral Theorem 8.4.4 are special cases of Stokes' Theorem 8.6.10 below.

Infinitesimal changes dx of a variable x are described by vectors in the tangent space at x to the space of those variables. Now let f be a function of x. The differential $df(x)$ of f at x is the linear part of the infinitesimal change in f when an infinitesimal change dx occurs in x. This implies that $df(x)$ should be interpreted as a linear function acting on tangent vectors at x. Thus df itself can be regarded as a function that gets evaluated along parametrized curves, by averaging over x on the curve the values of $df(x)$, acting on the tangent vector that is determined by the parametrization, at x to the curve. Indeed, we know that the rate of increase of a function along a curve depends linearly on the tangent vector to the curve. By integrating we obtain differences between values of the function as oriented line integrals. In the development of the calculus for differentials, one introduces differential forms, that is, linear combinations of differentials, having functions as coefficients; and an expression of that kind can only be the differential of one function if integrability conditions analogous to those from Definition 8.2.1 are met.

Higher-order differentials, products of differentials of functions, occur as obstructions against the integrability of the differential forms above. In fact, in the Remark following Proposition 8.1.12 we saw that $Af(x)$, the obstruction at x against the integrability of a vector field f, led to an antisymmetric bilinear mapping acting on two infinitesimal changes. In analogy with the situation for df, this property enables the function Af to be evaluated along parametrized surfaces, by averaging over x on the surface the values of $Af(x)$, acting on the pairs of tangent vectors that are determined by the parametrization, at x to the surface. Thus we see that higher-order differentials, that is, products of differentials of functions, occur in particular in the measurement of oriented volumes of infinitesimal parallelepipeds spanned by tangent vectors. One desires these volumes to linearly depend on those tangent vectors, and to vanish when the tangent vectors become linearly dependent. Together these considerations render it plausible that the higher-order differentials are generally defined as antisymmetric linear functions of the infinitesimal changes, that is, acting on collections of tangent vectors.

In order to arrive at a rigorous formulation of the foregoing we begin by defining differential forms. Then we give the proof of Stokes' Theorem, where we also introduce the exterior derivative of a differential form; this concept thus finds a first justification in the fact that it plays an important role in Stokes' Theorem. In addition we develop some *multilinear algebra* in order to derive the usual expressions for the exterior derivative. Our treatment heavily uses properties of determinants of a matrix, but other approaches are possible, at the cost of more abstract algebraic constructions.

Given a vector space V we write V^* for the dual linear space of V consisting of the linear functionals on V, that is, $V^* = \mathrm{Lin}(V, \mathbf{R})$. Note that $\mathbf{R}^{n*} := (\mathbf{R}^n)^*$ is linearly isomorphic with $\mathbf{R}^n$ in view of Lemma 2.6.1.

Definition 8.6.1. Let V be a vector space. Define $\bigwedge^0 V^* = \mathbf{R}$. For $k > 0$ we define $\bigwedge^k V^*$, the *k-th exterior power* of V^*, as the linear space of all *antisymmetric k-linear mappings* of the k-fold Cartesian product V^k of V with itself, into $\mathbf{R}$, that is,

ω in $\bigwedge^k V^*$ if and only if, in the notation of Definition 2.7.3,

$$\omega \in \mathrm{Lin}^k(V, \mathbf{R}), \qquad \omega(v_{\sigma(1)}, \dots, v_{\sigma(k)}) = \mathrm{sgn}(\sigma)\, \omega(v_1, \dots, v_k) \qquad (\sigma \in S_k).$$

Here S_k is the permutation group on k elements, and $\mathrm{sgn}(\sigma)$ is the sign of σ, that is, $\mathrm{sgn}(\sigma) = (-1)^m$ if $\sigma = \sigma_1 \cdots \sigma_m$ is the product of m transpositions. ❍

Note that $\bigwedge^1 V^* = V^*$. Considering $\omega(v_1, \dots, v_i + v_j, \dots, v_j + v_i, \dots, v_k)$ we see that the condition of $\omega \in \mathrm{Lin}^k(V, \mathbf{R})$ being antisymmetric is equivalent to the condition, where $v \in V$ occurs in i-th as well as in j-th position,

$$\omega(v_1, \dots, v, \dots, v, \dots, v_k) = 0 \qquad (1 \le i < j \le n).$$

Definition 8.6.2. Assume $\alpha_1, \dots, \alpha_k$ to be arbitrary elements in V^*. One readily sees that $\alpha_1 \wedge \cdots \wedge \alpha_k \in \bigwedge^k V^*$, if we define, for all $v_1, \dots, v_k \in V$,

$$(\alpha_1 \wedge \cdots \wedge \alpha_k)(v_1, \dots, v_k) = \det\left(\alpha_i(v_j)\right)_{1 \le i, j \le k}. \qquad ❍$$

Definition 8.6.3. Let $1 \le i_1 < \cdots < i_k \le n$ and $v_{i_1}, \dots, v_{i_k} \in V$, then we use the notation

$$I = (i_1, \dots, i_k), \qquad \mathcal{I}_k = \{\, (i_1, \dots, i_k) \mid 1 \le i_1 < \cdots < i_k \le n \,\},$$
$$v_I = (v_{i_1}, \dots, v_{i_k}) \in V^k.$$

From now on the notation $\widehat{\ }$ indicates that the variable underneath is omitted. In particular, if $k = n - 1$ we set

$$\begin{aligned} \widehat{i} &= (1, \dots, \widehat{i}, \dots, n) \in \mathcal{I}^{n-1}, \\ v_{\widehat{i}} &= (v_1, \dots, \widehat{v_i}, \dots, v_n) \in V^{n-1} \qquad (1 \le i \le n). \end{aligned} \qquad ❍$$

Lemma 8.6.4. *Let* $\dim V = n$ *and let* $(e_1, \dots, e_n)$ *be a basis for* V. *Further, let* $(e_1^*, \dots, e_n^*)$ *be the dual basis for* V^* *given by* $e_i^*(e_j) = \delta_{ij}$, *for* $1 \le i, j \le n$. *Then the elements*

$$e_I^* := e_{i_1}^* \wedge \cdots \wedge e_{i_k}^* \qquad (I \in \mathcal{I}_k) \tag{8.33}$$

form a basis for $\bigwedge^k V^*$. *In particular,* $\dim \bigwedge^k V^* = \binom{n}{k}$; *and so* $\dim \bigwedge^k V^* = 0$, *for* $k > n$. *Furthermore,* $\dim \bigwedge^n V^* = 1$; *and for* $\omega \in \bigwedge^n V^*$ *we have*

$$\omega = \omega(e_{(1,\dots,n)}) \det. \tag{8.34}$$

Proof. The linear independence of the vectors in (8.33) readily follows from the fact that, for I and $J \in \mathcal{I}_k$

$$e_I^*(e_J) = \begin{cases} 1, & \text{if } I = J; \\ 0, & \text{in all other cases.} \end{cases} \tag{8.35}$$

Now let $\omega \in \bigwedge^k V^*$ be arbitrary, and write $\omega_I = \omega(e_I)$. We assert that

$$\omega = \sum_{I \in \mathcal{I}_k} \omega_I \, e_I^*. \tag{8.36}$$

Indeed, both members of (8.36) have the same value on all k-tuples e_J, for $J \in \mathcal{I}_k$, and thus on V^k, since ω and the e_I^* are all contained in $\bigwedge^k V^*$. ❑

Example 8.6.5 (Relation to cross product of vectors). Let $V = \mathbf{R}^n$ and let $(e_1, \dots, e_n)$ be the standard basis. In this case, the cross product of $n - 1$ vectors in $\mathbf{R}^n$ (see the Remark on linear algebra in Example 5.3.11) is closely linked to the results in Lemma 8.6.4. Since $\dim \bigwedge^1 \mathbf{R}^{n*} = \dim \bigwedge^{n-1} \mathbf{R}^{n*} = n$, we can introduce an isomorphism between $\mathbf{R}^n$ and each of these two exterior powers. In fact, with $v \in \mathbf{R}^n$ we may associate $\flat_1 v \in \bigwedge^1 \mathbf{R}^{n*} = \mathbf{R}^{n*}$ as in Lemma 2.6.1, that is, we define $(\flat_1 v)h = \langle v, h \rangle$, for all $h \in \mathbf{R}^n$. For the other isomorphism we set

$$\flat_{n-1} v = i_v \det \in \bigwedge^{n-1} \mathbf{R}^{n*}, \qquad (i_v \det)(v_1, \dots, v_{n-1}) = \det(v \; v_1 \; \cdots \; v_{n-1}),$$

for any choice of vectors v_1, ..., $v_{n-1} \in \mathbf{R}^n$. Here $i_v \det$ is called the *contraction* of $\det \in \bigwedge^n \mathbf{R}^{n*}$ with the vector v. It is a direct consequence of Formula (8.36) that $\flat_{n-1} v$ is completely determined by its action on the $e_{\hat{i}}$, for $1 \le i \le n$. Expanding the $n \times n$ determinant by its first column we see $(\flat_{n-1} v)(e_{\hat{i}}) = (i_v \det)(e_{\hat{i}}) = \det(v \, e_1 \cdots \widehat{e_i} \cdots e_n) = (-1)^{i-1} v_i$. Thus we have obtained

$$\flat_{n-1} v = \sum_{1 \le i \le n} (-1)^{i-1} v_i \, e_{\hat{i}}^* \in \bigwedge^{n-1} \mathbf{R}^{n*}. \tag{8.37}$$

(The relation between $\flat_{n-1} v$ and contraction with v explains the minus signs in this formula. For more details on $\flat_1$ and $\flat_{n-1}$ see Example 8.8.2.)

Now we claim, for arbitrary v_1, ..., $v_{n-1} \in \mathbf{R}^n$,

$$\flat_1 v_1 \wedge \cdots \wedge \flat_1 v_{n-1} = \flat_{n-1}(v_1 \times \cdots \times v_{n-1}).$$

Indeed, testing both sides of this equality on $e_{\hat{i}}$, for $1 \le i \le n$, and applying Definition 8.6.2, we obtain

$$\det(\langle v_h, e_j \rangle)_{\substack{1 \le h < n \\ 1 \le j \le n,\, j \ne i}} = (-1)^{i-1} \det(e_i \; v_1 \; \cdots \; v_{n-1}) = (-1)^{i-1} (v_1 \times \cdots \times v_{n-1})_i,$$

which agrees with Formula (5.2) for the components of the cross product. This proves the claim. ☆

In the following we pointwise apply Definition 8.6.1.

Definition 8.6.6. Let $U \subset \mathbf{R}^n$ be an open subset and $0 \leq k \leq n$. We consider

$$\coprod_{x \in U} \bigwedge^k T_x^* U,$$

the disjoint union over all $x \in U$ of the k-th exterior powers $\bigwedge^k T_x^* U$ of the dual spaces $T_x^* U$ associated with the tangent spaces $T_x U = \mathbf{R}^n$ to U at x. The fact that $T_x U = \mathbf{R}^n$ follows from Theorem 5.1.2, because $\dim U = n$. Here we ignore the fact that all these exterior powers actually equal the same space $\bigwedge^k \mathbf{R}^{n*}$, taking for every point $x \in U$ a different copy of this. A *differential k-form* ω on U is a mapping

$$\omega : U \to \coprod_{x \in U} \bigwedge^k T_x^* U \qquad \text{with} \qquad \omega(x) \in \bigwedge^k T_x^* U \qquad (x \in U).$$

A differential k-form on U therefore acts on a k-tuple $(v_1, \dots, v_k)$ of vector fields $v_i : U \to \mathbf{R}^n$, if the v_i are considered as mappings $U \ni x \mapsto v_i(x) \in \mathbf{R}^n = T_x U$. Hereafter, a differential k-form on U will frequently also be made to act on a k-tuple of vectors $(v_1, \dots, v_k)$; these $v_i \in \mathbf{R}^n$ then are to be regarded as translation invariant vector fields on U, that is, as constant vector fields $U \ni x \mapsto v_i \in T_x U$.

We write

$$\Omega^k(U)$$

for the linear space with respect to the pointwise operations of differential k-forms on U. Further, ω is said to be a C^l differential k-form on U, with $l \geq 0$, if $x \mapsto \omega(x)(v_1, \dots, v_k) : U \to \mathbf{R}$ is a C^l function on U, for all $v_1, \dots, v_k \in \mathbf{R}^n$. The derivative of this function is an element of $\mathrm{Lin}(\mathbf{R}^n, \mathbf{R})$ that we denote by

$$v_0 \mapsto D\omega(x)(v_0)(v_1, \dots, v_k). \tag{8.38}$$

❍

Definition 8.6.7. Assume $D \subset \mathbf{R}^d$ and $U \subset \mathbf{R}^n$ are open subsets and $0 \leq k \leq n$. Let $\phi \in C^1(D, U)$ and $\omega \in \Omega^k(U)$. Then $\phi^* \omega \in \Omega^k(D)$ is defined by, for $y \in D$, $v_i \in T_y D \simeq \mathbf{R}^d$,

$$(\phi^* \omega)(y)(v_1, \dots, v_k) = \omega(\phi(y))(D\phi(y)v_1, \dots, D\phi(y)v_k).$$

The mapping $\phi^* \in \mathrm{Lin}\,(\Omega^k(U), \Omega^k(D))$ is said to be the operator of *pullback under the mapping* ϕ. ❍

Let $V \subset \mathbf{R}^p$ be open and let $f \in C^1(U, V)$. Applying the chain rule one readily proves

$$(f \circ \phi)^* = \phi^* \circ f^*. \tag{8.39}$$

A generalization of Definitions 8.1.1, 8.4.3 and 8.5.2 is the following:

Definition 8.6.8. Assume $0 \leq k \leq n$, and $D \subset \mathbf{R}^k$ and $U \subset \mathbf{R}^n$ are open subsets. Let $\phi \in C^1(D, U)$ and let $\omega \in \Omega^k(U)$ be a continuous differential k-form. Then define $\int_\phi \omega$, the *oriented integral* of ω over ϕ, by

$$\int_\phi \omega = \int_D (\phi^*\omega)(y)(e_{(1,\dots,k)})\,dy = \int_D \omega(\phi(y))(D_1\phi(y), \dots, D_k\phi(y))\,dy,$$

provided the integral in the last term converges. Here $(e_1, \dots, e_k)$ is the standard basis for $\mathbf{R}^k$. ❍

Remark. The value of the oriented integral of ω over ϕ does not change upon a *reparametrization with positive orientation* of the set D. That is, if $\Psi : D \to D$ is a C^1 diffeomorphism with $\det D\Psi > 0$, then

$$\int_\phi \omega = \int_{\phi\circ\Psi} \omega. \tag{8.40}$$

Indeed, application of Formula (8.34) to $\eta(\Psi(y)) \in \bigwedge^k T^*_{\Psi(y)} D$ yields

$$\begin{aligned} \Psi^*\eta(y)(e_{(1,\dots,k)}) &= \eta(\Psi(y))(D_1\Psi(y), \dots, D_k\Psi(y)) \\ &= \eta(\Psi(y))(e_{(1,\dots,k)})\,|\det D\Psi(y)|. \end{aligned}$$

Thus the assertion follows from the Change of Variables Theorem 6.6.1.

Further, let $V \subset \mathbf{R}^p$ be open, let $f \in C^1(U,\ V)$, and $\omega \in \Omega^k(V)$ a continuous differential k-form. An immediate consequence of Definition 8.6.8 and (8.39) then is

$$\int_{f\circ\phi} \omega = \int_\phi f^*\omega.$$

Finally, assume in particular that ϕ is a C^1 embedding. Then $X = \phi(D)$ is a C^1 submanifold in $\mathbf{R}^n$ of dimension d, on account of Corollary 4.3.2. If $\widetilde{\phi} : \widetilde{D} \to \mathbf{R}^n$ is another C^1 embedding with $X = \widetilde{\phi}(\widetilde{D})$, it follows from Lemma 4.3.3.(iii) that $\Psi = \phi^{-1} \circ \widetilde{\phi}$ is a C^1 diffeomorphism, with $\widetilde{\phi} = \phi \circ \Psi$. Under the assumption $\det D(\phi^{-1} \circ \widetilde{\phi}) > 0$ one verifies, analogous to the proof of Formula (8.40),

$$\int_\phi \omega = \int_{\widetilde{\phi}} \omega.$$

In this case, therefore, $\int_X \omega$, the *oriented integral* of ω over the manifold X, has been defined by

$$\int_X \omega = \int_\phi \omega.$$

We are now in a position to prepare the proof of Theorem 8.6.10 below, known as Stokes' Theorem; this proof is a direct generalization of the proof of Proposition 8.1.12.

Let $k \geq 1$; assume that $D \subset \mathbf{R}^k$ and $U \subset \mathbf{R}^n$ are open subsets, and that $\phi \in C^2(D, U)$; finally, let ω be a C^1 differential $(k-1)$-form on U. In the notation from Definition 8.6.3 consider $f_i : D \to \mathbf{R}$, for $1 \leq i \leq k$, given by

$$f_i(y) = (\phi^*\omega)(y)(e_{\hat{i}}) = \omega(\phi(y))(D_1\phi(y), \dots, \widehat{D_i\phi(y)}, \dots, D_k\phi(y)).$$

Then we have, with $D\omega$ defined in (8.38) and by using Proposition 2.7.6,

$$D_i f_i(y) = D\omega(\phi(y))(D_i\phi(y))(D_1\phi(y), \dots, \widehat{D_i\phi(y)}, \dots, D_k\phi(y))$$

$$+ \sum_{1\leq j\leq k,\, j\neq i} \omega(\phi(y))(D_1\phi(y), \dots, \widehat{D_i\phi(y)}, \dots, D_i D_j\phi(y), \dots, D_k\phi(y)).$$

In the last sum the summation is over $j \in \{1, \dots, i-1, i+1, \dots, k\}$, while $D_i D_j\phi(y)$ can also occur to the left of $\widehat{D_i\phi(y)}$, specifically, if $j < i$. Applying alternating summation over $1 \leq i \leq k$ in order to ensure antisymmetry, one obtains

$$\sum_{1\leq i\leq k} (-1)^{i-1} D_i f_i(y)$$

$$= \sum_{1\leq i\leq k} (-1)^{i-1} D\omega(\phi(y))(D_i\phi(y))(D_1\phi(y), \dots, \widehat{D_i\phi(y)}, \dots, D_k\phi(y))$$

$$+ \sum_{1\leq i\leq k} (-1)^{i-1} \sum_{1\leq j\leq k,\, j\neq i}$$

$$\omega(\phi(y))(D_1\phi(y), \dots, \widehat{D_i\phi(y)}, \dots, D_i D_j\phi(y), \dots, D_k\phi(y))$$

$$= \sum_{1\leq i\leq k} (-1)^{i-1} D\omega(\phi(y))(D_i\phi(y))(D_1\phi(y), \dots, \widehat{D_i\phi(y)}, \dots, D_k\phi(y))$$

$$+ \sum_{1\leq i<j\leq k} (-1)^{i+j-1}$$

$$\omega(\phi(y))(D_i D_j\phi(y) - D_j D_i\phi(y), D_1\phi(y), \dots, \widehat{D_i\phi(y)}, \dots, D_k\phi(y))$$

$$= \sum_{1\leq i\leq k} (-1)^{i-1} D\omega(\phi(y))(D_i\phi(y))(D_1\phi(y), \dots, \widehat{D_i\phi(y)}, \dots, D_k\phi(y)), \tag{8.41}$$

on the strength of Theorem 2.7.2. Note that

$$(v_1, \dots, v_k) \mapsto D\omega(x)(v_i)(v_1, \dots, \widehat{v_i}, \dots, v_k)$$

induces a mapping belonging to $\mathrm{Lin}^k(T_xU, \mathbf{R})$.

Definition 8.6.9. On account of Lemma 8.6.13 below, the mapping in $\mathrm{Lin}^k(T_xU, \mathbf{R})$ given by

$$(v_1, \dots, v_k) \mapsto \sum_{1 \le i \le k} (-1)^{i-1} D\omega(x)(v_i)(v_1, \dots, \widehat{v_i}, \dots, v_k)$$

belongs to $\bigwedge^k T_x^*U$; we denote it by

$$(v_1, \dots, v_k) \mapsto d\omega(x)(v_1, \dots, v_k).$$

Here $d\omega \in \Omega^k(U)$ is said to be the *exterior derivative* of $\omega \in \Omega^{k-1}(U)$. The operator

$$d \in \mathrm{Lin}\,(\Omega^{k-1}(U),\ \Omega^k(U))$$

is known as *exterior differentiation*. We say that $\omega \in \Omega^{k-1}(U)$ is *closed* on U if $d\omega = 0$. Furthermore, $\eta \in \Omega^k(U)$ is said to be *exact* on U if there exists an $\omega \in \Omega^{k-1}(U)$ with $\eta = d\omega$. ❍

By the use of Definitions 8.6.9 and 8.6.7, Formula (8.41) may be converted into the following equality of functions, valid for $y \in D$:

$$\begin{aligned} d(\phi^*\omega)(y)(e_{(1,\dots,k)}) &= \sum_{1 \le i \le k} (-1)^{i-1} D_i(\phi^*\omega)(y)(e_{\hat{i}}) \\ &= d\omega(\phi(y))(D_1\phi(y), \dots, D_k\phi(y)) = (\phi^*(d\omega))(y)(e_{(1,\dots,k)}). \end{aligned} \tag{8.42}$$

Note that (8.42) implies the following identity of differential k-forms on D:

$$d(\phi^*\omega) = \phi^*(d\omega); \tag{8.43}$$

in other words, exterior differentiation and pullback to $D \subset \mathbf{R}^k$ commute, acting on $\Omega^{k-1}(U)$.

Now let $I^k = [\,0, 1\,]^k$, a closed k-dimensional hypercube in $\mathbf{R}^k$, and assume that $\phi : I^k \to \mathbf{R}^n$ is the restriction of a C^2 mapping $D \to \mathbf{R}^n$ for an open neighborhood D of I^k. Integrating the equality between the second and the fourth term in (8.42) over I^k, changing the order of integration, and applying the Fundamental Theorem of Integral Calculus on $\mathbf{R}$ we obtain

$$\sum_{1 \le i \le k} \sum_{\pm} \pm(-1)^{i-1} I_{i,\pm} = \int_{I^k} (\phi^*(d\omega))(y)(e_{(1,\dots,k)})\,dy. \tag{8.44}$$

Here we have

$$I_{i,\pm} = \int_{I^{k-1}} ((\phi_{i,\pm})^*\omega)(y_{\hat{i}})(e_{\hat{i}})\, dy_{\hat{i}}, \qquad \text{where}$$

$$\phi_{i,\pm} : y_{\hat{i}} \mapsto \phi(y_1, \dots, y_{i-1}, \begin{smallmatrix}1\\0\end{smallmatrix}, y_{i+1}, \dots, y_k) : I^{k-1} \to U$$

is a parametrization of the smooth parts $\partial_{i,\pm}\phi$ of the i-th $(n-1)$-dimensional faces of $\partial(\phi(I^k))$. Further, the direction of the outer normal to $\partial_{i,\pm}\phi$ is given by

$$\pm \operatorname{sgn}(\det D\phi)(-1)^{i-1} D_1\phi \times \cdots \times \widehat{D_i\phi} \times \cdots \times D_k\phi.$$

Indeed, the inner product of this vector with $\pm D_i\phi$ is positive (compare with Exercise 7.59.(ii)). This enables us to recognize the left–hand side in (8.44) as $\int_{\partial\phi} \omega$, and the right–hand side as $\int_\phi d\omega$; thus we have now obtained:

Theorem 8.6.10 (Stokes' Theorem). *Let $0 \le k \le n$, let $I^k = [\,0, 1\,]^k \subset \mathbf{R}^k$, let U be open in $\mathbf{R}^n$, and let $\phi \in C^2(I^k, U)$; in addition, let $\omega \in \Omega^{k-1}(U)$ be a C^1-form. One then has*

$$\int_{\partial\phi} \omega = \int_\phi d\omega.$$

Note that, moreover, a limit argument yields:

Proposition 8.6.11. *The definition of $\omega \mapsto d\omega : \Omega^{k-1}(U) \to \Omega^k(U)$ is independent of the choice of coordinates in $\mathbf{R}^n$.*

The following theorem is a generalization of Formula (8.43). It is yet another expression of the fact that the exterior differentiation d is invariant under coordinate transformations; but it is more general, because the mapping f below need not be a diffeomorphism.

Theorem 8.6.12 (d and pullback commute). *Assume $U \subset \mathbf{R}^n$ and $V \subset \mathbf{R}^p$ are open subsets, and $f \in C^1(U, V)$. Let $\omega \in \Omega^{k-1}(V)$ be a C^1-form. Then we have the following identity of elements in $\Omega^k(U)$:*

$$d(f^*\omega) = f^*(d\omega).$$

Proof. It suffices to show that, for all $\phi \in C^2(I^k, U)$,

$$\int_\phi (d \circ f^*)\, \omega = \int_\phi (f^* \circ d)\, \omega.$$

By means of Definition 8.6.8 and Formulae (8.43) and (8.39) we find

$$\begin{aligned}\int_\phi (d\circ f^*)\,\omega &= \int_{I^k} \phi^*\circ(d\circ f^*)\,\omega = \int_{I^k} d\circ(\phi^*\circ f^*)\,\omega\\ &= \int_{I^k} d\circ(f\circ\phi)^*\omega = \int_{I^k}(f\circ\phi)^*\circ d\,\omega\\ &= \int_{I^k}\phi^*\circ(f^*\circ d)\,\omega = \int_\phi (f^*\circ d)\,\omega. \qquad \square\end{aligned}$$

The next lemma is a special case of Formula (8.50) below, which provides a procedure for antisymmetrizing multilinear forms.

Lemma 8.6.13. *Let $\eta \in \mathrm{Lin}^k(V, \mathbf{R})$ be antisymmetric in the last $k-1$ variables. Then $A\eta \in \bigwedge^k V^*$, if we define*

$$A\eta(v_1,\dots,v_k) = \sum_{1\le i\le k}(-1)^{i-1}\eta(v_i, v_1,\dots,\widehat{v_i},\dots,v_k). \tag{8.45}$$

Proof. We prove $A\eta(v_1,\dots,v_{l-1},v,v_{l+1},\dots,v_{m-1},v,v_{m+1},\dots,v_k) = 0$, for $1\le l < m \le k$. In view of the antisymmetry with respect to the last $k-1$ variables, the summands in (8.45) with index i different from l or m vanish in this case. The summands with indices l and m, respectively, are of the form

$$\begin{aligned}&(-1)^{l-1}\ \eta(v, v_1,\dots,v_{l-1},v_{l+1},v_{l+2},\dots,v_{m-1},v\ \ ,v_{m+1},\dots,v_k),\\ &(-1)^{m-1}\ \eta(v, v_1,\dots,v_{l-1},v\ \ ,v_{l+1},\dots,v_{m-2},v_{m-1},v_{m+1},\dots,v_k).\end{aligned} \tag{8.46}$$

Ignoring sign, the first term in (8.46) can be obtained from the second by transpositions of neighbors; to achieve this, the element v has to be carried from the l-th position to the $(m-1)$-th position. This requires $m-1-l$ transpositions of neighbors; consequently, the sign of the second term becomes $(-1)^{m-1-(m-1-l)} = -(-1)^{l-1}$. ❑

8.7 Properties of differential forms

We begin with a result in linear algebra. Let $A \in \mathrm{Mat}(n, \mathbf{R})$. The computation of $\det A$ using expansion by a row or a column of A is well-known. It is possible, however, to expand $\det A$ by several rows, or columns, simultaneously.

Indeed, let $0 \le k \le n$. For $I = (i_1,\dots,i_k) \in \mathcal{I}_k$ as in Definition 8.6.3, introduce $|I| = \sum_{1\le p\le k} i_p$, and $I' = (i'_1,\dots,i'_{n-k}) \in \mathcal{I}_{n-k}$ by $\{i_1,\dots,i_k\}\cup\{i'_1,\dots,i'_{n-k}\} = \{1,\dots,n\}$. Furthermore, for I and $J \in \mathcal{I}_k$, write $A_{IJ} \in \mathrm{Mat}(k, \mathbf{R})$ for the matrix whose (p,q)-th entry equals $a_{i_p j_q}$, for $1 \le p, q \le k$; then $A_{I'J'} \in \mathrm{Mat}(n-k, \mathbf{R})$. We now claim that there is the following expansion by the rows of A labeled by $I \in \mathcal{I}_k$:

$$\det A = (-1)^{|I|}\sum_{J\in\mathcal{I}_k}(-1)^{|J|}\det A_{IJ}\,\det A_{I'J'}. \tag{8.47}$$

In order to verify this, denote by $a_j \in \mathbf{R}^n$ the j-th column vector of A. Next define $a'_j \in \mathbf{R}^n$ by $a'_{ij} = a_{ij}$ (i.e. the i-th entry of a_j) if $i \in \{i_1, \dots, i_k\}$, and $a'_{ij} = 0$ otherwise. Finally, set $a''_j = a_j - a'_j \in \mathbf{R}^n$. Observe that the linear subspace spanned by the a'_j and the a''_j, for $1 \le j \le n$, is of dimension $\le k$ and $\le n-k$, respectively. Because $\det \in \mathrm{Lin}^n(\mathbf{R}^n, \mathbf{R})$ is antisymmetric, the decomposition $a_j = a'_j + a''_j$ leads to the decomposition $\det A = \sum_{J \in \mathfrak{I}_k} \det A^{IJ}$, where the entries a^{ij} of $A^{IJ} \in \mathrm{Mat}(n, \mathbf{R})$ are given by

$$a^{ij} = a_{ij} \qquad \text{if} \qquad (i, j) = \begin{cases} (i_p, j_q), & \text{for some } 1 \le p, q \le k; \\ (i'_r, j'_s), & \text{for some } 1 \le r, s \le n-k; \end{cases}$$

and $a^{ij} = 0$ otherwise. In order to compute $\det A^{IJ}$, let us shuffle the rows and columns of A^{IJ} so as to place $A_{IJ} \in \mathrm{Mat}(k, \mathbf{R})$ as defined above in the upper left corner. To this end we have to perform

$$(i_1 - 1) + \cdots + (i_k - k) + (j_1 - 1) + \cdots + (j_k - k) \equiv |I| + |J| \mod 2$$

permutations. The lower right corner of A^{IJ} then is made up by $A_{I'J'}$, while the other entries equal 0.

Lemma 8.7.1. *Let V be a vector space of finite dimension. Then there exists a unique bilinear mapping*

$$\bigwedge^k V^* \times \bigwedge^l V^* \to \bigwedge^{k+l} V^* \qquad \text{with} \qquad (\omega, \eta) \mapsto \omega \wedge \eta,$$

which satisfies, for $I \in \mathfrak{I}_k$ and $J \in \mathfrak{I}_l$,

$$e_I^* \wedge e_J^* = (e_{i_1}^* \wedge \cdots \wedge e_{i_k}^*) \wedge (e_{j_1}^* \wedge \cdots \wedge e_{j_l}^*) = e_{i_1}^* \wedge \cdots \wedge e_{i_k}^* \wedge e_{j_1}^* \wedge \cdots \wedge e_{j_l}^*.$$

The operation $\wedge$ is known as the exterior multiplication. *It is associative and* anticommutative, *which means that $\eta \wedge \omega = (-1)^{kl} \omega \wedge \eta$.*

Proof. In order to meet the conditions we define, for linear combinations of basis vectors (see Lemma 8.6.4),

$$\Big(\sum_{I \in \mathfrak{I}_k} \omega_I \, e_I^*\Big) \wedge \Big(\sum_{J \in \mathfrak{I}_l} \eta_J \, e_J^*\Big) = \sum_{I \in \mathfrak{I}_k,\, J \in \mathfrak{I}_l} \omega_I \eta_J \, e_I^* \wedge e_J^*.$$

The last assertion follows by successive application of $e_j^* \wedge e_i^* = -e_i^* \wedge e_j^*$. What is left to prove is that the definition of exterior multiplication does not depend on the choice of the basis in V. To this end we obtain Formula (8.48) below for $\omega \wedge \eta$, where $\omega = e_{i_1}^* \wedge \cdots \wedge e_{i_k}^*$ and $\eta = e_{j_1}^* \wedge \cdots \wedge e_{j_l}^*$. Consider $v_1, \dots, v_{k+l} \in V$. Then, by Definition 8.6.2,

$$(\omega \wedge \eta)(v_1, \dots, v_{k+l}) = \det A,$$

where the q-th column vector of $A \in \mathrm{Mat}(k+l, \mathbf{R})$ is given by

$$(v_{i_1 q}, \dots, v_{i_k q}, v_{j_1 q}, \cdots, v_{j_l q})^t \qquad (1 \le q \le k+l).$$

We compute $\det A$ using expansion by the top k rows as in Formula (8.47), thus we find, with $P = (1, \dots, k)$, and $1 \le q_1 < \cdots < q_k \le k+l$ if $Q \in \mathcal{I}_k$,

$$\begin{aligned}(\omega \wedge \eta)(v_1, \dots, v_{k+l}) &= (-1)^{\frac{1}{2}k(k+1)} \sum_{Q \in \mathcal{I}_k} (-1)^{|Q|} \det A_{PQ} \det A_{P'Q'} \\ &= (-1)^{\frac{1}{2}k(k+1)} \sum_{Q \in \mathcal{I}_k} (-1)^{|Q|} \omega(v_{q_1}, \dots, v_{q_k})\, \eta(v_{q'_1}, \dots, v_{q'_l}).\end{aligned} \tag{8.48}$$

This shows that the definition of $\omega \wedge \eta$ is in terms of ω and η only, and does not depend on the choice of the basis $(e_1, \dots, e_n)$ of V. ❑

Remark. The permutation

$$\sigma_Q = \begin{pmatrix} 1 & \dots & k & k+1 & \dots & k+l \\ q_1 & \dots & q_k & q'_1 & \dots & q'_l \end{pmatrix} \in S_{k+l},$$

where S_{k+l} is the permutation group on $k+l$ elements, is said to be a *shuffle*. Such a permutation describes a possible way of shuffling a deck of k cards through a deck of l cards, placing the cards of the first deck in order in the positions $q_1, \dots, q_k$ and those of the second deck in order in the positions $q'_1, \dots, q'_l$. The set of all shuffles in S_{k+l} is denoted by $S_{k,l}$ and consists of $\binom{k+l}{k}$ elements. Furthermore, $\mathrm{sgn}(\sigma_Q) = (-1)^{\frac{1}{2}k(k+1)+|Q|}$. Accordingly Formula (8.48) takes the form

$$(\omega \wedge \eta)(v_1, \dots, v_{k+l}) = \sum_{\sigma \in S_{k,l}} \mathrm{sgn}(\sigma)\, (\omega \circ \sigma)(v_1, \dots, v_k)\, (\eta \circ \sigma)(v_{k+1}, \dots, v_{k+l}). \tag{8.49}$$

Here $\sigma(v_1, \dots, v_k) = (v_{\sigma(1)}, \dots, v_{\sigma(k)})$, etc.

The results above are used in the preliminaries for the proof of the next theorem. We write $\bigwedge^{k,l} V^*$ for the linear subspace of $\mathrm{Lin}^{k+l}(V, \mathbf{R})$ consisting of the mappings that are antisymmetric with respect to the first k variables as well as the last l variables, and similarly we introduce $\bigwedge^{k,l,m} V^*$. Next define (see Lemma 8.6.13 for the case of $k = 1$, $l = k-1$ and $m = 0$)

$$A_{k,l} \in \mathrm{Lin}\Big(\bigwedge^{k,l,m} V^*, \bigwedge^{k+l,m} V^*\Big) \qquad \text{by} \qquad A_{k,l}\omega = \sum_{\sigma \in S_{k,l}} \mathrm{sgn}(\sigma)\, \omega \circ \sigma, \tag{8.50}$$

where $\sigma(k+l+i) = k+l+i$, for $1 \le i \le m$. The associativity of the exterior multiplication from Lemma 8.7.1 implies the commutativity of the following diagram:

$$\begin{array}{ccc} \bigwedge^{k,l,m} V^* & \xrightarrow{A_{l,m}} & \bigwedge^{k,l+m} V^* \\ {\scriptstyle A_{k,l}}\downarrow & & \downarrow{\scriptstyle A_{k,l+m}} \\ \bigwedge^{k+l,m} V^* & \xrightarrow{A_{k+l,m}} & \bigwedge^{k+l+m} V^* \end{array}. \qquad \text{That is,} \qquad A_{k+l,m} \circ A_{k,l} = A_{k,l+m} \circ A_{l,m}. \tag{8.51}$$

Theorem 8.7.2 **($d^2 = 0$).** *Let U be open in $\mathbf{R}^n$. For d acting on C^2 forms we have, for $k \in \mathbf{N}$,*

$$0 = d^2 : \Omega^{k-1}(U) \to \Omega^{k+1}(U).$$

In particular, in the terminology of Definition 8.6.9 we have that every exact differential form is also closed.

Proof. Consider a C^2 form $\omega \in \Omega^{k-1}(U)$ and $x \in U$. From $\omega(x) \in \bigwedge^{k-1} T_x^*U$ it follows that $D\omega(x)v_1 \in \bigwedge^{k-1} T_x^*U$ too, for a given $v_1 \in T_xU$. According to Definition 8.6.9 we find $d\omega(x) \in \bigwedge^k T_x^*U$ by application of $A_{1,k-1}$ to

$$D\omega(x)^\vee \in \bigwedge^{1,k-1} T_x^*U, \qquad D\omega(x)^\vee(v_1, \dots, v_k) = D\omega(x)(v_1)(v_2, \dots, v_k).$$

We may differentiate the resulting identity $d\omega(x) = A_{1,k-1}D\omega(x)^\vee$ at x in the direction of a fixed vector $v_1 \in T_xU$. In view of Lemma 2.4.7 we have that $D(d\omega)(x)v_1 \in \bigwedge^k T_x^*U$ is the result of application of $A_{1,k-1}$ to the mapping in $\bigwedge^{1,k-1} T_x^*U$ given by

$$(v_2, \dots, v_{k+1}) \mapsto (D^2\omega(x)(v_1)v_2)(v_3, \dots, v_{k+1}).$$

On account of Lemma 2.7.4 we may write this mapping as

$$D^2\omega(x)^\vee : (v_2, \dots, v_{k+1}) \mapsto D^2\omega(x)(v_1, v_2)(v_3, \dots, v_{k+1}),$$

$$\text{where} \qquad D^2\omega(x)^\vee \in \bigwedge^{1,1,k-1} T_x^*U.$$

Now it follows that $A_{1,k-1}D^2\omega(x)^\vee \in \bigwedge^{1,k}$ is the element which on evaluation on $(v_1, \cdots)$ equals $D(d\omega)(x)v_1$. Again by Definition 8.6.9, application of $A_{1,k}$ to this element gives $d(d\omega)(x)$, that is

$$d^2\omega(x) = A_{1,k} \circ A_{1,k-1}D^2\omega(x)^\vee = A_{2,k-1} \circ A_{1,1}D^2\omega(x)^\vee,$$

where we used Formula (8.51). Finally, Theorem 2.7.9 implies

$$\begin{aligned} &A_{1,1}D^2\omega(x)^\vee(v_1, \dots, v_{k+1}) \\ &\quad = D^2\omega(x)(v_1, v_2)(v_3, \dots, v_{k+1}) - D^2\omega(x)(v_2, v_1)(v_3, \dots, v_{k+1}) = 0. \end{aligned}$$ ❑

Proposition 8.7.3. *d is an* antiderivation, *that is, for C^1 forms $\omega \in \Omega^k(U)$ and $\eta \in \Omega^l(U)$ we have*

$$d(\omega \wedge \eta) = d\omega \wedge \eta + (-1)^k \omega \wedge d\eta.$$

Proof. As in Definition 8.6.9 and in the proof of Theorem 8.7.2 we have

$$d(\omega \wedge \eta)(x) = A_{1,k+l} D(\omega \wedge \eta)(x)^\vee, \qquad \text{where} \qquad D(\omega \wedge \eta)(x)^\vee \in \bigwedge^{1,k+l} T_x^* U \tag{8.52}$$

is given by $D(\omega \wedge \eta)(x)^\vee(v_1, \dots, v_{k+l+1}) = D(\omega \wedge \eta)(x)(v_1)(v_2, \dots, v_{k+l+1})$. Accordingly we now consider $D(\omega \wedge \eta)(x)v_1 \in \bigwedge^{k+l} T_x^* U$, for fixed $v_1 \in T_x^* U$, and note that, on account of Proposition 2.7.6,

$$\begin{aligned} D(\omega \wedge \eta)(x)v_1 &= D\omega(x)v_1 \wedge \eta(x) + \omega(x) \wedge D\eta(x)v_1 \\ &= D\omega(x)v_1 \wedge \eta(x) + (-1)^{kl} D\eta(x)v_1 \wedge \omega(x), \end{aligned}$$

the exterior multiplication being anticommutative by Lemma 8.7.1. Thus, we obtain

$$D(\omega \wedge \eta)(x)^\vee = D\omega(x)^\vee \wedge \eta(x) + (-1)^{kl} D\eta(x)^\vee \wedge \omega(x). \tag{8.53}$$

Because it is an exterior product of elements in $\bigwedge^{1,k} T_x^* U$ and $\bigwedge^l T_x^* U$, we have

$$D\omega(x)^\vee \wedge \eta(x) = A_{k,l}(D\omega(x)^\vee \cdot \eta(x)), \qquad \text{with} \qquad D\omega(x)^\vee \cdot \eta(x) \in \bigwedge^{1,k,l} T_x^* U.$$

Formula (8.51) now implies

$$\begin{aligned} &A_{1,k+l}(D\omega(x)^\vee \wedge \eta(x)) = A_{1,k+l} \circ A_{k,l}(D\omega(x)^\vee \cdot \eta(x)) \\ &\quad = A_{k+1,l} \circ A_{1,k}(D\omega(x)^\vee \cdot \eta(x)) = A_{k+1,l}((d\omega)(x) \cdot \eta(x)) = (d\omega)(x) \wedge \eta(x). \end{aligned}$$

Exactly the same arguments with the roles of k and l interchanged apply to the second summand in (8.53); hence we get, using (8.52),

$$d(\omega \wedge \eta) = (d\omega) \wedge \eta + (-1)^{kl} (d\eta) \wedge \omega = (d\omega) \wedge \eta + (-1)^{kl+k(l+1)} \omega \wedge (d\eta). \quad \square$$

Definition 8.7.4. We introduce a standard notation for differential k-forms on an open subset U of $\mathbf{R}^n$. If $x_i : U \to \mathbf{R}$ is the i-th coordinate function, Definition 8.6.9 implies that $dx_i(x) \in T_x^* U$ is constant, equaling the i-th standard basis vector $e_i^* \in \mathbf{R}^{n*}$, for $1 \le i \le n$ and $x \in U$. Formula 8.36 therefore implies that $\omega \in \Omega^k(U)$ can uniquely be written as

$$\omega = \sum_{I \in \mathcal{I}_k} \omega_I \, dx_I \qquad \text{with} \qquad \omega_I \in \Omega^0(U), \qquad x \mapsto \omega(x)(e_I) : U \to \mathbf{R},$$

$$dx_I = dx_{i_1} \wedge \cdots \wedge dx_{i_k} \qquad \text{if} \qquad I = (i_1, \dots, i_k).$$

Furthermore, we define (compare with Definition 8.6.3)

$$\begin{aligned}
e &= (e_1, \dots, e_n) \in (\mathbf{R}^n)^n, \qquad e_{\hat{i}} = (e_1, \dots, \widehat{e_i}, \dots, e_n) \in (\mathbf{R}^n)^{n-1},\\
dx &= dx_1 \wedge \cdots \wedge dx_n \in \Omega^n(U),\\
dx_{\hat{i}} &= dx_1 \wedge \cdots \wedge \widehat{dx_i} \wedge \cdots \wedge dx_n \in \Omega^{n-1}(U).
\end{aligned}$$

❍

Corollary 8.7.5. *For U open in $\mathbf{R}^n$ and $f \in \Omega^0(U)$ a C^1 differential form, one has*

$$df = \sum_{1 \le i \le n} D_i f \, dx_i \in \Omega^1(U).$$

More generally, for ω as in Definition 8.7.4 which is a C^1 differential form,

$$d\omega = \sum_{I \in \mathcal{I}_k} d\omega_I \wedge dx_I = \sum_{I \in \mathcal{I}_k} \sum_{1 \le i \le n} D_i \omega_I \, dx_i \wedge dx_I \in \Omega^{k+1}(U).$$

Here $d\omega_I \in \Omega^1(U)$ is the exterior derivative of $\omega_I \in \Omega^0(U)$.

Proof. According to Definition 8.6.9 one gets $df(x)(e_i) = D_i f(x)$, for $1 \le i \le n$, and this implies the formula for df. Successively using mathematical induction over $k \in \mathbf{N}$, Proposition 8.7.3 and Theorem 8.7.2, one derives $d(dx_I) = 0$, for all $I \in \mathcal{I}_k$. Hence the first equality in the formula for $d\omega$ follows from Proposition 8.7.3, while the second equality follows from the formula for $d\omega_I$. ❑

8.8 Applications of differential forms

Example 8.8.1 (Exterior derivative of differential 1- and $(n-1)$-form). For a C^1 form

$$\omega = \sum_{1 \le j \le n} \omega_j \, dx_j \in \Omega^1(U), \qquad \text{one has} \qquad d\omega = \sum_{1 \le i, j \le n} D_i \omega_j \, dx_i \wedge dx_j.$$

In the notation of Definition 8.7.4 we have $(-1)^{i-1} dx_i \wedge dx_{\hat{i}} = dx$. For $\omega \in \Omega^{n-1}(U)$ we define $(-1)^{i-1}\omega_i = \omega(e_{\hat{i}})$, which yields

$$\omega = \sum_{1 \le i \le n} (-1)^{i-1} \omega_i \, dx_{\hat{i}}, \qquad \text{and so} \qquad d\omega = \Big(\sum_{1 \le i \le n} D_i \omega_i \Big) dx.$$

☆

Example 8.8.2 (Relation between vector fields and differential forms). We apply the results from Example 8.6.5 in order to define $\flat_{n-1}$ and $\flat_1$ acting on vector fields. Thus, to a C^1 vector field $f : U \to \mathbf{R}^n$ we may, on the one hand, assign the C^1 form

$$\flat_{n-1} f = i_f \det \in \Omega^{n-1}(U), \qquad (i_f \det)(v_1, \dots, v_{n-1}) = \det(f \ v_1 \ \cdots \ v_{n-1}),$$

for arbitrary vector fields $v_1, \dots, v_{n-1}$ on U. As in Example 8.6.5, we find

$$\flat_{n-1} f = \sum_{1 \le i \le n} (-1)^{i-1} f_i \, dx_{\hat{i}} \in \Omega^{n-1}(U);$$

so

$$d(\flat_{n-1} f) = \flat_n(\operatorname{div} f) := \operatorname{div} f \, dx \in \Omega^n(U),$$

by the preceding example. Moreover, for $y \in \partial_{i,\pm} I^n$ (see the proof of Stokes' Theorem 8.6.10), we see that

$$\begin{aligned} \phi^*(\flat_{n-1} f)(y)(e_{\hat{i}}) &= (i_f \det)(\phi(y))(D_1\phi(y), \dots, \widehat{D_i\phi(y)}, \dots, D_n\phi(y)) \\ &= \det(f \circ \phi(y) \ D_1\phi(y) \ \cdots \ \widehat{D_i\phi(y)} \ \cdots \ D_n\phi(y)) \\ &= \langle \, f \circ \phi, \ D_1\phi \times \cdots \times \widehat{D_i\phi} \times \cdots \times D_n\phi \, \rangle(y). \end{aligned}$$

Furthermore, for $x \in I^n$,

$$\begin{aligned} &\phi^*(d(\flat_{n-1} f))(x)(e) = \operatorname{div} f(\phi(x)) \, dx(D_1\phi(x), \dots, D_n\phi(x)) \\ &\quad = (\operatorname{div} f) \circ \phi(x) \det(D_1\phi, \dots, D_n\phi)(x) = (\operatorname{div} f) \circ \phi(x) \det D\phi(x). \end{aligned}$$

Thus Gauss' Divergence Theorem 7.8.5 is seen to be a special case of Stokes' Theorem 8.6.10.

On the other hand, to a C^1 vector field $f : U \to \mathbf{R}^n$ we can assign the C^1 form

$$\flat_1 f = \sum_{1 \le i \le n} f_i \, dx_i \in \Omega^1(U).$$

Then, in the notation of Formula (8.4),

$$\begin{aligned} d(\flat_1 f) &= \sum_{1 \le i, j \le n} D_j f_i \, dx_j \wedge dx_i = \sum_{1 \le i < j \le n} (D_j f_i - D_i f_j) \, dx_j \wedge dx_i \\ &= \sum_{1 \le i < j \le n} A f_{ji} \, dx_i \wedge dx_j. \end{aligned}$$

In particular, for $n = 3$,

$$\begin{aligned} d(\flat_1 f) &= (D_1 f_2 - D_2 f_1) \, dx_1 \wedge dx_2 + (D_2 f_3 - D_3 f_2) \, dx_2 \wedge dx_3 \\ &\quad + (D_3 f_1 - D_1 f_3) \, dx_3 \wedge dx_1. \end{aligned}$$

That is

$$d(\flat_1 f) = \flat_2(\text{curl } f),$$

$$\text{because} \qquad \flat_2(g) = g_1\, dx_2 \wedge dx_3 + g_2\, dx_3 \wedge dx_1 + g_3\, dx_1 \wedge dx_2.$$

Consequently, Stokes' Integral Theorem 8.4.4 also is a special case of Stokes' Theorem 8.6.10.

We summarize the situation in $\mathbf{R}^3$ in the following table:

$$\begin{array}{llll} \text{function } g & \text{vector field grad } g & \flat_1(\text{grad } g) & = dg; \\ \text{vector field } f & \text{vector field curl } f & \flat_2(\text{curl } f) & = d(\flat_1 f); \\ \text{vector field } f & \text{function div } f & \flat_3(\text{div } f) & = d(\flat_2 f). \end{array} \tag{8.54}$$

These identifications are valid only in the standard coordinates on $\mathbf{R}^3$ and use the orientation of $\mathbf{R}^3$. We conclude that

$$\flat_2(\text{curl grad } g) = d^2 g = 0, \qquad \flat_3(\text{div curl } h) = d^2(\flat_1 h) = 0;$$

therefore the identities from Formula (8.17) follow from the identity $d^2 = 0$ (see Theorem 8.7.2).

The classical notation for a vector field is $v = v^i \frac{\partial}{\partial x^i}$, and for a differential form $\omega = \omega_I\, dx^I$, where according to the *Einstein summation convention* the summation is carried out over those indices that occur as subscript and superscript. Upon the transition from v to ω the indices ^{i} of the coefficients are lowered to $_I$; hence the notation $\flat$ (a character indicating a half step drop in pitch) for these *"musical" isomorphisms.* ☆

Example 8.8.3 (Special case of Brouwer's Fixed-point Theorem). Assume that U is open in $\mathbf{R}^n$, that $\phi \in C^2(I^n, U)$, and that $K := \phi(I^n)$ has a nonempty interior. Let $g : U \to \partial K$ be a C^2 mapping. Then the restriction $g|_{\partial K}$ of g to ∂K cannot equal the identity on ∂K. This result implies, among other things, that a membrane cannot be retracted onto its boundary without being punctured somewhere.

Indeed, let $x = (x_1, \dots, x_n)$ be the coordinate mapping on $\mathbf{R}^n$ and let $g = (g_1, \dots, g_n)$. Consider the integrals

$$\int_{\partial\phi} x_1\, dx_2 \wedge \cdots \wedge dx_n \qquad \text{and} \qquad \int_{\partial\phi} g_1\, dg_2 \wedge \cdots \wedge dg_n.$$

If $g(x) = x$, for all $x \in \partial K$, then the two integrals are equal. By Stokes' Theorem 8.6.10 and Corollary 8.7.5 it then follows that

$$\text{vol}(K) = \int_\phi dx = \int_\phi dx_1 \wedge \cdots \wedge dx_n = \int_\phi dg_1 \wedge \cdots \wedge dg_n.$$

This results in a contradiction, because the left–hand side does not vanish, whereas the right–hand side does. Indeed, we have $g_i = x_i \circ g = g^* x_i$, for $1 \le i \le n$. By Theorem 8.6.12 we find $dg_i = d(g^* x_i) = g^*(dx_i)$. Thus, for $x \in U$ and for any n-tuple of vectors $v_1, \dots, v_n \in T_x U \simeq \mathbf{R}^n$,

$$\begin{aligned}(dg_1 \wedge \cdots \wedge dg_n)(x)(v_1, \dots, v_n) &= (g^*(dx_1) \wedge \cdots \wedge g^*(dx_n))(x)(v_1, \dots, v_n)\\ &= (dx_1 \wedge \cdots \wedge dx_n)(g(x))(Dg(x)v_1, \dots, Dg(x)v_n)\\ &= \det(Dg(x)v_1 \ \cdots \ Dg(x)v_n) = 0,\end{aligned}$$

since the n vectors $Dg(x)v_1, \dots, Dg(x)v_n$ are elements of the $(n-1)$-dimensional linear subspace $T_{g(x)}(\partial K)$, and therefore linearly dependent.

Now assume K to be a convex open set, and let B be its closure. Every C^2 mapping $f : U \to \mathbf{R}^n$ which maps B into itself has a fixed point in B, in other words, there exists an $x \in B$ with $f(x) = x$. Indeed, if $x \neq f(x)$ for all $x \in B$, one can assign to x the unique point of intersection $g(x)$ with $\partial K = \partial B$ of the half-line from $f(x)$ to x. The mapping $g : B \to \partial K$ thus defined can be extended to a C^2 mapping $g : U \to \partial K$ for an open neighborhood U of B, but this leads to a contradiction with the foregoing. ☆

Example 8.8.4 (Jacobi's identity for minors). Let U and V be open in $\mathbf{R}^n$, let $\phi : U \to V$ be a C^2 mapping and denote the (i, j)-th minor of the matrix of $D\phi(x)$ by $\phi_{ij}(x)$. Then we have *Jacobi's identity for minors*:

$$\sum_{1 \le j \le n} (-1)^j D_j \phi_{ij} = 0 \qquad (1 \le i \le n).$$

Indeed, consider the standard volume form $dy = dy_1 \wedge \cdots \wedge dy_{n-1} \in \Omega^{n-1}(\mathbf{R}^{n-1})$. Fix $1 \le i \le n$ and write $\xi_i = (\phi_1, \dots, \widehat{\phi_i}, \dots, \phi_n) : U \to \mathbf{R}^{n-1}$. In the notation of Example 8.8.1 we have

$$\xi_i^*(dy) = \sum_{1 \le j \le n} \xi_i^*(dy)(e_{\hat{j}})\, dx_{\hat{j}} \in \Omega^{n-1}(U).$$

By Definitions 8.6.7 and 8.6.2,

$$\begin{aligned}\xi_i^*(dy)(e_{\hat{j}}) &= dy(D\xi_i e_1, \dots, \widehat{D\xi_i e_j}, \dots, D\xi_i e_n)\\ &= dy(D_1\xi_i, \dots, \widehat{D_j\xi_i}, \dots, D_n\xi_i) = \det(D_1\xi_i \ \cdots \ \widehat{D_j\xi_i} \ \cdots \ D_n\xi_i) = \phi_{ij}.\end{aligned}$$

Now $d(dy) = 0$, since it belongs to $\Omega^n(\mathbf{R}^{n-1})$, whence

$$0 = \xi_i^*(d(dy)) = d(\xi_i^*(dy)) \in \Omega^n(U)$$

by Theorem 8.6.12. Thus the assertion follows from Corollary 8.7.5 and the identity $dx_j \wedge dx_{\hat{j}} = (-1)^{j-1} dx$. Note that, for $n = 2$, Jacobi's identity takes the well-known form $D_1 D_2 \phi_i - D_2 D_1 \phi_i = 0$, for $1 \le i \le 2$. ☆

8.9 Homotopy Lemma

The formulation of Lemma 8.9.1 below requires a certain amount of preparation.

Let $U \subset \mathbf{R}^n$ be open. Assume $(\Phi^t)_{t\in\mathbf{R}}$ to be a one-parameter group of diffeomorphisms on U with C^1 tangent vector field $v : U \to \mathbf{R}^n$, that is (see Formula (5.21))

$$v(x) = \frac{d}{dt}\Big|_{t=0} \Phi^t(x) \qquad (x \in U).$$

Then we have the induced mapping L_v, the *Lie derivative* in the direction of the vector field v, acting on the C^1 forms $\omega \in \Omega^k(U)$ (see Definition 8.6.7)

$$L_v\omega = \frac{d}{dt}\Big|_{t=0} (\Phi^t)^*\omega \in \Omega^k(U). \tag{8.55}$$

It follows, by Theorem 8.6.12, that for all C^1 forms $\omega \in \Omega^k(U)$ and $\eta \in \Omega^l(U)$, and C^2 forms $\omega \in \Omega^k(U)$, respectively,

$$L_v(\omega \wedge \eta) = (L_v\omega) \wedge \eta + \omega \wedge L_v\eta, \qquad \text{and} \qquad L_v(d\omega) = d(L_v\omega). \tag{8.56}$$

We say that L_v is a *derivation*.

We further introduce i_v, the *contraction* with the vector field v, acting on $\Omega^k(U)$ as follows. If $f \in \Omega^0(U)$, then $i_v f = 0$, and if $1 \le k \le n$,

$$i_v : \Omega^k(U) \to \Omega^{k-1}(U) \qquad \text{with} \qquad (i_v\omega)(v_2, \dots, v_k) = \omega(v, v_2, \dots, v_k). \tag{8.57}$$

By expansion according to the upper row of the determinant $(df \wedge \eta)(v, v_1, \dots, v_k)$, for all $f \in C^1(U)$ and $\eta \in \Omega^k(U)$, where $v_1, \dots, v_k$ are vector fields on U, we see

$$i_v(df \wedge \eta) = (i_v df)\eta - df \wedge i_v\eta. \tag{8.58}$$

We claim that i_v is an antiderivation (see Proposition 8.7.3), that is, for C^0 forms $\omega \in \Omega^k(U)$ and $\eta \in \Omega^l(U)$ we have

$$i_v(\omega \wedge \eta) = i_v\omega \wedge \eta + (-1)^k\omega \wedge i_v\eta. \tag{8.59}$$

Using Formula (8.58) we can prove this by mathematical induction over $k \in \mathbf{N}$ as follows, for $f \in C^1(U)$:

$$\begin{aligned} i_v((df \wedge \omega) \wedge \eta) &= i_v(df \wedge (\omega \wedge \eta)) = i_v(df)\omega \wedge \eta - df \wedge i_v(\omega \wedge \eta) \\ &= i_v(df)\omega \wedge \eta - df \wedge i_v\omega \wedge \eta + (-1)^{k+1} df \wedge \omega \wedge i_v\eta \\ &= i_v(df \wedge \omega) \wedge \eta + (-1)^{k+1}(df \wedge \omega) \wedge i_v\eta. \end{aligned}$$

Lemma 8.9.1. *We have the following* homotopy formula *as identity of linear operators on C^1 forms in $\Omega^k(U)$:*

$$L_v = d \circ i_v + i_v \circ d.$$

In particular, the operators L_v and d commute, because $L_v \circ d = d \circ L_v = d \circ i_v \circ d$.

Note that for $\omega \in \Omega^k(U)$ one has $i_v\omega \in \Omega^{k-1}(U)$, and hence $di_v\omega \in \Omega^k(U)$, while $d\omega \in \Omega^{k+1}(U)$, from which $i_v d\omega \in \Omega^k(U)$.

Proof. d is an antiderivation according to Proposition 8.7.3 and i_v is an antiderivation too according to Formula (8.59). But then $D_v := di_v + i_v d$ is a derivation, since for C^1 forms $\omega \in \Omega^k(U)$ and $\eta \in \Omega^l(U)$,

$$\begin{aligned}
D_v(\omega \wedge \eta) &= d(i_v\omega \wedge \eta + (-1)^k \omega \wedge i_v\eta) + i_v(d\omega \wedge \eta + (-1)^k \omega \wedge d\eta) \\
&= di_v\omega \wedge \eta + (-1)^{k-1} i_v\omega \wedge d\eta + (-1)^k d\omega \wedge i_v\eta + (-1)^{2k}\omega \wedge di_v\eta \\
&\quad + i_v d\omega \wedge \eta + (-1)^k i_v\omega \wedge d\eta + (-1)^{k+1} d\omega \wedge i_v\eta + (-1)^{2k}\omega \wedge i_v d\eta \\
&= D_v\omega \wedge \eta + \omega \wedge D_v\eta.
\end{aligned}$$

Set $k = 0$; application of the chain rule to $t \mapsto f \circ \Phi^t$ then results in the homotopy formula, for $f \in C^1(U)$. Furthermore, for $\omega = df \in \Omega^1(U)$ one then shows, by (8.56) and using $d^2 = 0$ (see Theorem 8.7.2),

$$L_v\omega = d(L_v f) = d(d(i_v f) + i_v(df)) = d(i_v\omega) = D_v\omega. \tag{8.60}$$

Hence we know that L_v and D_v are derivations agreeing on the elements of $\Omega^0(U)$ and $\Omega^1(U)$, therefore they agree on those belonging to $\Omega^k(U)$, for $0 \le k \le n$. ❑

Example 8.9.2. Suppose $\alpha \in \Omega^n(\mathbf{R}^n)$ is a C^1 form that vanishes nowhere and let v be a C^1 vector field on $\mathbf{R}^n$. Then there exists a uniquely determined function $\operatorname{div}_\alpha v$ on $\mathbf{R}^n$, the *divergence of v with respect to the volume form α*, such that

$$L_v\alpha = \operatorname{div}_\alpha v\, \alpha \in \Omega^n(\mathbf{R}^n).$$

The homotopy formula from Lemma 8.9.1 then implies $d(i_v\alpha) = (di_v + i_v d)\alpha = L_v\alpha = \operatorname{div}_\alpha v\,\alpha$. Further, suppose that v is the tangent vector field of the one-parameter group of diffeomorphisms $(\Phi^t)_{t\in\mathbf{R}}$ on $\mathbf{R}^n$, and consider arbitrary $\phi \in C^2([\,0,1\,]^n, \mathbf{R}^n)$. Application of Stokes' Theorem 8.6.10 to $i_v\alpha \in \Omega^{n-1}(\mathbf{R}^n)$ now leads to the following analog of Formula (7.52):

$$\frac{d}{dt}\Big|_{t=0} \int_{\Phi^t\circ\phi} \alpha = \int_\phi \operatorname{div}_\alpha v\,\alpha = \int_{\partial\phi} i_v\alpha.$$

Here the first identity follows from

$$\frac{d}{dt}\Big|_{t=0} \int_{\Phi^t\circ\phi} \alpha = \frac{d}{dt}\Big|_{t=0} \int_\phi (\Phi^t)^*\alpha = \int_\phi \frac{d}{dt}\Big|_{t=0} (\Phi^t)^*\alpha = \int_\phi L_v\alpha. \qquad ☆$$

Example 8.9.3. The theory of differential forms sheds another light on Step IV on differentiation in the third proof of the Change of Variables Theorem in Section 6.13. As in Example 8.8.2, the function $g \circ \Xi^u \det D\Xi^u$ on V corresponds to the form $(\Xi^u)^*(g\,dy) \in \Omega^n(V)$. On account of the homotopy formula we then find, using that $d(g\,dy) \in \Omega^{n+1}(V) = \{0\}$,

$$\begin{aligned}\frac{d}{du}\Big|_{u=0} (\Xi^u)^*(g\,dy) &= L_\xi(g\,dy) = (d \circ i_\xi + i_\xi \circ d)(g\,dy) = d \circ i_\xi(g\,dy) \\ &= d(g\,i_\xi\,dy) = d\Big(\sum_{1\le i\le n} (-1)^{i-1} g\xi_i\,dy_{\hat{i}}\Big) = \operatorname{div}(g\xi)\,dy.\end{aligned} \tag{8.61}$$

Here we applied the formula from (8.37) that $i_\xi\,dy = \sum_{1\le i\le n}(-1)^{i-1}\xi_i\,dy_{\hat{i}} \in \Omega^{n-1}(V)$, and Example 8.8.1.

It is tempting to terminate the computation in (8.61) as soon as one encounters d applied to an $n-1$ form and to invoke Stokes' Theorem; this, however, would lead to a circular argument, because the proof of that theorem uses the Change of Variables Theorem. ☆

Definition 8.9.4. Let U and V be open in $\mathbf{R}^n$ and let Φ_0 and $\Phi_1 : U \to V$ be two C^1 mappings. Then Φ_0 and Φ_1 are said to be C^1 *homotopic* mappings if there exists a C^1 *homotopy* between Φ_0 and Φ_1, that is, a C^1 mapping $\Gamma : I \times U \to V$ satisfying

$$\Gamma(0,\cdot) = \Phi_0, \qquad \Gamma(1,\cdot) = \Phi_1.$$

Now assume that Φ_0 and Φ_1 are C^1-diffeomorphisms. A C^1 homotopy Γ between Φ_0 and Φ_1 is said to be a C^1 *isotopy* if $\Gamma(t,\cdot) : U \to V$ is a C^1 diffeomorphism, for every $t \in I$. ❍

Lemma 8.9.5 (Homotopy Lemma). *Let U and V be open in $\mathbf{R}^n$ and let Φ_0 and $\Phi_1 : U \to V$ be two homotopic C^1 mappings. Then, for $1 \le k \le n$, there exists an operator $H = H_k$, the* homotopy operator, *satisfying*

$$H_k \in \operatorname{Lin}(\Omega^k(V),\,\Omega^{k-1}(U)), \qquad \Phi_1^* - \Phi_0^* = d \circ H_k + H_{k+1} \circ d \qquad \text{on} \qquad \Omega^k(V).$$

Proof. Suppose $\Gamma : I \times U \to V$ is a C^1 homotopy between Φ_0 and Φ_1. Further, let $\iota_t : U \to I \times U$ be given by $\iota_t(x) = (t, x)$. Then $\Gamma \circ \iota_t = \Gamma(t,\cdot) \in C^1(U, V)$, for $t \in I$. Let $\omega \in \Omega^k(V)$. Application of the Fundamental Theorem of Integral Calculus on $\mathbf{R}$ to $t \mapsto (\Gamma \circ \iota_t)^*\omega \in \Omega^k(U)$ (evaluated on k arbitrary vector fields on U, if necessary, to convert it to a scalar function) now gives

$$\Phi_1^*\omega - \Phi_0^*\omega = (\Gamma \circ \iota_1)^*\omega - (\Gamma \circ \iota_0)^*\omega = \int_0^1 \frac{d}{dt}(\Gamma \circ \iota_t)^*\omega\,dt \in \Omega^k(U). \tag{8.62}$$

Next we introduce $\Phi^h : I \times U \to I \times U$ by $\Phi^h(t, x) = (t + h, x)$. Then

$$\left.\frac{d}{dh}\right|_{h=0} \Phi^h(t, x) = (1, 0, \dots, 0) =: e_0 \in \mathbf{R} \times \mathbf{R}^n.$$

Additionally we have $\iota_{t+h} = \Phi^h \circ \iota_t : U \to I \times U$, and so

$$\frac{d}{dt}\iota_t^* = \left.\frac{d}{dh}\right|_{h=0} \iota_{t+h}^* = \iota_t^* \circ \left.\frac{d}{dh}\right|_{h=0} (\Phi^h)^* = \iota_t^* \circ L_{e_0} \quad : \quad \Omega^k(I \times U) \to \Omega^k(U),$$

where L_{e_0} is the Lie derivative in the direction of the vector field e_0 on $I \times U$. Using the homotopy formula from Lemma 8.9.1 in $\Omega^k(V)$ and Theorem 8.6.12, we find

$$\begin{aligned}\frac{d}{dt}(\Gamma \circ \iota_t)^* &= \left(\frac{d}{dt}\iota_t^*\right) \circ \Gamma^* = \iota_t^* \circ L_{e_0} \circ \Gamma^* = \iota_t^* \circ (d \circ i_{e_0} + i_{e_0} \circ d) \circ \Gamma^* \\ &= d \circ (\iota_t^* \circ i_{e_0} \circ \Gamma^*) + (\iota_t^* \circ i_{e_0} \circ \Gamma^*) \circ d \quad : \quad \Omega^k(V) \to \Omega^k(U).\end{aligned} \tag{8.63}$$

Formulae (8.62) and (8.63) then show that H is as desired, if we define $H_k \in \mathrm{Lin}\,(\Omega^k(V),\ \Omega^{k-1}(U))$ by

$$H_k = \int_0^1 (\iota_t^* \circ i_{e_0} \circ \Gamma^*)\, dt. \tag{8.64}$$

❑

For application later on, in the proof of Theorem 8.11.2, we note a consequence of the Homotopy Lemma (compare this result with Step IV on differentiation in the third proof of the Change of Variables Theorem in Section 6.13).

Proposition 8.9.6. *Let U and V be open in $\mathbf{R}^n$ and let Φ_0 and $\Phi_1 : U \to V$ be two homotopic C^1 mappings, by means of $\Gamma \in C^1(I \times U, V)$. Consider a C^1 form $\omega \in \Omega^n(V)$ and a compact set $K \subset U$ such that* $\mathrm{supp}\,(\Gamma(t, \cdot)^*\omega) \subset K$, *for all $t \in I$. Then we have*

$$\int_U \Phi_0^*\omega = \int_U \Phi_1^*\omega.$$

Proof. In view of the Homotopy Lemma 8.9.5 we have $\Phi_1^*\omega - \Phi_0^*\omega = d(H\omega)$, with $H\omega \in \Omega^{k-1}(U)$. Hence Stokes' Theorem 8.6.10 gives

$$\int_U \Phi_1^*\omega - \int_U \Phi_0^*\omega = \int_U d(H\omega) = \int_{\partial U} H\omega = 0,$$

since $\mathrm{supp}(H\omega) \cap \partial U = \emptyset$. Indeed, the condition $\mathrm{supp}\,((\Gamma \circ \iota_t)^*\omega) \subset K$ and Formula (8.64) imply $\mathrm{supp}(H\omega) \subset K$, while $K \cap \partial U = \emptyset$. ❑

8.10 Poincaré's Lemma

Next we prove the following generalization of Lemma 8.2.6: on contractible open sets (see Definition 8.10.1 below), closed and exact differential forms coincide. In addition, this result yields a different proof for the formulae from Lemma 8.2.6.

Definition 8.10.1. An open set $U \subset \mathbf{R}^n$ is said to be *contractible* if there exists a point $x^0 \in U$ such that the mapping $x \mapsto x^0$ on U and the identity mapping on U are C^1 homotopic. In other words, there is a C^1 mapping $\Gamma : I \times U \to U$ such that $\Gamma(0, x) = x^0$ and $\Gamma(1, x) = x$, for all $x \in U$. ❍

Note that a star-shaped open set is contractible. A bounded open subset in $\mathbf{R}^3$ bounded by two concentric spheres of different radii is simply connected (see Definition 8.2.7) but not contractible.

Theorem 8.10.2 (Poincaré's Lemma for differential forms). *Assume $U \subset \mathbf{R}^n$ to be a contractible open set, and $\omega \in \Omega^k(U)$ to be a closed C^1 form, that is, $d\omega = 0$. Then ω is exact on U, that is, there exists a C^1 form $\eta \in \Omega^{k-1}(U)$ with $\omega = d\eta$.*

Assume in particular that U is star-shaped and that $x^0 = 0$, in the definition of being star-shaped. If

$$\omega = \sum_{1 \le i_1 < \cdots < i_k \le n} \omega_{i_1,\dots,i_k}\, dx_{i_1} \wedge \cdots \wedge dx_{i_k}, \tag{8.65}$$

then $\eta = H\omega$ is an example of a C^1 form having this property, where, for $x \in U$,

$$\begin{aligned} H\omega(x) \;=\; & \sum_{1 \le i_1 < \cdots < i_k \le n} \left(\int_0^1 t^{k-1} \omega_{i_1,\dots,i_k}(tx)\, dt \right) \\ & \cdot \sum_{1 \le j \le k} (-1)^{j-1} x_{i_j}\, dx_{i_1} \wedge \cdots \wedge \widehat{dx_{i_j}} \wedge \cdots \wedge dx_{i_k}. \end{aligned}$$

Proof. We apply the Homotopy Lemma 8.9.5. In the case under discussion $\Gamma \circ \iota_t \in C^1(U, U)$, for $t \in I$; in particular, $\Gamma \circ \iota_1$ is the identity on U, while $\Gamma \circ \iota_0$ is the mapping on U with constant value x^0. Consequently,

$$\omega = (\Gamma \circ \iota_1)^* \omega - (\Gamma \circ \iota_0)^* \omega = d(H\omega) + H(d\omega) = d(H\omega),$$

as $d\omega = 0$. Hence we see that $\eta = H\omega$ satisfies $d\eta = \omega$.

If U is star-shaped with $x^0 = 0$, we can choose $\Gamma(t, x) = tx$. For ω as in Formula (8.65) we have

$$\begin{aligned} (\Gamma^* \omega_{i_1,\dots,i_k})(t, x) &= \omega_{i_1,\dots,i_k}(tx), \\ \Gamma^*(dx_{i_j}) = d(\Gamma^* x_{i_j}) &= d(t x_{i_j}) = t\, dx_{i_j} + x_{i_j}\, dt. \end{aligned}$$

It follows that

$$(i_{e_0} \circ \Gamma^*)\omega(t, x)$$

$$= i_{e_0} \sum_{1 \le i_1 < \cdots < i_k \le n} \omega_{i_1,\dots,i_k}(tx) \sum_{1 \le j \le k} (t\, dx_{i_1}) \wedge \cdots \wedge (x_{i_j}\, dt) \wedge \cdots \wedge (t\, dx_{i_k})$$

$$= \sum_{1 \le i_1 < \cdots < i_k \le n} \omega_{i_1,\dots,i_k}(tx) \sum_{1 \le j \le k} (-1)^{j-1} t^{k-1} x_{i_j}\, dx_{i_1} \wedge \cdots \wedge \widehat{dx_{i_j}} \wedge \cdots \wedge dx_{i_k}.$$

Thus we find the desired formula for $H\omega$. ❑

Example 8.10.3 (De Rham cohomology). Differential forms are tools in the study of topological properties of a submanifold in $\mathbf{R}^n$. Here we take no more than a first step towards elaborating this observation.

Let $f \in C^1(V)$, with V a submanifold in $\mathbf{R}^n$. If the values of f do not change under small variations of the point at which the function is evaluated, then f is constant on the connected components of V, see Proposition 1.9.8.(iv); and this is the case if and only if $df = 0$ on V. Consequently, the number of connected components of V equals the dimension of the vector space

$$H^0(V) := \{ f \in \Omega^0(V) \mid df = 0 \}.$$

One dimension higher, an analog of a function f as described above is a function acting on curves, with the property that the values of the function do not change under small variations of the curve along which the function is evaluated. In other words, this now involves a C^1 form $\omega \in \Omega^1(V)$ whose integral along a mapping γ does not change under small variations of γ. Natural differential 1-forms ω are those of the form df with $f \in \Omega^0(V)$, that is, ω is exact; and for these

$$\int_\gamma \omega = \int_\gamma df = f(\text{beginning } \gamma) - f(\text{end } \gamma).$$

This value does not change under variations of γ that leave the end points of $\mathrm{im}(\gamma)$ invariant. Henceforth we shall, for that reason, consider variations with fixed end points. Or, rephrasing, we require $\int_\gamma \omega = 0$ for "small" closed curves γ. And if we choose $\gamma = \partial\phi$, Stokes' Theorem 8.6.10 yields $0 = \int_{\partial\phi} \omega = \int_\phi d\omega$ for "small" surfaces ϕ; and this is true of ω if and only if $d\omega = 0$, that is, if ω is closed. In view of Example 8.8.1, exact forms ω are always closed. As a consequence, the vector space that codifies the information of interest with respect to this problem is the quotient space

$$H^1(V) := \{\omega \in \Omega^1(V) \mid d\omega = 0\} / \{\omega \in \Omega^1(V) \mid \omega = df \text{ for } f \in \Omega^0(V)\}.$$

Continuing in this way, we define $H^k(V)$, the k-th *de Rham cohomology* of V, for $k \in \mathbf{N}_0$, as the quotient vector space

$$H^k(V) := \{\omega \in \Omega^k(V) \mid d\omega = 0\} / \{\omega \in \Omega^k(V) \mid \omega = d\eta \text{ for } \eta \in \Omega^{k-1}(V)\}.$$

The $H^k(V)$ give a measure of the topological complexity of V. Thus, Poincaré's Lemma 8.10.2 asserts that $H^k(V) = (0)$ if V is a contractible open set. On the other hand, $\dim H^d(V) > 0$ if V is a compact orientable submanifold of dimension d without boundary. Indeed, if $\omega = d\eta \in \Omega^d(V)$, then

$$\int_V \omega = \int_{\partial V} \eta = \int_{\emptyset} \eta = 0;$$

but we also have $\text{vol}_d(V) > 0$. ☆

8.11 Degree of mapping

Definition 8.11.1. Let U and V be open in $\mathbf{R}^n$. A mapping $\Phi : U \to V$ is said to be *proper* if the inverse image of every compact set in V is also compact in U (compare with Definition 1.8.5). ❍

We want to prove the following theorem.

Theorem 8.11.2 (Degree of mapping). *Let U and V be open in $\mathbf{R}^n$ and V be connected, let $\Phi : U \to V$ be a proper C^2 mapping. Then there exists an integer* $\deg(\Phi) \in \mathbf{Z}$, *the* degree of the mapping Φ, *with the property that for every C^1 form $\omega \in \Omega^n(V)$ with compact support*

$$\int_U \Phi^*\omega = \deg(\Phi) \int_V \omega. \tag{8.66}$$

We say that $y \in V$ is a *regular value* for Φ if Φ is regular at every point $x \in \Phi^{-1}(\{y\}) \subset U$, that is, if $D\Phi(x) \in \text{Aut}(\mathbf{R}^n)$ (see Definition 3.2.6). A local version of Formula (8.66) in a neighborhood of a regular value is easy to prove; this proof also makes it clear why $\deg(\Phi)$ is an integer. Note that y is a regular value if $y \notin \Phi(U)$, and that, according to Sard's Theorem (see Exercise 6.36), the set of singular values for Φ is a negligible subset of V.

Lemma 8.11.3. *Let the notation be as in Theorem 8.11.2. Let $y \in V$ be a regular value for Φ. Then there exists a neighborhood V_0 of y in V such that Formula (8.66) holds for every ω with compact support contained in V_0.*

Proof. On account of the Submersion Theorem 4.5.2, the set $\Phi^{-1}(\{y\})$ is a submanifold of dimension ≤ 0, and therefore a discrete set in U; and this set is finite (say $m \in \mathbf{N}_0$ elements) owing to Φ being proper. Because of the Local Inverse

Function Theorem 3.2.4 there exist disjoint connected open sets $U_i \subset U$ and an open set $V_0 \subset V$ such that

$$\Phi^{-1}(V_0) = \bigcup_{1\le i\le m} U_i, \qquad \Phi|_{U_i} : U_i \to V_0 \quad C^2 \text{ diffeomorphism} \quad (1 \le i \le m). \tag{8.67}$$

If the support of ω is contained in V_0, the support of $\Phi^*\omega$ is contained in $\Phi^{-1}(V_0)$, thus

$$\int_U \Phi^*\omega = \sum_{1\le i\le m} \int_U (\Phi|_{U_i})^*\omega.$$

But in view of (8.67), the Remark following Definition 8.6.8 yields $\int_U (\Phi|_{U_i})^*\omega = \sigma_i \int_V \omega$, where $\sigma_i = \pm 1$, according as $\det D(\Phi|_{U_i}) \gtrless 0$. Hence we find

$$\deg(\Phi) = \sum_{1\le i\le m} \sigma_i \in \mathbf{Z}. \tag{8.68}$$

In particular, therefore, $\deg(\Phi) = 0$ if $y \notin \Phi(U)$. ❑

Next we show, by deformation arguments and a partition of unity, that the general case of Formula (8.66) can be reduced to the special case from Lemma 8.11.3. To show this we first derive the Isotopy Lemma; here we recall the notion of isotopy from Definition 8.9.4.

Theorem 8.11.4 (Isotopy Lemma). *Assume U to be a connected open set in $\mathbf{R}^n$, and let x^0 and $x^1 \in U$. Then there exists a C^2 isotopy $\Gamma : I \times U \to U$ with $\Gamma(0, x) = x$, for all $x \in U$, and $\Gamma(1, x^0) = x^1$, while outside a fixed compact subset of U the $\Gamma(t, \cdot)$, for all $t \in I$, equal the identity.*

Proof. If the assertion holds for a pair x^0 and $x^1 \in U$, we say these are isotopic. This defines an equivalence relation between the elements of U. We shall demonstrate that every equivalence class is an open set. Then U is a disjoint union of open sets, and in view of the connectedness of U there can only be one such class. As a result, it suffices to show that the assertion of the lemma holds for x^1 lying in a sufficiently small ball $B \subset U$ about x^0.

We may assume that $x^0 = 0$ and $x^1 = (x_1^1, 0, \dots, 0)$, which may require a prior translation and a rotation in $\mathbf{R}^n$. Let $\chi : U \to I$ be a C^2 function which is 1 at x^0 and which vanishes outside $\frac{1}{2}B$. Define, for $x \in U$ and $t \in I$,

$$\Gamma(t, x) = x + t\chi(x)(x^1 - x^0) = (x_1 + t\chi(x)x_1^1, x_2, \dots, x_n) \in U.$$

That is, $\Gamma(0, \cdot)$ is the identity on U; in addition, $\Gamma(t, \cdot)$ is the identity on U outside B; and $\Gamma(1, x^0) = x^1$. We now have

$$D_1\Gamma_1(t, x) = 1 + tD_1\chi(x)x_1^1,$$

while $x \mapsto D_1\chi(x)$ is a bounded function on B. It follows that we can arrange, by taking $|x_1^1|$ small enough, that $x_1 \mapsto \Gamma_1(t, x)$ is monotonically strictly increasing, for all $t \in I$ and $x_2, \dots, x_n$. Consequently, all $\Gamma(t, \cdot)$ are C^2 diffeomorphisms. ❑

Proof. (of Theorem 8.11.2.) Assume $y^0 \in V$ to be a regular value for the mapping $\Phi : U \to V$ and V_0 to be an open neighborhood of y^0, as in Lemma 8.11.3. Application of the Isotopy Lemma 8.11.4 to the connected set V yields, for every $y \in V$, a C^2 diffeomorphism $\Psi_y : V \to V$ such that $\Psi_y(y^0) = y$ and Ψ_y is C^2 isotopic to the identity on V. The collection $\{ \Psi_y(V_0) \mid y \in V \}$ forms an open covering of $\mathrm{supp}(\omega)$. In view of the compactness of $\mathrm{supp}(\omega)$ there exists a C^2 partition of unity $\{ \chi_j \mid 1 \le j \le l \}$ subordinate to this covering. By changing over to $\chi_j\omega$ we may assume the support of ω to be contained in an open set $\Psi_y(V_0)$, for some $y \in V$. Because the identity on V is C^2 isotopic to Ψ_y, the mappings Φ and $\Psi_y \circ \Phi : U \to V$ are C^2 homotopic. Furthermore, it follows from the Isotopy Lemma and the properness of Φ that the condition on the supports in Proposition 8.9.6 is satisfied. Hence, this proposition yields

$$\int_U \Phi^*\omega = \int_U (\Psi_y \circ \Phi)^*\omega = \int_U \Phi^*(\Psi_y^*\omega).$$

The support of $\Psi_y^*\omega$ is contained in V_0, and so we find, by Lemma 8.11.3,

$$\int_U \Phi^*(\Psi_y^*\omega) = \deg(\Phi) \int_V \Psi_y^*\omega.$$

For the diffeomorphism $\Psi_y : V \to V$ one observes that $\det D\Psi_y > 0$, because Ψ_y is C^2 isotopic to the identity on V; hence, by the Remark following Definition 8.6.8,

$$\int_V \Psi_y^*\omega = \int_V \omega.$$

❑

Example 8.11.5 (Degree of polynomial and Fundamental Theorem of Algebra). Let $p : \mathbf{C} \to \mathbf{C}$ be a complex polynomial function of degree n, that is

$$p(z) = \sum_{0 \le k \le n} c_k z^k, \qquad c_k \in \mathbf{C}, \qquad c_n \neq 0.$$

Then the degree of p as a polynomial equals the degree of p considered as a mapping $\mathbf{R}^2 \to \mathbf{R}^2$. This is obvious for $p_0(z) = c_n z^n$, on account of Formula (8.68) and $\det Dp_0(x) = |p_0'(z)|^2 > 0$ (use the Cauchy–Riemann equation), for $z = x_1 + ix_2 \neq 0$; and

$$\Gamma(t, z) = c_n z^n + t \sum_{0 \le k < n} c_k z^k$$

gives a C^∞ homotopy between p_0 and p. An immediate consequence of this is the *Fundamental Theorem of Algebra.* According to this theorem there exists, for every polynomial function $p : \mathbf{C} \to \mathbf{C}$ with positive degree, a $z \in \mathbf{C}$ with $p(z) = 0$.

Indeed, p not surjective implies that the degree of the mapping p equals 0, but then the degree of the polynomial p also equals 0. In fact, p possesses precisely n different roots if 0 is a regular value for p. See Exercises 3.48 and 8.13 for other proofs. ☆

Definition 8.11.6. Assume X and Y are C^k submanifolds in $\mathbf{R}^n$ of dimension d, and let $\Phi : X \to Y$ be a *mapping of manifolds*. We say that Φ is a C^k *mapping* if, for every $x \in X$, there exist C^k embeddings $\phi : U \to \mathbf{R}^n$ and $\psi : V \to \mathbf{R}^n$ with U and V open sets in $\mathbf{R}^d$, for which

$$x \in \operatorname{im}(\phi), \qquad \Phi(x) \in \operatorname{im}(\psi), \qquad \widetilde{\Phi} := \psi^{-1} \circ \Phi \circ \phi : U \to V \quad \text{a } C^k \text{ mapping.}$$

Analogously we say that a bijective mapping $\Phi : X \to Y$ is a C^k *diffeomorphism* if, for every $x \in X$, the mapping $\widetilde{\Phi} : U \to V$ is a C^k diffeomorphism. Let $\omega \in \Omega^k(\mathbf{R}^n)$ be a C^1 form with $\operatorname{supp}(\omega) \cap Y \subset \operatorname{im}(\psi)$, then ω_Y, the restriction of ω to Y, is defined as the differential form $\psi^*\omega \in \Omega^k(V)$. Finally we define $\int_Y \omega_Y := \int_V \psi^*\omega$, the *integral* of ω over the submanifold Y. ❍

Using Lemma 4.3.3 one readily verifies that the choices of ϕ and ψ are irrelevant in the first two definitions above. Furthermore, $\int_Y \omega_Y = \int_V \psi^*\omega$ is independent of the choice of ψ, provided $\det D(\psi^{-1} \circ \widetilde{\psi}) > 0$ for a $\widetilde{\psi}$ with the same properties as ψ.

From the foregoing one immediately derives:

Theorem 8.11.7 (Degree of mapping). *Let W be open in $\mathbf{R}^n$, let X and Y be compact oriented C^2 submanifolds of dimension k, and assume $Y \subset W$ to be connected. Let $\Phi : X \to Y$ be a C^2 mapping. Then there exists an integer* $\deg(\Phi) \in \mathbf{Z}$, *the* degree of the mapping Φ, *with the property that, for every C^1 form $\omega \in \Omega^k(W)$,*

$$\int_X \Phi^*\omega_Y = \deg(\Phi) \int_Y \omega_Y. \tag{8.69}$$

In particular, $\deg(\Phi) \neq 0$ *implies that Φ is surjective. Furthermore,* $\deg(\Phi)$ *is invariant under C^2 homotopy of Φ.*

In the notation of Definition 8.11.6, the compactness of X and Y implies that $\widetilde{\Phi} : U \to V$ is proper. Formula (8.69) now means that, in fact, for $\eta = \psi^*\omega \in \Omega^k(V)$,

$$\int_U \widetilde{\Phi}^*\eta = \deg(\Phi) \int_V \eta.$$

Example 8.11.8 (Hairy sphere of even dimension has a cowlick). There exists a C^2 tangent vector field to S^{n-1} without any zeros if and only if $n \in \mathbf{N}$ is even (see Exercise 6.22 for another proof).

Indeed, assume that $f : S^{n-1} \to \mathbf{R}^n$ is a C^2 mapping with $f(x) \in T_x S^{n-1}$ and $f(x) \neq 0$, for all $x \in S^{n-1}$. Then let $g(x) = \frac{1}{\|f(x)\|} f(x)$, and define

$$\Gamma : I \times S^{n-1} \to S^{n-1} \qquad \text{by} \qquad \Gamma(t, x) = (\cos \pi t)\, x + (\sin \pi t)\, g(x).$$

One sees that indeed we have $\Gamma(t, x) \in S^{n-1}$, since $\|x\| = \|g(x)\| = 1$ and $\langle x, g(x) \rangle = 0$, for $x \in S^{n-1}$. Furthermore, $\Gamma(0, x) = x$ and $\Gamma(1, x) = -x$. Thus $Id|_{S^{n-1}}$ is C^2 homotopic with $-Id|_{S^{n-1}}$, which implies that the two mappings are of the same degree. The degree of $-Id|_{S^{n-1}} : x \mapsto -x$ is $(-1)^n$, because $(-Id)^*\omega = (-1)^n\omega$, with ω as in (8.71). From $(-1)^n = 1$ it follows that n is even.

An example of a tangent vector field f to S^{2n-1} without any zeros is the following:

$$f(x) = (-x_2, x_1, -x_4, x_3, \dots, -x_{2n}, x_{2n-1}) \qquad (x \in S^{2n-1}). \qquad ☆$$

Example 8.11.9 (Winding number and Kronecker's integral). Let $X \subset \mathbf{R}^n$ denote a connected compact and oriented C^2 submanifold of codimension 1 (it can be shown that orientability of X is a consequence of the other conditions), let $a \in \mathbf{R}^n$ and let $\phi : X \to U := \mathbf{R}^n \setminus \{a\}$ be a C^2 mapping. Note that the compact set $Y := \phi(X)$ need not be a manifold. The number $w(Y, a) \in \mathbf{Z}$, the *winding number* of Y with respect to a in $\mathbf{R}^n$, that is, the number of times the set Y winds around a in $\mathbf{R}^n$, is defined as $\deg(\Phi)$ with $\Phi = \pi \circ \phi : X \to S^{n-1}$. Here $\pi : U \to S^{n-1}$ denotes the radial projection with respect to a given by $\pi(x) = \frac{1}{\|x-a\|}(x - a)$. For computing $w(Y, a)$ we have the following generalization of Example 7.9.4, known as *Kronecker's integral*:

$$w(Y, a) = \frac{1}{|S^{n-1}|} \int_Y \frac{1}{\|x - a\|^n} i_{x-a} dx \in \mathbf{Z}. \tag{8.70}$$

Here the integrand is a closed differential $(n-1)$-form on U.

Indeed, we may assume $a = 0$ and we will apply Theorem 8.11.7 with $\omega \in \Omega^{n-1}(U)$ equal to a C^∞ differential form whose restriction to S^{n-1} determines the hypersurface area on S^{n-1}. Formula (7.15) implies that we can take $\omega = i_x dx$, the contraction of $dx \in \Omega^n(U)$ with the vector field $x \mapsto x$ on U, which satisfies

$$\int_{S^{n-1}} \omega = |S^{n-1}|. \tag{8.71}$$

The properties above of ω also can be seen as follows. In the notation of Example 8.8.1,

$$\omega = \sum_{1 \le i \le n} (-1)^{i-1} x_i \, dx_{\hat{i}} \in \Omega^{n-1}(U); \qquad \text{and} \qquad d\omega = n \, dx \in \Omega^n(U).$$

(The second identity also follows from the homotopy formula in Lemma 8.9.1.) Because $S^{n-1} = \partial B^n$, we find by means of Stokes' Theorem and Example 7.9.1

$$\int_{S^{n-1}} \omega = \int_{B^n} d\omega = n\ \mathrm{vol}_n(B^n) = \mathrm{hyperarea}(S^{n-1}) = |S^{n-1}|.$$

Now, on the strength of Example 2.4.8,

$$D\pi(x)v = \frac{1}{\|x\|}v + \text{some multiple of } x \qquad (x \in U,\ v \in \mathbf{R}^n). \tag{8.72}$$

Using the definition of ω, Formula (8.72) and the antisymmetry of dx, we obtain for $x \in U$ and any $(n-1)$-tuple of vectors $v_1, \dots, v_{n-1} \in T_xU \simeq \mathbf{R}^n$

$$(\pi^*\omega)(x)(v_1, \dots, v_{n-1}) = \omega(\pi(x))(D\pi(x)v_1, \dots, D\pi(x)v_{n-1})$$
$$= dx\Big(\frac{1}{\|x\|}x, \frac{1}{\|x\|}v_1, \dots, \frac{1}{\|x\|}v_{n-1}\Big) = \frac{1}{\|x\|^n}\omega(x)(v_1, \dots, v_{n-1}).$$

In other words,

$$\sigma := \pi^*\omega = \frac{1}{\|x\|^n}i_x dx = |S^{n-1}|\flat_{n-1}f \in \Omega^{n-1}(\mathbf{R}^n \setminus \{0\}), \tag{8.73}$$

where $\flat_{n-1}f$ is the differential form associated as in Example 8.8.2 with the Newton vector field $f(x) = \frac{1}{|S^{n-1}|\,\|x\|^n}x$ from Example 7.8.4. A direct computation shows that σ is a closed differential form, contrary to ω. We may prove this also by means of the homotopy formula from Lemma 8.9.1, using that $I : \mathbf{R}^n \to \mathbf{R}^n$ is the tangent vector field of $(\Phi^t)_{t\in\mathbf{R}}$ with $\Phi^t : \mathbf{R}^n \to \mathbf{R}^n$ given by $\Phi^t(x) = tx$. Formula (8.70) now follows from (8.73) since we have, on account of (8.71) and Theorem 8.11.7,

$$\deg(\Phi) = \frac{\deg(\Phi)}{|S^{n-1}|}\int_{S^{n-1}} \omega = \frac{1}{|S^{n-1}|}\int_X \phi^*(\pi^*\omega) = \frac{1}{|S^{n-1}|}\int_X \phi^*\sigma. \tag{8.74}$$

For Y fixed the function $a \mapsto w(Y, a)$ is a continuous function $\mathbf{R}^n \setminus Y \to \mathbf{Z}$, and therefore it is locally constant. So it is constant on the connected components of $\mathbf{R}^n \setminus Y$; and since $\lim_{a\to\infty} w(Y, a) = 0$, we have $w(Y, a) = 0$ for all a in the unbounded component of $\mathbf{R}^n \setminus Y$. ☆

Example 8.11.10 (Special case of Jordan–Brouwer Separation Theorem). Let $V \subset \mathbf{R}^n$ be a connected compact C^2 submanifold of codimension 1 which is orientable. (It can be shown that the last condition is a consequence of the preceding ones.) Then the complement $\mathbf{R}^n \setminus V$ consists of two nonempty disjoint open connected sets, and V is the boundary of both these sets. We will now prove this result.

Let $a \notin V$ and consider $w(V, a)$, the winding number of V with respect to a, that is, in the notation of the preceding example we take $X = Y = V$ and

$\phi = I$. More precisely, we study what happens to Kronecker's integral in (8.70) when a crosses V. Write $a_\pm = \mp\epsilon\, e_1$, for $\epsilon > 0$ arbitrary but fixed. On account of Theorem 4.7.1.(iv) we then may assume that V, locally near the $a_\pm$, is given by $\{\,(0, y) \mid y \in \mathbf{R}^{n-1} \text{ near } 0\,\}$. The main contribution to the difference

$$\sum_\pm \pm w(V, a_\pm) = \frac{1}{|S^{n-1}|}\int_V \sum_\pm \frac{\pm 1}{\|x - a_\pm\|^n} i_{x-a_\pm} dx$$

comes from the $x \in V$ closest to the $a_\pm$, that is the y near 0. In fact, for x far away from the $a_\pm$, the integrand, being a difference, is small and the integral is small too, the integration being over the compact submanifold V. Hence we may as well assume that $V = \{0\} \times \mathbf{R}^{n-1}$, manifestly ignoring the compactness of V. Now, for $x = (0, y) \in V$, we have $x - a_\pm = (\pm\epsilon, y_2, \dots, y_n)^t \in \mathbf{R}^n$ and so

$$\|x - a_\pm\| = (\epsilon^2 + \|y\|^2)^{\frac{1}{2}}, \qquad i_{x-a_\pm} dx(e_2, \dots, e_n) = \det(x - a_\pm \;\; e_2 \;\cdots\; e_n) = \pm\epsilon.$$

We obtain, with the substitution $y = \epsilon x$ in the third term,

$$\sum_\pm \pm w(V, a_\pm) = \sum_\pm \frac{\pm 1}{|S^{n-1}|}\int_V \frac{1}{\|x - a_\pm\|^n} i_{x-a_\pm}\, dx$$

$$= \frac{2\epsilon}{|S^{n-1}|}\int_{\mathbf{R}^{n-1}} \frac{1}{(\epsilon^2 + \|y\|^2)^{\frac{n}{2}}}\, dy = \frac{2}{|S^{n-1}|}\int_{\mathbf{R}^{n-1}} \frac{1}{(1 + \|x\|^2)^{\frac{n}{2}}}\, dx = 1,$$

on the strength of Exercise 7.23. In this case, of V being a manifold, we obtain that the function $a \mapsto w(V, a)$ jumps by ± 1 when a crosses V. (These approximate calculations are rigorous because we are dealing with a $\mathbf{Z}$-valued function.) In turn, this implies that $\mathbf{R}^n \setminus V$ has at least two connected components. Furthermore, suppose l is a ray emanating from $a \notin V$, and let $a' \in l \setminus V$. Suppose that k is the number of times that l intersects V between a and a'. Conclude that $w(V, a) - w(V, a') = k \mod 2$.

Next we prove that there are exactly two connected components. To this end note, analogously to Exercise 7.35.(i), that there exists a number $\delta > 0$ such that

$$\Phi : \,]-\delta, \delta[\, \times V \to \mathbf{R}^n \qquad \text{with} \qquad \Phi(t, x) = x + t\,\nu(x),$$

is a C^2 diffeomorphism onto an open neighborhood V_δ of V in $\mathbf{R}^n$. Here ν is a continuous choice for a normal to V. Then it follows from Theorem 1.9.4 that V_δ is a connected subset of $\mathbf{R}^n$, and this implies that $V_\delta \setminus V = V_+ \cup V_-$, where the $V_\pm$ both are connected subsets of $\mathbf{R}^n$. It now suffices to show that every connected component C of $\mathbf{R}^n \setminus V$ intersects either V_+ or V_-. To this end, consider $x \in \partial C$. If $x \notin V$, then $x \in \mathbf{R}^n \setminus V$, which is open; so x cannot be a boundary point of any connected component of $\mathbf{R}^n \setminus V$. Therefore $\partial C \subset V$, which implies that C must intersect a $V_\pm$.

Observe that of the two connected components of $\mathbf{R}^n \setminus V$, precisely one is unbounded, call it C_0; and the other is bounded, call it C_1. If V is given the outward orientation it gets as ∂C_1, then

$$a \in C_i \qquad \Longleftrightarrow \qquad w(V, a) = i \qquad (0 \le i \le 1).$$

Because V is differentiable, infinitesimally (via the normal to V) one has a ready criterion for deciding at which side of V a given point lies. The winding number of V with respect to a point makes this criterion into a global one. The proof given above can be adapted to the case of V being closed but not compact. The general formulation of the Jordan–Brouwer Separation Theorem is valid for a set $V \subset \mathbf{R}^n$ that is a homeomorphic image of S^{n-1}. The proof must take into account that V may then be much more irregular. ☆

Example 8.11.11 (Number of solutions of an equation). As a further application of Example 8.11.9, we show how the number of solutions of an equation $\phi(x) = 0$, counted with signs, within an open set can be computed by means of Kronecker's integral taken over its boundary.

Let $\Omega \subset \mathbf{R}^n$ be a connected bounded open set for which $\partial\Omega$ is a compact submanifold of dimension $n-1$. Suppose that Ω_0 is an open neighborhood in $\mathbf{R}^n$ of the closure $\overline{\Omega}$ and that $\phi : \Omega_0 \to \mathbf{R}^n$ is a C^2 mapping. Assume that $\phi(x) \neq 0$, for $x \in \partial\Omega$, and that 0 is a regular value for ϕ. Then $\phi^{-1}(\{0\}) \cap \Omega$ is a finite set, say $\{a_i \mid 1 \le i \le m\}$; and $D\phi(a_i) \in \mathrm{Aut}(\mathbf{R}^n)$, for $1 \le i \le m$. If $\pi : \mathbf{R}^n \setminus \{0\} \to S^{n-1}$ denotes the radial projection and $\sigma \in \Omega^{n-1}(\mathbf{R}^n \setminus \{0\})$ is as in Formula (8.73), we have

$$\frac{1}{|S^{n-1}|}\int_{\partial\Omega} \phi^*\sigma = \sum_{1\le i\le m} \mathrm{sgn}\,(\det D\phi(a_i)). \tag{8.75}$$

In fact, $\phi^{-1}(\{0\}) \cap \Omega$ is a discrete set in Ω. If this set were infinite, it would have a cluster point in $\partial\Omega$; by continuity, in this case ϕ would vanish at such a point, contrary to the assumptions. Select an open ball V_0 about 0 of radius $\delta > 0$ such that $V_0 \cap \phi(\partial\Omega) = \emptyset$, and further disjoint connected open neighborhoods $U_i \subset \Omega$ of a_i, with $1 \le i \le m$, for which the conditions in (8.67) are satisfied. Next, introduce $W_i = U_i \cap \phi^{-1}(\frac{1}{2}V_0)$; then the restriction of ϕ to each ∂W_i is a C^2 diffeomorphism from this manifold onto $\{x \in \mathbf{R}^n \mid \|x\| = \frac{\delta}{2}\}$. The set $U := \Omega \setminus \cup_{1\le i\le m}\overline{W_i}$ is open in $\mathbf{R}^n$ and its boundary is the disjoint union of the $(n-1)$-dimensional submanifolds $\partial\Omega$ and ∂W_i, for $1 \le i \le m$. We may assume that ϕ has no zeros on $\Omega_0 \setminus \overline{\Omega}$. Hence the differential $(n-1)$-form $\phi^*\sigma$ on the open neighborhood $\Omega_0 \setminus \{a_i \mid 1 \le i \le m\}$ of U is closed, on account of σ being closed and Theorem 8.6.12. Therefore we may apply Stokes' Theorem to $\int_U \phi^*\sigma$ and Formula (8.74) in order to obtain

$$\frac{1}{|S^{n-1}|}\int_{\partial\Omega} \phi^*\sigma = \deg(\pi \circ (\phi|_{\partial\Omega})) = -\sum_{1\le i\le m} \deg(\pi \circ (\phi|_{\partial W_i})).$$

But the orientation of ∂W_i is the outward one with respect to Ω, while $\phi|_{\partial W_i}$: $W_i \to \{x \in \mathbf{R}^n \mid \|x\| = \frac{\delta}{2}\}$ is a diffeomorphism where the sphere is oriented by the outward normal. Therefore there is an extra minus sign when applying Formula (8.68), and this proves Formula (8.75). Note the similarity with the arguments in Example 7.9.4. ☆

Exercises

Exercises for Chapter 6

Exercise 6.1 (Not Jordan measurable, compact set). Let $\{ r_n \mid n \in \mathbf{N} \}$ be an enumeration of $\mathbf{Q} \cap \,]\, 0, 1\, [$, and let $0 < \epsilon < 1$ be chosen arbitrarily. For each $n \in \mathbf{N}$ we select an open interval $I_n \subset \,]\, 0, 1\, [$ such that $r_n \in I_n$ and $\text{length}(I_n) = \frac{\epsilon}{2^n}$. Now define $A = \bigcup_{n \in \mathbf{N}} I_n$ and $K = [\, 0, 1\,] \setminus A$.

(i) Show that the inner measure of A cannot exceed ϵ.

(ii) Prove that A is a dense subset of $[\, 0, 1\,]$, that is, $\overline{A} = [\, 0, 1\,]$. Conclude that the outer measure of A equals 1.

(iii) Prove that A is a not Jordan measurable, open set in $\mathbf{R}$.

(iv) Prove that K is a compact set in $\mathbf{R}$ which is not Jordan measurable.

Exercise 6.2. Let $B = [\, 0, 1\,] \times [\, 1, 2\,]$. Prove

$$\int_B (x_1 + x_2)^{-2} \, dx = \log \left(\frac{4}{3} \right).$$

Exercise 6.3. Demonstrate that $\frac{2}{3}$ is the area of the bounded set in $\mathbf{R}^2$ bounded by the line $\{ x \in \mathbf{R}^2 \mid x_1 = x_2 \}$ and the parabola $\{ x \in \mathbf{R}^2 \mid x_2^2 = 2x_1 \}$.

Exercise 6.4. Show that the volume of the solid in $\mathbf{R}^3$ under the paraboloid $\{ x \in \mathbf{R}^3 \mid x_1^2 + x_2^2 - x_3 = 0 \}$ and above the square $K = [\, 0, 1\,] \times [\, 0, 1\,]$ equals $\frac{2}{3}$.

Exercise 6.5. Verify that the volume of the bounded solid in $\mathbf{R}^3$ bounded by the parabolic cylinder $\{x \in \mathbf{R}^3 \mid x_1^2 + x_3 = 4\}$ and the planes $\{x \in \mathbf{R}^3 \mid x_1 = 0\}$, $\{x \in \mathbf{R}^3 \mid x_2 = 0\}$, $\{x \in \mathbf{R}^3 \mid x_2 = 6\}$, $\{x \in \mathbf{R}^3 \mid x_3 = 0\}$ equals 32.

Illustration for Exercise 6.6

Exercise 6.6. Prove that the volume of the bounded solid in $\mathbf{R}^3$ bounded by the paraboloid $\{ x \in \mathbf{R}^3 \mid x_1^2 + x_2^2 - x_3 = 0 \}$, the cylinder $\{ x \in \mathbf{R}^3 \mid x_1^2 + x_2^2 = a^2 \}$, for $a > 0$, and the plane $\{ x \in \mathbf{R}^3 \mid x_3 = 0 \}$ equals $\frac{1}{2}\pi a^4$.

Exercise 6.7. Let B be the unit disk in $\mathbf{R}^2$. Prove $\int_B x_1^2 x_2^2 \, dx = \frac{\pi}{24}$.

Exercise 6.8. Let B be the ball of radius R about the origin in $\mathbf{R}^3$. Calculate $\int_B x_1 x_2 x_3 \|x\|^2 \, dx$.
Hint: Do not plunge into the calculation straight away.

Exercise 6.9. Let $B = \{ x \in \mathbf{R}^2 \mid |x_1| \le 1, \ |x_2| \le 1 \}$. Prove $\int_B \|x\|^{-1} \, dx = 4 \log(1 + \sqrt{2})$.

Exercise 6.10. Define $B = \{ x \in \mathbf{R}^3 \mid 1 \le x_1 \le e^{x_3}, \ x_2 \ge x_3, \ x_2^2 + x_3^2 \le 4 \}$. Prove

$$\int_B \frac{1}{x_1} \, dx = \frac{8 - 4\sqrt{2}}{3}.$$

Exercise 6.11. Let $f \in C([\, a, b\,])$, and define $V = \{ x \in \mathbf{R}^n \mid a \le x_1 \le x_2 \le \cdots \le x_n \le b \}$. Show

$$\int_V \prod_{1 \le j \le n} f(x_j) \, dx = \frac{1}{n!} \Big(\int_a^b f(t) \, dt \Big)^n.$$

Exercise 6.12. Let B be a rectangle or a ball in $\mathbf{R}^n$ and suppose $f \in C(B)$. Prove that there exists a point $x_0 \in B$ such that $\int_B f(x) \, dx = f(x_0) \operatorname{vol}_n(B)$.

Exercise 6.13. Let $U \subset \mathbf{R}^2$ be the open disk about 0 and of radius $a > 0$. Prove

$$\int_U \|x\|\, dx = \frac{2\pi a^3}{3}; \qquad \int_U e^{-\|x\|^2}\, dx = \pi(1 - e^{-a^2}).$$

Exercise 6.14 (Needed for Exercise 6.28). Let a and $b > 0$. Using polar coordinates prove that πab is the area of the ellipse $\{\, x \in \mathbf{R}^2 \mid \frac{x_1^2}{a^2} + \frac{x_2^2}{b^2} \le 1 \,\}$.

Exercise 6.15 (Sequel to Exercise 2.73). Now we can give the background for that exercise.

(i) Prove that, for all $a \in \mathbf{R}$,

$$\Big(\int_0^a e^{-x^2}\, dx\Big)^2 = 2\int_0^{\frac{\pi}{4}} \int_0^{\frac{a}{\cos\alpha}} e^{-r^2} r\, dr\, d\alpha = \frac{\pi}{4} - \int_0^{\frac{\pi}{4}} e^{-\frac{a^2}{\cos^2\alpha}}\, d\alpha.$$

(ii) Conclude by means of the substitution $\alpha = \arctan t$ that, for all $a \in \mathbf{R}$,

$$\Big(\int_0^a e^{-x^2}\, dx\Big)^2 + \int_0^1 \frac{e^{-a^2(1+t^2)}}{1+t^2}\, dt = \frac{\pi}{4}.$$

Exercise 6.16. $U \subset \mathbf{R}^2$ denotes the interior of the triangle having vertices $(0, 0)$, $(1, 0)$ and $(0, 1)$. Prove

$$\int_U e^{(x_1 - x_2)/(x_1 + x_2)}\, dx = \frac{\sinh(1)}{2}.$$

Hint: Consider the C^∞ diffeomorphism $\Phi : x \mapsto (x_1 - x_2,\ x_1 + x_2)$, and find the triangle V for which $V = \Phi(U)$.

Exercise 6.17. Assume $0 < a < b$ and let $U = \{\, x \in \mathbf{R}^3 \mid a < \|x\| < b \,\}$. Prove

$$\int_U \|x\|^{-3}\, dx = 4\pi \log\Big(\frac{b}{a}\Big).$$

Exercise 6.18. Prove that 16π is the volume of the open bounded set in $\mathbf{R}^3$ bounded by the paraboloids $\{\, x \in \mathbf{R}^3 \mid x_1^2 + x_2^2 - x_3 = 0 \,\}$ and $\{\, x \in \mathbf{R}^3 \mid x_1^2 + x_2^2 + x_3 = 8 \,\}$.

Exercise 6.19. Let $C = \{x \in \mathbf{R}^3 \mid \|x\|^2 \le 2,\ x_1^2 + x_2^2 \ge 1,\ x_3 \ge 0\}$. Calculate $\int_C \sqrt{3x_3 - x_3^3}\, dx$, by means of substitution of cylindrical coordinates $x = (r\cos\alpha,\ r\sin\alpha,\ x_3)$.

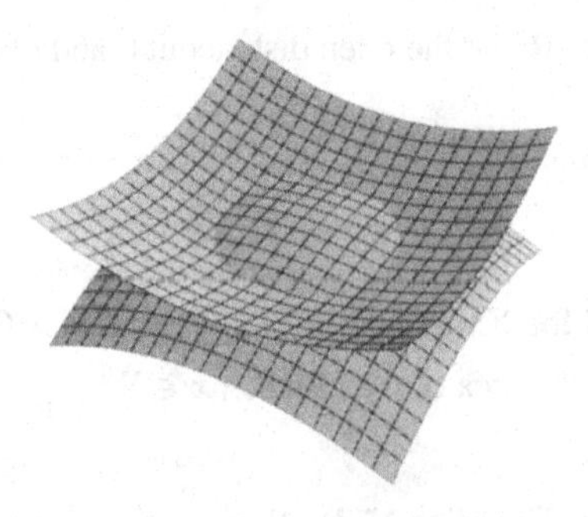

Illustration for Exercise 6.18

Exercise 6.20. Let $U \subset \mathbf{R}^3$ be the open unit ball. Prove, for all $y \in \mathbf{R}^3$,

$$\int_U \cos\langle x, y\rangle\, dx = \frac{4\pi}{\|y\|^2}\Big(\frac{\sin\|y\|}{\|y\|} - \cos\|y\|\Big).$$

Conclude that $\mathrm{vol}_3(U) = \frac{4}{3}\pi$.
Hint: Use the fact that the integral is invariant under rotations acting on y.

Exercise 6.21. Let $U = \{\, x \in \mathbf{R}^2_+ \mid x_1 < 1,\ x_2 < 1\,\}$. Prove

$$\int_U e^{x_1^2+x_2^2-x_1^2x_2^2}\sqrt{1-x_1^2}\, dx = \frac{\pi(e-1)}{4}.$$

Hint: Consider $\Phi(x) = (x_1,\ x_2\sqrt{1-x_1^2})$.

Exercise 6.22 (No C^1 field of unit tangent vectors on even-dimensional sphere – sequel to Exercise 3.27). We use the notation from that exercise. Deduce from the definition of Φ_t that $t \mapsto \det \Phi_t(x)$, for $x \in \mathbf{R}^n$, is a polynomial function, which is strictly positive if $x \in A$ and $|t|$ is sufficiently small. More precisely, show $\det \Phi_t(x) = 1 + \sum_{1\le i\le n} t^i \alpha_i(x)$, where the $\alpha_i \in C(\mathbf{R}^n)$. Conclude by integration over A that

$$\mathrm{vol}_n\,(\Phi_t(A)) = \mathrm{vol}_n(A) + \sum_{1\le i\le n} t^i \int_A \alpha_i(x)\, dx.$$

On the other hand, deduce from Exercise 3.27.(iv) that

$$\mathrm{vol}_n\,(\Phi_t(A)) = (1+t^2)^{\frac{n}{2}}\,\mathrm{vol}_n(A).$$

For $n \in \mathbf{N}$ odd, conclude that a mapping f satisfying all the conditions in Exercise 3.27 does not exist.
Background. We have proved the result from Example 8.11.8 that a hairy sphere of even dimension has a cowlick, that is, on the unit sphere in $\mathbf{R}^n$ there exists a C^1 field of unit tangent vectors if and only if n is even.

Exercise 6.23 (Sequel to Exercise 1.15 – needed for Exercise 8.37). Let $B^n = \{ y \in \mathbf{R}^n \mid \|y\| < 1 \}$, and define $\Psi : B^n \to \mathbf{R}^n$ by $\Psi(y) = (1 - \|y\|^2)^{-1/2}\, y$.

(i) Using Exercise 1.15 prove that $\Psi : B^n \to \mathbf{R}^n$ is a C^∞ diffeomorphism, with inverse $\Phi := \Psi^{-1} : \mathbf{R}^n \to B^n$ given by $\Phi(x) = (1 + \|x\|^2)^{-1/2}\, x$.

(ii) Prove the following equality of functions on B^n:

$$D_j \Psi_i = (1 - \|\cdot\|^2)^{-1/2}\, (\delta_{ij} + \Psi_i \Psi_j).$$

If we regard $x \in \mathbf{R}^n$ as an element of $\mathrm{Mat}(n \times 1, \mathbf{R})$, then $xx^t = (x_i x_j)_{1 \le i, j \le n} \in \mathrm{Mat}(n, \mathbf{R})$.

(iii) Let $A \in \mathbf{O}(n, \mathbf{R})$ (hence $A^t A = I$), let $x \in \mathbf{R}^n$, and write $z = Ax$. Then prove $\det(I + xx^t) = \det(I + zz^t)$, and conclude, making a suitable choice for A, that

$$\det(I + xx^t) = 1 + \|x\|^2 \qquad (x \in \mathbf{R}^n).$$

(iv) Show that, for every continuous function f or g with compact support on $\mathbf{R}^n$ or B^n, respectively,

$$\int_{\mathbf{R}^n} f(x)\, dx = \int_{B^n} f\Big(\frac{1}{(1 - \|y\|^2)^{\frac{1}{2}}}\, y\Big) \frac{dy}{(1 - \|y\|^2)^{\frac{n}{2}+1}},$$

$$\int_{B^n} g(y)\, dy = \int_{\mathbf{R}^n} g\Big(\frac{1}{(1 + \|x\|^2)^{\frac{1}{2}}}\, x\Big) \frac{dx}{(1 + \|x\|^2)^{\frac{n}{2}+1}}.$$

Exercise 6.24. Let $n \ge 2$, let $K \in \mathcal{J}(\mathbf{R}^n)$ and let $\Phi : \mathbf{R}^n \to \mathbf{R}^n$ be the C^1 mapping defined by

$$\begin{aligned} \Phi_1(x) &= x_1 + a_1 && (a_1 \in \mathbf{R}), \\ \Phi_i(x) &= x_i + a_i(x_1, \dots, x_{i-1}) && (1 < i \le n,\ a_i \in C^1(\mathbf{R}^{i-1})). \end{aligned}$$

Prove that $\Phi(K) \in \mathcal{J}(\mathbf{R}^n)$, and that $\mathrm{vol}_n(\Phi(K)) = \mathrm{vol}_n(K)$.

Exercise 6.25 (Solids of revolution – sequel to Exercise 4.6). Let $f \in C([a, b])$ be nonnegative and let $\gamma : [a, b] \to \mathbf{R}^3$ be the curve given by $\gamma(t) = (f(t),\ 0,\ t)$. Let V be the *surface of revolution* in $\mathbf{R}^3$ obtained by revolving $\mathrm{im}(\gamma)$ in $\mathbf{R}^3$ about the x_3-axis, as in Exercise 4.6. Let L be the compact set in $\mathbf{R}^3$ bounded by V and the planes $\{ x \in \mathbf{R}^3 \mid x_3 = a \}$ and $\{ x \in \mathbf{R}^3 \mid x_3 = b \}$.

(i) Prove that L is Jordan measurable in $\mathbf{R}^3$, and that $\mathrm{vol}_3(L) = \pi \int_a^b f(t)^2\, dt$.

Assume $K \subset \mathbf{R}^3$ to be a compact subset in the half-plane $\{ x \in \mathbf{R}^3 \mid x_2 = 0,\ x_1 \geq 0 \}$ and which is Jordan measurable in $\mathbf{R}^2$, if the plane $\{ x \in \mathbf{R}^3 \mid x_2 = 0 \}$ is identified with $\mathbf{R}^2$. Construct, in a way analogous to that in Exercise 4.6, the *solid of revolution* L in $\mathbf{R}^3$ obtained by revolving K in $\mathbf{R}^3$ about the x_3-axis.

(ii) Prove that L is Jordan measurable in $\mathbf{R}^3$ and that $\mathrm{vol}_3(L) = 2\pi \int_K x_1 \, dx_1 \, dx_3$.

Exercise 6.26 (Archimedes' Theorem). Prove that the volumes of an inscribed cone, a half ball, and a circumscribed cylinder, all having the same base plane and radius, are in the ratios $1 : 2 : 3$.

Exercise 6.27. Two cylinders are inscribed inside a half ball in the following fashion. The lower face of one of the cylinders is in the plane face of the half ball, and the circumference of the top face of that cylinder lies on the round surface of the half ball. The lower face of the other cylinder lies in the upper face of the first cylinder, and the circumference of its top face lies on the round surface of the half ball. How should the heights of the two cylinders be chosen for the sum of their volumes to be maximal ?

Exercise 6.28 (From Newton to Kepler – sequel to Exercises 5.24 and 6.14). Let the notation be as in Example 6.6.8 on Kepler's second law. In particular, suppose $t \mapsto x(t)$ is a C^2 curve in $\mathbf{R}^2$ such that the position $x(t)$ and the acceleration $x''(t)$ are linearly dependent vectors, and more precisely, that there exists $0 \neq k \in \mathbf{R}$ such that

$$x''(t) = -k\|x(t)\|^{-3}x(t) = -\operatorname{grad}\Big(-\frac{k}{\|x(t)\|}\Big) \qquad (t \in \mathbf{R}).$$

This corresponds with an *inverse-square law* (for the magnitude of $x''(t)$) as studied by Newton. The acceleration is *centripetal* for $k > 0$, and *centrifugal* for $k < 0$. In the following we often write x instead of $x(t)$ to simplify the notation, and similarly x' and x''. We will determine the geometric properties of the orbit $\{ x(t) \mid t \in \mathbf{R} \}$ and we begin by computing the velocity of the normalized position vector.

(i) Assuming $x \neq 0$, use Example 2.4.8 to prove

$$\begin{aligned}(\|x\|^{-1}x)' &= -\|x\|^{-3}\langle x, x'\rangle x + \|x\|^{-1}x' \\ &= \|x\|^{-3}(-\langle x, x'\rangle x + \langle x, x\rangle x') =: \|x\|^{-3}v.\end{aligned}$$

Show $\langle v, x\rangle = 0$ in order to obtain $v = \lambda Jx$, with $\lambda \in \mathbf{R}$ and $J \in \mathbf{SO}(2, \mathbf{R})$ as in Lemma 8.1.5.

(ii) On account of Example 6.6.8 there exists $l \in \mathbf{R}$ with $l = \det(x\ x') = \langle Jx, x'\rangle = -\langle Jx', x\rangle$. With λ as in part (i), verify $\lambda \langle x, x\rangle = \det(x\ \lambda Jx) = \det(x\ v) = \langle x, x\rangle \det(x\ x') = l\langle x, x\rangle$, and conclude $-\langle x, x'\rangle x + \langle x, x\rangle x' = v = lJx$.

(iii) Combine parts (i) and (ii) to find

$$(\|x\|^{-1}x)' = l\|x\|^{-3}Jx = -\frac{l}{k}Jx''.$$

Integrating once with respect to t, obtain a constant vector $\epsilon \in \mathbf{R}^2$ such that

$$(\star) \qquad \|x(t)\|^{-1}x(t) = \epsilon - \frac{l}{k}Jx'(t) \qquad (t \in \mathbf{R}).$$

Verify that $\|x' + \frac{k}{l}J\epsilon\| = \left|\frac{k}{l}\right|$. In other words, the *hodograph* (ἡ ὁδός = way), the curve traced by the velocity vector x', is the circle of center $-\frac{k}{l}J\epsilon$ and of radius $|\frac{k}{l}|$. Next, take the inner product of the equality $(\star)$ with x and apply (ii) once more to conclude that

$$\|x\| = \langle \epsilon, x\rangle - \frac{l}{k}\langle Jx', x\rangle = \langle \epsilon, x\rangle + \frac{l^2}{k}.$$

Deduce from Exercise 5.24.(iii) that the orbit of x is a (branch of a) conic section with eccentricity vector ϵ and $d = \frac{l^2}{k}$, the sign of which depends on that of k. This is *Kepler's first law*.

(iv) By a translation in $\mathbf{R}$ we may assume $\|x(0)\| = \min\{\,\|x(t)\| \mid t \in \mathbf{R}\,\}$. Deduce from part (iii) and (ii), respectively,

$$\|x(0)\| = \frac{l^2}{k(1+e)} \qquad \text{and} \qquad \|x(0)\|\,\|x'(0)\| = l.$$

(v) According to Exercise 3.47.(i) the total energy per unit mass $H = \frac{\|x'\|^2}{2} - \frac{k}{\|x\|}$ is constant along orbits. Starting from $(\star)$ in part (iii) prove that this constant value equals $H = \frac{k^2}{2l^2}(e^2-1)$. Conclude that the orbit is an ellipse, a parabola, or a hyperbola, as $H < 0$, $H = 0$, or $H > 0$, respectively.

(vi) Assume that $k > 0$, and that the orbit is an ellipse, with semimajor axis a and semiminor axis b. This implies that there exists minimal $T \in \mathbf{R}_+$ with $x(t) = x(t+T)$, for all $t \in \mathbf{R}$. After a time T the area swept out by the radius vector equals the area of the ellipse. Deduce from Example 6.6.8, Exercise 6.14, Exercise 5.24.(ii), and part (iii), successively, that T satisfies

$$\frac{l}{2}T = \pi ab = \pi d^2(1-e^2)^{-\frac{3}{2}}, \qquad \text{or} \qquad \frac{T^2}{a^3} = \frac{4\pi^2}{k},$$

where the constant at the right–hand side is independent of the particular orbit. This is the assertion of *Kepler's third law*. Again on account of Exercise 5.24.(ii), prove $l = (ka(1-e^2))^{\frac{1}{2}}$. Furthermore, show

$$H = -\frac{k}{2a}, \qquad \|x'\|^2 = k\Big(\frac{2}{\|x\|} - \frac{1}{a}\Big), \qquad T = 2\pi k(-2H)^{-\frac{3}{2}}.$$

Demonstrate $\|x'(0)\| = \max\{\,\|x'(t)\| \mid t \in \mathbf{R}\}$ as well as $\|x'(\frac{T}{2})\| = \min\{\,\|x'(t)\| \mid t \in \mathbf{R}\,\}$, and also

$$\|x'(0)\|^2 = \frac{k}{a}\frac{1+e}{1-e}, \qquad \left\|x'\Big(\frac{T}{2}\Big)\right\|^2 = \frac{k}{a}\frac{1-e}{1+e},$$

$$\|x'(0)\|\,\left\|x'\Big(\frac{T}{2}\Big)\right\| = \frac{k}{a} = -2H.$$

Exercise 6.29 (Volumes, Bernoulli and Euler numbers – sequel to Exercise 0.16 – needed for Exercises 6.30 and 6.40). Define $\Theta^n = \{x \in \mathbf{R}^n_+ \mid x_j + x_{j+1} < 1 \ (1 \le j < n)\}$.

(i) Prove

$$\theta_n = \mathrm{vol}_n(\Theta^n) = \int_0^1 \int_0^{1-x_1} \cdots \int_0^{1-x_{n-1}} dx_n \cdots dx_2\, dx_1.$$

For the calculation of this integral we introduce polynomial functions p_k on $\mathbf{R}$ by

$$p_0(x) = 1, \qquad p_k(x) = \int_0^{1-x} p_{k-1}(t)\, dt \quad (k \in \mathbf{N}).$$

(ii) Prove

$$p_n(0) = \theta_n \quad (n \in \mathbf{N}), \qquad p_0(1) = 1, \qquad p_k(1) = 0,$$
$$p_k'(x) = -p_{k-1}(1-x) \qquad (k \in \mathbf{N}).$$

Next, we introduce the *formal power series* in y (that is, without considering convergence)

$$f(x, y) = \sum_{k \in \mathbf{N}_0} p_k(x) y^k.$$

(iii) Prove that f satisfies the differential equation

$$(\star) \qquad \frac{\partial f}{\partial x}(x, y) + y f(1-x, y) = 0, \qquad \text{and} \qquad (\star\star) \qquad f(1, y) = 1.$$

(iv) Prove that $\frac{\partial^2 f}{\partial x^2}(x, y) + y^2 f(x, y) = 0$, and conclude

$$f(x, y) = a(y)\cos(xy) + b(y)\sin(xy).$$

Substitute $x = 0$ into $(\star)$ and prove $b(y) = -1$, and, using $(\star\star)$, show that $a(y) = \tan y + \frac{1}{\cos y}$. Conclude

$$\tan y + \sec y = f(0, y) = 1 + \sum_{n \in \mathbf{N}} p_n(0) y^n = 1 + \sum_{n \in \mathbf{N}} \theta_n y^n.$$

For all $n \in \mathbf{N}$ we find that θ_n equals the coefficient of y^n in the power series expansion of $\tan y + \sec y$. On account of Exercise 0.16.(xiii) we have

$$\tan y = \sum_{n\in\mathbf{N}} t_n y^{2n-1} = \sum_{n\in\mathbf{N}} (-1)^{n-1} 2^{2n}(2^{2n}-1)\frac{B_{2n}}{(2n)!}\, y^{2n-1} \qquad \Big(|y| < \frac{\pi}{2}\Big).$$

Here the B_{2n} are the *Bernoulli numbers*. Moreover, we have

$$\sec y = \sum_{n\in\mathbf{N}_0} \frac{E_n}{(2n)!}\, y^{2n} \qquad \Big(|y| < \frac{\pi}{2}\Big).$$

Here the E_n are known as the *Euler numbers*. One has

$$E_0 = 1, \qquad E_1 = 1, \qquad E_2 = 5, \qquad E_3 = 61, \qquad E_4 = 1385, \qquad \dots$$

(v) Deduce, for all $n \in \mathbf{N}$,

$$\theta_{2n-1} = t_n = (-1)^{n-1} 2^{2n}(2^{2n}-1)\frac{B_{2n}}{(2n)!}, \qquad \theta_{2n} = \frac{E_n}{(2n)!}.$$

Exercise 6.30 (Sequel to Exercise 6.29). We use the notation from that exercise. The formulae from part (v) of that exercise can be proved in a somewhat more direct way.

(i) Prove, for $i \in \mathbf{N}_0$ and $x \in \mathbf{R}$,

$$p_{i+2}(x) = \int_0^1 p_{i+1}(t)\,dt - \int_{1-x}^1 p_{i+1}(t)\,dt$$

$$= p_{i+2}(0) - \int_0^x p_{i+1}(1-t)\,dt = p_{i+2}(0) - \int_0^x \int_0^t p_i(s)\,ds\,dt.$$

(ii) Use part (i) and mathematical induction on $n \in \mathbf{N}_0$ to show, for $x \in \mathbf{R}$,

$$p_{2n+1}(x) = \sum_{0\le i\le n} p_{2n+1-2i}(0)\frac{(-1)^i}{(2i)!}x^{2i} - \frac{(-1)^n}{(2n+1)!}x^{2n+1}.$$

Conclude with $p_{2n+1}(1) = 0$ that

$$\sum_{0\le i\le n} p_{2n+1-2i}(0)\frac{(-1)^i}{(2i)!} = \frac{(-1)^n}{(2n+1)!}.$$

(iii) Now derive the following identity of formal power series:

$$\Big(\sum_{n\in\mathbf{N}_0} p_{2n+1}(0)x^{2n+1}\Big)\Big(\sum_{n\in\mathbf{N}_0}\frac{(-1)^n}{(2n)!}x^{2n}\Big)=\sum_{n\in\mathbf{N}_0}\frac{(-1)^n}{(2n+1)!}x^{2n+1}.$$

That is,

$$\sum_{n\in\mathbf{N}}\theta_{2n-1}x^{2n-1}=\sum_{n\in\mathbf{N}}p_{2n-1}(0)x^{2n-1}=\tan x=\sum_{n\in\mathbf{N}}t_n x^{2n-1}.$$

(iv) Now discuss the case of the function sec.

Exercise 6.31. Let $(\Psi^t)_{t\in\mathbf{R}}$ be the one-parameter group of C^∞ diffeomorphisms of $\mathbf{R}$ given by $\Psi^t(x)=e^{2t}x$, set $L=[\,0,1\,]$ and suppose $f\in C^1(\mathbf{R})$. Prove

$$\frac{d}{dt}\int_{\Psi^t(L)} f(x)\,dx = 2e^{2t}f(e^{2t})\qquad (t\in\mathbf{R})$$

by means of the chain rule, the substitution $x=e^{2t}y$, and the transport equation from Example 6.6.9, respectively.

Exercise 6.32 (Special case of Urysohn's Lemma – needed for Exercise 6.33). Let $K\subset\mathbf{R}^n$ be a compact set, let $U\subset\mathbf{R}^n$ be an open set, and assume $K\subset U$.

(i) Prove that a continuous function $f:\mathbf{R}^n\to[\,0,1\,]$ exists such that $f(x)=1$, for $x\in K$, and $\operatorname{supp}(f)\subset U$.
Hint: Consider the open covering $\{U\}$ of K.

(ii) Verify that, for given $a,b\in\mathbf{R}$ with $a<b$, the function $a+(b-a)f:\mathbf{R}^n\to[\,a,b\,]$ has the value b on K, and a on $\mathbf{R}^n\setminus U$.

Exercise 6.33 (Special case of Tietze's Extension Theorem – sequel to Exercise 6.32). Let $K\subset\mathbf{R}^n$ be a compact set, assume $a,b\in\mathbf{R}$ with $a<b$, and let $f:K\to[\,a,b\,]$ be a continuous function. Then there exists a continuous function $F:\mathbf{R}^n\to[\,a,b\,]$ such that $F|_K=f$.

(i) Replace f by $(f-a)/(b-a)$ and conclude that one may assume $[\,a,b\,]=[\,0,1\,]$.

Let $t=\frac{1}{3}$. We assert that there exists a sequence $(f_k)_{k\in\mathbf{N}}$ of continuous functions on $\mathbf{R}^n$ such that

$$f_k:\mathbf{R}^n\to[\,0,\,t\,(2t)^{k-1}\,];\qquad 0\le f(x)-\sum_{1\le i\le k}f_i(x)\le(2t)^k\qquad(x\in K).$$

(ii) Let $A = f^{-1}([\,0, t\,])$ and $B = f^{-1}([\,2t, 1\,])$. Prove by using Exercise 6.32 that there exists a continuous function $f_1 : \mathbf{R}^n \to [\,0, t\,]$ such that $f_1(x) = 0$ for $x \in A$, and $f_1(x) = t$ for $x \in B$. Verify that $0 \le f(x) - f_1(x) \le 2t$, for $x \in K$.

(iii) Now prove, by induction on $k \in \mathbf{N}$, that one can find f_k as above, such that $f_k = 0$ on the set where $f - \sum_{1 \le i < k} f_i(x) \le t\,(2t)^{k-1}$, and that $f_k = t\,(2t)^{k-1}$ on the set where $f - \sum_{1 \le i < k} f_i(x) \ge (2t)^k$.

(iv) Prove that $F = \sum_{k \in \mathbf{N}} f_k$ satisfies the requirements.

Exercise 6.34 (Every closed set is zero-set – needed for Exercise 6.37).

(i) Let $g \in C(\mathbf{R}^n)$. Prove that $C = N(0) = \{\, x \in \mathbf{R}^n \mid g(x) = 0 \,\}$ is a closed subset of $\mathbf{R}^n$.

We shall now prove the converse of the assertion in part (i) in five steps. Let C be an arbitrary closed subset of $\mathbf{R}^n$.

(ii) Prove that there exists a countable collection $\{B_k\}_{k \in \mathbf{N}}$ of open balls $B_k \subset \mathbf{R}^n \setminus C$ with

$$\mathbf{R}^n \setminus C = \bigcup_{k \in \mathbf{N}} B_k.$$

(iii) Show that for every $k \in \mathbf{N}$ there exists $g_k \in C^\infty(\mathbf{R}^n)$ with the properties (see the proof of Theorem 6.7.4) $g_k \ge 0$, and $g_k(x) > 0$ if and only if $x \in B_k$.

(iv) Verify that for every $k \in \mathbf{N}_0$ the number $m_k = \sup\{\, |D^\alpha g_k(x)| \in \mathbf{R} \mid x \in \mathbf{R}^n,\ \alpha \in \mathbf{N}_0^n,\ |\alpha| \le k \,\}$ is well-defined.

(v) Prove that, for every $\alpha \in \mathbf{N}_0^n$, the series $\sum_{k \in \mathbf{N}} \frac{1}{m_k\, 2^k} D^\alpha g_k$ converges uniformly on $\mathbf{R}^n$; and use termwise differentiation to conclude that $g \in C^\infty(\mathbf{R}^n)$, if

$$g = \sum_{k \in \mathbf{N}} \frac{1}{m_k\, 2^k}\, g_k.$$

(vi) Verify that $C = \{\, x \in \mathbf{R}^n \mid g(x) = 0 \,\}$.

Exercise 6.35. Let K and K_δ, for $\delta > 0$, be as in Lemma 6.8.1. Let $f \in C(\mathbf{R}^n)$ and assume $N(f) \cap K = \emptyset$. Prove that a number $\delta > 0$ exists with $N(f) \cap K_\delta = \emptyset$.

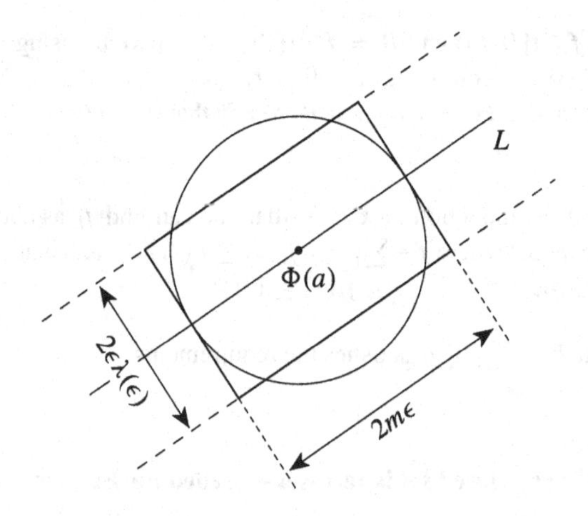

Illustration for Exercise 6.36: Sard's Theorem

Exercise 6.36 (Sard's Theorem – sequel to Exercise 2.27 – needed for Exercise 6.37). The following is an (easy) version of this theorem. Let $U \subset \mathbf{R}^n$ be open and let $\Phi : U \to \mathbf{R}^n$ be a C^1 mapping, then

$$\mathrm{vol}_n(\{\, \Phi(x) \mid x \in U \text{ is singular point for } \Phi \,\}) = 0.$$

We will prove this in a number of steps.

(i) Verify that it is sufficient to prove the assertion for every rectangle B in $\mathbf{R}^n$ with $B \subset U$.

Let B be a fixed rectangle thus chosen.

(ii) Prove that a number $m > 0$ exists such that $\|\Phi(x) - \Phi(a)\| \leq m\|x - a\|$, for all $x, a \in B$.

Now define

$$K = \{\, a \in B \mid \Phi \text{ is singular at } a \,\}.$$

Let $a \in K$ be fixed for the moment. Because $D\Phi(a)(\mathbf{R}^n)$ is a proper linear subspace of $\mathbf{R}^n$, there exists an affine submanifold L of codimension 1 in $\mathbf{R}^n$ going through $\Phi(a)$, such that, for all $x \in \mathbf{R}^n$,

$$\Phi(a) + D\Phi(a)(x - a) \in L.$$

Now let $\epsilon > 0$ be chosen arbitrarily.

(iii) For every $x \in B$ with $\|x - a\| \leq \epsilon$, prove that $\Phi(x)$ belongs to

the ball in $\mathbf{R}^n$ about $\Phi(a)$ of radius $m\epsilon$;

the tubular neighborhood in $\mathbf{R}^n$ about L of half thickness $\epsilon\lambda(\epsilon)$,

where λ is as in Exercise 2.27.(ii).

(iv) Conclude for such x that $\Phi(x)$ is contained in a rectangular parallelepiped in $\mathbf{R}^n$ which can be written as a product of a cube in L of dimension $n-1$ having edges of length $2m\epsilon$, and of an interval in $\mathbf{R}$ of length $2\epsilon\lambda(\epsilon)$. Then prove

$$\text{vol}_n(\Phi\{\, x \in B \mid \|x-a\| \leq \epsilon\,\}) \leq (2m\epsilon)^{n-1}\,2\epsilon\lambda(\epsilon).$$

Note that the constants on the right–hand side of this inequality are independent of $a \in K$.

(v) Verify that the number of balls in $\mathbf{R}^n$ of radius ϵ about $a \in K$ (where the point a now is considered to be variable), required to cover K is of the order $\mathcal{O}(\epsilon^{-n})$, $\epsilon \downarrow 0$; and conclude that

$$\text{vol}_n\,(\Phi(K)) = \mathcal{O}(\lambda(\epsilon)), \qquad \epsilon \downarrow 0.$$

(vi) Why has Sard's Theorem now been proved ?

Exercise 6.37 (Functional dependence – sequel to Exercises 6.34 and 6.36). (See also Exercise 4.33.) Let $U \subset \mathbf{R}^n$ be open and let $\Phi \in C^1(U,\ \mathbf{R}^n)$. We assume that the rank of $D\Phi(x)$ is lower than n, for all $x \in U$.

(i) Prove by Exercise 6.36 that $\text{vol}_n\,(\,\text{im}(\Phi)) = 0$.

(ii) Let $C \subset U$ be a compact subset. Conclude by Exercise 6.34 that a submersion $g \in C^\infty(\mathbf{R}^n)$ exists such that $g \circ \Phi(x) = 0$, for all $x \in C$.

(iii) Next, take $n = 2$ and assume $D\Phi(x)$ has constant rank 1, for all $x \in U$. Prove that, locally on U, one of the two component functions of Φ can always be written as a C^1 function of the other one.

Exercise 6.38 (Volume of a neighborhood of a parallelepiped). Define $N = \{\,1, 2, \dots, n\,\}$. Consider linearly independent vectors $v_j \in \mathbf{R}^n$, for $j \in N$, and let $P = \{\,\sum_{1\leq j\leq n} t_j\, v_j \mid 0 \leq t_j \leq 1\,\}$ be the parallelepiped spanned by the v_j. Suppose $\delta > 0$ and write

$$P_\delta = \{\, y \in \mathbf{R}^n \mid \text{ there exists } x \in P \text{ with } \|y-x\| \leq \delta\}$$

as in Lemma 6.8.1. Then, with the notation $\#K$ for the number of elements in $K \subset N$,

$$\text{vol}_n(P_\delta) = \sum_{0\leq m\leq n}\Big(\sum_{K\subset N,\,\#K=n-m} \text{vol}_{n-m}(P^K)\Big)\, c_m\delta^m.$$

Here P^K denotes the $(n-m)$-dimensional parallelepiped spanned by the $n-m$ of the v_k satisfying $k \in K$,

$$(\star) \qquad \text{vol}_{n-m}(P^K) = (\,\det(\langle v_k,\, v_{k'}\rangle)_{k,\,k'\in K})^{1/2},$$

while c_m is the m-dimensional volume of the unit ball in $\mathbf{R}^m$ (see, for instance, Exercise 6.50.(viii)), which makes the factor $c_m\,\delta^m$ equal to the volume of an m-dimensional ball of radius δ. The details of the following proof are left for the reader to check.

Proof. Let $x \in \mathbf{R}^n$. Because P is a closed subset of $\mathbf{R}^n$, there exists $p \in P$ such that $\|x - p\| \le \|x - p'\|$, for every $p' \in P$. The convexity of P implies that we have a strict inequality here if $p' \ne p$. In other words, the element $p = \pi(x)$ is unique, hence we have a mapping $\pi : \mathbf{R}^n \to P$, the projection to the nearest point in P.

The relative interiors of the faces of P are the sets $F_{I,J,K}$ of the form

$$\{ \sum_{1\le j\le n} t_j\, v_j \mid j \in I \Rightarrow t_j = 0;\ \ j \in J \Rightarrow t_j = 1;\ \ j \in K \Rightarrow 0 < t_j < 1 \},$$

where (I, J, K) denotes any partition of N into three disjoint subsets. Now suppose $p \in F_{I,J,K}$, then $y := x - p \in \mathbf{R}^n$ has the following properties.

$$\text{(i) } i \in I \quad \Rightarrow \quad -\langle y, v_i\rangle = \frac{d}{dt}\Big|_{t=0} \frac{1}{2}\|x - (p + t\,v_i)\|^2 \ge 0;$$

$$\text{(ii) } j \in J \quad \Rightarrow \quad \langle y, v_j\rangle = \frac{d}{dt}\Big|_{t=0} \frac{1}{2}\|x - (p - t\,v_j)\|^2 \ge 0;$$

$$\text{(iii) } k \in K \quad \Rightarrow \quad \langle y, v_k\rangle = \frac{d}{dt}\Big|_{t=0} \frac{1}{2}\|x - (p + t\,v_k)\|^2 = 0.$$

Conversely, a convexity argument yields that (i), (ii) and (iii) imply that $p = \pi(x)$. Condition (iii) means that y belongs to C_K, the orthogonal complement of the v_k, for $k \in K$, which is a linear subspace of $\mathbf{R}^n$ of dimension equal to $m = n - \#K$. The conditions (i) and (ii) then describe a polyhedral cone $C_{I,J,K}$ in C_K. The condition $x \in P_\delta$ now corresponds to the condition $\|y\| < \delta$. Write $B_K(\delta)$ for the open ball in C_K with center at the origin and radius equal to δ. The fact that y is perpendicular to the face $F_{I,J,K}$ then implies that the n-dimensional volume of $\pi^{-1}(F_{I,J,K}) \cap P_\delta$ is equal to the $\#K$-dimensional volume of $F_{I,J,K}$ times the m-dimensional volume of $C_{I,J,K} \cap B_K(\delta)$. Furthermore, the $\#K$-dimensional volume of $F_{I,J,K}$ is equal to $(\star)$, which is independent of the partition (I, J) of $N \setminus K$. If, for given $K \subset N$, we let (I, J) run over all partitions of $N \setminus K$ into two disjoint subsets, then the union of the $C_{I,J,K}$ is equal to C_K. On the other hand, the intersection of $C_{I,J,K}$ and $C_{I',J',K}$, if $(I, J) \ne (I', J')$, is contained in a linear subspace of C_K of positive codimension, and therefore has m-dimensional volume equal to 0. It follows that the sum of the m-dimensional volumes of the sets $C_{I,J,K} \cap B_K(\delta)$, over all partitions (I, J) of $N \setminus K$, is equal to the m-dimensional volume of $B_K(\delta)$, which in turn is $c_m\,\delta^m$. ❑

Exercise 6.39 (zeta function and dilogarithm – sequel to Exercise 3.20 – needed for Exercises 6.40 and 7.6). As in Exercise 0.25 *Riemann's zeta function* is defined by $\zeta(s) = \sum_{k\in\mathbf{N}} k^{-s}$, for $s \in \mathbf{C}$ with $\operatorname{Re} s > 1$.

(i) Demonstrate that one has, with $U = \,]\,0, 1\,[\,^2 \subset \mathbf{R}^2$,

$$\begin{aligned}\frac{3\zeta(2)}{4} &= \sum_{n\in\mathbf{N}} \frac{1}{n^2} - \sum_{n\in\mathbf{N}} \frac{1}{(2n)^2} = \sum_{n\in\mathbf{N}_0} \frac{1}{(2n+1)^2} = \sum_{n\in\mathbf{N}_0} \Big(\int_0^1 x^{2n}\,dx\Big)^2 \\ &= \sum_{n\in\mathbf{N}_0} \int_0^1 \int_0^1 (x_1 x_2)^{2n}\,dx_1 dx_2 = \int_U \frac{1}{1 - x_1^2 x_2^2}\,dx.\end{aligned}$$

(ii) Deduce by means of parts (i) and (ii) of Exercise 3.20 that $\zeta(2) = \frac{\pi^2}{6} = 1.644\,934\,066\,848\cdots$.

(iii) Perform one of the integrations in the double integral in part (i) and demonstrate by partial fraction decomposition that

$$\int_0^1 \frac{1}{x} \log\Big(\frac{1+x}{1-x}\Big)\,dx = \frac{\pi^2}{4}.$$

This result can also be obtained by series expansion of the integrand according to powers of x, leading to

$$\int_0^1 \frac{1}{x} \log\Big(\frac{1+x}{1-x}\Big)\,dx = 2 \sum_{n\in\mathbf{N}_0} \frac{1}{(2n+1)^2}.$$

And series expansion also gives

$$\int_0^1 \frac{\log(1-x)}{x}\,dx = -\frac{\pi^2}{6}.$$

Background. The function $\mathrm{Li}_2 : \,]-\infty, 1\,[\, \to \mathbf{R}$ with

$$\mathrm{Li}_2(x) = -\int_0^x \frac{\log(1-t)}{t}\,dt$$

is called the *dilogarithm*. Thus $\mathrm{Li}_2(1) = \frac{\pi^2}{6}$, and one has $\mathrm{Li}_2(x) = \sum_{n\in\mathbf{N}} \frac{x^n}{n^2}$, for $|x| \le 1$. By means of differentiation one obtains the following *functional equation for the dilogarithm*:

$$\mathrm{Li}_2(x) + \mathrm{Li}_2(1-x) + (\log x)(\log(1-x)) = \frac{\pi^2}{6} \qquad (0 < x < 1).$$

(iv) Prove in a similar way (recall that $\arctan x = \sum_{n\in\mathbf{N}_0} (-1)^n \frac{x^{2n+1}}{2n+1}$)

$$\begin{aligned}\sum_{n\in\mathbf{N}_0} \frac{(-1)^n}{(2n+1)^2} &= \int_U \frac{1}{1 + x_1^2 x_2^2}\,dx = \int_0^1 \frac{\arctan x}{x}\,dx \\ &= 0.915\,965\,594\,177\cdots.\end{aligned}$$

The integral on the right–hand side cannot be calculated by antidifferentiation; its value is known as *Catalan's constant* $\mathbf{G}$.

Background. We have the function $\mathrm{Ti}_2 : \mathbf{R} \to \mathbf{R}$ with

$$\mathrm{Ti}_2(x) = \int_0^x \frac{\arctan t}{t}\, dt.$$

So $\mathrm{Ti}_2(1) = \mathbf{G}$, and $\mathrm{Ti}_2(x) = \sum_{n\in\mathbf{N}_0} (-1)^n \frac{x^{2n+1}}{(2n+1)^2}$, for $|x| \leq 1$. Furthermore, the *Clausen function* $\mathrm{Cl}_2 : \mathbf{R} \to \mathbf{R}$, and the *Lobachevsky function* $Л : \mathbf{R} \to \mathbf{R}$ are defined by

$$\mathrm{Cl}_2(x) = -\int_0^x \log\left|2\sin\frac{1}{2}t\right| dt = \sum_{n\in\mathbf{N}} \frac{\sin nx}{n^2}, \qquad Л(x) = -\int_0^x \log|2\sin t|\, dt.$$

Therefore $Л(x) = \frac{1}{2}\mathrm{Cl}_2(2x)$. Substitution of $x = \pi$ and $x = \frac{\pi}{2}$ in $\mathrm{Cl}_2(x)$ gives

$$\int_0^{\frac{\pi}{2}} \log(\sin t)\, dt = -\frac{\pi}{2}\log 2 \qquad \text{and} \qquad \mathrm{Cl}_2\left(\frac{\pi}{2}\right) = \mathbf{G}.$$

(v) Conclude that $\displaystyle\int_U \frac{1}{1 - x_1^4 x_2^4}\, dx = \frac{\mathbf{G}}{2} + \frac{\pi^2}{16}.$

(vi) Prove as in part (i)

$$\begin{aligned}\int_U \frac{1}{1 - x_1 x_2}\, dx &= \int_U \sum_{n\in\mathbf{N}_0} (x_1 x_2)^n\, dx = \sum_{n\in\mathbf{N}_0} \left(\int_0^1 x^n\, dx\right)^2 \\ &= \sum_{n\in\mathbf{N}_0} \frac{1}{(n+1)^2} = \zeta(2).\end{aligned}$$

Furthermore, show by means of integration by parts in the third equality

$$\begin{aligned}\int_U \frac{\log x_1 x_2}{1 - x_1 x_2}\, dx &= 2\int_U \sum_{n\in\mathbf{N}_0} x_1^n x_2^n \log x_1\, dx \\ &= 2\sum_{n\in\mathbf{N}_0} \int_0^1 x_1^n \log x_1\, dx_1 \int_0^1 x_2^n\, dx_2 = -2\zeta(3).\end{aligned}$$

Exercise 6.40 ($\zeta(2n)$ and Euler's series – sequel to Exercises 3.20, 6.29 and 6.39 – needed for Exercise 6.52).

(i) Prove, as in Exercise 6.39.(i), with $\square^n = \,]0,1[^{\,n} \subset \mathbf{R}^n$,

$$\left(1 - \frac{1}{2^{2n}}\right)\zeta(2n) = \int_{\square^{2n}} \frac{1}{1 - x_1^2 \cdots x_{2n}^2}\, dx,$$

$$\sum_{k\in\mathbf{N}_0} \frac{(-1)^k}{(2k+1)^{2n+1}} = \int_{\square^{2n+1}} \frac{1}{1 + x_1^2 \cdots x_{2n+1}^2}\, dx.$$

Define

$$v_n = \mathrm{vol}_n(\Upsilon^n), \qquad \Upsilon^n = \{\, y \in \mathbf{R}^n_+ \mid y_1{+}y_2 < 1,\ y_2{+}y_3 < 1, \dots,\ y_n{+}y_1 < 1 \,\}.$$

(ii) Conclude by Exercise 3.20.(iii) that

$$\zeta(2n) = \frac{\pi^{2n}}{2^{2n}-1}\, v_{2n}, \qquad \sum_{k\in\mathbf{N}_0} \frac{(-1)^k}{(2k+1)^{2n+1}} = \Big(\frac{\pi}{2}\Big)^{2n+1} v_{2n+1}.$$

Let $a = \frac{1}{2}(1, 1, \dots, 1) \in \overline{\Upsilon^n}$. The $2n$-tope Υ^n has n-fold symmetry about the axis $\mathbf{R}a$; and Υ^n is the union of n congruent pyramids, each having a as its apex, and as basis the intersection of $\overline{\Upsilon^n}$ with the j-th coordinate plane for $1 \le j \le n$, respectively. Indeed, let Γ^n be the collection of the $x \in \Upsilon^n$ satisfying $x_n < x_j$, for $1 \le j < n$. In view of the cyclic symmetry in our problem we have

$$v_n = n\, \mathrm{vol}_n(\Gamma^n).$$

Note that

$$\Gamma^n = \{\, x \in \mathbf{R}^n_+ \mid x_n < x_j \ (1 \le j < n),\ x_j + x_{j+1} < 1 \ (1 \le j \le n-2) \,\}.$$

Indeed, the two "missing" equations for $x \in \Gamma^n$ are a consequence of the other

$$x_{n-1} + x_n < x_{n-1} + x_{n-2} < 1, \qquad x_n + x_1 < x_2 + x_1 < 1.$$

Define $p : \mathbf{R}^n \to \mathbf{R}^{n-1}$ as the projection onto the first $n-1$ coordinates, and $\Psi : \mathbf{R}^n \to \mathbf{R}^n$ by

$$\Psi(y) = (1 - y_n)\, p(y) + y_n\, a.$$

Now introduce the $(2n-1)$-tope $\Theta^n = \{\, y \in \mathbf{R}^n_+ \mid y_j + y_{j+1} < 1 \ (1 \le j < n) \,\}$.

(iii) Verify $\Psi : \Theta^{n-1} \times \square \to \Gamma^n$ is a C^∞ diffeomorphism and $\det D\Psi(y) = \frac{1}{2}(1 - y_n)^{n-1}$. Conclude that

$$\mathrm{vol}_n(\Gamma^n) = \frac{1}{2n}\, \mathrm{vol}_{n-1}(\Theta^{n-1}).$$

(iv) Now prove by Exercise 6.29.(v)

$$v_{2n} = (-1)^{n-1} 2^{2n-1} (2^{2n} - 1) \frac{B_{2n}}{(2n)!}, \qquad v_{2n+1} = \frac{1}{2} \frac{E_n}{(2n)!};$$

and derive from this (cf. Exercise 0.20)

$$\zeta(2n) = (-1)^{n-1} \frac{1}{2} (2\pi)^{2n} \frac{B_{2n}}{(2n)!},$$

$$\sum_{k\in\mathbf{N}_0} \frac{(-1)^k}{(2k+1)^{2n+1}} = \frac{1}{2} \Big(\frac{\pi}{2}\Big)^{2n+1} \frac{E_n}{(2n)!}.$$

The series on the left–hand side of the second identity are known as *Euler's series*.

Example. For $1 \le n \le 4$ and $0 \le n \le 3$ the sums of the series above are, respectively,

$$\frac{\pi^2}{6}, \qquad \frac{\pi^4}{90}, \qquad \frac{\pi^6}{945}, \qquad \frac{\pi^8}{9450},$$

$$\frac{\pi}{4}, \qquad \frac{\pi^3}{32}, \qquad \frac{5\pi^5}{1536}, \qquad \frac{61\pi^7}{184\,320}.$$

(v) Verify that the results above can be summarized as follows:

$$\sum_{k\in\mathbf{N}_0} \frac{(-1)^{nk}}{(2k+1)^n} = \upsilon_n \Big(\frac{\pi}{2}\Big)^n \qquad (n \in \mathbf{N}).$$

Here each $\upsilon_n \in \mathbf{Q}$ is found as the volume of an n-dimensional convex polytope Υ^n with rational vertices. We have found by direct computation of $\mathrm{vol}_n\, \Upsilon^n$

$$\sum_{n\in\mathbf{N}} \upsilon_n\, t^{n-1} = \frac{1}{2}(\sec t + \tan t) \qquad \Big(|t| < \frac{\pi}{2}\Big).$$

(vi) From part (iv), derive the following estimates:

$$2\frac{1}{(2\pi)^{2n}} < \frac{|B_{2n}|}{(2n)!} \le \frac{\pi^2}{3}\frac{1}{(2\pi)^{2n}}, \qquad \frac{4}{3}\Big(\frac{2}{\pi}\Big)^{2n+1} < \frac{E_n}{(2n)!} < 2\Big(\frac{2}{\pi}\Big)^{2n+1}.$$

(vii) Verify that the solid generated by symmetrizing Υ^3 with respect to the origin is a *rhombic dodecahedron* (δώδεκα = twelve). That is, it is a dodecahedron whose faces are congruent rhombi having diagonals of lengths $\sqrt{2}$ and 1, respectively. Verify that its volume equals 2.

Exercise 6.41 (Another proof for $\int_{\mathbf{R}} e^{-x^2}\,dx = \sqrt{\pi}$). (Compare with Exercises 2.73 and 6.50.(i).) From Example 6.10.8 we know

$$\Big(\int_{\mathbf{R}_+} e^{-x^2}\,dx\Big)^2 = \int_{\mathbf{R}_+}\int_{\mathbf{R}_+} e^{-(x^2+y^2)}\,dy\,dx.$$

(i) Introduce the new variable t via $y = xt$, and conclude that

$$\Big(\int_{\mathbf{R}_+} e^{-x^2}\,dx\Big)^2 = \int_{\mathbf{R}_+}\int_{\mathbf{R}_+} e^{-x^2(1+t^2)}x\,dt\,dx = \int_{\mathbf{R}_+}\int_{\mathbf{R}_+} e^{-x^2(1+t^2)}x\,dx\,dt.$$

(ii) Now finish the proof, using $\int_{\mathbf{R}_+} e^{-x^2(1+t^2)}x\,dx = \frac{1}{2(1+t^2)}$.

Exercise 6.42 (Probability density, expectation vector and covariance matrix of distribution – needed for Exercises 6.43, 6.44, 6.51, 6.96, 6.97). In *stochastics* a function $f : \mathbf{R}^n \to \mathbf{R}$ is said to be the *probability density of a distribution on* $\mathbf{R}^n$ if

$$f(x) \geq 0 \quad (x \in \mathbf{R}^n), \qquad \int_{\mathbf{R}^n} f(x)\,dx = 1.$$

In the case of convergence of the following integrals, the vector $\mu \in \mathbf{R}^n$ and the matrix $C \in \mathrm{Mat}(n, \mathbf{R})$ given by

$$\mu_j = \int_{\mathbf{R}} x_j f(x)\,dx \qquad (1 \leq j \leq n),$$

$$C_{ij} = \int_{\mathbf{R}^n} (x_i - \mu_i)(x_j - \mu_j) f(x)\,dx \qquad (1 \leq i,\ j \leq n)$$

are said to be the *expectation vector* and the *covariance matrix* of that distribution, respectively.

(i) Prove that $C \in \mathrm{Mat}^+(n, \mathbf{R})$ is a positive semidefinite matrix.

In the case where $n = 1$, the vector $\mu \in \mathbf{R}$ is said to be the *expectation* and the number $C \geq 0$ is known as the *variance* of the distribution; the usual notation then is σ^2 instead of C. The number $\sigma \geq 0$ itself is known as the *standard deviation* of the distribution.

(ii) Now set $n = 1$, and verify $\sigma^2 = \int_{\mathbf{R}} x^2 f(x)\,dx - \mu^2$.

Exercise 6.43 (Sequel to Exercise 6.42). Let $\mu \in \mathbf{R}$ and $\sigma > 0$. Let $f = f(\mu, \sigma) \in C^\infty(\mathbf{R})$ be defined by

$$f(x) = \frac{1}{\sigma\sqrt{2\pi}}\, e^{-\frac{(x-\mu)^2}{2\sigma^2}}.$$

Demonstrate

$$\int_{\mathbf{R}} f(x)\,dx = 1, \qquad \int_{\mathbf{R}} (x - \mu) f(x)\,dx = 0, \qquad \int_{\mathbf{R}} (x - \mu)^2 f(x)\,dx = \sigma^2.$$

Hint: The last identity follows by differentiation with respect to σ of the first identity.

Background. In the terminology of Exercise 6.42 the function $f(\mu, \sigma)$ is said to be the *probability density of the normal distribution* on $\mathbf{R}$ with *expectation* μ and *variance* σ^2.

Exercise 6.44 (Sequel to Exercise 6.42 – needed for Exercises 6.92 and 6.93). Let $Q \in \mathrm{Mat}^+(n, \mathbf{R})$ be positive definite. Let E be the ellipsoid, and B the unit ball in $\mathbf{R}^n$,

$$E = \{\, x \in \mathbf{R}^n \mid \langle Qx, x\rangle \leq 1 \,\}, \qquad \text{and} \qquad B = \{\, x \in \mathbf{R}^n \mid \|x\| \leq 1 \,\}.$$

(i) Prove that E is Jordan measurable in $\mathbf{R}^n$, and that $\mathrm{vol}_n(E) = \frac{\mathrm{vol}_n(B)}{\sqrt{\det Q}}$.
Hint: See Formula (2.29).

(ii) Prove

$$\int_{\mathbf{R}^n} e^{-\frac{1}{2}\langle Qx, x\rangle}\, dx = \frac{(2\pi)^{\frac{n}{2}}}{\sqrt{\det Q}}.$$

Background. Now replace Q by C^{-1} in the expression above, for a symmetric positive–definite matrix $C \in \mathrm{Mat}(n, \mathbf{R})$. In the terminology of Exercise 6.42 the function on $\mathbf{R}^n$

$$x \mapsto \frac{1}{(2\pi)^{\frac{n}{2}}\sqrt{\det C}}\, e^{-\frac{1}{2}\langle C^{-1}x, x\rangle}$$

is said to be the *probability density of the normal distribution* on $\mathbf{R}^n$ with the origin as expectation vector and C as covariance matrix (the latter property will be proved in Exercise 6.93).

Exercise 6.45. Let $u \in \mathbf{R}$ be arbitrary, define $U \subset \mathbf{R}^2$ and $\Psi : \mathbf{R}^2 \to \mathbf{R}^2$ by

$$U = \{\, x \in \mathbf{R}^2 \mid x_1 + x_2 < u \,\} \qquad \text{and} \qquad \Psi(y) = (y_1,\ y_2 - y_1).$$

(i) Find $V \subset \mathbf{R}^2$ such that $\Psi : V \to U$ is a C^∞ diffeomorphism.

Assume f and g belong to $C_c(\mathbf{R})$.

(ii) Prove

$$\int_U f(x_1)g(x_2)\, dx = \int_{-\infty}^{u} \int_{\mathbf{R}} f(y_1)g(y_2 - y_1)\, dy_1\, dy_2.$$

(iii) Now assume $f(x) = e^{-x^2}$ for $x \in \mathbf{R}$. Prove

$$\int_U f(x_1)f(x_2)\, dx = \sqrt{\pi} \int_{-\infty}^{u/\sqrt{2}} f(x)\, dx.$$

Hint: The present function f is absolutely Riemann integrable over $\mathbf{R}$, but does not have a compact support. Verify that the result from (ii) is valid for this function f.

Exercise 6.46. Define the C^∞ function $f_p : \mathbf{R}^4 \setminus \{0\} \to \mathbf{R}$ by $f_p(x) = \|x\|^p$, for $p \in \mathbf{R}$. Calculate the values of p for which f_p is absolutely Riemann integrable over $U = \{\, x \in \mathbf{R}^4 \mid 0 < \|x\| < 1 \,\}$ and $V = \{\, x \in \mathbf{R}^4 \mid \|x\| > 1 \,\}$, respectively; for these values of p, calculate the integrals of f_p over U and V, respectively.

Exercise 6.47. Define in $\mathbf{R}^3$ the sets $V = \{ x \in \mathbf{R}^3 \mid x_3 \geq a \}$, where $a \in \mathbf{R}_+$, and $B(R) = \{ x \in \mathbf{R}^3 \mid \|x\| \leq R \}$, where $R > a$.

(i) Calculate $\int_{B(R)\cap V} \frac{1}{\|x\|^6}\,dx$ and $\int_V \frac{1}{\|x\|^6}\,dx$.

The set $W \subset \mathbf{R}^3$ is defined by $W = \{ x \in \mathbf{R}^3 \mid x_1 + x_2 + x_3 \geq 1 \}$.

(ii) Calculate $\int_W \frac{1}{\|x\|^6}\,dx$.
Hint: The answer can be very easily found by using the result of $\int_V \frac{1}{\|x\|^6}\,dx$; why?

Exercise 6.48 (Sequel to Exercise 5.51). Let $V \subset \mathbf{R}^3$ be the pseudosphere from Exercise 5.51. Prove that the volume of the set in $\mathbf{R}^3$ which contains the x_3-axis and which is bounded by V equals $\frac{2}{3}\pi$.

Exercise 6.49 (Sequel to Exercise 3.8 – needed for Exercise 8.35). Suppose $f \in C^1(\mathbf{R}^2)$ has compact support, and let $i = \sqrt{-1}$. In four steps we now prove that

$$(\star) \qquad \int_{\mathbf{R}^2} \frac{1}{z}\frac{\partial f}{\partial \bar{z}}(x)\,dx := \int_{\mathbf{R}^2} \frac{1}{x_1 + i\,x_2}\frac{1}{2}(D_1 f(x) + i\,D_2 f(x))\,dx = -\pi f(0).$$

(i) Prove the convergence of the integral in $(\star)$ by means of $\frac{1}{|x_1 + i\,x_2|} = \frac{1}{\|x\|}$, for $x \neq 0$.

Let $\Psi(r, \alpha) = r(\cos\alpha,\ \sin\alpha)$, for $(r, \alpha) \in [\,0,\ \infty\,[\, \times \mathbf{R}$, and let $\widetilde{f} = f \circ \Psi$.

(ii) Prove by Exercise 3.8.(iii) that the integral in $(\star)$ equals

$$\frac{1}{2}\int_{\mathbf{R}_+}\int_{-\pi}^{\pi}\Big(\frac{\partial \widetilde{f}}{\partial r} + \frac{i}{r}\frac{\partial \widetilde{f}}{\partial \alpha}\Big)(r, \alpha)\,d\alpha\,dr.$$

(iii) Prove

$$\int_{\mathbf{R}_+}\int_{-\pi}^{\pi}\frac{\partial \widetilde{f}}{\partial r}(r, \alpha)\,d\alpha\,dr = \int_{-\pi}^{\pi}\int_{\mathbf{R}_+}\frac{\partial \widetilde{f}}{\partial r}(r, \alpha)\,dr\,d\alpha = -2\pi f(0, 0).$$

(iv) Prove $(\star)$ by means of the 2π-periodicity of $\alpha \mapsto \widetilde{f}(r, \alpha)$.

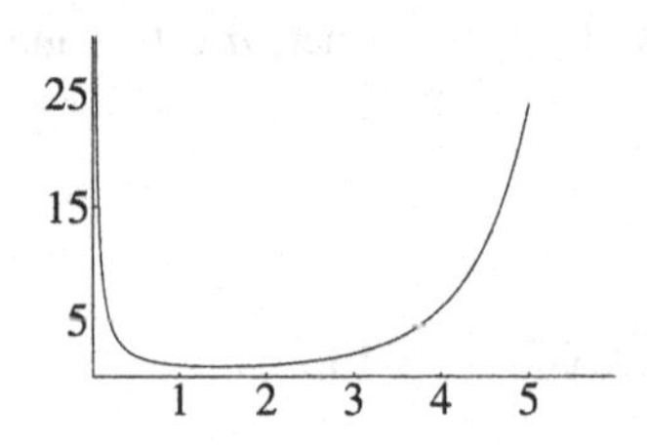

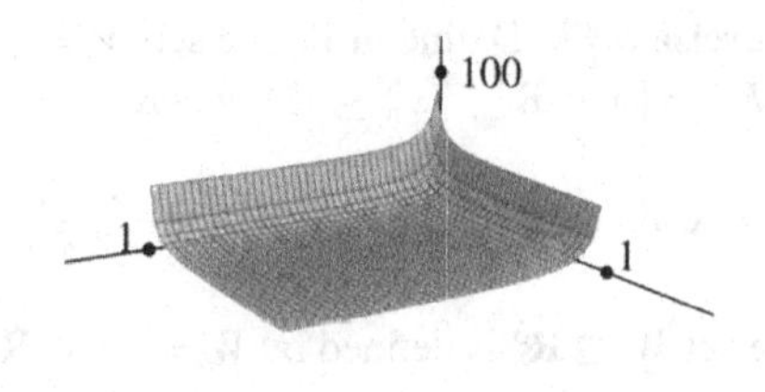

Illustration for Exercise 6.50: Euler's Gamma and Beta functions
Gamma function, and Beta function on $]0, 1[\times]0, 1[$

Exercise 6.50 (Euler's Gamma and Beta functions – sequel to Exercise 2.79 – needed for Exercises 6.51, 6.52, 6.53, 6.55, 6.56, 6.57, 6.58, 6.59, 6.60, 6.63, 6.64, 6.65, 6.69, 6.89, 6.96, 6.98, 6.104, 7.5, 7.21, 7.23, 7.24 and 7.28). Define the *Gamma function* $\Gamma : \mathbf{R}_+ \to \mathbf{R}$ by

$$\Gamma(p) = \int_{\mathbf{R}_+} e^{-t}\, t^{p-1}\, dt.$$

(i) Prove

$$\Gamma(p+1) = p\,\Gamma(p) \qquad (p \in \mathbf{R}_+), \qquad \Gamma(n+1) = n! \qquad (n \in \mathbf{N}_0).$$

Further, compare with Example 6.10.8 and Exercises 2.73, 6.15 and 6.41,

$$\Gamma(p) = 2\int_{\mathbf{R}_+} e^{-u^2}\, u^{2p-1}\, du; \qquad \text{in particular} \qquad \Gamma\Big(\frac{1}{2}\Big) = \sqrt{\pi}.$$

(ii) Show, for p_1 and $p_2 \in \mathbf{R}_+$,

$$\Gamma(p_1)\Gamma(p_2) = \Gamma(p_1+p_2)\, 2\int_0^{\frac{\pi}{2}} \cos^{2p_1-1}\alpha\, \sin^{2p_2-1}\alpha\, d\alpha.$$

Now define $\mathrm{B} : \mathbf{R}_+^2 \to \mathbf{R}$, the *Beta function*, by

$$\mathrm{B}(p_1, p_2) = \frac{\Gamma(p_1)\Gamma(p_2)}{\Gamma(p_1+p_2)}.$$

(iii) Prove

$$\int_0^{\frac{\pi}{2}} \cos^{p_1}\alpha\, \sin^{p_2}\alpha\, d\alpha = \frac{1}{2}\mathrm{B}\Big(\frac{p_1+1}{2}, \frac{p_2+1}{2}\Big) \qquad (p_1, p_2 \in \mathbf{R}_+).$$

(iv) Show, using the substitution $t = \frac{u}{u+1}$, for p_1 and $p_2 \in \mathbf{R}_+$,

$$\begin{aligned} \mathrm{B}(p_1, p_2) &= \int_0^1 t^{p_1-1}(1-t)^{p_2-1}\,dt = \int_{\mathbf{R}_+} \frac{u^{p_1-1}}{(1+u)^{p_1+p_2}}\,du \\ &= 2\int_{\mathbf{R}_+} \frac{v^{2p_1-1}}{(1+v^2)^{p_1+p_2}}\,dv. \end{aligned}$$

(v) Setting $p_1 = \frac{m}{n}$ and $p_2 = \frac{1}{2}$, and substituting $t = u^n$, prove

$$\int_0^1 \frac{u^{m-1}}{\sqrt{1-u^n}}\,du = \frac{\sqrt{\pi}\,\Gamma(\frac{m}{n})}{n\,\Gamma(\frac{m}{n}+\frac{1}{2})} \qquad (m,\ n \in \mathbf{N}).$$

(vi) Prove

$$\mathrm{B}(p, p) = 2^{1-2p}\,\mathrm{B}\Big(p, \frac{1}{2}\Big),$$

and in addition *Legendre's duplication formula* (see Exercise 6.53 for a different proof)

$$\Gamma(2p) = 2^{2p-1}\,\pi^{-1/2}\,\Gamma(p)\Gamma\Big(p+\frac{1}{2}\Big) \qquad (p \in \mathbf{R}_+).$$

Conclude, using part (v) (see Exercise 3.44), that

$$\frac{\varpi}{2} := \int_0^1 \frac{1}{\sqrt{1-t^4}}\,dt = \frac{1}{4\sqrt{2\pi}}\Gamma\Big(\frac{1}{4}\Big)^2.$$

(vii) Prove

$$\Gamma\Big(n+\frac{1}{2}\Big) = \frac{(2n)!\sqrt{\pi}}{2^{2n}\,n!} \qquad (n \in \mathbf{N}_0).$$

Conclude that

$$\sum_{0\le k\le n} \frac{(-1)^k}{2k+1}\binom{n}{k} = \int_0^1 (1-u^2)^n\,du = \frac{(n!)^2\,2^{2n}}{(2n+1)!} \qquad (n \in \mathbf{N}).$$

Prove

$$\int_0^\pi \cos^{2n}\alpha\,d\alpha = \int_0^\pi \sin^{2n}\alpha\,d\alpha = \frac{1\cdot 3\cdots(2n-1)}{2\cdot 4\cdots 2n}\pi =: \frac{(2n-1)!!}{(2n)!!}\pi.$$

Show by expanding the logarithm in Exercise 2.79 in a power series in $x\cos\alpha$ and integrating term-by-term

$$\arcsin x = x + \sum_{n\in\mathbf{N}} \frac{(2n-1)!!}{(2n)!!}\,\frac{x^{2n+1}}{2n+1} \qquad (|x| < 1).$$

(viii) Let $B^n = \{ x \in \mathbf{R}^n \mid \|x\| \le 1 \}$. Prove

$$\mathrm{vol}_n(B^n) = \frac{\pi^{\frac{n}{2}}}{\Gamma(\frac{n}{2}+1)} = \begin{cases} \dfrac{\pi^k}{k!}, & n = 2k; \\ \dfrac{2^{2k}\,\pi^{k-1}\,k!}{(2k)!}, & n = 2k-1. \end{cases}$$

Hint: Apply mathematical induction and the formula

$$\mathrm{vol}_{n+1}(B^{n+1}) = \int_{-1}^{1} \mathrm{vol}_n(\text{ball of radius}\sqrt{1-h^2})\,dh.$$

Exercise 6.51 (Sequel to Exercises 6.42 and 6.50 – needed for Exercises 6.83 and 7.30). In *statistical mechanics* the following integrals play a role. Let $a \in \mathbf{R}_+$, then

$$\int_{\mathbf{R}_+} x^n e^{-ax^2}\,dx = \begin{cases} \left(\dfrac{n-1}{2}\right)!\,\dfrac{1}{2(\sqrt{a})^{n+1}}, & n \text{ odd}; \\ \dfrac{n!\,\sqrt{\pi}}{(\frac{n}{2})!\,(2\sqrt{a})^{n+1}}, & n \text{ even}. \end{cases}$$

Prove these formulae in the following two ways.

(i) Using parts (i) and (vii) of Exercise 6.50.

(ii) Consider the functions on $\mathbf{R}_+$

$$a \mapsto \int_{\mathbf{R}_+} x\,e^{-ax^2}\,dx = \frac{1}{2a} \qquad \text{and} \qquad a \mapsto \int_{\mathbf{R}_+} e^{-ax^2}\,dx = \frac{1}{2}\sqrt{\frac{\pi}{a}},$$

respectively, and apply the Differentiation Theorem 2.10.13.

Let $n \in \mathbf{N}$. In the terminology of Exercise 6.42 the function $f_n : \mathbf{R} \to \mathbf{R}$ with

$$f_n(x) = \begin{cases} 0, & x \le 0; \\ 2\dfrac{x^{n-1}e^{-\frac{1}{2}x^2}}{2^{\frac{n}{2}}\Gamma(\frac{n}{2})}, & x > 0, \end{cases}$$

is said to be the *probability density of the* χ*-distribution with n degrees of freedom.*

(iii) Verify that $\int_{\mathbf{R}} f_n(x)\,dx = 1$, for $n \in \mathbf{N}$.

In particular, the function $f_3(x) = \sqrt{\frac{2}{\pi}}\,x^2 e^{-\frac{1}{2}x^2}$, for $x \in \mathbf{R}_+$, occurs in *kinetic gas theory.*

Exercise 6.52 (Sequel to Exercises 6.40 and 6.50). *Riemann's zeta function* is defined by $\zeta(s) = \sum_{k\in\mathbf{N}} k^{-s}$, for $s \in \mathbf{C}$ with $\mathrm{Re}\, s > 1$, as in Exercise 0.25.

(i) Prove, for $\mathrm{Re}\, s > 1$,

$$\int_{\mathbf{R}_+} \frac{x^{s-1}}{e^x - 1}\, dx = \Gamma(s)\zeta(s), \qquad \int_{\mathbf{R}_+} \frac{x^{s-1}}{e^x + 1}\, dx = (1 - 2^{1-s})\Gamma(s)\zeta(s).$$

Hint: Use $\frac{1}{e^x-1} = \sum_{k\in\mathbf{N}} e^{-kx}$ and interchange the order of summation and integration. Then prove by Exercise 6.50 that $\int_{\mathbf{R}_+} x^{s-1}e^{-kx}\, dx = \frac{\Gamma(s)}{k^s}$.

(ii) Using Exercise 0.20 or 6.40.(iv), show that

$$\int_{\mathbf{R}_+} \frac{x^{2n-1}}{e^x - 1}\, dx = \int_0^1 \frac{(\log x)^{2n-1}}{x - 1}\, dx = (-1)^{n-1}(2\pi)^{2n}\frac{B_{2n}}{4n} \qquad (n \in \mathbf{N}).$$

Verify that for $n = 1$ the answer is in agreement with that in Exercise 6.39.(iii).

(iii) Replace x by ax, for $a > 0$, in the first integral in part (i), and deduce by means of differentiation with respect to a of the resulting formula

$$\int_{\mathbf{R}_+} \frac{x^s e^x}{(e^x - 1)^2}\, dx = \Gamma(s+1)\zeta(s) \qquad (\mathrm{Re}\, s > 1).$$

(iv) Imitate the proof in part (ii) to show

$$\int_{\mathbf{R}_+} \frac{x^{2n}}{\cosh x}\, dx = 2\int_0^1 \frac{(\log x)^{2n}}{x^2 + 1}\, dx = \left(\frac{\pi}{2}\right)^{2n+1} E_n \qquad (n \in \mathbf{N}_0),$$

$$\int_{\mathbf{R}_+} \frac{x^{2n-1}}{\sinh x}\, dx = (-1)^{n-1}(2^{2n} - 1)\pi^{2n}\frac{B_{2n}}{2n} \qquad (n \in \mathbf{N}).$$

Background. The integral $\int_{\mathbf{R}_+} \frac{x^3}{e^x-1}\, dx = \frac{\pi^4}{15}$ occurs in the theory of the energy intensity of *black body radiation*, and the one in part (iii) in the quantum theory of transport effects.

Exercise 6.53 (Another proof of Legendre's duplication formula – sequel to Exercises 2.87 and 6.50). Multiply both sides of the identity in Exercise 2.87.(v) by x^{p-1} for $p \in \mathbf{R}_+$, integrate with respect to x over $\mathbf{R}_+$, and change the order of integration on the right–hand side. Conclude that *Legendre's duplication formula* from Exercise 6.50.(vi) follows.

Exercise 6.54 (Asymptotic behavior of some integrals and Stirling's formula). We make the following assumptions: (i) $a < 0 < b$, and g and $h \in C([\,a, b\,])$; (ii) there exists $c > 0$ such that $h(t) \geq ct^2$, for $t \in [\,a, b\,]$; (iii) there exists $H > 0$ such that $\lim_{t\to 0} \frac{h(t)}{t^2} = \frac{H}{2}$. Then we have

$$\lim_{x\to\infty} \sqrt{x} \int_a^b g(t)e^{-xh(t)}\, dt = g(0)\sqrt{\frac{2\pi}{H}}.$$

Indeed, in the following we may assume $x \in \mathbf{R}_+$. The substitution of variables $t = \frac{s}{\sqrt{x}}$ implies

$$I(x) := \sqrt{x} \int_a^b g(t) e^{-xh(t)}\, dt = \int_{\mathbf{R}} f(x, s)\, ds,$$

where $f : \mathbf{R}_+ \times \mathbf{R} \to \mathbf{R}$ is given by $f(x, s) := 1_{[\sqrt{x}\, a,\, \sqrt{x}\, b]}\, g(s/\sqrt{x}) e^{-xh(s/\sqrt{x})}$. Set $m = \max\{\, |g(t)| \mid t \in [\, a, b\,]\,\}$ and deduce from (ii)

$$|f(x, s)| \le m e^{-xc\,(s/\sqrt{x})^2} = m e^{-cs^2} \qquad (x \in \mathbf{R}_+,\ s \in \mathbf{R}).$$

Note that the function $s \mapsto m e^{-cs^2}$ is absolutely Riemann integrable over $\mathbf{R}$. Furthermore, $I(x)$ is uniformly convergent for $x \in \mathbf{R}_+$ on account of De la Vallée-Poussin's test from Lemma 2.10.10. Conclude from (i) and (iii) that we have $\lim_{x\to\infty} f(x, s) = g(0) e^{-\frac{1}{2} H s^2}$, for $s \in \mathbf{R}$. Use the Continuity Theorem 2.10.12 and Example 6.10.8 to verify

$$\lim_{x\to\infty} I(x) = g(0) \int_{\mathbf{R}} e^{-\frac{1}{2} H s^2}\, ds = g(0) \sqrt{\frac{2\pi}{H}}.$$

For the asymptotic behavior of the Gamma function $\Gamma(x) = \int_{\mathbf{R}_+} e^{-y}\, y^{x-1}\, dy$, for $x \to \infty$, from Exercise 6.50 substitute $y = x e^t$ and then apply the preceding result with $g(t) = 1$ and $h(t) = e^t - 1 - t$ to obtain *Stirling's formula*

$$\lim_{x\to\infty} \frac{\Gamma(x)}{e^{-x} x^{x-\frac{1}{2}}} = \sqrt{2\pi}, \qquad \text{that is} \qquad \Gamma(x) \sim \sqrt{2\pi}\, x^{x-\frac{1}{2}} e^{-x}, \quad x \to \infty.$$

Background. For more precise results, see Exercise 6.55.

Exercise 6.55 (Asymptotic expansion of Gamma function – sequel to Exercise 6.50). For the complete asymptotic behavior of the Gamma function $\Gamma(x) = \int_{\mathbf{R}_+} e^{-y}\, y^{x-1}\, dy$, for $x \to \infty$, from Exercise 6.50 introduce a new variable t by means of $y = x e^t$. This gives

$$\Gamma(x) = x^x e^{-x} \int_{\mathbf{R}} e^{-x(e^t - 1 - t)}\, dt.$$

The function $h : \mathbf{R} \to \mathbf{R}$ with $h(t) = e^t - 1 - t$ attains its absolute minimum 0 at 0 and has quadratic approximation $\frac{1}{2} t^2$ near 0, that is, $h(t) = \frac{1}{2} t^2 (1 + \mathcal{O}(t)),\quad t \to 0$. Therefore apply the change of variable $u = u(t)$ given by $h(t) = \frac{1}{2} u^2$, in other words, $u(t) = \operatorname{sgn}(t) \sqrt{2h(t)}$. Show that u is a bijection from $\mathbf{R}$ into itself. Further, deduce

$$\lim_{t\to 0} u'(t) = \lim_{t\to 0} \operatorname{sgn}(t) \frac{2(e^t - 1)}{2\sqrt{2h(t)}} = \lim_{t\to 0} \operatorname{sgn}(t) \frac{t + \mathcal{O}(t^2)}{\operatorname{sgn}(t)(t + \mathcal{O}(t^2))} = 1.$$

This implies that $u'(0) = 1$. Prove that $u'(t) \neq 0$, for all $t \neq 0$. Deduce from the Global Inverse Function Theorem 3.2.8 that the inverse mapping $t = t(u)$ is a C^∞

function on $\mathbf{R}$ with $t(0) = 0$. Differentiate the equality $h(t) = \frac{1}{2}u^2$ with respect to the variable u to find

$$t'(u)(e^t - 1) = u, \qquad \text{hence} \qquad (\star) \qquad t'(u)(\frac{1}{2}u^2 + t(u)) = u.$$

Next consider the MacLaurin expansion of t in powers of u; that is, write, without assuming convergence

$$t(u) = u + a_2 u^2 + \cdots + a_k u^k + \cdots .$$

Verify that this implies

$$t'(u)(\frac{1}{2}u^2 + t(u)) = u + (\frac{1}{2} + 3a_2)u^2 + (a_2 + 2a_2^2 + 4a_3)u^3 + \cdots$$
$$+(\frac{1}{2}(k-1)a_{k-1} + ka_k + (k-1)a_2 a_{k-1} + \cdots + 2a_{k-1}a_2 + a_k)u^k + \cdots .$$

Hence, equating like powers of u in $(\star)$ conclude that $a_2 = -\frac{1}{6}$, $a_3 = \frac{1}{36}$, $a_4 = -\frac{1}{270}$, and in general, for $k \geq 3$, obtain the following *recursion relation* for the coefficients a_k:

$$(k+1)a_k = -\frac{1}{2}(k-1)a_{k-1} - (k-1)a_2 a_{k-1} - (k-2)a_3 a_{k-2} - \cdots - 2a_{k-1}a_2.$$

Show that accordingly

$$\begin{aligned}\Gamma(x) \quad &\sim x^x e^{-x} \int_{\mathbf{R}} e^{-\frac{1}{2}xu^2} \sum_{k \in \mathbf{N}_0} (k+1)a_{k+1} u^k \, du \\ &= x^x e^{-x} \sum_{k \in \mathbf{N}_0} (k+1)a_{k+1} \int_{\mathbf{R}} e^{-\frac{1}{2}xu^2} u^k \, du.\end{aligned}$$

Prove that the integrals on the right–hand side vanish for k odd, and use Exercise 6.51 to conclude that, for $x \to \infty$,

$$\begin{aligned}\Gamma(x) &\sim 2x^x e^{-x} \sum_{k \in \mathbf{N}_0} (2k+1)a_{2k+1} \frac{(2k)!\,\sqrt{\pi}}{k!\,(\sqrt{2x})^{2k+1}} \\ &= \sqrt{2\pi}\, x^{x-\frac{1}{2}} e^{-x} \sum_{k \in \mathbf{N}_0} \frac{(2k+1)!}{2^k\, k!} a_{2k+1} \frac{1}{x^k} \\ &= \sqrt{2\pi}\, x^{x-\frac{1}{2}} e^{-x} \left(1 + \frac{1}{12x} + \frac{1}{288x^2} - \frac{139}{51\,840x^3} - \frac{571}{2\,488\,320x^4} + \cdots\right).\end{aligned}$$

In the computation above we ignored the fact that the MacLaurin expansion for t actually terminates and then contains an error term. It is therefore that we use the sign $\sim$ to indicate that we have only equality in the following sense. Let there be given a function $f : \mathbf{R}_+ \to \mathbf{R}$. A formal power series in $\frac{1}{x}$, for $x \in \mathbf{R}_+$, with coefficients $b_k \in \mathbf{R}$,

$$\sum_{k\in\mathbf{N}_0} b_k \frac{1}{x^k},$$

is said to be an *asymptotic expansion* for f if the following condition is met. Write $s_n(x)$ for the sum of the first $n+1$ terms of the series, and $r_n(x) = x^n(f(x) - s_n(x))$. Then one must have, for each $n \in \mathbf{N}$,

$$\lim_{x\to\infty} r_n(x) = 0.$$

Note that no condition is imposed on $\lim_{n\to\infty} r_n(x)$, for x fixed. When the foregoing definition is satisfied, we write

$$f(x) \sim \sum_{k\in\mathbf{N}_0} b_k \frac{1}{x^k}, \qquad x \to \infty.$$

Background. See Exercise 0.24 for related results.
Finally an application. Using Exercise 6.50.(viii), verify

$$\mathrm{vol}_n(B^n) \sim \frac{1}{\sqrt{\pi n}}\Big(\frac{2\pi e}{n}\Big)^{\frac{n}{2}}, \qquad n \to \infty,$$

and deduce $\lim_{n\to\infty} \mathrm{vol}_n(B^n) = 0$.

Exercise 6.56 (Product formula for Gamma function and Wallis' product – sequel to Exercise 6.50). The well-known formula $\lim_{n\to\infty}(1-\frac{t}{n})^n = e^{-t}$, for all $t \in \mathbf{R}$, makes it plausible that, for all $p > 0$,

$$(\star) \qquad \Gamma(p) = \int_{\mathbf{R}_+} e^{-t}\, t^{p-1}\, dt = \lim_{n\to\infty} \int_0^n t^{p-1}\Big(1-\frac{t}{n}\Big)^n dt.$$

In order to prove $(\star)$ we define, for $n \in \mathbf{N}$ and $p > 0$,

$$I_n = \int_0^n t^{p-1}\Big(1-\frac{t}{n}\Big)^n dt,$$

$$E_n = \Big|\int_0^n t^{p-1}e^{-t}\, dt - I_n\Big| = \Big|\int_0^n t^{p-1}e^{-t}\Big(1-e^t\Big(1-\frac{t}{n}\Big)^n\Big)\, dt\Big|.$$

(i) Prove that $e^x \geq 1+x$, for all $x \in \mathbf{R}$. Use this result to show that, for $0 \leq t \leq n$,

$$e^t\Big(1-\frac{t}{n}\Big)^n \leq 1, \qquad -e^t \leq -\Big(1+\frac{t}{n}\Big)^n,$$

hence

$$E_n \leq \int_0^n t^{p-1}e^{-t}\Big(1-\Big(1-\frac{t^2}{n^2}\Big)^n\Big)\, dt.$$

(ii) By mathematical induction over $n \in \mathbf{N}$ verify that $(1-x)^n \geq 1-nx$ if $x \leq 1$. Use this to show, for $n \in \mathbf{N}$ and $p > 0$,

$$E_n \leq \frac{1}{n}\int_0^n t^{p+1}e^{-t}\,dt \leq \frac{1}{n}\Gamma(p+2).$$

Conclude that $\lim_{n\to\infty} E_n = 0$, and that $(\star)$ is valid.

Background. Write χ_n for the characteristic function of the interval $[\,0, n\,]$ and $f_n(t) = \chi_n(t)(1-\frac{t}{n})^n$. Part (i) asserts $0 \leq f_n(t) \leq e^{-t}$, while $\lim_{n\to\infty} f_n(t) = e^{-t}$, for all $t \in \mathbf{R}$. Arzelà's Dominated Convergence Theorem 6.12.3 therefore immediately implies the identity in $(\star)$.

(iii) Prove, by means of the substitution $t = nu$ and Exercise 6.50.(iv),

$$\Gamma(p) = \lim_{n\to\infty}\frac{n^p\,n!}{p(p+1)\cdots(p+n)} = \lim_{n\to\infty}\frac{n^p\,n!}{(p)_{n+1}} \qquad (p>0),$$

in the notation of Exercise 0.11. Use this to derive the following *product formulae*:

$$\frac{1}{\Gamma(p)} = pe^{\gamma p}\prod_{n\in\mathbf{N}}\Big(1+\frac{p}{n}\Big)e^{-\frac{p}{n}}, \qquad \Gamma(p) = \frac{1}{p}\prod_{n\in\mathbf{N}}\frac{\big(1+\frac{1}{n}\big)^p}{1+\frac{p}{n}}, \qquad (p>0),$$

where $\gamma = \lim_{n\to\infty}\big(\sum_{1\leq k\leq n}\frac{1}{k} - \log n\big)$ is *Euler's constant.*

(iv) Conclude that

$$\lim_{n\to\infty}\frac{n^{\frac{1}{2}}n!}{\frac{1}{2}(\frac{1}{2}+1)\cdots(\frac{1}{2}+n)} = \lim_{n\to\infty}\frac{2\sqrt{n}\,2^n n!}{1\cdot 3\cdot 5\cdots(2n-1)(2n+1)} = \sqrt{\pi}.$$

From this, show

$$\lim_{n\to\infty}\frac{2^n n!\,\sqrt{2n+1}}{1\cdot 3\cdot 5\cdots(2n-1)(2n+1)} = \sqrt{\frac{\pi}{2}},$$

and verify that in this way we have obtained *Wallis' product* from Exercise 0.13.(iii)

$$\lim_{n\to\infty}\frac{2}{1}\,\frac{2}{3}\,\frac{4}{3}\,\frac{4}{5}\,\frac{6}{5}\,\frac{6}{7}\cdots\frac{2n}{2n-1}\,\frac{2n}{2n+1} = \frac{\pi}{2}, \qquad \text{or} \qquad \frac{2}{\pi} = \prod_{n\in\mathbf{N}}\Big(1-\frac{1}{4n^2}\Big).$$

Background. Part (iii), and therefore Wallis's product also, amounts to the following property, which is an asymptotic variant of the functional equation for the Gamma function. For $p > 0$,

$$n^p\Gamma(n) \sim \Gamma(n+p), \quad n\to\infty.$$

Exercise 6.57 (Analytic continuation of Gamma function – sequel to Exercise 6.50 – needed for Exercises 6.58, 6.62, 6.89 and 8.19). We prove that the Gamma function can be extended to a complex-differentiable function on the subset $\Omega = \mathbf{C} \setminus (-\mathbf{N}_0)$ of the complex plane $\mathbf{C}$.

(i) Prove that the integral $\Gamma(z) = \int_{\mathbf{R}_+} e^{-t} t^{z-1}\, dt$ is well-defined, for every $z \in \mathbf{C}$ with $\operatorname{Re} z > 0$.

(ii) Now let $z \in \Omega$, then there exists $n \in \mathbf{N}$ with $\operatorname{Re}(z+n) > 0$. Define

$$\Gamma(z) = \frac{\Gamma(z+n)}{(z+n-1)\cdots(z+1)z}.$$

Verify that this definition is independent of the choice of $n \in \mathbf{N}$.

(iii) Prove $\Gamma(z) \neq 0$, for all $z \in \Omega$.

(iv) Verify that Γ is a complex-differentiable function on Ω.

Exercise 6.58 (Reflection formula for Gamma function – sequel to Exercises 6.50 and 6.57 – needed for Exercises 6.60, 6.61, 6.74, 6.89 and 6.108). Define $f : \mathbf{R} \setminus \mathbf{Z} \to \mathbf{R}$ by (see Exercise 6.57)

$$f(x) = \Gamma(x)\,\Gamma(1-x)\,\sin \pi x.$$

(i) Define $f(0) = \pi$. Prove by $f(x) = \Gamma(1+x)\,\Gamma(1-x)\,\frac{\sin \pi x}{x}$ that f is continuous at 0, and also that f is infinitely differentiable at 0.

(ii) Demonstrate that $f(x+1) = x\Gamma(x)\,\frac{\Gamma(1-x)}{-x}\,\sin(-\pi x) = f(x)$. Conclude that $f : \mathbf{R} \to \mathbf{R}$ is continuous and periodic, and therefore bounded on $\mathbf{R}$.

(iii) By means of Legendre's duplication formula from Exercise 6.50.(vi), prove

$$f\Big(\frac{x}{2}\Big) f\Big(\frac{x+1}{2}\Big) = \pi f(x).$$

(iv) Let $g = (\log \circ f)^{(2)}$. Verify that g is a bounded function on $\mathbf{R}$ satisfying

$$g\Big(\frac{x}{2}\Big) + g\Big(\frac{x+1}{2}\Big) = 4g(x).$$

Show that this implies $g = 0$ on $\mathbf{R}$. Prove that $\log \circ f$ is a linear periodic function on $\mathbf{R}$, and that therefore f is constant and equals π. Thus we find the following *reflection formula* for the Gamma function (see Exercise 6.59 for another proof):

$$\Gamma(x)\,\Gamma(1-x) = \frac{\pi}{\sin \pi x} \qquad (x \in \mathbf{R} \setminus \mathbf{Z}).$$

(v) Using Exercise 6.50.(iv), prove (see Exercise 0.15.(iv) for a different proof)

$$\int_0^1 t^{p-1}(1-t)^{-p}\,dt = \int_{\mathbf{R}_+} \frac{x^{p-1}}{1+x}\,dx = \frac{\pi}{\sin \pi p} \qquad (0 < p < 1).$$

Show that

$$\int_{\mathbf{R}_+} \frac{dx}{1+x^n} = \frac{\pi}{n \sin \frac{\pi}{n}} \qquad (n \in \mathbf{N}).$$

(vi) Conclude by means of Exercise 6.50.(v) and (vi), and by using part (iv), that

$$\int_0^1 \frac{1}{\sqrt{1-t^3}}\,dt = \frac{1}{2\pi\sqrt{3}\sqrt[3]{2}}\Gamma\Big(\frac{1}{3}\Big)^3 = 1.402\,182\,105\,325\cdots.$$

Background. The reflection formula can also be proved via $\Gamma(1-x) = -x\Gamma(-x)$ using the product formulae for the Gamma function from Exercise 6.56.(iii) and for the sine function from Exercise 0.13.(ii). Combination of Exercise 6.57 and the formula $\frac{1}{\Gamma(z)} = \frac{1}{\pi}\Gamma(1-z)\sin \pi z$ shows that $\frac{1}{\Gamma}$ is complex-differentiable on $\mathbf{C}$. This also follows from the product formula for $\frac{1}{\Gamma}$ in Exercise 6.56.(iii), which is valid on all of $\mathbf{C}$.

Exercise 6.59 (Another proof of reflection formula for Gamma function – sequel to Exercises 0.15, 3.1 and 6.50). Using Corollary 6.4.3 verify, for $0 < x < 1$,

$$\Gamma(x)\Gamma(1-x) = \int_{\mathbf{R}_+^2} e^{-(t_1+t_2)}\Big(\frac{t_1}{t_2}\Big)^x \frac{1}{t_1}\,dt.$$

By means of Exercises 3.1 and 0.15.(iv) deduce (see Exercise 6.58.(iv) for another proof)

$$\Gamma(x)\Gamma(1-x) = \int_{\mathbf{R}_+^2} e^{-s_1}\frac{s_2^{x-1}}{s_2+1}\,ds = \int_{\mathbf{R}_+} \frac{s_2^{x-1}}{s_2+1}\,ds_2 = \frac{\pi}{\sin \pi x} \qquad (0 < x < 1).$$

Exercise 6.60 (Fresnel's integral – sequel to Exercises 0.8, 6.50 and 6.58 – needed for Exercises 6.61 and 6.62). One has

$$(\star) \qquad \int_{\mathbf{R}_+} \frac{\sin x}{x^p}\,dx = \frac{\Gamma(\frac{p}{2})\,\Gamma(1-\frac{p}{2})}{2\Gamma(p)} = \frac{\pi}{2\Gamma(p)\sin(p\frac{\pi}{2})} \qquad (0 < p < 2).$$

In particular,

$$\int_{\mathbf{R}_+} \frac{\sin x}{x}\,dx = \frac{\pi}{2}, \qquad \int_{\mathbf{R}_+} \frac{\sin x}{\sqrt{x}}\,dx = \sqrt{\frac{\pi}{2}}, \qquad \int_{\mathbf{R}_+} \frac{\sin x}{x\sqrt{x}}\,dx = \sqrt{2\pi},$$

and *Fresnel's integral*

$$\int_{\mathbf{R}_+} \sin(x^2)\,dx = \sqrt{\frac{\pi}{8}}.$$

Similarly,

$$\int_{\mathbf{R}_+} \frac{\cos x}{x^p}\,dx = \frac{\pi}{2\Gamma(p)\cos\left(p\frac{\pi}{2}\right)} \qquad (0 < p < 1).$$

We now prove (⋆) in two steps.

(i) Prove

$$\frac{1}{x^p} = \frac{1}{\Gamma(p)}\int_{\mathbf{R}_+} t^{p-1}\,e^{-xt}\,dt \qquad (x \in \mathbf{R}_+),$$

and show

$$\int_{\mathbf{R}_+} \frac{\sin x}{x^p}\,dx = \frac{1}{\Gamma(p)}\int_{\mathbf{R}_+} t^{p-1}\int_{\mathbf{R}_+} e^{-xt}\,\sin x\,dx\,dt.$$

(ii) Using Exercise 0.8, prove

$$\int_{\mathbf{R}_+} \frac{\sin x}{x^p}\,dx = \frac{1}{\Gamma(p)}\int_{\mathbf{R}_+} \frac{t^{p-1}}{1+t^2}\,dt.$$

Now apply Exercise 6.58.(v).

(iii) Show (see Exercise 8.19 for a different proof)

$$\int_{\mathbf{R}_+} x^{s-1}\left\{\begin{matrix}\sin\\ \cos\end{matrix}\right. x\,dx = \Gamma(s)\left\{\begin{matrix}\sin\\ \cos\end{matrix}\right. \left(s\frac{\pi}{2}\right) \qquad \begin{matrix}-1 < s < 1;\\ 0 < s < 1.\end{matrix}$$

Exercise 6.61 (Values of Airy function – sequel to Exercises 2.90, 6.58 and 6.60). We use the notation from Exercise 2.90. Prove directly using Exercises 6.60 or 8.19, or deduce from Exercise 2.90.(vi) and the reflection formula from Exercise 6.58.(iv),

$$\mathrm{Ai}(0) = \frac{1}{3^{\frac{2}{3}}\Gamma\left(\frac{2}{3}\right)}, \qquad \mathrm{Ai}'(0) = -\frac{1}{3^{\frac{1}{3}}\Gamma\left(\frac{1}{3}\right)}.$$

Exercise 6.62 (Functional equation for zeta function – sequel to Exercises 0.16, 0.18, 0.25, 6.57 and 6.60 – needed for Exercise 7.31). The notation is as in Exercise 0.25. Deduce from (⋆) in that exercise

$$\zeta(s) = \frac{1}{s-1} + \frac{1}{2} + \frac{s}{12} - \frac{s(s+1)}{2}\int_1^\infty \frac{b_2(x-[x])}{x^{s+2}}\,dx \qquad (-1 < \mathrm{Re}\,s).$$

Using integration by parts and Exercise 0.16.(i) and (iii), show that the integral also converges for $-2 < \operatorname{Re} s$, and prove

$$-\frac{s(s+1)}{2}\int_0^1 \frac{b_2(x-[x])}{x^{s+2}}\,dx = \frac{1}{s-1}+\frac{1}{2}+\frac{s}{12} \qquad (\operatorname{Re} s < -1).$$

Conclude that

$$\zeta(s) = -\frac{s(s+1)}{2}\int_{\mathbf{R}_+} \frac{b_2(x-[x])}{x^{s+2}}\,dx \qquad (-2 < \operatorname{Re} s < -1).$$

Next use the identity $b_2(x-[x]) = 4\sum_{k\in\mathbf{N}} \frac{\cos 2k\pi x}{(2k\pi)^2}$ from Exercise 0.18.(ii) and then Exercises 6.60 and 6.57 to show, for $-2 < \operatorname{Re} s < -1$,

$$\begin{aligned}\zeta(s) &= -2s(s+1)\int_{\mathbf{R}_+} \frac{\cos t}{t^{s+2}}\,dt \sum_{k\in\mathbf{N}} (2k\pi)^{s-1} \\ &= -2^{s-1}\pi^s \frac{s(s+1)}{\Gamma(s+2)}\,\frac{1}{\cos(\frac{\pi}{2}(s+2))}\zeta(1-s).\end{aligned}$$

Deduce the following *functional equation for the zeta function* (see Exercise 6.89 for another proof):

$$2^{s-1}\pi^s\zeta(1-s) = \cos\Big(\frac{s}{2}\pi\Big)\Gamma(s)\,\zeta(s) \qquad (-2 < \operatorname{Re} s < -1).$$

Background. It follows from complex analysis that the functional equation actually is valid on the common domain of the left- and the right–hand side. It is seemingly easier to work with b_1 instead of b_2, but in that case the interchange of integration and summation is a more delicate matter. One can prove that the partial sums $\sum_{1\le k\le N} \frac{\sin 2k\pi x}{k\pi}$ are bounded on $\mathbf{R}$ and then use Arzelà's Dominated Convergence Theorem 6.12.3.

Exercise 6.63 (Sequel to Exercises 0.4 and 6.50). Let $l \in \mathbf{N}_0$ and let P_l be the Legendre polynomial as in Exercise 0.4. Use repeated integration by parts to show

$$\begin{aligned}2^{2l}(l!)^2\int_{-1}^1 P_l(x)^2\,dx &= (-1)^l\int_{-1}^1 (x^2-1)^l\Big(\frac{d}{dx}\Big)^{2l}(x^2-1)^l\,dx \\ &= (-1)^l(2l)!\int_{-1}^1 (x^2-1)^l\,dx.\end{aligned}$$

Now apply Exercise 6.50.(vii) to deduce

$$\int_{-1}^1 P_l(x)^2\,dx = \frac{2}{2l+1}.$$

Exercise 6.64 (Dirichlet's formula – sequel to Exercises 3.2 and 6.50 – needed for Exercises 7.26 and 7.37). Let $p \in \mathbf{R}_+^2$, and let $\Delta^2 \subset \mathbf{R}^2$ be the triangle with $\Delta^2 = \{ x \in \mathbf{R}_+^2 \mid x_1 + x_2 < 1 \}$.

(i) Prove the following formula:

$$\int_{\Delta^2} x_1^{p_1-1} x_2^{p_2-1}\, dx = \frac{\Gamma(p_1)\Gamma(p_2)}{\Gamma(1+p_1+p_2)}.$$

Hint: Apply Theorem 6.4.5, and subsequently Exercise 6.50.(iv).

(ii) Let $p_3 \in \mathbf{R}_+$. Prove

$$\int_{\Delta^2} x_1^{p_1-1} x_2^{p_2-1} (1 - x_1 - x_2)^{p_3-1}\, dx = \frac{\Gamma(p_1)\Gamma(p_2)\Gamma(p_3)}{\Gamma(p_1+p_2+p_3)}.$$

Hint: Begin as in part (i); substitute $x_2 = (1 - x_1)t$ in the inner integral.

(iii) Let $B_+^2 = \{ y \in \mathbf{R}_+^2 \mid \|y\| \le 1 \}$. Prove that we then have the following analog of the formula from Exercise 6.50.(iii):

$$\int_{B_+^2} y_1^{p_1} y_2^{p_2} (\sqrt{1-\|y\|^2})^{p_3} \frac{dy}{\sqrt{1-\|y\|^2}} = \frac{\Gamma(\frac{p_1+1}{2})\Gamma(\frac{p_2+1}{2})\Gamma(\frac{p_3+1}{2})}{2^2\Gamma(\frac{p_1+p_2+p_3+3}{2})}.$$

(iv) Suppose $f \in C([\,0,1\,])$. Prove the following, known as *Dirichlet's formula*:

$$\int_{\Delta^2} x_1^{p_1-1} x_2^{p_2-1} f(x_1+x_2)\, dx = \frac{\Gamma(p_1)\Gamma(p_2)}{\Gamma(p_1+p_2)} \int_0^1 x^{p_1+p_2-1} f(x)\, dx.$$

Show the formula from part (i) to be a special case.
Hint: Use Exercise 3.2.

Exercise 6.65 (Generalization of Dirichlet's formula – sequel to Exercises 3.19, 6.50 and 6.64 – needed for Exercise 7.52). Let Δ^n be as in Exercise 3.19.

(i) Prove the following generalization of Exercise 6.64.(i):

$$\int_{\Delta^n} \prod_{1\le j\le n} x_j^{p_j-1}\, dx = \frac{\prod_{1\le j\le n}\Gamma(p_j)}{\Gamma(1+\sum_{1\le j\le n} p_j)} \qquad (p \in \mathbf{R}_+^n).$$

Hint: Use Exercises 3.19.(i) and 6.50.(iv).

(ii) In particular, conclude that $\mathrm{vol}_n(\Delta^n) = \frac{1}{n!}$.

(iii) Consider $B_k^n = \{ x \in \mathbf{R}^n \mid \sum_{1\le j\le n} |x_j|^k \le 1 \}$, for $k \in \mathbf{N}$. Prove

$$\mathrm{vol}_n(B_k^n) = \Big(\frac{2}{k}\Big)^n \frac{\Gamma(\frac{1}{k})^n}{\Gamma(\frac{n}{k}+1)}.$$

Verify that, on geometrical grounds, $\lim_{k\to\infty} \mathrm{vol}_n(B_k^n) = 2^n$.

(iv) Suppose $f \in C([0,1])$. Prove the following generalization of *Dirichlet's formula*, for $p \in \mathbf{R}_+^n$:

$$\int_{\Delta^n} \prod_{1\le j\le n} x_j^{p_j-1} f\Big(\sum_{1\le j\le n} x_j\Big)\,dx = \frac{\prod_{1\le j\le n}\Gamma(p_j)}{\Gamma(\sum_{1\le j\le n} p_j)} \int_0^1 x^{-1+\sum_{1\le j\le n} p_j} f(x)\,dx.$$

(v) Let a and $d \in \mathbf{R}_+^n$ and define $\Theta^n = \{ x \in \mathbf{R}_+^n \mid \sum_{1\le j\le n} \big(\frac{x_j}{a_j}\big)^{d_j} < 1 \}$. Using part (i) prove

$$\int_{\Theta^n} \prod_{1\le j\le n} x_j^{p_j-1}\,dx = \frac{\prod_{1\le j\le n} \frac{a_j^{p_j}}{d_j}\Gamma\big(\frac{p_j}{d_j}\big)}{\Gamma\big(1+\sum_{1\le j\le n}\frac{p_j}{d_j}\big)}.$$

Hint: Consider $\Psi : \mathbf{R}_+^n \to \mathbf{R}_+^n$ given by $x_i = \Psi_i(y) = a_i y_i^{\frac{1}{d_i}}$, for $1 \le i \le n$, and note that $\Theta^n = \Psi(\Delta^n)$.
In particular, the n-dimensional volume of Θ^n and of the ellipsoid $\{ x \in \mathbf{R}^n \mid \sum_{1\le j\le n} \big(\frac{x_j}{a_j}\big)^2 < 1 \}$, respectively, are equal to, see Exercise 6.50.(viii)

$$\frac{\prod_{1\le j\le n} a_j \Gamma\big(1+\frac{1}{d_j}\big)}{\Gamma\big(1+\sum_{1\le j\le n}\frac{1}{d_j}\big)} \qquad \text{and} \qquad \frac{\pi^{\frac{n}{2}}}{\Gamma(\frac{n}{2}+1)} \prod_{1\le j\le n} a_j = \mathrm{vol}_n(B^n) \prod_{1\le j\le n} a_j.$$

(vi) Suppose that $p_j \in 2\mathbf{N}_0$, for $1 \le j \le n$. Deduce from part (v) that the average of the monomial function $x \mapsto x^p = \prod_{1\le j\le n} x_j^{p_j}$ over B^n satisfies, in the notation of Exercise 6.50.(vii) and with $(-1)!! = 1$

$$\frac{1}{\mathrm{vol}_n(B^n)} \int_{B^n} x^p\,dx = \frac{\prod_{1\le j\le n}(p_j-1)!!}{\prod_{1\le l\le \frac{|p|}{2}}(n+2l)} \in \mathbf{Q}.$$

Exercise 6.66 (Bessel function – sequel to Exercise 6.50 – needed for Exercises 6.67, 6.68, 6.70, 6.98, 6.102, 6.107, 7.22, 7.30 and 8.20). Let $\lambda > -\frac{1}{2}$. Define the function

$$f_\lambda : \mathbf{R} \to \mathbf{C} \qquad \text{by} \qquad f_\lambda(x) = \int_0^\pi e^{-ix\cos\alpha} \sin^{2\lambda}\alpha\,d\alpha.$$

(i) Prove by the substitution of variables $\alpha \mapsto \pi - \alpha$, for $x \in \mathbf{R}$,

$$f_\lambda(x) = 2\int_0^{\frac{\pi}{2}} \cos(x\cos\alpha)\sin^{2\lambda}\alpha\,d\alpha = 2\int_0^{\frac{\pi}{2}} \cos(x\sin\alpha)\cos^{2\lambda}\alpha\,d\alpha.$$

Conclude that f_λ is real-valued.

(ii) Show that

$$f_\lambda^{(n)}(x) = (-i)^n \int_0^\pi \cos^n \alpha \, e^{-ix\cos\alpha} \sin^{2\lambda} \alpha \, d\alpha \qquad (n \in \mathbf{N},\ x \in \mathbf{R}),$$

and conclude that $f_\lambda'' = f_{\lambda+1} - f_\lambda$ on $\mathbf{R}$.

(iii) Verify by integration by parts that $(2\lambda + 1) f_\lambda'(x) = -x\, f_{\lambda+1}(x)$, for $x \in \mathbf{R}$.

(iv) Prove that f_λ satisfies the following differential equation for $u \in C^2(\mathbf{R})$:

$$x\, u''(x) + (2\lambda + 1)\, u'(x) + x\, u(x) = 0 \qquad (x \in \mathbf{R}).$$

The *Bessel function of order* λ is defined as the function

$$J_\lambda : \mathbf{R}_+ \to \mathbf{R} \qquad \text{with} \qquad J_\lambda(x) = \frac{1}{\Gamma(\frac{1}{2})\Gamma(\lambda + \frac{1}{2})} \Big(\frac{x}{2}\Big)^\lambda \int_0^\pi e^{-ix\cos\alpha} \sin^{2\lambda} \alpha \, d\alpha.$$

(v) Demonstrate by integration that $J_{\frac{1}{2}}(x) = \sqrt{\frac{2}{\pi}}\frac{\sin x}{\sqrt{x}}$, for $x \in \mathbf{R}_+$.

(vi) Show by part (iii) that $J_{\lambda+1}(x) = -J_\lambda'(x) + \frac{\lambda}{x} J_\lambda(x)$, for $x \in \mathbf{R}_+$. Conclude that

$$J_{n+\frac{1}{2}}(x) = (-1)^n \sqrt{\frac{2}{\pi}} x^{n+\frac{1}{2}} \Big(\frac{1}{x}\frac{d}{dx}\Big)^n \Big(\frac{\sin x}{x}\Big) \qquad (n \in \mathbf{N},\ x \in \mathbf{R}_+).$$

(vii) Using part (iv), prove that on $\mathbf{R}_+$ the function J_λ satisfies the following, known as *Bessel's equation*:

$$x^2\, u''(x) + x\, u'(x) + (x^2 - \lambda^2)\, u(x) = 0 \qquad (x \in \mathbf{R}_+).$$

(viii) Prove by means of Exercise 6.50.(iii) that $\lim_{x \downarrow 0} \big(\frac{2}{x}\big)^\lambda J_\lambda(x) = \frac{1}{\Gamma(\lambda+1)}$.

(ix) For fixed $x \in \mathbf{R}$ prove that the following power series expansion is convergent uniformly in $\alpha \in \mathbf{R}$:

$$e^{-ix\cos\alpha} = \sum_{k \in \mathbf{N}_0} (-i)^k \frac{x^k}{k!} \cos^k \alpha.$$

Next apply the theorem on the termwise integration of a uniformly convergent series, then Exercise 6.50.(iii) and (vii), to conclude that

$$J_\lambda(x) = \Big(\frac{x}{2}\Big)^\lambda \sum_{k \in \mathbf{N}_0} \frac{(-1)^k}{k!\,\Gamma(\lambda + k + 1)} \Big(\frac{x}{2}\Big)^{2k} \qquad (x \in \mathbf{R}_+).$$

Now let $\lambda \leq -\frac{1}{2}$. Then the defining integral for J_λ diverges. Nonetheless it follows by Exercise 6.57 that the right–hand side in part (ix) is well-defined if $\lambda \neq -n$, for $n \in \mathbf{N}$. Therefore we define the Bessel function J_λ for those values of λ by means of the series in (ix).

(x) Using Exercise 6.50.(vii), prove $J_{-\frac{1}{2}}(x) = \sqrt{\frac{2}{\pi}}\frac{\cos x}{\sqrt{x}}$, for $x \in \mathbf{R}_+$.

(xi) Verify that J_λ and $J_{-\lambda}$ both satisfy Bessel's differential equation from (vii).

(xii) Prove by part (viii) that J_λ and $J_{-\lambda}$ are linearly independent functions over $\mathbf{R}$.

Background. In texts on ordinary differential equations it is proved that the dimension over $\mathbf{R}$ of the solution space of Bessel's differential equation equals 2. Bessel's differential equation often occurs when spherical coordinates are used, see for example Exercise 6.68.

(xiii) For $n \in \mathbf{N}_0$ and $x > 0$, show that

$$J_n(x) = \frac{1}{2\pi}\int_{-\pi}^{\pi} e^{ix\sin\alpha - in\alpha}\,d\alpha = \frac{1}{\pi}\int_0^{\pi}\cos(n\alpha - x\sin\alpha)\,d\alpha.$$

Hint: Write

$$e^{ix\sin\alpha} = \sum_{k\in\mathbf{N}_0}\frac{1}{k!}\Big(\frac{x}{2}\Big)^k(e^{i\alpha} - e^{-i\alpha})^k,$$

and prove by means of Newton's Binomial Theorem, for $m \in \mathbf{N}_0$,

$$\frac{1}{2\pi}\int_{-\pi}^{\pi}(e^{i\alpha} - e^{-i\alpha})^{n+2m}e^{-in\alpha}\,d\alpha = (-1)^m\binom{n+2m}{m}.$$

Exercise 6.67 (Kepler's equation – sequel to Exercises 0.19, 3.32 and 6.66). This equation reads, for unknown $x \in \mathbf{R}$ with $y \in \mathbf{R}^2$ as a parameter,

$$x = y_1 + y_2\sin x.$$

Geometrically this comes down to a point of intersection of the graph of the affine function $x \mapsto x - y_1$ with the graph of the sine $x \mapsto y_2\sin x$. If $x = \psi(y)$ with $\psi : \mathbf{R}^2 \to \mathbf{R}$ is a solution, then $f_{y_2} : \mathbf{R} \to \mathbf{R}$ with $f_{y_2}(y_1) = \psi(y) - y_1 = y_2\sin\psi(y)$ is an odd 2π-periodic function. Therefore $y_1 \mapsto D_1\psi(y)$ is even and 2π-periodic. Prove by implicit differentiation

$$D_1\psi(y) = \frac{1}{1 - y_2\cos x}.$$

Prove by Fourier series expansion according to the variable y_1, compare with Exercise 0.19, with coefficients depending on y_2,

$$D_1\psi(y) = \frac{1}{2\pi}\int_{-\pi}^{\pi}\frac{1}{1 - y_2\cos x}\,dy_1 + \sum_{n\in\mathbf{N}}\frac{\cos ny_1}{\pi}\int_{-\pi}^{\pi}\frac{\cos ny_1}{1 - y_2\cos x}\,dy_1.$$

By means of the substitution $y_1 = y_1(x) = x - y_2 \sin x$ and of Exercise 6.66.(xiii), show that

$$\begin{aligned} D_1\psi(y) &= \frac{1}{2\pi}\int_{-\pi}^{\pi} dx + \sum_{n\in\mathbf{N}} \frac{\cos ny_1}{\pi}\int_{-\pi}^{\pi} \cos n(x - y_2 \sin x)\,dx \\ &= 1 + 2\sum_{n\in\mathbf{N}} J_n(ny_2)\cos ny_1. \end{aligned}$$

Hence conclude that

$$x = \psi(y) = y_1 + 2\sum_{n\in\mathbf{N}} \frac{\sin(ny_1)J_n(ny_2)}{n}.$$

Here we do not consider the problem for which values of y the series actually converges.

Exercise 6.68 (Helmholtz' equation and Bessel function – sequel to Exercises 3.9, 3.17 and 6.66 – needed for Exercise 8.35). We use the notations from those exercises. We look for $f \in C^2(\mathbf{R}^3)$ that satisfy the following, known as *Helmholtz' equation*:

$$(\Delta + \mu^2)f(x) = 0 \qquad (x \in \mathbf{R}^3,\ \mu \geq 0),$$

which occurs, for example, in the eigenvalue problem for the Laplacian, see Exercise 7.64.(iii). Here we try to find a solution f of the form

$$f \circ \Psi : (r, \alpha, \theta) \mapsto g(r)\, Y_l^m(\alpha, \theta);$$

where $(r, \alpha, \theta) \in \mathbf{R}_+ \times\ ]-\pi, \pi[\ \times \left]-\frac{\pi}{2}, \frac{\pi}{2}\right[$, while $g \in C^2(\mathbf{R}_+)$ and $l \in \mathbf{N}_0$ and $m \in \mathbf{Z}$ with $|m| \leq l$. By means of Exercises 3.9.(vi) and 3.17.(i), prove that g must satisfy the differential equation

$$(\star) \qquad r^2 g''(r) + 2r\, g'(r) + (\mu^2 r^2 - l(l+1))\, g(r) = 0.$$

Verify that $h : r \mapsto \sqrt{r}\, g(r)$ must then satisfy the differential equation

$$r^2 h''(r) + r\, h'(r) + \left(\mu^2 r^2 - \left(l + \frac{1}{2}\right)^2\right) h(r) = 0.$$

Show that $u : x \mapsto h(\frac{x}{\mu})$ must be a solution of Bessel's differential equation

$$x^2 u''(x) + x\, u'(x) + \left(x^2 - \left(l + \frac{1}{2}\right)^2\right) u(x) = 0.$$

Conclude by Exercise 6.66 that, with a and $b \in \mathbf{C}$, every solution g of $(\star)$ is given by

$$g(r) = \frac{1}{\sqrt{r}}\big(a\, J_{l+\frac{1}{2}}(\mu r) + b\, J_{-l-\frac{1}{2}}(\mu r)\big).$$

In particular, $f_\pm(x) = -\frac{1}{4\pi}\frac{e^{\pm i\mu\|x\|}}{\|x\|}$, for $x \in \mathbf{R}^3 \setminus \{0\}$, satisfy $(\Delta + \mu^2)f(x) = 0$.

Exercise 6.69 (Hypergeometric differential equation – sequel to Exercises 0.11 and 6.50 – needed for Exercises 6.70, 6.106 and 7.5). Assume a, b and c in $\mathbf{R}$. Let $Z \subset \mathbf{C}$ be the complex plane excluding the interval $[\,1, \infty\,[$ along the real axis. For every $z \in Z$ the function $[\,0, 1\,] \to \mathbf{C}$ with $t \mapsto (1 - zt)^{-a}$ is well-defined, and uniquely determined by the requirement that its value be 1, for $z = 0$. Consequently, for a, b and c in $\mathbf{R}$, $z \in Z$, the following function is well-defined on $[\,0, 1\,]$ by

$$t \mapsto t^{b-1}(1 - t)^{c-b-1}(1 - zt)^{-a}.$$

Moreover, the integral from 0 to 1 of this function converges, for $a, b, c \in \mathbf{R}$, $0 < b < c$, $z \in Z$; thus one defines the *hypergeometric function* ${}_2F_1(a, b; c : \cdot) = F(a, b; c : \cdot) : Z \to \mathbf{C}$ by

$$(\star) \qquad F(a, b; c : z) = \frac{\Gamma(c)}{\Gamma(b)\Gamma(c - b)} \int_0^1 t^{b-1}(1 - t)^{c-b-1}(1 - zt)^{-a}\, dt,$$

with Γ as in Exercise 6.50.

(i) In the notation from Exercise 0.11 prove

$$(a)_n = \frac{\Gamma(a + n)}{\Gamma(a)} \qquad (n \in \mathbf{N}),$$

and conclude from that exercise that, if $z \in Z$, $t \in \,]\,0, 1\,[$, with $|zt| < 1$,

$$(1 - zt)^{-a} = \sum_{n \in \mathbf{N}_0} \frac{\Gamma(a + n)}{\Gamma(a)\, n!} (zt)^n.$$

(ii) Now use integration term-by-term of the power series in (i), then Exercise 6.50.(iv) to conclude that $F(a, b; c : z)$, for all $z \in \mathbf{C}$ with $|z| < 1$, is given by the following *hypergeometric series*:

$$\begin{aligned} F(a, b; c : z) &= \frac{\Gamma(c)}{\Gamma(a)\Gamma(b)} \sum_{n \in \mathbf{N}_0} \frac{\Gamma(a + n)\Gamma(b + n)}{\Gamma(c + n)\, n!} z^n \\ &= 1 + \frac{a\, b}{c\, 1} z + \frac{a(a + 1)\, b(b + 1)}{c(c + 1)\, 1 \cdot 2} z^2 \\ &\quad + \frac{a(a + 1)(a + 2)\, b(b + 1)(b + 2)}{c(c + 1)(c + 2)\, 1 \cdot 2 \cdot 3} z^3 + \cdots . \end{aligned}$$

The terminology "hypergeometric" is explained by, inter alia, the relation with the geometric series

$$F(1, 1; 1 : z) = \sum_{n \in \mathbf{N}_0} z^n = \frac{1}{1 - z}.$$

(iii) Assume $0 < a < c$ and $z \in Z$. Prove $F(a, b; c : z) = F(b, a; c : z)$, that is,

$$F(a, b; c : z) = \frac{\Gamma(c)}{\Gamma(a)\Gamma(c-a)} \int_0^1 t^{a-1}(1-t)^{c-a-1}(1-zt)^{-b}\, dt.$$

(iv) Demonstrate (compare with Exercise 6.50.(vii)), for $|z| < 1$,

$$(1+z)^n = F(-n, 1; 1 : -z); \qquad \log(1+z) = zF(1, 1; 2 : -z);$$

$$\arcsin z = zF\Big(\frac{1}{2}, \frac{1}{2}; \frac{3}{2} : z^2\Big) = z + \sum_{n\in\mathbf{N}} \frac{(2n-1)!!}{(2n)!!} \frac{z^{2n+1}}{2n+1}$$

$$:= z + \sum_{n\in\mathbf{N}} \frac{1\cdot 3\cdots(2n-1)}{2\cdot 4\cdots 2n} \frac{z^{2n+1}}{2n+1};$$

$$\arctan z = zF\Big(\frac{1}{2}, 1; \frac{3}{2} : -z^2\Big) = \sum_{n\in\mathbf{N}_0} (-1)^n \frac{z^{2n+1}}{2n+1}.$$

For the moment, write $f(z) = \frac{\Gamma(a)\Gamma(b)}{\Gamma(c)} F(a, b; c : z)$.

(v) Prove that, for $|z| < 1$,

$$\Big(z\frac{d}{dz} + a\Big) f(z) = \sum_{n\in\mathbf{N}_0} \frac{\Gamma(a+n+1)\Gamma(b+n)}{\Gamma(c+n)\, n!} z^n,$$

and conclude that the function f satisfies

$$z\Big(z\frac{d}{dz} + a\Big)\Big(z\frac{d}{dz} + b\Big) f(z) = z\frac{d}{dz}\Big(z\frac{d}{dz} + c - 1\Big) f(z).$$

(vi) Verify, for every $p \in \mathbf{R}$,

$$\frac{d}{dz}\Big(z\frac{d}{dz} + p\Big) = z\frac{d^2}{dz^2} + (p+1)\frac{d}{dz};$$

and conclude from (v) that $u = F(a, b; c : \cdot)$ satisfies the following *hypergeometric differential equation*:

$$(\star\star) \qquad z(1-z)\frac{d^2u}{dz^2}(z) + (c-(a+b+1)z)\frac{du}{dz}(z) - ab\, u(z) = 0 \qquad (z \in \mathbf{C}).$$

(vii) Let $l \in \mathbf{N}_0$. Assume the function u satisfies the hypergeometric differential equation with

$$a = -l, \qquad b = l+1, \qquad c = 1.$$

Define $\Psi : \mathbf{R} \to \mathbf{R}$ by $z = \Psi(t) = \frac{1-t}{2}$. Verify that $v = u \circ \Psi$ satisfies Legendre's differential equation from Exercise 0.4.(i). Note that in this case the parameters a, b and c do not satisfy the restrictions mentioned above.

We now want to prove directly from its integral representation ($\star$) that $F(a, b; c : \cdot)$ satisfies the differential equation in ($\star\star$). To do this we define

$$g(z,t) := t^b(1-t)^{c-b}(1-zt)^{-a-1};$$

then

$$\frac{\partial g}{\partial t}(z,t) = t^{b-1}(1-t)^{c-b-1}(1-zt)^{-a-2}h(z,t),$$

with

$$\begin{aligned} h(z,t) &= b\,(1-t)(1-zt) + (b-c)\,t(1-zt) + (a+1)\,zt(1-t) \\ &= b\,(1-zt)^2 + ((a+b+1)z - c)\,t(1-zt) + (a+1)\,z(z-1)t^2. \end{aligned}$$

So now we have

$$\begin{aligned} \frac{\partial g}{\partial t}(z,t) &= b\,t^{b-1}(1-t)^{c-b-1}(1-zt)^{-a} \\ &\quad + ((a+b+1)z - c)\,t^b(1-t)^{c-b-1}(1-zt)^{-a-1} \\ &\quad + (z^2 - z)(a+1)\,t^{b+1}(1-t)^{c-b-1}(1-zt)^{-a-2}. \end{aligned}$$

In addition we write for the integrand in ($\star$)

$$k(z,t) := t^{b-1}(1-t)^{c-b-1}(1-zt)^{-a}.$$

Note that then

$$\begin{aligned} \frac{\partial k}{\partial z}(z,t) &= a t^b(1-t)^{c-b-1}(1-zt)^{-a-1}; \\ \frac{\partial^2 k}{\partial z^2}(z,t) &= a(a+1)t^{b+1}(1-t)^{c-b-1}(1-zt)^{-a-2}; \\ a\frac{\partial g}{\partial t}(z,t) &= (z^2 - z)\frac{\partial^2 k}{\partial z^2}(z,t) + ((a+b+1)z - c)\frac{\partial k}{\partial z}(z,t) + ab\,k(z,t). \end{aligned}$$

(viii) Prove that the results above do indeed imply that $F(a, b; c : \cdot)$ satisfies ($\star\star$).

Exercise 6.70 (Confluent hypergeometric differential equation – sequel to Exercises 6.66 and 6.69 – needed for Exercise 6.71). Assume a and c in $\mathbf{R}$ are such that $0 < a < c$. We define the *confluent hypergeometric function* or the *Kummer function* ${}_1F_1(a; c : \cdot) = K(a; c : \cdot) : \mathbf{C} \to \mathbf{C}$ by

$$K(a; c : z) = \frac{\Gamma(c)}{\Gamma(a)\Gamma(c-a)} \int_0^1 t^{a-1}(1-t)^{c-a-1}e^{zt}\,dt.$$

(i) Let f_λ be as in Exercise 6.66. Prove, by the substitution of variables $\sin^2 \frac{\alpha}{2} = s$,

$$u(t) := e^{\frac{t}{2}} f_\lambda\Big(\frac{t}{2i}\Big) = 2^{2\lambda} \frac{\Gamma(\lambda + \frac{1}{2})^2}{\Gamma(2\lambda + 1)} K(\lambda + \frac{1}{2};\ 2\lambda + 1 : t) \qquad (t \in \mathbf{R}).$$

Next, use Legendre's duplication formula from Exercise 6.50.(vi) to show that the following formula holds for J_λ, the *Bessel function of order* λ:

$$J_\lambda(x) = \frac{1}{\Gamma(\lambda + 1)} e^{-ix} \Big(\frac{x}{2}\Big)^\lambda K(\lambda + \frac{1}{2};\ 2\lambda + 1 : 2ix) \qquad (x \in \mathbf{R}_+).$$

(ii) Use termwise integration of a power series, then Exercise 6.50.(iv), to conclude that, for all $z \in \mathbf{C}$,

$$K(a; c : z) = \frac{\Gamma(c)}{\Gamma(a)} \sum_{n \in \mathbf{N}_0} \frac{\Gamma(a + n)}{\Gamma(c + n)\, n!} z^n$$

$$= 1 + \frac{a}{c\,1} z + \frac{a(a+1)}{c(c+1)\,1 \cdot 2} z^2 + \frac{a(a+1)(a+2)}{c(c+1)(c+2)\,1 \cdot 2 \cdot 3} z^3 + \cdots.$$

Hint: Use the ratio test to prove the convergence of the series for $z \in \mathbf{C}$.

(iii) Show that part (ii) implies $K(a; a : z) = e^z$, for all $a > 0$ and $z \in \mathbf{C}$.

(iv) Let a, b, c, z and $F(a, b; c : \cdot)$ be as in Exercise 6.69. Note that

$$\lim_{b \to \infty} (1 - \frac{zt}{b})^{-b} = e^{zt},$$

and use Exercise 6.69.(iii) to prove the following equality for the integral representations:

$$(\star) \qquad \lim_{b \to \infty} F\Big(a, b; c : \frac{z}{b}\Big) = K(a; c : z).$$

With a, c and z fixed, show the series for $F(a, b; c : \frac{z}{b})$ from Exercise 6.69.(ii) to be uniformly convergent for $b \in [\,2|z|, \infty\,[$, and conclude that the identity $(\star)$ also holds for the series representations.

For the moment, write $f(z) = \frac{\Gamma(a)}{\Gamma(c)} K(a; c : z)$.

(v) Prove, for $z \in \mathbf{C}$,

$$\Big(z\frac{d}{dz} + a\Big) f(z) = \sum_{n \in \mathbf{N}_0} \frac{\Gamma(a + n + 1)}{\Gamma(c + n)\, n!} z^n,$$

to conclude that the function f satisfies

$$z\Big(z\frac{d}{dz} + a\Big) f(z) = z\frac{d}{dz}\Big(z\frac{d}{dz} + c - 1\Big) f(z).$$

(vi) Verify that part (v), in the same manner as in Exercise 6.69.(vi), implies that $u = K(a; c : \cdot)$ satisfies the following *confluent hypergeometric differential equation*:

$$(\star) \qquad z\,\frac{d^2u}{dz^2}(z) + (c - z)\,\frac{du}{dz}(z) - a\,u(z) = 0 \qquad (z \in \mathbf{C}).$$

(vii) Let $F(a, b; c : \cdot)$ be as in Exercise 6.69. Verify that $u : z \mapsto F(a, b; c : \frac{z}{b})$ satisfies the differential equation

$$z\Big(1 - \frac{z}{b}\Big)\frac{d^2u}{dz^2}(z) + \Big(c - z - \frac{a+1}{b}z\Big)\frac{du}{dz}(z) - a\,u(z) = 0.$$

Conclude that $(\star)$ is obtained from this equation by taking the limit for $b \to \infty$, as expected on the basis of part (iv).

(viii) Assume u satisfies $(\star)$ and write $w(z) = e^{-\frac{1}{2}z} z^{\frac{1}{2}c} u(z)$, $k = \frac{1}{2}c - a$ and $m = \frac{1}{2}c - \frac{1}{2}$. Verify that w then satisfies the following, known as *Whittaker's equation*:

$$\frac{d^2w}{dz^2}(z) - \Big(\frac{1}{4} - \frac{k}{z} + \frac{m^2 - \frac{1}{4}}{z^2}\Big)w(z) = 0.$$

(ix) Prove by Exercise 6.66.(iv) that the function u from part (i) satisfies the following confluent hypergeometric differential equation:

$$t\,u''(t) + (2\lambda + 1 - t)\,u'(t) - (\lambda + \frac{1}{2})\,u(t) = 0 \qquad (t \in \mathbf{R}).$$

Background. One observes that we have now proved that the confluent hypergeometric differential equation is a limiting case of the hypergeometric differential equation, and that Whittaker's and Bessel's differential equations transform, by a substitution of variables among other things, into a confluent hypergeometric differential equation (see also Exercise 6.71).

Exercise 6.71 (Further confluence – sequel to Exercise 6.70).

(i) As in Exercise 6.70.(vii), determine a differential equation for $u : z \mapsto K(a; c : \frac{z}{a})$, and by taking the limit for $a \to \infty$ derive the following differential equation:

$$(\star) \qquad z\,\frac{d^2u}{dz^2}(z) + c\,\frac{du}{dz}(z) - u(z) = 0 \qquad (z \in \mathbf{C}).$$

(ii) Verify that $(\star)$ is satisfied by ${}_0F_1(c : z) := \Gamma(c)\sum_{n\in\mathbf{N}_0} \frac{1}{\Gamma(c+n)\,n!}\,z^n$.

(iii) Assume u satisfies ($\star$), and write $v(z) = u(-\frac{z^2}{4})$. Show that v satisfies the differential equation

$$z\,v''(z) + (2c-1)\,v'(z) + z\,v(z) = 0.$$

Note that, with $c = \lambda + 1$, this is the differential equation from Exercise 6.66.(iv) associated with the Bessel function.

(iv) Finally, examine the effect of the substitution $w(z) = u(cz)$, and thus derive the differential equation $w'(z) - w(z) = 0$, for $w = {}_0F_0 : z \mapsto e^z = \sum_{n\in\mathbf{N}_0} \frac{1}{n!} z^n$.

Exercise 6.72 (Cauchy–Schwarz inequality – needed for Exercise 6.76). Here we give two proofs of this integral inequality. It is assumed that f and g are contained in $C_c(\mathbf{R}^n)$.

(i) Prove that the following numbers are well-defined:

$$\|f\|_2 = \Big(\int_{\mathbf{R}^n} |f(x)|^2\,dx\Big)^{1/2}, \qquad \langle f, g\rangle = \int_{\mathbf{R}^n} f(x)g(x)\,dx.$$

(ii) Verify that Proposition 1.1.6 as well as its proof apply and directly yield $|\langle f, g\rangle| \le \|f\|_2\|g\|_2$, known as the *Cauchy–Schwarz inequality*.

(iii) Prove $ab \le \frac{1}{2}(a^2 + b^2)$, for all $a, b \in \mathbf{R}$. Substitute $a = \frac{f(x)}{\|f\|_2}$ and $b = \frac{g(x)}{\|g\|_2}$ and integrate, to conclude again that the Cauchy–Schwarz inequality holds.

Exercise 6.73 (Hölder's and Minkowski's inequalities – sequel to Exercise 5.41 – needed for Exercises 6.74, 6.75 and 6.79). For $p \ge 1$ and f in $C_c(\mathbf{R}^n)$, we define

$$\|f\|_p = \Big(\int_{\mathbf{R}^n} |f(x)|^p\,dx\Big)^{1/p}.$$

(i) Assume $p > 1$ and $q > 1$ satisfy $\frac{1}{p} + \frac{1}{q} = 1$, and consider $f, g \in C_c(\mathbf{R}^n)$. Prove the following, known as *Hölder's inequality*:

$$\int_{\mathbf{R}^n} |f(x)g(x)|\,dx \le \|f\|_p\,\|g\|_q.$$

Hint: Compare with Exercise 5.41.(v).

(ii) Next, assume $p_k > 1$, for $1 \le k \le r$, and $\sum_{1\le k\le r} \frac{1}{p_k} = 1$, and let $f_k \in C_c(\mathbf{R}^n)$, for $1 \le k \le r$. Prove

$$\int_{\mathbf{R}^n} \prod_{1\le k\le r} |f_k(x)|\,dx \le \prod_{1\le k\le r} \|f_k\|_{p_k}.$$

(iii) Let $p \geq 1$ and $f, g \in C_c(\mathbf{R}^n)$. Prove the following, known as *Minkowski's inequality*:

$$\|f+g\|_p \leq \|f\|_p + \|g\|_p.$$

Hint: The inequality readily follows for $p = 1$. Therefore assume $p > 1$. One has

$$\begin{aligned}\int_{\mathbf{R}^n} |f(x)+g(x)|^p\,dx \quad &\leq \int_{\mathbf{R}^n} |f(x)||f(x)+g(x)|^{p-1}\,dx \\ &\quad + \int_{\mathbf{R}^n} |g(x)||f(x)+g(x)|^{p-1}\,dx.\end{aligned}$$

Now apply part (i), use $(p-1)q = p$, divide by $\left(\int_{\mathbf{R}^n} |f(x)+g(x)|^p\,dx\right)^{1/q}$, and make use of $1 - \frac{1}{q} = \frac{1}{p}$.

Exercise 6.74 (Hilbert's inequality – sequel to Exercises 6.58 and 6.73). Let $I = \mathbf{R}_+$ and let $k : I^2 \to I$ be homogeneous of degree -1. Let $p > 1$ and $q > 1$ with $\frac{1}{p} + \frac{1}{q} = 1$, and define $c \geq 0$ by

$$c = \int_I k(x, 1) x^{-1/p}\,dx.$$

(i) Prove that $c = \int_I k(1, y) y^{-1/q}\,dy$.

(ii) Prove that, for all f and $g \in C_c(I)$ one has, in the notation of Exercise 6.73,

$$\left| \int_{I^2} k(x, y) f(x) g(y)\,dx dy \right| \leq c\,\|f\|_p \|g\|_q.$$

Hint: We have

$$\begin{aligned}\int_I f(x) \int_I k(x, y) g(y)\,dy\,dx \quad &= \int_I f(x) \int_I x k(x, xw) g(xw)\,dw\,dx \\ &= \int_I f(x) \int_I k(1, w) g(xw)\,dw\,dx \\ &= \int_I k(1, w) \int_I f(x) g(xw)\,dx\,dw.\end{aligned}$$

Now apply Hölder's inequality from Exercise 6.73.(i) to the inner integral and use the fact that $\int_I |g(xw)|^q\,dx = \frac{1}{w} \int_I |g(y)|^q\,dy$.

(iii) Consider in particular $k(x, y) = \frac{1}{x+y}$. By means of Exercise 6.58.(v), prove the following, known as *Hilbert's inequality*:

$$\int_{I^2} \frac{|f(x)g(y)|}{x+y}\,dx dy \leq \frac{\pi}{\sin(\frac{\pi}{p})} \|f\|_p \|g\|_q.$$

Exercise 6.75 (Sequel to Exercises 5.41 and 6.73). Let $k \in C(\mathbf{R}^{n+p})$ and assume that $c_1 > 0$ and $c_2 > 0$ satisfy

$$\int_{\mathbf{R}^n} |k(x, y)|\, dx \leq c_1 \quad (y \in \mathbf{R}^p), \qquad \int_{\mathbf{R}^p} |k(x, y)|\, dy \leq c_2 \quad (x \in \mathbf{R}^n).$$

Let $f \in C_c(\mathbf{R}^p)$, $g \in C_c(\mathbf{R}^n)$, $p > 1$ and $q > 1$ with $\frac{1}{p} + \frac{1}{q} = 1$. One then has, in the notation of Exercise 6.73, the following inequality:

$$\left| \int_{\mathbf{R}^{n+p}} k(x, y) f(y) g(x)\, dx dy \right| \leq c_1^{1/p} c_2^{1/q} \, \|f\|_p \|g\|_q.$$

Hint: On account of Young's inequality from Exercise 5.41 one has, for all $t > 0$,

$$|f(y)g(x)| \leq \frac{1}{p} t^p |f(y)|^p + \frac{1}{q} t^{-q} |g(x)|^q.$$

Now minimize the resulting upper bound for the integral with respect to all $t > 0$.

Exercise 6.76 (Poincaré's inequality – sequel to Exercise 6.72 – needed for Exercise 6.77). Let $f \in C_c(\mathbf{R}^n)$ be a C^1 function.

(i) Prove $\int_{\mathbf{R}^n} f(x)^2\, dx = -2 \int_{\mathbf{R}^n} x_j\, f(x)\, D_j f(x)\, dx$, for $1 \leq j \leq n$.

(ii) Use (i) and the Cauchy–Schwarz inequality from Exercise 6.72 to show that

$$\int_{\mathbf{R}^n} f(x)^2\, dx \leq 2 \Big(\int_{\mathbf{R}^n} (x_j\, f(x))^2\, dx \Big)^{1/2} \Big(\int_{\mathbf{R}^n} D_j f(x)^2\, dx \Big)^{1/2}.$$

(iii) Let $U \subset \mathbf{R}^n$ be a bounded open set. Prove the existence of a constant $c = c(U) > 0$ such that, for all f having the additional property that $\mathrm{supp}(f) \subset U$, we have the following, known as *Poincaré's inequality*:

$$\int_U f(x)^2\, dx \leq c \int_U \| \operatorname{grad} f(x) \|^2\, dx.$$

Exercise 6.77 (Heisenberg's uncertainty relations – sequel to Exercise 6.76). In *quantum physics* the state of a particle is described by means of a *wave function*; this will be understood to be a C^2 function with compact support $f : \mathbf{R}^n \times \mathbf{R} \to \mathbf{C}$, for which $\int_{\mathbf{R}^n} |f(x, t)|^2\, dx = 1$, for all $t \in \mathbf{R}$. Here $|f(x, t)|^2$ gives the *probability density* (see also Exercise 6.42) of the particle being found at the point $x \in \mathbf{R}^n$ at time $t \in \mathbf{R}$. The collection of these wave functions is acted upon by the *position operators* Q_j and the *momentum operators* P_j, for $1 \leq j \leq n$, as follows:

$$Q_j f(x, t) = x_j f(x, t), \qquad P_j f(x, t) = \frac{1}{\sqrt{-1}} \frac{\partial f}{\partial x_j}(x, t).$$

(i) Verify that the Q_j and P_j are self-adjoint operators with respect to the *Hermitian inner product* on the collection of wave functions

$$\langle f, g \rangle = \int_{\mathbf{R}^n} f(x,t)\,\overline{g(x,t)}\,dx.$$

Then the *expectation vectors* $x^0 \in \mathbf{C}^n$ for the position and $p^0 \in \mathbf{C}^n$ for the momentum, respectively, of the particle described by the wave function f are given by

$$x_j^0 = \langle\, Q_j f,\ f\,\rangle \qquad \text{and} \qquad p_j^0 = \langle\, P_j f,\ f\,\rangle \qquad (1 \le j \le n).$$

(ii) Check that in fact we have $x^0 \in \mathbf{R}^n$ and $p^0 \in \mathbf{R}^n$.

The *uncertainties* or *standard deviations* $\Delta x_j \ge 0$ and $\Delta p_j \ge 0$, in the j-th coordinate of the position x and of the momentum p, respectively, of the particle described by the wave function f are defined by

$$(\Delta x_j)^2 = \|(Q_j - x_j^0)f\|^2 = \int_{\mathbf{R}^n} x_j^2\, f(x,t)\,\overline{f(x,t)}\,dx - (x_j^0)^2,$$

$$(\Delta p_j)^2 = \|(P_j - p_j^0)f\|^2 = \int_{\mathbf{R}^n} -\frac{\partial^2 f}{\partial x_j^2}(x,t)\,\overline{f(x,t)}\,dx - (p_j^0)^2.$$

(iii) Check that one may assume $x^0 = 0$ and $p^0 = 0$, using the substitutions where $f(x)$ is replaced by $f(x + x^0)$ or by $e^{-\sqrt{-1}\langle p^0, x\rangle} f(x)$, respectively.

(iv) Prove the following, known as *Heisenberg's uncertainty relations*:

$$\Delta x_j\, \Delta p_j \ge \frac{1}{2} \qquad (1 \le j \le n).$$

That is, the smaller the uncertainty in the position x of the particle, the larger the uncertainty in its momentum p, and vice versa.
Hint: See Exercise 6.76.(ii).

Exercise 6.78 (Sobolev's inequality). Let $C_c^1(\mathbf{R}^n) = C_c(\mathbf{R}^n) \cap C^1(\mathbf{R}^n)$. In general, the constant occurring on the right–hand side of Poincaré's inequality from Exercise 6.76.(iii) depends on the open set $U \subset \mathbf{R}^n$. In contrast with this, *Sobolev's inequality* asserts the following. Let $p \in \mathbf{R}$ with $1 \le p < n$ and define $p^\star \in \mathbf{R}$ by

$$\frac{1}{p^\star} = \frac{1}{p} - \frac{1}{n} \qquad \left(\text{that is, } p^\star = \frac{n}{n-p}\, p > p\right).$$

Then there exists a constant $c = c(n,p) > 0$ such that in the notation from Exercise 6.73, for all $f \in C_c^1(\mathbf{R}^n)$,

$$(\star) \qquad \|f\|_{p^\star} \le c \sum_{1 \le j \le n} \|D_j f\|_p.$$

This is an estimate for a function f in terms of its partial derivatives which is valid regardless of the support of f. This generality is possible owing to the occurrence, on the left–hand side, of an integral expression containing $p^\star$ instead of p. In estimates of this kind the rate at which f converges to 0 is of importance, and $|f|^{p^\star}$ converges to 0 more rapidly than $|f|^p$. We now prove ($\star$) in five steps.

(i) Apply the Fundamental Theorem of Integral Calculus 2.10.1 twice, to prove, for $1 \le j \le n$

$$2|f(x)| \le \int_{\mathbf{R}} |D_j f(x)|\, dx_j =: f_j(x)^{n-1} \qquad (x \in \mathbf{R}^n).$$

Conclude that

$$|2f(x)|^{n/(n-1)} \le \Big(\prod_{1 \le j \le n-1} f_j(x) \Big) \Big(\int_{\mathbf{R}} |D_n f(x)|\, dx_n \Big)^{1/(n-1)}.$$

(ii) Now integrate with respect to the variable x_n, and apply Exercise 6.73.(ii), with

$$\frac{1}{p_1} = \cdots = \frac{1}{p_{n-1}} = \frac{1}{n-1}, \qquad \text{and} \qquad f_j \qquad (1 \le j < n).$$

Conclude that

$$\int_{\mathbf{R}} |2f(x)|^{n/(n-1)}\, dx_n$$

$$\le \prod_{1 \le j \le n-1} \Big(\int_{\mathbf{R}^2} |D_j f(x)|\, dx_j\, dx_n \Big)^{1/(n-1)} \Big(\int_{\mathbf{R}} |D_n f(x)|\, dx_n \Big)^{1/(n-1)}.$$

(iii) Integrate with respect to the variable x_{n-1}, and again apply Exercise 6.73.(ii), this time with the functions, for $1 \le j \le n-2$,

$$\Big(\int_{\mathbf{R}^2} |D_j f(x)|\, dx_j\, dx_n \Big)^{1/(n-1)}, \qquad \Big(\int_{\mathbf{R}} |D_n f(x)|\, dx_n \Big)^{1/(n-1)}.$$

Repeat these operations to prove

$$\int_{\mathbf{R}^n} |2f(x)|^{n/(n-1)}\, dx \le \prod_{1 \le j \le n} \Big(\int_{\mathbf{R}^n} |D_j f(x)|\, dx \Big)^{1/(n-1)},$$

that is,

$$(\star\star) \qquad \Big(\int_{\mathbf{R}^n} |f(x)|^{n/(n-1)}\, dx \Big)^{(n-1)/n} \le \frac{1}{2} \Big(\prod_{1 \le j \le n} \int_{\mathbf{R}^n} |D_j f(x)|\, dx \Big)^{1/n}.$$

(iv) In the case $p = 1$, the inequality $(\star)$ immediately follows from $(\star\star)$ by means of the elementary inequality

$$(\star\star\star) \qquad \prod_{1\le j\le n} g_j \le \frac{1}{n!}\Big(\sum_{1\le j\le n} g_j\Big)^n \qquad (g_j \ge 0,\ 1 \le j \le n).$$

(v) Now let $p > 1$. Then

$$\frac{n-1}{n}p^\star = \frac{n-1}{n-p}p > 1;$$

and as a consequence we have $f^{\frac{n-1}{n}p^\star} \in C_c^1(\mathbf{R}^n)$. Apply $(\star\star)$, with f replaced by $f^{\frac{n-1}{n}p^\star}$. Then use the fact that, for q obeying $\frac{1}{p} + \frac{1}{q} = 1$,

$$\int_{\mathbf{R}^n} |f(x)|^{\frac{n-1}{n}p^\star - 1}|D_j f(x)|\,dx$$

$$\le \Big(\int_{\mathbf{R}^n} |f(x)|^{(\frac{n-1}{n}p^\star - 1)q}\,dx\Big)^{1/q}\Big(\int_{\mathbf{R}^n} |D_j f(x)|^p\,dx\Big)^{1/p}.$$

One has

$$(\frac{n-1}{n}p^\star - 1)q = p^\star, \qquad \frac{n-1}{n} - \frac{1}{q} = \frac{1}{p^\star}, \qquad \frac{n-1}{n}p^\star = \frac{n-1}{n-p}p.$$

Therefore, $(\star\star)$ leads to

$$\Big(\int_{\mathbf{R}^n} |f(x)|^{p^\star}\,dx\Big)^{1/p^\star} \le \frac{p}{2}\frac{n-1}{n-p}\Big(\prod_{1\le j\le n}\Big(\int_{\mathbf{R}^n} |D_j f(x)|^p\,dx\Big)^{1/p}\Big)^{1/n}.$$

Now use $(\star\star\star)$ once again to prove the validity of $(\star)$, for $p > 1$.

Exercise 6.79 (Hardy's inequality – sequel to Exercise 6.73). On the left–hand side in Sobolev's inequality from Exercise 6.78 an exponent $p^\star$ occurs which differs from p on the right–hand side. In that exercise the point was made that this is related to the rate at which the integrands converge to 0. On $\mathbf{R}$ the desired increase in the rate of convergence is obtained if on the left–hand side one writes $\frac{1}{x}f(x)$ instead of $f(x)$; in this context, see also Exercise 7.72. That is, for functions $f \in C_c^1(I)$ where $I = \mathbf{R}_+$, one has *Hardy's inequality*, which reads

$$(\star) \qquad \Big(\int_I \Big|\frac{1}{x}f(x)\Big|^p\,dx\Big)^{1/p} \le \frac{p}{p-1}\Big(\int_I |f'(x)|^p\,dx\Big)^{1/p}.$$

We prove $(\star)$ in three steps.

(i) Write $F(x) = \frac{1}{x}f(x)$, and prove $\int_I F(x)^p\,dx = -p\int_I F(x)^{p-1}\,x\,F'(x)\,dx$.

(ii) Note that $x\,F'(x) = -F(x) + f'(x)$, and deduce $(p-1)\int_I F(x)^p\,dx = p\int_I F(x)^{p-1} f'(x)\,dx$.

(iii) Check that there is no loss of generality in assuming that $F(x) \geq 0$. Then use Exercise 6.73.(i) and arguments such as those in Exercise 6.73.(iii) to prove $(\star)$.

Exercise 6.80 (Inequality by Fourier transformation). For $f \in \mathcal{S}(\mathbf{R}^n)$, write (compare with Exercise 6.73)

$$\|f\|_\infty = \sup\{\,|f(x)| \mid x \in \mathbf{R}^n\,\}, \qquad \|f\|_1 = \int_{\mathbf{R}^n} |f(x)|\,dx.$$

Show that there exists a constant $c = c_n > 0$ such that, for all $f \in \mathcal{S}(\mathbf{R}^n)$,

$$\|f\|_\infty \leq c_n \sum_{|\alpha|\leq n+1} \|D^\alpha f\|_1.$$

Hint: Prove

$$(1+\|\xi\|^2)^{\frac{n+1}{2}} \leq \Big(1 + \sum_{1\leq j\leq n} |\xi_j|\Big)^{n+1} \leq c_n' \sum_{|\alpha|\leq n+1} |\xi^\alpha| \qquad (\xi \in \mathbf{R}^n),$$

and deduce

$$|\widehat{f}(\xi)| \leq c_n'(1+\|\xi\|^2)^{-\frac{n+1}{2}} \sum_{|\alpha|\leq n+1} |\xi^\alpha \widehat{f}(\xi)|$$

$$= c_n'(1+\|\xi\|^2)^{-\frac{n+1}{2}} \sum_{|\alpha|\leq n+1} |\widehat{D^\alpha f}(\xi)| \leq c_n'(1+\|\xi\|^2)^{-\frac{n+1}{2}} \sum_{|\alpha|\leq n+1} \|D^\alpha f\|_1.$$

Verify $\|f\|_\infty \leq (2\pi)^{-n}\|\widehat{f}\|_1$ by means of the Fourier Inversion Theorem, and conclude that

$$\|f\|_\infty \leq c_n'' \int_{\mathbf{R}^n} (1+\|\xi\|^2)^{-\frac{n+1}{2}}\,d\xi \sum_{|\alpha|\leq n+1} \|D^\alpha f\|_1.$$

Exercise 6.81 (Eigenfunction for translation). Suppose $f \in C(\mathbf{R}^n)$ is a nontrivial eigenfunction for translation, that is, for every $y \in \mathbf{R}^n$ there exists $\lambda(y) \in \mathbf{R}$ with

$$(\star) \qquad f(x+y) = \lambda(y) f(x) \qquad (x \in \mathbf{R}^n), \qquad f(0) = 1.$$

Show $\lambda(y) = f(y)$, for $y \in \mathbf{R}^n$. Select $g \in C_c(\mathbf{R}^n) \cap C^\infty(\mathbf{R}^n)$ satisfying $\int_{\mathbf{R}^n} f(y)g(y)\,dy = 1$. Deduce

$$f(x) = \int_{\mathbf{R}^n} f(x+y)g(y)\,dy = \int_{\mathbf{R}^n} g(y-x)f(y)\,dy,$$

and prove $f \in C^\infty(\mathbf{R}^n)$ by differentiation under the integral sign. Now differentiate $(\star)$ with respect to the variable y and show $f(x) = e^{Df(0)x}$, for all $x \in \mathbf{R}^n$ (compare with Exercise 0.2.(iii)).

Exercise 6.82 (Another proof of Theorem 6.11.3.(iii)). The assertion can be proved without recourse to the Differentiation Theorem 2.10.13.

(i) Using Taylor expansion of $t \mapsto e^{-it}$ show that there exists a constant $c > 0$ such that for all $t \in \mathbf{R}$ we have $|r(t)| \le ct^2$ if $e^{-it} = 1 - it + r(t)$.

(ii) Deduce, for $\xi, \eta \in \mathbf{R}^n$,

$$\begin{aligned}\widehat{f}(\xi + \eta) - \widehat{f}(\xi) &= \int_{\mathbf{R}^n} e^{-i\langle x,\xi\rangle}(e^{-i\langle x,\eta\rangle} - 1) f(x)\, dx \\ &= \int_{\mathbf{R}^n} e^{-i\langle x,\xi\rangle}(-i\langle x, \eta\rangle + r(\langle x, \eta\rangle)) f(x)\, dx \\ &=: -i \sum_{1\le j\le n} \eta_j \widehat{(\cdot_j f)}(\xi) + R(\xi, \eta).\end{aligned}$$

Conclude by part (i) and the Cauchy–Schwarz inequality that

$$\begin{aligned}|R(\xi, \eta)| &\le \int_{\mathbf{R}^n} |r(\langle x, \eta\rangle)||f(x)|\, dx \le c\|\eta\|^2 \int_{\mathbf{R}^n} \|x\|^2 |f(x)|\, dx \\ &= \mathcal{O}(\|\eta\|^2), \qquad \|\eta\| \downarrow 0.\end{aligned}$$

Deduce that $\widehat{f}$ is differentiable in ξ; apply this argument repeatedly to conclude that $\widehat{f} \in C^\infty(\mathbf{R}^n)$, and that Theorem 6.11.3.(iii) holds.

Exercise 6.83 (Another proof of Example 6.11.4 – sequel to Exercises 2.84 and 6.51).

(i) By taking the real and imaginary parts of $\widehat{g}$, for $n = 1$, show that the result from Example 6.11.4 is equivalent to the formulae, valid for $\xi \in \mathbf{R}$,

$$\sqrt{\frac{\pi}{2}} e^{-\frac{1}{2}\xi^2} = \int_{\mathbf{R}_+} e^{-\frac{1}{2}x^2} \cos(\xi x)\, dx, \qquad 0 = \int_{\mathbf{R}} e^{-\frac{1}{2}x^2} \sin(\xi x)\, dx.$$

(ii) Demonstrate that the first identity follows from Exercise 2.84 or by series expansion of the cosine and by Exercise 6.51, and that the second one is trivial.

Exercise 6.84. Assume $0 \ne g \in \mathcal{S}(\mathbf{R}^n)$ and that g vanishes in a neighborhood in $\mathbf{R}^n$ of 0. Let $f = \widehat{g} \in \mathcal{S}(\mathbf{R}^n)$. Prove that all *moments* of f vanish, that is,

$$\int_{\mathbf{R}^n} x^\alpha f(x)\, dx = 0 \qquad (\alpha \in \mathbf{N}_0^n).$$

Exercise 6.85. Prove that the convolution $f * g \in \mathcal{S}(\mathbf{R}^n)$, if f and $g \in \mathcal{S}(\mathbf{R}^n)$.
Hint: Use that $|x^\beta| \le 2^{|\beta|} \sum_{0 \le j \le |\beta|} \|x - y\|^j \, \|y\|^{|\beta|-j}$, for all $x, y \in \mathbf{R}^n$ and $\beta \in \mathbf{N}_0^n$.

Exercise 6.86 (Parseval–Plancherel's identity – needed for Exercises 6.94, 6.100 and 7.31). Let $f, h \in \mathcal{S}(\mathbf{R}^n)$ and $y \in \mathbf{R}^n$.

(i) Prove $\int_{\mathbf{R}^n} \widehat{f}(\xi) h(\xi) e^{i\langle y, \xi \rangle} \, d\xi = \int_{\mathbf{R}^n} f(x) \widehat{h}(x - y) \, dx$.

(ii) Conclude that $\int_{\mathbf{R}^n} f(x) \widehat{h}(x) \, dx = \int_{\mathbf{R}^n} \widehat{f}(\xi) h(\xi) \, d\xi$.

Now let $g \in \mathcal{S}(\mathbf{R}^n)$, and introduce $h = (2\pi)^{-n} \overline{\widehat{g}}$.

(iii) Verify that $h \in \mathcal{S}(\mathbf{R}^n)$ and $\widehat{h} = \overline{g}$.

(iv) Conclude that we have the following, known as *Parseval–Plancherel's identity*:

$$\int_{\mathbf{R}^n} f(x) \overline{g(x)} \, dx = (2\pi)^{-n} \int_{\mathbf{R}^n} \widehat{f}(\xi) \overline{\widehat{g}(\xi)} \, d\xi.$$

Exercise 6.87 (Poisson's summation formula for R – sequel to Exercise 0.18). Let $f \in \mathcal{S}(\mathbf{R})$. Multiply both sides of the identity in Exercise 0.18.(ii) for $n = 2$ by the second derivative $f''(x)$, integrate over $\mathbf{R}$, and use integration by parts twice, keeping in mind that $\lim_{x \to \pm\infty} f(x) = \lim_{x \to \pm\infty} f'(x) = 0$. Verify *Poisson's summation formula* (see Exercise 6.88.(ii) for a different proof)

$$\sum_{k \in \mathbf{Z}} f(k) = \sum_{k \in \mathbf{Z}} \widehat{f}(2\pi k),$$

and derive

$$\sum_{k \in \mathbf{Z}^n} f(x + k) = \sum_{k \in \mathbf{Z}^n} \widehat{f}(2\pi k) \, e^{2\pi i \langle k, x \rangle} \qquad (x \in \mathbf{R}^n).$$

Exercise 6.88 (Poisson's summation formula and Jacobi's functional equation – sequel to Exercise 0.19 – needed for Exercises 6.89, 6.90, 6.91 and 6.96). For $f \in \mathcal{S}(\mathbf{R}^n)$ the function $F : \mathbf{R}^n \to \mathbf{R}$ is well-defined by $F(x) = \sum_{j \in \mathbf{Z}^n} f(x + j)$. Check that this series, as indeed the series of the α-th derivatives also, is uniformly convergent on $\mathbf{R}^n$, for $\alpha \in \mathbf{Z}^n$. Thus one defines a C^∞ function F on $\mathbf{R}^n$ which is invariant under translation with elements from $\mathbf{Z}^n$. Moreover, $(f_k)_{k \in \mathbf{Z}^n}$ with $f_k(x) = e^{2\pi i \langle k, x \rangle}$ is a maximal orthonormal system on $[\,0, 1\,]^n$; expansion according to Exercise 0.19 of F in the corresponding Fourier series therefore yields $F = \sum_{k \in \mathbf{Z}^n} \langle F, f_k \rangle f_k$.

(i) Demonstrate, for $k \in \mathbf{Z}^n$,

$$\begin{aligned}\langle F, f_k\rangle &= \int_{[0,1]^n} \sum_{j\in\mathbf{Z}^n} f(x+j)e^{-2\pi i\langle k,x\rangle}\,dx = \int_{\mathbf{R}^n} e^{-i\langle 2\pi k,x\rangle} f(x)\,dx \\ &= \widehat{f}(2\pi k).\end{aligned}$$

(ii) Prove the following two formulae, the latter of which is known as *Poisson's summation formula*, for $x \in \mathbf{R}^n$:

$$\sum_{k\in\mathbf{Z}^n} f(x+k) = \sum_{k\in\mathbf{Z}^n} \widehat{f}(2\pi k)\, e^{2\pi i\langle k,x\rangle}, \qquad \sum_{k\in\mathbf{Z}^n} f(k) = \sum_{k\in\mathbf{Z}^n} \widehat{f}(2\pi k).$$

Define ψ and $\phi : \mathbf{R}_+ \to \mathbf{R}$ by

$$\psi(x) = \sum_{n\in\mathbf{Z}} e^{-\pi n^2 x}, \qquad \phi(x) = \frac{1}{2}(\psi(x) - 1) = \sum_{n\in\mathbf{N}} e^{-\pi n^2 x}.$$

(iii) Using part (ii) and Example 6.11.4, prove the following, known as *Jacobi's functional equation*:

$$\psi\Big(\frac{1}{x}\Big) = x^{\frac{1}{2}}\psi(x) \qquad (x \in \mathbf{R}_+).$$

(iv) Conclude that $\phi(\frac{1}{x}) = x^{\frac{1}{2}}\,\phi(x) + \frac{1}{2}(x^{\frac{1}{2}} - 1)$ for $x \in \mathbf{R}_+$.

(v) Demonstrate $0 < \phi(x) \le \sum_{n\in\mathbf{N}} e^{-\pi n x} = \frac{e^{-\pi x}}{1-e^{-\pi x}} = \mathcal{O}(e^{-\pi x}), \quad x \to \infty.$

(vi) Conclude by parts (iv) and (v) that $\phi(x) = \mathcal{O}(x^{-\frac{1}{2}}), \quad x \downarrow 0.$

Exercise 6.89 (Functional equation for zeta function – sequel to Exercises 0.25, 6.50, 6.57, 6.58 and 6.88). For $s \in \mathbf{C}$ with $\operatorname{Re} s > 0$ one has

$$\pi^{-\frac{s}{2}}\Gamma\Big(\frac{s}{2}\Big)\frac{1}{n^s} = \int_{\mathbf{R}_+} e^{-\pi n^2 x}x^{\frac{s}{2}-1}\,dx \qquad (n \in \mathbf{N}).$$

Sum over $n \in \mathbf{N}$ and interchange the order of summation and integration. Let $\phi : \mathbf{R}_+ \to \mathbf{R}$ be as in Exercise 6.88. Using parts (v) and (vi) from that exercise one obtains, for $\operatorname{Re} s > 1$,

$$\Lambda(s) := \pi^{-\frac{s}{2}}\Gamma\Big(\frac{s}{2}\Big)\zeta(s) = \int_{\mathbf{R}_+} \sum_{n\in\mathbf{N}} e^{-\pi n^2 x}x^{\frac{s}{2}-1}\,dx = \int_{\mathbf{R}_+} \phi(x)x^{\frac{s}{2}-1}\,dx.$$

Prove with Exercise 6.88.(iv)

$$\int_0^1 \phi(x)x^{\frac{s}{2}-1}\,dx = \int_1^\infty \phi(x)x^{\frac{1-s}{2}-1}\,dx - \frac{1}{1-s} - \frac{1}{s} \qquad (\mathrm{Re}\, s > 1).$$

Therefore

$$\Lambda(s) = \int_1^\infty \phi(x)x^{\frac{1-s}{2}-1}\,dx - \frac{1}{1-s} + \int_1^\infty \phi(x)x^{\frac{s}{2}-1}\,dx - \frac{1}{s} \qquad (\mathrm{Re}\, s > 1).$$

The integral

$$\int_1^\infty \phi(x)x^{\frac{s}{2}-1}\,dx$$

converges for all $s \in \mathbf{C}$, and moreover it defines a complex-differentiable function of $s \in \mathbf{C}$. This implies that $\Lambda(s)$ can be defined for all $s \in \mathbf{C}$ with the exception of $s = 1$ and $s = 0$. Because $\Gamma(\frac{s}{2}) \neq 0$ for all $s \in \mathbf{C}$, the preceding result defines $\zeta(s)$ for those values of s also. In this connection, see Exercise 6.57 for the definition of $\Gamma(s)$, with $s \in \mathbf{C}$. Furthermore, we find that $s \mapsto \Lambda(s)$ is invariant under the substitution $s \mapsto 1 - s$; that is,

$$\Lambda(s) = \Lambda(1-s); \qquad \text{and} \qquad \Gamma\Big(\frac{s}{2}\Big)\,\zeta(s) = \pi^{s-\frac{1}{2}}\Gamma\Big(\frac{1-s}{2}\Big)\,\zeta(1-s).$$

Multiply both sides by $\Gamma(1 - \frac{1}{2}s)$, then apply the reflection formula from Exercise 6.58.(iv), and Legendre's duplication formula from Exercise 6.50.(vi); this yields the *functional equation for the zeta function*

$$2^{s-1}\pi^s\zeta(1-s) = \cos\Big(\frac{s}{2}\pi\Big)\Gamma(s)\,\zeta(s) \qquad (s \in \mathbf{C}).$$

In particular we find, by Exercise 0.25, the result from Exercise 0.20 or 6.40.(iv)

$$\zeta(2n) = (-1)^{n-1}\frac{1}{2}(2\pi)^{2n}\frac{B_{2n}}{(2n)!}.$$

Exercise 6.90 (Poisson's summation formula and Fourier Inversion Theorem – sequel to Exercise 6.88). For $f \in \mathcal{S}(\mathbf{R}^n)$ and $x \in [\,0, 2\pi\,]^n$ define $g \in \mathcal{S}(\mathbf{R}^n)$ by $g(t) = e^{-i\langle t, x\rangle} f(t)$.

(i) Verify $\widehat{g}(\xi) = \widehat{f}(\xi + x)$. Next apply Exercise 6.88.(ii) to g and deduce

$$\sum_{k\in\mathbf{Z}^n} f(k)e^{-i\langle k, x\rangle} = \sum_{k\in\mathbf{Z}^n} \widehat{f}(2\pi k + x).$$

(ii) Integrate this equality termwise with respect to the variable x over $[\,0, 2\pi\,]^n$ in order to obtain

$$(2\pi)^n f(0) = \sum_{k\in\mathbf{Z}^n} \int_{[\,0,2\pi\,]^n} \widehat{f}(2\pi k + x)\,dx = \int_{\mathbf{R}^n} \widehat{f}(\xi)\,d\xi.$$

Obtain the Fourier Inversion Theorem 6.11.6 for $f \in \mathcal{S}(\mathbf{R}^n)$.

Exercise 6.91 (Partial-fraction decomposition of hyperbolic functions – sequel to Exercise 6.88). For $t > 0$ we define $f : \mathbf{R} \to \mathbf{R}$ by $f(x) = e^{-t|x|}$. Then we have (see also Exercise 6.99.(ii))

$$\widehat{f}(\xi) = 2\int_{\mathbf{R}_+} e^{-tx}\cos\xi x\,dx = \frac{2t}{t^2+\xi^2} \qquad (\xi \in \mathbf{R}).$$

(i) Show by summation of the geometric series, for $t > 0$ and $0 \le x < 1$,

$$\sum_{k\in\mathbf{Z}} e^{-t|x+k|} = e^{-tx} + \sum_{k\in\mathbf{N}} e^{-t(x+k)} + \sum_{k\in\mathbf{N}} e^{t(x-k)} = \frac{\cosh t(x-\frac{1}{2})}{\sinh\frac{t}{2}}.$$

(ii) Prove by using Exercise 6.88.(ii), part (i), and the fact that $\widehat{f}$ is an even function

$$\frac{\cosh t(x-\frac{1}{2})}{\sinh\frac{t}{2}} = \frac{2}{t} + 2\sum_{k\in\mathbf{N}} \frac{2t}{t^2+4\pi^2k^2}\cos(2\pi kx) \qquad (t > 0,\ 0 \le x < 1).$$

(iii) Substitute $x = 0$ and $x = \frac{1}{2}$, in the identity in (ii) and derive the following *partial-fraction decomposition* of the hyperbolic cotangent and the hyperbolic cosecant, respectively, for $x \in \mathbf{R} \setminus \{0\}$,

$$\pi\,\mathrm{cotanh}(\pi x) = \frac{1}{x} + 2x\sum_{k\in\mathbf{N}} \frac{1}{x^2+k^2}, \qquad \frac{\pi}{\sinh(\pi x)} = \frac{1}{x} + 2x\sum_{k\in\mathbf{N}} \frac{(-1)^k}{x^2+k^2}.$$

(iv) Use the equality $\mathrm{cotanh}(\pi x) - \frac{1}{\sinh(\pi x)} = \tanh(\pi\frac{x}{2})$ to obtain from (iii)

$$\frac{\pi}{2}\tanh(\frac{\pi}{2}x) = 2x\sum_{k\in\mathbf{N}} \frac{1}{x^2+(2k-1)^2} \qquad (x \in \mathbf{R}).$$

(v) Differentiate the identity in (ii) with respect to x, take $x = \frac{1}{4}$, and use the equality $\sinh\frac{t}{2} = 2\sinh\frac{t}{4}\cosh\frac{t}{4}$ to find

$$\frac{1}{\cosh\frac{t}{4}} = 16\pi\sum_{k\in\mathbf{N}} \frac{k}{t^2+4\pi^2k^2}\sin\Big(\frac{\pi}{2}k\Big).$$

Prove

$$\frac{\pi}{2\cosh(\frac{\pi}{2}x)} = 2\sum_{k\in\mathbf{N}} (-1)^{k-1}\frac{2k-1}{x^2+(2k-1)^2} \qquad (x \in \mathbf{R}).$$

Background. By replacing x by ix and using the identities $\cosh ix = \cos x$ and $\sinh ix = i \sin x$, we see that the identities in parts (iii) – (v) go over in their counterparts in Exercise 0.13.(i).

(vi) From parts (iii) – (v) deduce the following formulae, by expanding the denominators of the integrands in a geometric series and integrating termwise, for $\xi \in \mathbf{R}$:

$$\int_{\mathbf{R}_+} \frac{\sin \xi x}{e^x - 1}\,dx = \frac{\pi}{2}\Big(\operatorname{cotanh} \pi\xi - \frac{1}{\pi\xi}\Big),$$

$$\int_{\mathbf{R}_+} \frac{\sin \xi x}{\sinh x}\,dx = \frac{\pi}{2} \tanh\Big(\frac{\pi}{2}\xi\Big),$$

$$\int_{\mathbf{R}_+} \frac{\cos \xi x}{\cosh x}\,dx = \frac{\pi}{2\cosh\big(\frac{\pi}{2}\xi\big)},$$

$$\int_{\mathbf{R}} \frac{e^{-ix\xi}}{\cosh x}\,dx = \frac{\pi}{\cosh\big(\frac{\pi}{2}\xi\big)}.$$

Exercise 6.92 (Solution of heat equation – sequel to Exercise 6.43 – needed for Exercises 6.103 and 8.35). Let $k > 0$ and define $g : \mathbf{R}^{n+1} \setminus \{0\} \to \mathbf{R}$ by (see Formula (6.48))

$$g(x,t) = g_t(x) = \begin{cases} (4\pi kt)^{-\frac{n}{2}}\, e^{-\frac{\|x\|^2}{4kt}}, & t > 0; \\ 0, & t \le 0. \end{cases}$$

(i) Prove that $g : \mathbf{R}^{n+1} \setminus \{0\} \to \mathbf{R}$ is a C^∞ function (see the proof of Theorem 6.7.4 for the points $(x, 0) \in \mathbf{R}^{n+1}$ with $x \neq 0$).

(ii) Using Exercise 6.43, check that $\int_{\mathbf{R}^n} g_t(x)\,dx = 1$, for $t \in \mathbf{R}_+$.

(iii) Prove, for $x \in \mathbf{R}^n \setminus \{0\}$, $t \in \mathbf{R}_+$ and $1 \le j \le n$,

$$D_t g(x,t) = \Big(\frac{\|x\|^2}{4kt^2} - \frac{n}{2t}\Big) g(x,t); \qquad D_j g(x,t) = -\frac{x_j}{2kt} g(x,t).$$

Conclude that g on $\mathbf{R}^{n+1} \setminus \{0\}$ satisfies the heat equation (6.46).

(iv) Conclude by application of the Differentiation Theorem 2.10.13 that the function u from Formula (6.49) satisfies the heat equation; so that one has

$$u(x,t) = (f * g_t)(x) = \int_{\mathbf{R}^n} f(y) g_t(x-y)\,dy$$

$$= \pi^{-n/2} \int_{\mathbf{R}^n} f(x - 2\sqrt{kt}\,y) e^{-\|y\|^2}\,dy.$$

Further show that $u(x,0) = f(x)$.

Exercise 6.93 (Fourier transform of probability density of normal distribution and covariance matrix – sequel to Exercise 6.44 – needed for Exercise 6.94). Let $C \in \mathrm{Mat}^+(n, \mathbf{R})$ be positive definite, and define $f : \mathbf{R}^n \to \mathbf{R}$ by

$$f(x) = \frac{1}{(2\pi)^{\frac{n}{2}}\sqrt{\det C}} e^{-\frac{1}{2}\langle C^{-1}x,\, x\rangle}.$$

(i) (Compare with Example 6.11.4.) Verify $(D_k f)(x) = -(C^{-1}x)_k f(x)$ for $1 \le k \le n$. Now write $D = (D_1, \dots, D_n)$. Fourier transformation then yields

$$-\xi_k \widehat{f}(\xi) = ((C^{-1}D)_k \widehat{f})(\xi) \qquad (1 \le k \le n).$$

Next, multiply by C_{jk} and sum over $1 \le k \le n$ to obtain $-(C\xi)_j \widehat{f}(\xi) = (D_j \widehat{f})(\xi)$, for $1 \le j \le n$. Now conclude that $\widehat{f}(\xi) = e^{-\frac{1}{2}\langle C\xi,\, \xi\rangle}$.

(ii) Expand both sides of the identity

$$\int_{\mathbf{R}^n} e^{-i\langle x,\, \xi\rangle} f(x)\, dx = e^{-\frac{1}{2}\langle C\xi,\, \xi\rangle}$$

in a power series in $\xi \in \mathbf{R}^n$, and compare the coefficients. This leads to

$$\int_{\mathbf{R}^n} f(x)\, dx = 1; \qquad \int_{\mathbf{R}^n} x_j f(x)\, dx = 0, \qquad \int_{\mathbf{R}^n} x_j x_k f(x)\, dx = C_{jk},$$

for $1 \le j, k \le n$.

Now let $\mu \in \mathbf{R}^n$ and define

$$f(x) = \frac{1}{(2\pi)^{\frac{n}{2}}\sqrt{\det C}} e^{-\frac{1}{2}\langle C^{-1}(x-\mu),\, (x-\mu)\rangle}.$$

(iii) Prove, for $1 \le j, k \le n$,

$$\int_{\mathbf{R}^n} f(x)\, dx = 1, \qquad \int_{\mathbf{R}^n} (x_j - \mu_j) f(x)\, dx = 0,$$

$$\int_{\mathbf{R}^n} (x_j - \mu_j)(x_k - \mu_k) f(x)\, dx = C_{jk}.$$

Background. In the terminology of Exercise 6.42 the function $f = f(\mu, C)$ is said to be the *probability density of the normal distribution* on $\mathbf{R}^n$ with *expectation vector* μ and *covariance matrix* C.

Exercise 6.94 (Asymptotic expansion for oscillatory integral – sequel to Exercises 6.86 and 6.93). Let $V \in \mathrm{Mat}^+(n, \mathbf{R})$ be positive definite.

(i) Prove by Exercises 6.86.(ii) and 6.93.(i) that, for all $f \in \mathcal{S}(\mathbf{R}^n)$ and $t \in \mathbf{R}_+$,

$$\int_{\mathbf{R}^n} f(x) e^{-\frac{1}{2}t\langle Vx,x\rangle}\,dx = (2\pi t)^{-\frac{n}{2}}(\det V)^{-\frac{1}{2}} \int_{\mathbf{R}^n} \widehat{f}(\xi) e^{-\frac{1}{2t}\langle V^{-1}\xi,\xi\rangle}\,d\xi.$$

With the use of the theory of functions of one complex variable it can be shown that a similar identity holds for the following *oscillatory integral* with *amplitude function* f and quadratic *phase function*, which is obtained by replacing t with $-i\omega$, where $i = \sqrt{-1}$ and $\omega > 0$:

$$\int_{\mathbf{R}^n} f(x) e^{\frac{1}{2}i\omega\langle Vx,x\rangle}\,dx = (2\pi\omega)^{-\frac{n}{2}}|\det V|^{-\frac{1}{2}} e^{i\frac{\pi}{4}\operatorname{sgn} V} \int_{\mathbf{R}^n} \widehat{f}(\xi) e^{-\frac{i}{2\omega}\langle V^{-1}\xi,\xi\rangle}\,d\xi.$$

Here sgn V equals the signature of V, the difference between the number of positive and the number of negative eigenvalues of V, all counted with multiplicities. Introduce the differential operator

$$\langle V^{-1}D,\, D\rangle = \sum_{1\le j,k\le n} (V^{-1})_{jk} D_j D_k.$$

(ii) Prove by means of Theorem 6.11.3.(ii) and Theorem 6.11.6 that, for $k \in \mathbf{N}_0$,

$$\int_{\mathbf{R}^n} \widehat{f}(\xi)\Big(-\frac{i}{2\omega}\langle V^{-1}\xi,\,\xi\rangle\Big)^k\,d\xi = (2\pi)^n\Big(\frac{i}{2\omega}\Big)^k(\langle V^{-1}D,\,D\rangle^k f)(0).$$

(iii) Prove that we have the following asymptotic expansion (see Exercise 6.55) for the oscillatory integral:

$$\int_{\mathbf{R}^n} f(x)\, e^{\langle \frac{i}{2}\omega Vx,\,x\rangle}\,dx$$

$$\sim \Big(\frac{2\pi}{\omega}\Big)^{\frac{n}{2}}|\det V|^{-\frac{1}{2}} e^{i\frac{\pi}{4}\operatorname{sgn} V}\sum_{k\in\mathbf{N}_0}\frac{1}{k!}\Big(\Big\langle\frac{i}{2}\,(\omega V)^{-1}D,\,D\Big\rangle^k f\Big)(0), \qquad \omega\to\infty.$$

Exercise 6.95 (Principle of stationary phase). Let $\phi \in C^\infty(\mathbf{R}^n)$ and $f \in C_c^\infty(\mathbf{R}^n)$. Define the *oscillatory integral* $I : \mathbf{R}_+ \to \mathbf{C}$ with *phase function* ϕ and *amplitude function* f by

$$I(\omega) = \int_{\mathbf{R}^n} e^{i\omega\phi(x)} f(x)\,dx.$$

We want to study the asymptotic behavior of $I(\omega)$, for $\omega\to\infty$, under the assumption $D\phi(x) \neq 0$, for all $x \in \operatorname{supp}(f)$, that is, ϕ has no stationary points in $\operatorname{supp}(f)$. We define the differential operator L on $\mathbf{R}^n$ by $L = \|D\phi\|^{-2}\sum_{1\le j\le n}(D_j\phi)\,D_j$.

(i) Prove $I(\omega) = \frac{1}{i\omega}\int_{\mathbf{R}^n} L(e^{i\omega\phi})(x) f(x)\,dx$.

(ii) Show that there exists a differential operator $L^\dagger$ on $\mathbf{R}^n$ with

$$I(\omega) = \frac{1}{i\omega} \int_{\mathbf{R}^n} e^{i\omega\phi(x)} (L^\dagger f)(x)\, dx.$$

(iii) Demonstrate that $I(\omega) = \mathcal{O}(\omega^{-k}), \quad \omega \to \infty$, for all $k \in \mathbf{N}$.

Background. The result in part (iii) shows that the function $\omega \mapsto I(\omega)$ may not decrease arbitrarily rapidly, for $\omega \to \infty$, only if the phase function ϕ has stationary points in $\operatorname{supp}(f)$. In Exercise 6.94 we have a quadratic phase function ϕ, with a stationary point at 0. In that case, $I(\omega) \sim c\,\omega^{-\frac{n}{2}}$, for $\omega \to \infty$, with $c \neq 0$ if $f(0) \neq 0$.

Exercise 6.96 (Gamma distribution, Lipschitz' formula and Eisenstein series – sequel to the Exercises 0.20, 6.42, 6.50 and 6.88). Let α and $\lambda > 0$. In the terminology of Exercise 6.42, the function $f_{\alpha,\lambda} : \mathbf{R} \to \mathbf{R}$ with

$$f_{\alpha,\lambda}(x) = \begin{cases} 0, & x \leq 0; \\ \dfrac{\lambda^\alpha}{\Gamma(\alpha)} x^{\alpha-1} e^{-\lambda x}, & x > 0, \end{cases}$$

is said to be the *probability density of the Gamma distribution* with parameters α and λ. In particular, for $n \in \mathbf{N}$, the function

$$f_{\frac{n}{2},\frac{1}{2}}(x) = \frac{1}{2^{\frac{n}{2}}\Gamma(\frac{n}{2})} x^{\frac{n}{2}-1} e^{-\frac{1}{2}x} \qquad (x \in \mathbf{R}_+)$$

is said to be the *probability density of Pearson's χ^2 distribution with n degrees of freedom.*

(i) Prove

$$\int_{\mathbf{R}} f_{\alpha,\lambda}(x)\, dx = 1, \qquad \int_{\mathbf{R}} x f_{\alpha,\lambda}(x)\, dx = \frac{\alpha}{\lambda},$$

$$\int_{\mathbf{R}} (x-\mu)^2 f_{\alpha,\lambda}(x)\, dx = \frac{\alpha}{\lambda^2}.$$

Thus $\frac{\alpha}{\lambda}$ is the expectation and $\frac{\alpha}{\lambda^2}$ the variance of this Gamma distribution. In many cases the convolution and the Fourier transform are well-defined for functions not contained in $\mathcal{S}(\mathbf{R})$, and this is also true in the particular case of the $f_{\alpha,\lambda}$.

(ii) By means of Exercise 6.50.(iv) prove $f_{\alpha_1,\lambda} * f_{\alpha_2,\lambda} = f_{\alpha_1+\alpha_2,\lambda}$, for α_1, α_2 and $\lambda > 0$.

(iii) Show by Exercise 6.69.(i) that $\widehat{f_{\alpha,\lambda}}(\xi) = \left(\frac{\lambda}{\lambda+i\xi}\right)^\alpha$, for $|\xi| < \lambda$.

In fact, this formula is true for all $\xi \in \mathbf{R}$. Moreover, the Fourier Inversion Theorem is valid in this case. Let $\mathcal{H}$ be the upper half-plane $\{ z \in \mathbf{C} \mid \operatorname{Im} z > 0 \}$.

(iv) For any $z \in \mathcal{H}$, show that the Fourier transform of $x \mapsto (x - z)^{-\alpha}$ equals

$$\xi \mapsto \begin{cases} 0, & \xi \geq 0; \\ 2\pi i \dfrac{(-i\xi)^{\alpha-1}}{\Gamma(\alpha)} e^{-iz\xi}, & \xi < 0. \end{cases}$$

(v) Now use Poisson's summation formula from Exercise 6.88.(ii) to derive *Lipschitz' formula*

$$\sum_{n\in\mathbf{Z}} \frac{1}{(z+n)^k} = \frac{(-2\pi i)^k}{(k-1)!} \sum_{n\in\mathbf{N}} n^{k-1} e^{2\pi i n z} \qquad (z \in \mathcal{H},\ k \in \mathbf{N} \setminus \{1\}).$$

Furthermore, prove the following identity for the *Lerch function* Λ, valid for $z \in \mathcal{H}$, $x \in \mathbf{R}$ and $\alpha > 0$:

$$\Lambda(z, x, \alpha - 1) := \sum_{\{n\in\mathbf{Z} \mid n+x>0\}} (n+x)^{\alpha-1} e^{2\pi i n z} = \frac{\Gamma(\alpha)}{(-2\pi i)^\alpha} \sum_{n\in\mathbf{Z}} \frac{e^{-2\pi i (z+n)x}}{(z+n)^\alpha}.$$

(vi) Lipschitz' formula also can be obtained by expanding the right–hand side of the following formula from Exercise 0.13.(i):

$$\sum_{n\in\mathbf{Z}}{}' \frac{1}{z+n} = \pi \cot(\pi z) = -\pi i - 2\pi i \frac{e^{2\pi i z}}{1 - e^{2\pi i z}}$$

as a geometric series in $e^{2\pi i z}$ (this is where the condition $z \in \mathcal{H}$ is necessary) and differentiating $k - 1$ times with respect to z. Conversely, the partial-fraction decomposition of the cotangent can be derived from Lipschitz' formula.

The *Eisenstein series* G_k *of index* $k > 1$ is the function $G_k : \mathcal{H} \to \mathbf{C}$ given by

$$G_k(z) = \sum_{(0,0)\neq(m,n)\in\mathbf{Z}\times\mathbf{Z}} \frac{1}{(mz+n)^{2k}} = 2\zeta(2k) + 2 \sum_{m\in\mathbf{N}} \sum_{n\in\mathbf{Z}} \frac{1}{(mz+n)^{2k}}.$$

(vii) Apply Lipschitz' formula with z replaced by mz, to get the so-called Fourier expansion of G_k at infinity

$$\begin{aligned} G_k(z) &= 2\zeta(2k) + \frac{2(-2\pi i)^{2k}}{(2k-1)!} \sum_{m\in\mathbf{N}} \sum_{n\in\mathbf{N}} n^{2k-1} e^{2\pi i m n z} \\ &= 2\zeta(2k) + \frac{2(2\pi i)^{2k}}{(2k-1)!} \sum_{j\in\mathbf{N}} \sigma_{2k-1}(j) e^{2\pi i j z} \\ &= 2\zeta(2k)\Big(1 - \frac{4k}{B_{2k}} \sum_{j\in\mathbf{N}} \sigma_{2k-1}(j) e^{2\pi i j z}\Big). \end{aligned}$$

Here $\sigma_{2k-1}(j)$ denotes the sum of the $(2k-1)$-th powers of positive divisors of j, while in the last equality we used Exercise 0.20.

Exercise 6.97 (One-sided stable distribution – sequel to Exercises 2.87 and 6.42). In the terminology of Exercise 6.42, the function $f : \mathbf{R} \to \mathbf{R}$ with

$$f(x) = \begin{cases} 0, & x \le 0; \\ \dfrac{1}{\sqrt{\pi}} x^{-\frac{3}{2}} e^{-\frac{1}{x}}, & x > 0, \end{cases}$$

is said to be the *probability density of the one-sided stable distribution of order* $\frac{1}{2}$. Define, for $t > 0$,

$$f_t(x) = \frac{1}{t} f\Big(\frac{x}{t}\Big) = \sqrt{\frac{t}{\pi}}\, x^{-\frac{3}{2}} e^{-\frac{t}{x}}.$$

(i) Using Exercise 2.87.(vii), prove that $\int_{\mathbf{R}} f_t(x)\, dx = 1$, for all $t > 0$.

(ii) The convolution is well-defined for the functions f_t. Show $f_{t_1^2} * f_{t_2^2} = f_{(t_1+t_2)^2}$, for t_1 and $t_2 > 0$.
Hint: Verify that in $x > 0$ the left–hand side equals

$$\frac{t_1 t_2}{\pi} \int_0^x ((x-y)y)^{-\frac{3}{2}} e^{-\frac{t_1^2}{x-y} - \frac{t_2^2}{y}}\, dy,$$

and introduce the new variable $z > 0$ via $z = \frac{y}{x-y}$. Finally, use Exercise 2.87.(vii).

Exercise 6.98 (Bessel function as Fourier transform – sequel to Exercises 6.50 and 6.66). The Fourier transform $\widehat{f}$ is often well-defined for functions $f : \mathbf{R}^n \to \mathbf{R}$ which are not contained in $\mathcal{S}(\mathbf{R}^n)$; in particular this applies to the characteristic function f of the unit ball B^n in $\mathbf{R}^n$; we therefore define

$$(\star) \qquad \widehat{f}(\xi) = \int_{B^n} e^{-i\langle x, \xi \rangle}\, dx \qquad (\xi \in \mathbf{R}^n).$$

(i) Check that $\widehat{f} : \mathbf{R}^n \to \mathbf{C}$ is a continuous function, and that $\widehat{f}(0) = \frac{\pi^{\frac{n}{2}}}{\Gamma(\frac{n}{2}+1)}$.

(ii) Note that the expression on the right–hand side of $(\star)$ is independent of the direction of $\xi \in \mathbf{R}^n$. Therefore, set $\xi = (0, \dots, 0, \|\xi\|)$, and use the result and the hint from Exercise 6.50.(viii) to prove that

$$\widehat{f}(\xi) = \frac{\pi^{\frac{n-1}{2}}}{\Gamma(\frac{n+1}{2})} \int_{-1}^{1} e^{-ih\|\xi\|} (1-h^2)^{\frac{n-1}{2}}\, dh.$$

(iii) Substitute $h = \cos\alpha$ and conclude that

$$\widehat{f}(\xi) = \frac{\pi^{\frac{n-1}{2}}}{\Gamma(\frac{n+1}{2})} \int_0^{\pi} e^{-i\|\xi\|\cos\alpha} \sin^n\alpha \, d\alpha.$$

In particular, for $\xi \neq 0$,

$$(\star\star) \qquad \widehat{f}(\xi) = \Big(\frac{2\pi}{\|\xi\|}\Big)^{\frac{n}{2}} J_{\frac{n}{2}}(\|\xi\|).$$

(iv) Prove by means of Exercise 6.66.(viii) that the expression on the right in $(\star\star)$ does indeed approach the value of $\widehat{f}(0)$ for $\xi \to 0$.

Exercise 6.99 (Fourier transform of $x \mapsto e^{-\|x\|}$ and Poisson's integral – sequel to Exercise 2.87 – needed for Exercise 7.30). The Fourier transform $\widehat{f}$ is often well-defined for functions $f : \mathbf{R}^n \to \mathbf{R}$ not contained in $\mathcal{S}(\mathbf{R}^n)$; in particular this applies to the function $f : \mathbf{R}^n \to \mathbf{R}$ given by $f(x) = e^{-\|x\|}$.

(i) Prove

$$\widehat{f}(\xi) = 2^n \pi^{\frac{n-1}{2}} \Gamma\Big(\frac{n+1}{2}\Big)(1 + \|\xi\|^2)^{-\frac{n+1}{2}} \qquad (\xi \in \mathbf{R}^n).$$

Hint: Prove by Exercise 2.87.(v) that

$$\widehat{f}(\xi) = \frac{1}{\sqrt{\pi}} \int_{\mathbf{R}^n} e^{-i\langle x, \xi\rangle} \int_{\mathbf{R}_+} \frac{e^{-y}}{\sqrt{y}} e^{-\frac{\|x\|^2}{4y}} \, dy \, dx,$$

interchange the order of integration, and use Example 6.11.4.

(ii) Deduce, for all $\xi \in \mathbf{R}^n$ and $t > 0$,

$$\int_{\mathbf{R}^n} e^{-i\langle x, \xi\rangle - t\|x\|} \, dx = 2^n \pi^{\frac{n-1}{2}} \Gamma\Big(\frac{n+1}{2}\Big) \frac{t}{(\|\xi\|^2 + t^2)^{\frac{n+1}{2}}}.$$

Let $\mathbf{R}^{n+1}_* = \mathbf{R}^{n+1} \setminus \{0\}$. We define *Poisson's kernel* $P : \mathbf{R}^{n+1}_* \to \mathbf{R}$ by (see also Exercise 7.70.(ii))

$$P(x, t) = \frac{\Gamma(\frac{n+1}{2})}{\pi^{\frac{n+1}{2}}} \frac{|t|}{\|(x, t)\|^{n+1}}.$$

In the terminology of Exercise 6.42 the function $\mathbf{R}^n \to \mathbf{R}$ with

$$x \mapsto \frac{\Gamma(\frac{n+1}{2})}{\pi^{\frac{n+1}{2}}} \frac{1}{(x^2 + 1)^{\frac{n+1}{2}}}$$

is said to be the *probability density of the Cauchy distribution*. Note that $P(x, -t) = P(x, t)$. Assume that the properties of the Fourier transformation are also valid for the function on $\mathbf{R}^n$ given by $x \mapsto e^{-t\|x\|}$, where $t > 0$, as can be shown by approximation arguments.

(iii) Conclude that $\int_{\mathbf{R}^n} P(x,t)\,dx = 1$, for all $t \neq 0$. More generally, prove

$$\int_{\mathbf{R}^n} \frac{\cos\langle \xi, x\rangle}{(\|x\|^2+t^2)^{\frac{n+1}{2}}}\,dx = \frac{\pi^{\frac{n+1}{2}}}{\Gamma\big(\frac{n+1}{2}\big)}\frac{e^{-t\|\xi\|}}{t} \qquad (\xi \in \mathbf{R}^n,\ t>0),$$

and show that one therefore has the following, known as *Laplace's integrals*: (compare with Exercise 2.85):

$$\int_{\mathbf{R}_+} \frac{\cos \xi x}{x^2+t^2}\,dx = \frac{\pi}{2}\frac{e^{-t\xi}}{t} \qquad (\xi \geq 0,\ t>0),$$

$$\int_{\mathbf{R}_+} \frac{x \sin \xi x}{x^2+t^2}\,dx = \frac{\pi}{2}e^{-t\xi} \qquad (\xi > 0,\ t \geq 0).$$

Use the identity $\int_{\mathbf{R}_+} \frac{\sin x}{x}\,dx = \frac{\pi}{2}$ from, for instance, Example 2.10.14 to obtain the validity of the latter formula for $t=0$.

(iv) Verify that we have the following *semigroup property*, for all t_1 and $t_2 > 0$:

$$P(x,\ t_1+t_2) = \int_{\mathbf{R}^n} P(x-y,\ t_1)\,P(y,\ t_2)\,dy.$$

(v) Let $\delta > 0$ be arbitrary. Check that $\int_{\{x\in\mathbf{R}^n \mid \|x\|>\delta\}} \|x\|^{-n-1}\,dx < \infty$, and use this to prove

$$\lim_{t\to 0}\int_{\{x\in\mathbf{R}^n \mid \|x\|>\delta\}} P(x,t)\,dx = 0.$$

Define $\mathbf{R}^{n+1}_\pm = \{(x,t) \in \mathbf{R}^{n+1} \mid \pm t > 0\}$ (note that in the remainder of this exercise the meaning of $\mathbf{R}^{n+1}_+$ differs from the usual one).

(vi) Prove by Example 7.8.4 that the functions

$$(x,t) \mapsto \log\|(x,t)\| \qquad (n=1); \qquad (x,t)\mapsto \frac{1}{\|(x,t)\|^{n-1}} \qquad (n>1),$$

are harmonic on $\mathbf{R}^{n+1}_*$. Verify that, leaving scalars aside, P on $\mathbf{R}^{n+1}_\pm$ is the partial derivative with respect to t of the preceding function, and conclude that P is a harmonic function on $\mathbf{R}^{n+1}_\pm$.

Now let $h \in C(\mathbf{R}^n)$ be bounded, and define *Poisson's integral* $\mathcal{P}h : \mathbf{R}^{n+1}_\pm \to \mathbf{R}$ of h by

$$(\mathcal{P}h)(x,t) = (h * P(\cdot,t))(x) = \int_{\mathbf{R}^n} h(x-y)P(y,t)\,dy.$$

(vii) Prove that $\mathcal{P}h$ is a well-defined harmonic function on $\mathbf{R}^{n+1}_\pm$.

(viii) Assume $(x, t) \in \mathbf{R}^{n+1}_{\pm}$ and prove by means of part (iii) that

$$|(\mathcal{P}h)(x,t) - h(x)| \le \int_{\|y\|\le\delta} |h(x-y) - h(x)| P(y,t)\, dy$$

$$+2 \sup_{x\in\mathbf{R}^n} |h(x)| \int_{\|y\|>\delta} P(y,t)\, dy.$$

Using part (v) show that, uniformly for x in compact sets in $\mathbf{R}^n$,

$$\lim_{\pm t\downarrow 0} (\mathcal{P}h)(x,t) = h(x).$$

Background. Evidently, in the terminology of Example 7.9.7, the function $f = \mathcal{P}h$ is a solution of the following Dirichlet problem on $\Omega = \mathbf{R}^{n+1}_{+}$ or $\mathbf{R}^{n+1}_{-}$:

$$\Delta f = 0 \qquad \text{with} \qquad f|_{\partial\Omega} = h.$$

(ix) On the basis of the results from Exercise 7.21.(ii), show

$$(\mathcal{P}h)(x,t) = \frac{2}{\text{hyperarea}_n(S^n)} \int_{\mathbf{R}^n} \frac{h(x+ty)}{\|(y,1)\|^{n+1}}\, dy \qquad ((x,t) \in \mathbf{R}^{n+1}_{\pm}).$$

(x) Prove that there exists a harmonic function u on $\mathbf{R}^{n+1}_{\pm}$ with

$$\lim_{t\downarrow 0} u(x,t) - \lim_{t\uparrow 0} u(x,t) = h(x) \qquad (x \in \mathbf{R}^n).$$

Hint: Write $h = \frac{1}{2}h - (-\frac{1}{2}h)$.

Background. Evidently, a bounded continuous function on $\mathbf{R}^n$ can be represented by means of the jump along $\mathbf{R}^n \times \{0\}$ made by a suitably chosen harmonic function on $\mathbf{R}^{n+1}_{\pm}$. This point of view is of importance in the theory of *hyperfunctions*.

Exercise 6.100 (Mellin transformation – sequel to Exercise 6.86 – needed for Exercise 6.101). Let $\mathcal{S}^*(\mathbf{R}_+)$ be the linear space of functions $f : \mathbf{R}_+ \to \mathbf{C}$ with the property that the function $f^* : \mathbf{R} \to \mathbf{C}$ given by $f^*(x) = f(e^x)$ is contained in $\mathcal{S}(\mathbf{R})$. For $f \in \mathcal{S}^*(\mathbf{R}_+)$ we define the function $\mathcal{M}f : \mathbf{R} \to \mathbf{C}$, the *Mellin transform* of f, by

$$(\mathcal{M}f)(\xi) = \int_{\mathbf{R}_+} x^{-i\xi} f(x)\, \frac{dx}{x} \qquad (\xi \in \mathbf{R}).$$

We now prove the following, known as *Mellin's formula*, valid for all $f \in \mathcal{S}^*(\mathbf{R}_+)$ and $x \in \mathbf{R}_+$:

$$(\star) \qquad f(x) = \frac{1}{2\pi} \int_{\mathbf{R}} x^{i\xi} (\mathcal{M}f)(\xi)\, d\xi.$$

(i) Check that $\widehat{f^*}(\xi) = (\mathcal{M}f)(\xi)$, for all $f \in \mathcal{S}^*(\mathbf{R}_+)$ and $\xi \in \mathbf{R}$. Conclude that the *Mellin transformation* $\mathcal{M}$ belongs to $\mathrm{Lin}\,(\mathcal{S}^*(\mathbf{R}_+),\ \mathcal{S}(\mathbf{R}))$, and also that ($\star$) is obtained. Hence, for $g \in \mathcal{S}(\mathbf{R})$ and $x \in \mathbf{R}_+$,

$$(\mathcal{M}^{-1}g)(x) = \frac{1}{2\pi}\int_{\mathbf{R}} x^{i\xi}\, g(\xi)\, d\xi.$$

Define Hermitian inner products on $\mathcal{S}^*(\mathbf{R}_+)$ and $\mathcal{S}(\mathbf{R})$, respectively, by $\langle\, f, g\,\rangle = \int_{\mathbf{R}_+} f(x)\overline{g(x)}\,\frac{dx}{x}$ and $\langle\, f, g\,\rangle = \frac{1}{2\pi}\int_{\mathbf{R}} f(x)\overline{g(x)}\, dx$.

(ii) Prove by the Parseval–Plancherel identity from Exercise 6.86, for all $f \in \mathcal{S}^*(\mathbf{R}_+)$,

$$\int_{\mathbf{R}_+} |f(x)|^2\,\frac{dx}{x} = \frac{1}{2\pi}\int_{\mathbf{R}} |(\mathcal{M}f)(\xi)|^2\, d\xi.$$

Conclude that $\mathcal{M} \in \mathrm{Lin}\,(\mathcal{S}^*(\mathbf{R}_+),\ \mathcal{S}(\mathbf{R}))$ is *unitary*, that is, $\mathcal{M}$ preserves the Hermitian inner products.

(iii) Introduce the differential operator $\partial_x = \frac{1}{i}x\frac{d}{dx}$, and let $\xi \in \mathbf{R}$. Verify that the function $u : x \mapsto x^{i\xi}$ is the unique solution of the eigenvalue problem $(\partial_x u)(x) = \xi\, u(x)$, for $x \in \mathbf{R}_+$, and $u(1) = 1$.

(iv) Introduce the inner product $\langle f, g\rangle = \int_{\mathbf{R}_+} f(x)\overline{g(x)}\,\frac{dx}{x}$ and prove that ∂_x is a self-adjoint linear operator with respect to it; in other words, verify that, for all $f \in \mathcal{S}^*(\mathbf{R}_+)$ and all bounded C^1 functions g,

$$\int_{\mathbf{R}_+} (\partial_x f)(x)\overline{g(x)}\,\frac{dx}{x} = \int_{\mathbf{R}_+} f(x)\overline{(\partial_x g)(x)}\,\frac{dx}{x}.$$

(v) Using parts (iii) and (iv), show that, for all $f \in \mathcal{S}^*(\mathbf{R}_+)$ and $\xi \in \mathbf{R}$,

$$\mathcal{M}(\partial.f)(\xi) = \xi\,(\mathcal{M}f)(\xi), \qquad \mathcal{M}(-f\log\cdot)(\xi) = \Big(\frac{1}{i}\frac{d}{d\xi}(\mathcal{M}f)\Big)(\xi).$$

Now assume f and $g \in \mathcal{S}^*(\mathbf{R}_+)$. Define the *convolution* $f\Box g : \mathbf{R}_+ \to \mathbf{C}$ of f and g by

$$(f\Box g)(x) = \int_{\mathbf{R}_+} f\Big(\frac{x}{y}\Big)g(y)\,\frac{dy}{y}.$$

(vi) Demonstrate that $\mathcal{M}(f\Box g) = (\mathcal{M}f)\,(\mathcal{M}g)$.

Background. The formula $(\mathcal{M}\circ\partial.\circ\mathcal{M}^{-1})f(\xi) = \xi\, f(\xi)$ shows that the differential operator $\frac{1}{i}x\frac{d}{dx}$ acting on $\mathcal{S}^*(\mathbf{R}_+)$ can be diagonalized by conjugation with the Mellin transformation $\mathcal{M}$, and that the action of the conjugated operator in $\mathcal{S}(\mathbf{R})$ is that of multiplication by the coordinate ξ. In number theory one usually calls

$\int_{\mathbf{R}_+} x^s f(x) \frac{dx}{x}$, that is $(\mathcal{M}f)(is)$, the Mellin transform of f evaluated at s. In that discipline one studies functions $f(x) = \sum_{n\in\mathbf{N}} a_n e^{-nx}$ and computes

$$\frac{1}{\Gamma(s)} \int_{\mathbf{R}_+} x^{s-1} f(x)\,dx = \sum_{n\in\mathbf{N}} \frac{a_n}{n^s},$$

which is called a *Dirichlet series*. In this case the inversion formula takes a different form.

Exercise 6.101 (Harmonic function on sector – sequel to Exercises 3.8 and 6.100). Assume $0 < \alpha_0 < \pi$, let $V \subset \mathbf{R}^2$ be the half-strip $\{\,(r,\alpha) \in \mathbf{R}^2 \mid r \in \mathbf{R}_+,\ |\alpha| < \alpha_0\,\}$, and let $\Psi : V \to U := \Psi(V)$ be the substitution of polar coordinates as in Example 3.1.1. Then U is an open sector in $\mathbf{R}^2$. We look for a solution $f \in C^2(U) \cap C(\overline{U})$ for the following *partial differential equation* with *boundary condition*:

$$\Delta f = 0 \quad \text{on } U, \qquad f|_{\partial U} = h,$$

where $h(x_1, x_2) = h(x_1, -x_2)$, for $x \in \partial U$ and $(h \circ \Psi)(\cdot, \alpha_0) \in \mathcal{S}^*(\mathbf{R}_+)$, in the notation of Exercise 6.100.

(i) Conclude by Exercise 3.8.(v) that $f \circ \Psi$ on V must satisfy

$$\Big(\Big(r\frac{\partial}{\partial r}\Big)^2 + \frac{\partial^2}{\partial \alpha^2}\Big)(f \circ \Psi) = 0.$$

We now try to find f with the property that $f \circ \Psi(\cdot, \alpha) \in \mathcal{S}^*(\mathbf{R}_+)$, for all $|\alpha| \le \alpha_0$; we then write $g(\cdot, \alpha) \in \mathcal{S}(\mathbf{R})$ for the Mellin transform $\mathcal{M}(f \circ \Psi)(\cdot, \alpha)$.

(ii) Use Exercise 6.100.(v) to verify that, with $|\alpha| \le \alpha_0$ fixed for the moment,

$$(\star) \qquad \Big(-\xi^2 + \frac{\partial^2}{\partial \alpha^2}\Big) g(\xi, \alpha) = 0 \qquad (\xi \in \mathbf{R}).$$

(iii) Now interpret $(\star)$ as a differential equation with respect to the variable α, and infer the existence of functions a and $b : \mathbf{R} \to \mathbf{C}$ such that $g(\xi, \alpha) = a(\xi)\cosh(\xi\alpha) + b(\xi)\sinh(\xi\alpha)$, for all $|\alpha| \le \alpha_0$. Now choose in particular $\alpha = \pm\alpha_0$ and derive

$$g(\xi, \alpha) = \mathcal{M}(h \circ \Psi)(\xi, \alpha_0)\, \frac{\cosh(\xi\alpha)}{\cosh(\xi\alpha_0)}.$$

(iv) Prove by Mellin's formula from Exercise 6.100

$$f \circ \Psi(r, \alpha) = \frac{1}{2\pi} \int_{\mathbf{R}} r^{i\xi} \frac{\cosh(\xi\alpha)}{\cosh(\xi\alpha_0)} \mathcal{M}(h \circ \Psi)(\xi, \alpha_0)\, d\xi \qquad ((r, \alpha) \in V).$$

Conclude that, for $x \in U$,

$$f(x) = \frac{1}{2\pi} \int_{\mathbf{R}} \|x\|^{i\xi} \cosh\Big(2\xi \arctan(\frac{x_2}{x_1 + 1})\Big) \frac{\mathcal{M}(h \circ \Psi)(\xi,\, \alpha_0)}{\cosh(\xi\alpha_0)}\, d\xi.$$

Exercise 6.102 (Hankel transformation – sequel to Exercises 6.66 and 8.20 – needed for Exercise 7.30). We employ the notation from Exercise 6.66. Let $\mathcal{S}(\mathbf{R}_+)$ be the linear subspace of $\mathcal{S}(\mathbf{R})$ consisting of the even functions in $\mathcal{S}(\mathbf{R})$. For $\lambda > -\frac{1}{2}$ and $f \in \mathcal{S}(\mathbf{R}_+)$ we define the function $\mathcal{H}_\lambda f : \mathbf{R}_+ \to \mathbf{R}$, the *Hankel transform* of f of order λ, by

$$(\mathcal{H}_\lambda f)(\xi) = \int_{\mathbf{R}_+} f(x) \frac{J_\lambda(x\xi)}{(x\xi)^\lambda} x^{2\lambda+1}\, dx \qquad (\xi \in \mathbf{R}_+).$$

This integral is well-defined, see Exercise 8.20. We now prove the following, known as *Hankel's formula*:

$$\mathcal{H}_\lambda^2 = I,$$

that is, for all $f \in \mathcal{S}(\mathbf{R}_+)$ and $x \in \mathbf{R}_+$ we have

$$f(x) = \int_{\mathbf{R}_+} \int_{\mathbf{R}_+} f(y) \frac{J_\lambda(y\xi)}{(y\xi)^\lambda} y^{2\lambda+1}\, dy\ \frac{J_\lambda(x\xi)}{(x\xi)^\lambda} \xi^{2\lambda+1}\, d\xi.$$

Under the substitution $g(x) = x^\lambda f(x)$ we obtain the following variant of Hankel's formula:

$$g(x) = \int_{\mathbf{R}_+} \int_{\mathbf{R}_+} g(y) J_\lambda(y\xi) y\, dy\ J_\lambda(x\xi)\xi\, d\xi.$$

See Exercise 7.30 for the relation between the Hankel and Fourier transformations.

(i) Introduce the differential operator with variable coefficients

$$\Delta_{x,\lambda} = -\frac{1}{x^{2\lambda+1}} \frac{d}{dx}\Big(x^{2\lambda+1} \frac{d}{dx}\Big).$$

Let $\xi \in \mathbf{R}_+$. Use Exercise 6.66.(iv) to verify that the function $u : x \mapsto \frac{J_\lambda(x\xi)}{(x\xi)^\lambda}$ is a solution of the eigenvalue problem

$$(\Delta_{x,\lambda} u)(x) = \xi^2 u(x) \qquad (x \in \mathbf{R}_+).$$

(ii) Prove that $\Delta_{x,\lambda}$ is a self-adjoint linear operator with respect to the inner product $\langle f, g\rangle = \int_{\mathbf{R}_+} f(x)g(x)x^{2\lambda+1}\,dx$; that is, prove for all $f \in \mathcal{S}(\mathbf{R}_+)$, and all bounded C^2 functions g,

$$\int_{\mathbf{R}_+} (\Delta_{x,\lambda} f)(x)g(x)x^{2\lambda+1}\,dx = \int_{\mathbf{R}_+} f(x)(\Delta_{x,\lambda} g)(x)x^{2\lambda+1}\,dx.$$

(iii) Verify, by parts (i) and (ii), for all $f \in \mathcal{S}(\mathbf{R}_+)$ and $\xi \in \mathbf{R}_+$,

$$\mathcal{H}_\lambda(\Delta_{\cdot,\lambda} f)(\xi) = \xi^2(\mathcal{H}_\lambda f)(\xi), \qquad \mathcal{H}_\lambda(\cdot^2 f)(\xi) = (\Delta_{\xi,\lambda}(\mathcal{H}_\lambda f))(\xi).$$

Conclude that $\mathcal{H}_\lambda : \mathcal{S}(\mathbf{R}_+) \to \mathcal{S}(\mathbf{R}_+)$.

(iv) Verify $\mathcal{H}_\lambda k = k$ if $k(x) = e^{-\frac{1}{2}x^2}$.
Hint: There are two possibilities: apply the series expansion of J_λ from the Exercise 6.66.(ix); alternatively, check that both k and $\mathcal{H}_\lambda k$ (use part (iii)) satisfy the differential equation

$$(\Delta_{x,\lambda} f)(x) = (2\lambda + 2 - x^2)f(x) \qquad (x \in \mathbf{R}_+).$$

Prove by Exercise 6.66.(vi) that $(\mathcal{H}_\lambda f)'(\xi) = -\xi(\mathcal{H}_{\lambda+1} f)(\xi)$, and conclude by part (viii) of the same exercise that

$$k(0) = \lim_{\xi\downarrow 0}(\mathcal{H}_\lambda k)(\xi) = 1, \qquad k'(0) = \lim_{\xi\downarrow 0}(\mathcal{H}_\lambda k)'(\xi) = 0.$$

Assume $f \in \mathcal{S}(\mathbf{R}_+)$ and $x_0 \in \mathbf{R}_+$, and write $h(x) = f(x) - f(x_0)e^{\frac{1}{2}x_0^2}\,k(x)$. On account of part (iv), to prove the formula $(\mathcal{H}_\lambda^2 f)(x_0) = f(x_0)$ it suffices to show $(\mathcal{H}_\lambda^2 h)(x_0) = 0$.

(v) Check that $h(x_0) = h(-x_0) = 0$, and write $h(x) = (x^2 - x_0^2)g(x)$, with $g \in \mathcal{S}(\mathbf{R}_+)$ suitably chosen. Use part (iii) to prove

$$(\mathcal{H}_\lambda h)(\xi) = (\Delta_{\xi,\lambda} - x_0^2)(\mathcal{H}_\lambda g)(\xi).$$

Conclude by parts (ii) and (i) that

$$\begin{aligned}(\mathcal{H}_\lambda^2 h)(x_0) &= \int_{\mathbf{R}_+} (\mathcal{H}_\lambda h)(\xi)\frac{J_\lambda(x_0\xi)}{(x_0\xi)^\lambda}\xi^{2\lambda+1}\,d\xi \\ &= \int_{\mathbf{R}_+} (\mathcal{H}_\lambda g)(\xi)(\Delta_{\xi,\lambda} - x_0^2)\Big(\frac{J_\lambda(x_0\xi)}{(x_0\xi)^\lambda}\Big)\xi^{2\lambda+1}\,d\xi = 0.\end{aligned}$$

Background. The formula $(\mathcal{H}_\lambda \circ \Delta_{\cdot,\lambda} \circ \mathcal{H}_\lambda^{-1})f(\xi) = \xi^2 f(\xi)$ shows that the differential operator $-\frac{1}{x^{2\lambda+1}}\frac{d}{dx}(x^{2\lambda+1}\frac{d}{dx})$ acting on $\mathcal{S}(\mathbf{R}_+)$ is diagonalized by conjugation with the Hankel transformation $\mathcal{H}_\lambda$, and that the action of the conjugated operator in $\mathcal{S}(\mathbf{R}_+)$ is that of multiplication by ξ^2.

Exercise 6.103 (Weierstrass' Approximation Theorem – sequel to Exercise 6.92 – needed for Exercise 8.36). This theorem asserts that a differentiable function on a compact set can be uniformly approximated by a polynomial function, and that a finite number of derivatives of the function can then also be uniformly approximated by the corresponding derivatives of that polynomial function. In a more exact formulation, let $\Omega \subset \mathbf{R}^n$ be open and let $k \in \mathbf{N}_0$; then for every function $f \in C^k(\Omega)$, for every compact set $K \subset \Omega$, and for every $\epsilon > 0$, there exists a polynomial function $p : \mathbf{R}^n \to \mathbf{R}$ such that, for every multi-index $\alpha = (\alpha_1, \ldots, \alpha_n) \in \mathbf{N}_0^n$ with $|\alpha| \leq k$, one has

$$\sup\{\, |D^\alpha f(x) - D^\alpha p(x)| \mid x \in K \,\} < \epsilon.$$

For a proof in the case of a continuous function on an interval in $\mathbf{R}$, see Exercise 1.55.

We now give an outline of the proof. It is left to the reader to check and add its details. Let $\chi \in C_c^\infty(\mathbf{R}^n)$ with $\mathrm{supp}(\chi) \subset \Omega$ and $\chi = 1$ on a neighborhood of K. Then $f\chi \in C_c^k(\mathbf{R}^n)$, and thus we may henceforth assume $f \in C_c^k(\mathbf{R}^n)$. In particular, the integrations below are over the compact set $\mathrm{supp}(f)$. Now employ the notation from Exercise 6.92 with $k = 1$, and use the arguments applied in that exercise, with f replaced by $D^\alpha f$. One finds

$$u_\alpha(x, t) := (D^\alpha f) * g_t(x) = (4\pi t)^{-n/2} \int_{\mathbf{R}^n} D^\alpha f(y) e^{-\|x-y\|^2/4t}\, dy.$$

Now, from the Continuity Theorem 2.10.2, it follows that $u_\alpha : K \times [\,0, 1\,] \to \mathbf{R}$ is a continuous function on a compact set. But then u_α is uniformly continuous on that set, on account of Theorem 1.8.15. Consequently, for every $\epsilon > 0$ there exists a number $t > 0$ such that, for every multi-index $\alpha \in \mathbf{N}_0^n$ with $|\alpha| \leq k$, and for all $x \in K$,

$$|D^\alpha f(x) - u_\alpha(x, t)| = |u_\alpha(x, 0) - u_\alpha(x, t)| < \frac{\epsilon}{2}.$$

One has

$$e^{-\|x-y\|^2/4t} = \sum_{j \in \mathbf{N}_0} \frac{\|x - y\|^{2j}}{j!\,(-4t)^j}.$$

But, for x and y in compact sets in $\mathbf{R}^n$, this series can be uniformly approximated by its partial sums $p_N(x, y)$, where the summation runs from 0 to $N \in \mathbf{N}$. By choosing N sufficiently large, we can find a polynomial function $p := p_N : \mathbf{R}^n \to \mathbf{R}$ with

$$p(x) := (4\pi t)^{-n/2} \sum_{0 \leq j \leq N} \frac{1}{j!\,(-4t)^j} \int_{\mathbf{R}^n} f(y)((x_1 - y_1)^2 + \cdots + (x_n - y_n)^2)^j\, dy,$$

such that, for every multi-index $\alpha \in \mathbf{N}_0^n$ with $|\alpha| \leq k$ and for all $x \in K$,

$$|u_\alpha(x, t) - D^\alpha p(x)| \leq (4\pi t)^{-n/2} \int_{\mathbf{R}^n} D^\alpha f(y) \sum_{N<j} \frac{\|x - y\|^{2j}}{j!\,(4t)^j}\, dy < \frac{\epsilon}{2}.$$

Exercise 6.104 (Sequel to Exercise 6.50 – needed for Exercises 6.105 and 6.106). In $\mathcal{S}(\mathbf{R}^n)$ the convolution operation $*$ (see Example 6.11.5) defines a multiplication $*$ by

$$(\star) \qquad (f, g) \mapsto f * g \qquad \text{with} \qquad f * g(x) = \int_{\mathbf{R}^n} f(x-y)g(y)\,dy.$$

(i) Show this multiplication to be commutative and associative.

For $s > 0$, define the function $\phi_s : \mathbf{R} \to \mathbf{R}$ by

$$\phi_s(x) = \frac{x_+^{s-1}}{\Gamma(s)}, \qquad \text{where} \qquad x_+^{s-1} = \begin{cases} x^{s-1}, & x > 0; \\ 0, & x \le 0. \end{cases}$$

(ii) Verify that, for $s, t > 0$, the convolution $\phi_s * \phi_t$ is also well-defined by $(\star)$, and that it is given by

$$\phi_s * \phi_t(x) = \frac{1}{\Gamma(s)\Gamma(t)} \int_0^x (x-y)^{s-1} y^{t-1}\,dy.$$

Conclude by Exercise 6.50.(iv) that $\phi_s * \phi_t = \phi_{s+t}$, for $s, t > 0$.

Exercise 6.105 (Fractional integration and differentiation – sequel to Exercises 2.75 and 6.104 – needed for Exercises 6.106, 6.107, 6.108 and 8.35). Define

$$C_+^\infty(\mathbf{R}) = \{\, f \in C^\infty(\mathbf{R}) \mid \operatorname{supp}(f) \subset \mathbf{R}_+ \,\},$$

and let $D \in \operatorname{End}(C_+^\infty(\mathbf{R}))$ be the operator of differentiation. Define

$$D^{-1} \in \operatorname{End}(C_+^\infty(\mathbf{R})) \qquad \text{by} \qquad D^{-1}f(x) = \int_0^x f(t)\,dt.$$

(i) Prove that D^{-1} is in fact the inverse of D.

Define $D^{-n} = (D^{-1})^n$, for $n \in \mathbf{N}$.

(ii) From Exercise 2.75.(i) deduce that $D^{-n} f = f * \phi_n$, for $n \in \mathbf{N}$ and $f \in C_+^\infty(\mathbf{R})$, with ϕ_n as in Exercise 6.104.

(iii) Prove by Taylor's formula from Lemma 2.8.2 or Exercise 2.75.(iii) that D^{-n} is the inverse of D^n, that is, $D^{-n} = (D^n)^{-1}$.

For $s > 0$ we define the operator $D^{-s} \in \operatorname{End}(C_+^\infty(\mathbf{R}))$ of *fractional integration* from 0 *of order* s by (compare with part (ii))

$$D^{-s} f = f * \phi_s; \qquad \text{and so} \qquad D^{-s} f(x) = \frac{1}{\Gamma(s)} \int_0^x f(y)(x-y)^{s-1}\,dy.$$

(iv) By Exercise 6.104 prove that $D^{-s} \circ D^{-t} = D^{-t} \circ D^{-s} = D^{-(s+t)}$, for all s, $t > 0$.

For $t \geq 0$ we define $D^t \in \text{End}\,(C_+^\infty(\mathbf{R}))$ of *fractional differentiation* of order t as follows. We have the unique decomposition $t = n - s$ with $n \in \mathbf{N}$ and $0 < s \leq 1$. Now define

$$D^t = D^n \circ D^{-s}; \qquad \text{hence} \qquad D^t f = (f * \phi_s)^{(n)}.$$

(v) Prove $D^0 = I$, and demonstrate by integration by parts that $D^t = D^m \circ D^{-r}$, if $t = m - r$ with $m \in \mathbf{N}$ and $0 < r \leq m$.

(vi) Examine the validity of the following assertions. For $t \geq 0$ one has $D^t = D^{-s} \circ D^n$; therefore $D^t f = f^{(n)} * \phi_s$, and for all $s, t \in \mathbf{R}$ one has the group property $D^s \circ D^t = D^t \circ D^s = D^{s+t}$.

Background. In *distribution theory* the arguments above are generalized in a far-reaching manner and the *one-parameter group* $(D^s)_{s \in \mathbf{R}}$ of linear operators, together with its generalizations, becomes an aid in solving differential equations.

Exercise 6.106 (Hypergeometric function and fractional integration – sequel to Exercises 6.69, 6.104 and 6.105). Let the notation be as in these exercises and set $\phi_{a,b}(x) = (1-x)^{-a}\phi_b(x)$. Now substitute $zt = y$ in the integral for the hypergeometric function to obtain on $]\,0, 1\,[$

$$F(a, b; c : \cdot) = \frac{\phi_{a,b} * \phi_{c-b}}{\phi_c} = \frac{D^{b-c}\phi_{a,b}}{\phi_c}.$$

Exercise 6.107 (Bessel function and fractional integration and differentiation – sequel to Exercises 6.66 and 6.105).

(i) Verify that, for $\lambda > -\frac{1}{2}$ and $x \in \mathbf{R}_+$,

$$\begin{aligned} J_\lambda(x) &= \frac{2}{\Gamma(\frac{1}{2})\Gamma(\lambda + \frac{1}{2})} \Big(\frac{x}{2}\Big)^\lambda \int_0^1 (1 - t^2)^{\lambda - \frac{1}{2}} \cos xt \, dt \\ &= \frac{2}{\Gamma(\frac{1}{2})\Gamma(\lambda + \frac{1}{2})} \frac{1}{(2x)^\lambda} \int_0^x (x^2 - y^2)^{\lambda - \frac{1}{2}} \cos y \, dy. \end{aligned}$$

(ii) Conclude that, in the terminology of Exercise 6.105, this formula may be recognized as a fractional integration from 0 of order $\lambda + \frac{1}{2}$:

$$\begin{aligned} \Gamma\Big(\frac{1}{2}\Big)(2\sqrt{x})^\lambda J_\lambda(\sqrt{x}) &= \frac{1}{\Gamma(\lambda + \frac{1}{2})} \int_0^x (x - z)^{\lambda - \frac{1}{2}} \frac{\cos\sqrt{z}}{\sqrt{z}} \, dz \\ &= D^{-\lambda - \frac{1}{2}}\Big(\frac{\cos\sqrt{\cdot}}{\sqrt{\cdot}}\Big)(x). \end{aligned}$$

(iii) Prove

$$(\star) \qquad (2\sqrt{x})^\lambda J_\lambda(\sqrt{x}) = D^{-\lambda}\Big(J_0(\sqrt{\cdot})\Big)(x).$$

Let $\mu > 0$. From Exercise 6.105.(vi) conclude that

$$(2\sqrt{x})^{\lambda+\mu} J_{\lambda+\mu}(\sqrt{x}) = D^{-\mu}\Big((2\sqrt{\cdot})^\lambda J_\lambda(\sqrt{\cdot})\Big)(x).$$

Demonstrate that in integral form this becomes

$$2^\mu(\sqrt{x})^{\lambda+\mu} J_{\lambda+\mu}(\sqrt{x}) = \frac{1}{\Gamma(\mu)}\int_0^x (x-y)^{\mu-1}(\sqrt{y})^\lambda J_\lambda(\sqrt{y})\,dy.$$

Now substitute $x = z^2$, $y = z^2\sin^2\alpha$ and subsequently $z = x$. This gives the following, known as *Sonine's formula*, valid for $\lambda > -\frac{1}{2}$, $\mu > 0$ and $x \in \mathbf{R}_+$:

$$2^{\mu-1}\Gamma(\mu) J_{\lambda+\mu}(x) = x^\mu \int_0^{\frac{\pi}{2}} J_\lambda(x\sin\alpha)\sin^{\lambda+1}\alpha\cos^{2\mu-1}\alpha\,d\alpha.$$

Note that we can use the left–hand side of formula $(\star)$ to define the Bessel function J_λ, for $\lambda \le -\frac{1}{2}$. Indeed, for these values of λ the expression on the right–hand side in $(\star)$ can be interpreted as the fractional derivative of order $-\lambda$ of the function $x \mapsto J_0(\sqrt{x})$.

(iv) Let $\lambda \ge 0$. Prove, by means of Exercise 6.105.(vi), that both J_λ and $J_{-\lambda}$ satisfy Bessel's differential equation from Exercise 6.66.(vii).

(v) Prove, by means of Exercise 6.66.(vi), that

$$(-1)^n(2\sqrt{x})^{-n} J_n(\sqrt{x}) = D^n\Big(J_0(\sqrt{\cdot})\Big)(x) \qquad (n \in \mathbf{N},\ x \in \mathbf{R}_+).$$

Conclude that

$$J_{-n} = (-1)^n J_n \qquad (n \in \mathbf{N}).$$

Thus, in this case the second solution J_{-n} is linearly dependent on the first solution J_n.

Exercise 6.108 (Abel's integral equation – sequel to Exercises 0.6, 2.75 and 6.58). Let $a > 0$ and let $g \in C^1([0, a])$ be given with $g(0) = 0$. We want to solve the following, known as *Abel's integral equation*, for an unknown continuous function $f : [0, a] \to \mathbf{R}$:

$$g(x) = \int_0^x \frac{f(t)}{\sqrt{x-t}}\,dt \qquad (x \in [0, a]).$$

We prove in four steps that a solution f is given by

$$(\star) \qquad f(y) = \frac{1}{\pi}\int_0^y \frac{g'(x)}{\sqrt{y-x}}\,dx \qquad (y \in [0, a]).$$

(i) Multiply both parts of the integral equation by $\frac{1}{\sqrt{y-x}}$ to conclude that, for $y \in [\,0, a\,]$,

$$\begin{aligned}\int_0^y \frac{g(x)}{\sqrt{y-x}}\,dx &= \int_0^y \int_0^x \frac{f(t)}{\sqrt{(y-x)(x-t)}}\,dt\,dx \\ &= \int_0^y f(t) \int_t^y \frac{dx}{\sqrt{(x-t)(y-x)}}\,dt.\end{aligned}$$

(ii) The value of the inner integral follows from Exercise 0.6.(v). For $y \in [\,0, a\,]$, conclude that

$$\int_0^y \frac{g(x)}{\sqrt{y-x}}\,dx = \pi \int_0^y f(t)\,dt;$$

whence

$$\pi f(y) = \frac{d}{dy}\int_0^y \frac{g(x)}{\sqrt{y-x}}\,dx.$$

Straightforward application of Exercise 2.74 is not possible, because it leads to expressions of the form $\infty - \infty$.

(iii) Substitute $x = yt$ to prove that

$$\begin{aligned}\pi f(y) &= \frac{d}{dy}\int_0^1 \frac{\sqrt{y}\,g(yt)}{\sqrt{1-t}}\,dt = \int_0^1 \frac{\frac{1}{2\sqrt{y}}\,g(yt) + t\sqrt{y}\,g'(yt)}{\sqrt{1-t}}\,dt \\ &= \int_0^y \frac{g(x) + 2xg'(x)}{2y\sqrt{y-x}}\,dx.\end{aligned}$$

(iv) Finally, use integration by parts to prove

$$\frac{1}{y}\int_0^y \frac{g(x)}{2\sqrt{y-x}}\,dx = \int_0^y \frac{g'(x)}{\sqrt{y-x}}\,dx - \int_0^y \frac{xg'(x)}{y\sqrt{y-x}}\,dx.$$

Now conclude that $(\star)$ holds.

Let $0 < s < 1$ and consider the following generalization of Abel's integral equation:

$$g(x) = \int_0^x \frac{f(t)}{(x-t)^s}\,dt \qquad (x \in [\,0, a\,]).$$

(v) Now multiply by $\frac{1}{(y-x)^{1-s}}$, and conclude by Exercise 6.58.(iv) that the inner integral becomes

$$\int_0^1 u^{1-s-1}(1-u)^{s-1}\,du = \Gamma(1-s)\Gamma(s) = \frac{\pi}{\sin(s\pi)}.$$

Conclude that

$$f(y) = \frac{\sin(s\pi)}{\pi}\int_0^y \frac{g'(x)}{(y-x)^{1-s}}\,dx \qquad (y \in [\,0, a\,]).$$

It turns out a posteriori that the problem above can be elegantly formulated as follows. For $0 < s < 1$, define

$$\phi_s(x) = \frac{x_+^{s-1}}{\Gamma(s)}, \qquad \text{where} \qquad x_+^{s-1} = \begin{cases} x^{s-1}, & x > 0; \\ 0, & x \leq 0. \end{cases}$$

Then the *convolution equation* for f (see Example 6.11.5; and our present treatment is formal, that is, it is assumed that here, too, convolution is well-defined) $g = f * \phi_s$ has the solution $f = g' * \phi_{1-s}$.

(vi) Examine the relationship between this solution technique and that of Exercise 2.75.

Exercise 6.109 (Convolution and spline – needed for Exercise 7.39). The definition

$$f * g(x) = \int_{\mathbf{R}^n} f(x-y)g(y)\,dy$$

of convolution $*$ from Example 6.11.5 turns out to be meaningful for a much larger class of functions f and g than those from $\mathcal{S}(\mathbf{R}^n)$. The convolution operation "improves smoothness properties" and can be used to construct C^k functions with compact support, for $k \in \mathbf{N}$.

Let χ be the characteristic function of the interval $[-\frac{1}{2}, \frac{1}{2}] \subset \mathbf{R}$ and let $\chi_n = \chi * \ldots * \chi$ be the n-fold convolution product, for $n \in \mathbf{N}$, according to the definition above.

(i) Prove

$$\int_{\mathbf{R}} \chi(x)\,dx = 1, \qquad \int_{\mathbf{R}} x\chi(x)\,dx = 0, \qquad \int_{\mathbf{R}} x^2\chi(x)\,dx = \frac{1}{12},$$

$$\chi_n(x) = \chi_n(-x) \qquad (x \in \mathbf{R}), \qquad \int_{\mathbf{R}} \chi_n(x)\,dx = 1 \qquad (n \in \mathbf{N}).$$

(ii) Prove that χ_2 is the continuous function on $\mathbf{R}$ given by

$$\chi_2(x) = \begin{cases} 0, & x \leq -1; \\ x+1, & -1 \leq x \leq 0; \\ -x+1, & 0 \leq x \leq 1; \\ 0, & 1 \leq x. \end{cases}$$

Hint: $\chi_2(x) = \int_{\{y \in \mathbf{R} \mid x-\frac{1}{2} \leq y \leq x+\frac{1}{2}\}} \chi(y)\,dy$.

(iii) Prove that χ_3 is the C^1 function on $\mathbf{R}$ given by $\chi_3(x)$ is equal to

$$\begin{array}{lllrcl}
0, & & & & x \le & -\frac{3}{2};\\
\frac{1}{2!}(\frac{3}{2}+x)^2 & = & \frac{(x+\frac{3}{2})^2}{2!}, & -\frac{3}{2} \le & x \le & -\frac{1}{2};\\
\frac{3}{4}-x^2 & = & -2\frac{(x+\frac{1}{2})^2}{2!}+(x+\frac{1}{2})+\frac{1}{2!}, & -\frac{1}{2} \le & x \le & \frac{1}{2};\\
\frac{1}{2!}(\frac{3}{2}-x)^2 & = & \frac{(x-\frac{1}{2})^2}{2!}-(x-\frac{1}{2})+\frac{1}{2!}, & \frac{1}{2} \le & x \le & \frac{3}{2};\\
0, & & & \frac{3}{2} \le & x. &
\end{array}$$

(iv) Prove that χ_4 is the C^2 function on $\mathbf{R}$ given by $\chi_4(x)$ is equal to

$$\begin{array}{lllrcl}
0, & & & & x \le & -2;\\
\frac{1}{3!}(2+x)^3 & = & \frac{(x+2)^3}{3!}, & -2 \le & x \le & -1;\\
\frac{2}{3}-x^2-\frac{1}{2}x^3 & = & -3\frac{(x+1)^3}{3!}+\frac{(x+1)^2}{2!}+\frac{1}{2!}(x+1)+\frac{1}{3!}, & -1 \le & x \le & 0;\\
\frac{2}{3}-x^2+\frac{1}{2}x^3 & = & 3\frac{x^3}{3!}-2\frac{x^2}{2!}+\frac{4}{3!}, & 0 \le & x \le & 1;\\
\frac{1}{3!}(2-x)^3 & = & -\frac{(x-1)^3}{3!}+\frac{(x-1)^2}{2!}-\frac{1}{2!}(x-1)+\frac{1}{3!}, & 1 \le & x \le & 2;\\
0, & & & 2 \le & x. &
\end{array}$$

We now prove by mathematical induction on $n \in \mathbf{N}$ that χ_n is a piecewise polynomial C^{n-2} function on $\mathbf{R}$ whose support is the interval $[-\frac{n}{2}, \frac{n}{2}]$. The intervals

$$I_{n,0} := \big]-\infty, -\frac{n}{2}\big], \qquad I_{n,k} := \big[-\frac{n}{2}+k-1, -\frac{n}{2}+k\big] \qquad (1 \le k \le n),$$
$$I_{n,n+1} := \big[\frac{n}{2}, \infty\big[$$

are the maximal intervals I such that $\chi_n\big|_I$ is a polynomial function. Let

$$p_{n,k} := \chi_n\big|_{I_{n,k}} \qquad (0 \le k \le n+1);$$

then $p_{n,k}$, for $1 \le k \le n$, is a polynomial function of degree $n-1$, while $p_{n,0} = p_{n,n+1} = 0$. Now write

$$(\star) \qquad p_{n,k}(x) = \sum_{0 \le i \le n-1} p_{n,k,i}\,\frac{(x-(-\frac{n}{2}+k-1))^i}{i!}.$$

We shall want to prove, for $1 \le k \le n$ and $0 \le i \le n-1$,

$$(\star\star) \qquad p_{n,k,i} = \frac{1}{(n-1-i)!}\sum_{0 \le j \le k-1} (-1)^j \binom{n}{j}(k-1-j)^{n-1-i}.$$

(v) Verify that, for $\leq k \leq n+1$ and $x \in I_{n+1,\,k}$,

$$p_{n+1,\,k}(x) = \int_{x-\frac{1}{2}}^{-\frac{n}{2}+k-1} p_{n,\,k-1}(y)\,dy + \int_{-\frac{n}{2}+k-1}^{x+\frac{1}{2}} p_{n,\,k}(y)\,dy.$$

Conclude by means of ($\star$) that

$$p_{n+1,\,k}(x) =$$

$$\sum_{0\leq i\leq n-1} \frac{p_{n,\,k-1,\,i}}{(i+1)!} + \sum_{1\leq i\leq n} (p_{n,\,k,\,i-1} - p_{n,\,k-1,\,i-1}) \frac{(x-(-\frac{n+1}{2}+k-1))^i}{i!}.$$

(vi) Derive the following recursion relations between the coefficients $p_{n+1,\,k,\,i}$:

$$p_{n+1,\,k,\,0} = \sum_{0\leq i\leq n-1} \frac{p_{n,\,k-1,\,i}}{(i+1)!} \qquad (1 \leq k \leq n+1);$$

$$p_{n+1,\,k,\,i} = p_{n,\,k,\,i-1} - p_{n,\,k-1,\,i-1} \qquad (1 \leq k \leq n+1,\ 1 \leq i \leq n).$$

Then go on to show that ($\star\star$) is satisfied, using $\binom{n}{j-1} + \binom{n}{j} = \binom{n+1}{j}$.

(vii) Prove

$$\chi_n(x) = \frac{1}{(n-1)!} \sum_{0\leq j\leq [x+\frac{n}{2}]} (-1)^j \binom{n}{j} (x + \frac{n}{2} - j)^{n-1}.$$

(viii) The function χ_n is not $(n-1)$ times differentiable on $[-\frac{n}{2}, \frac{n}{2}]$; more precisely, prove, for $1 \leq k \leq n+1$,

$$\lim_{x\downarrow -\frac{n}{2}+k-1} \frac{d^{n-1} p_{n,\,k}}{d^{n-1}x}(x) - \lim_{x\uparrow -\frac{n}{2}+k-1} \frac{d^{n-1} p_{n,\,k}}{d^{n-1}x}(x) = (-1)^{k-1} \binom{n}{k-1}.$$

(ix) We want to give an independent proof that the following does indeed hold, for $0 \leq i < n-1$:

$$0 = \chi_n^{(i)}\Big(\frac{n}{2}\Big) = \frac{1}{(n-1-i)!} \sum_{0\leq j\leq n-1} (-1)^j \binom{n}{j} (n-j)^{n-1-i}.$$

For this we consider

$$f(t) = \sum_{0\leq j\leq n-1} \binom{n}{j} (-1)^j e^{t(n-j)} = (e^t - 1)^n - (-1)^n \qquad (t \in \mathbf{R});$$

note that $0 = f^{(n-1-i)}(0) = (n-1-i)!\, \chi_n^{(i)}(\frac{n}{2})$, for $0 \leq i < n-1$.

Background. In *numerical mathematics*, C^{n-1} functions on $\mathbf{R}$ which are piecewise polynomial of degree n are known as *splines* of degree n, after the flexible metal strips used to draw smooth curves.

Exercise 6.110. Define

$$f_k : \mathbf{R}_+ \to \mathbf{R} \qquad \text{by} \qquad f_k(x) = \begin{cases} \dfrac{1}{k}, & 0 < x < k^2; \\ 0, & k^2 \le x. \end{cases}$$

Show $\lim_{k\to\infty} f_k(x) = 0$, for all $x \in \mathbf{R}_+$, and this is even true uniformly on $\mathbf{R}_+$, while we have $\lim_{k\to\infty} \int_{\mathbf{R}_+} f_k(x)\,dx = \infty$.
Background. Note that majorizing functions like $g(x) = 1$ for $0 < x < 1$, and $g(x) = \frac{1}{\sqrt{x}}$, for $1 \le x$ are not absolutely Riemann integrable over $\mathbf{R}_+$. Hence Arzelà's Dominated Convergence Theorem 6.12.3 does not apply.

Exercises for Chapter 7

Exercise 7.1 (Formula of Binet–Cauchy and Pythagoras' Theorem). Let $d \leq n$, and suppose $A \in \mathrm{Mat}(n \times d, \mathbf{R})$ and $B \in \mathrm{Mat}(d \times n, \mathbf{R})$. Denote by $a_i \in \mathbf{R}^d$, for $1 \leq i \leq d$, the row vectors of A, and by $b_j \in \mathbf{R}^d$, for $1 \leq j \leq d$, the column vectors of B. Write K for the d-tuple $(k_1, \dots, k_d)$ from the set $\{1, 2, \dots, n\}$; and further $\mathcal{K}_d$ for the set of all such d-tuples; write A^K and B_K in $\mathrm{Mat}(d, \mathbf{R})$ for the matrix with $a_{k_1}, \dots, a_{k_d}$ as its row vectors and with $b_{k_1}, \dots, b_{k_d}$ as its column vectors, respectively. Finally, set

$$a^K = \det A^K, \qquad b_K = \det B_K.$$

(i) Prove the following *formula of Binet–Cauchy*:

$$\det(BA) = \sum_{K \in \mathcal{I}_d} b_K\, a^K.$$

Here $\mathcal{I}_d$ is the subset of $\mathcal{K}_d$ consisting of the strictly ascending d-tuples, that is, of the $K = (k_1, \dots, k_d)$ satisfying $1 \leq k_1 < \cdots < k_d \leq n$ (compare with Definition 8.6.3).

Hint: If $C = BA \in \mathrm{Mat}(d, \mathbf{R})$, then $c_{ij} = \sum_{1 \leq k \leq n} b_{ik} a_{kj}$, for $1 \leq i, j \leq d$. For $\sigma \in S_d$, the permutation group on d elements, write $\mathrm{sgn}(\sigma)$ for the sign of σ. Then, by a well-known formula for the determinant,

$$\begin{aligned}
\det(BA) &= \sum_{\sigma \in S_d} \mathrm{sgn}(\sigma) \prod_{1 \leq i \leq d} \sum_{1 \leq k_i \leq n} b_{ik_i} a_{k_i \sigma(i)} \\
&= \sum_{1 \leq k_1, \dots, k_d \leq n} \prod_{1 \leq i \leq d} b_{ik_i} \sum_{\sigma \in S_d} \mathrm{sgn}(\sigma) \prod_{1 \leq j \leq d} a_{k_j \sigma(j)} \\
&= \sum_{K \in \mathcal{K}_d} a^K \prod_{1 \leq i \leq d} b_{ik_i}.
\end{aligned}$$

Note that $a^K \neq 0$ only if K consists of mutually distinct numbers, therefore the summation can be performed only over such K. Since $a^{(\sigma(k_1), \dots, \sigma(k_d))} = \mathrm{sgn}(\sigma)\, a^{(k_1, \dots, k_d)}$, for $\sigma \in S_d$, we obtain

$$\det(BA) = \sum_{K \in \mathcal{I}_d} a^K \sum_{\sigma \in S_d} \mathrm{sgn}(\sigma) \prod_{1 \leq i \leq d} b_{i\sigma(k_i)} = \sum_{K \in \mathcal{I}_d} b_K\, a^K.$$

(ii) Let v and $w \in \mathbf{R}^n$; and let $A \in \mathrm{Mat}(n \times 2, \mathbf{R})$ and $B \in \mathrm{Mat}(2 \times n, \mathbf{R})$ have v and w as column and row vectors, respectively. From part (i) deduce the equality in Exercise 5.26.(i).

(iii) Suppose $B = A^t$. Using part (i) conclude the following, known as *Pythagoras' Theorem*, which generalizes the identity $\|a\|^2 = \sum_{1\leq j\leq n} a_j^2$, valid for a vector $a \in \mathbf{R}^n$:

$$\det(A^tA) = \sum_{K\in\mathcal{I}_d} (\det A^K)^2 \qquad (A \in \mathrm{Mat}(n \times d, \mathbf{R})).$$

Note that the column vectors of A^K, for $K \in \mathcal{I}_d$, are the projections of the column vectors of A onto the d-dimensional linear subspace of $\mathbf{R}^n$ spanned by the standard basis vectors $e_{k_1},\dots,e_{k_d}$. In other words, in the terminology of Section 7.3, the square of the d-volume of a d-dimensional parallelepiped in $\mathbf{R}^n$ is equal to the sum of the squares of the d-volumes of its projections onto all the d-dimensional linear subspaces of $\mathbf{R}^n$ given by the vanishing of some of the coordinate functions.

Exercise 7.2. Calculate the length of that part of the graph of the function log which lies between $(1, 0)$ and $(x, \log x)$, for $x > 0$.
Hint: Substitute $t = \tan\alpha$ to obtain

$$\int \frac{\sqrt{1+t^2}}{t}\,dt = \sqrt{1+t^2} + \log t - \log(1+\sqrt{1+t^2}) \qquad (t > 0).$$

Exercise 7.3 (Sequel to Exercise 4.8). Consider the part $V = \phi(\,]\,0, 2\pi\,[\,)$ of the helix from Exercise 4.8 for which $\phi(t) = (\cos t,\ \sin t,\ t)$. Prove

$$\int_V \|x\|^2\,d_1x = \frac{2\pi\sqrt{2}\,(3+4\pi^2)}{3}.$$

Exercise 7.4 (Semicubic parabola). This is defined as the zero-set in $\mathbf{R}^2$ of $g(x) = x_1^3 - x_2^2$. Calculate the length of the part of this curve lying between the points $(t,\ -t^{3/2})$ and $(t,\ t^{3/2})$, for $t > 0$.

Exercise 7.5 (Lemniscate, $\Gamma(\frac{1}{4})$ and elliptic integrals of first kind – sequel to Exercises 3.44, 4.34, 6.50 and 6.69 – needed for Exercise 7.6). Consider the *lemniscate* from Example 6.6.4, which in polar coordinates (r, α) for $\mathbf{R}^2$ is given by $r^2 = \cos 2\alpha$. Let $0 \leq \phi \leq \frac{\pi}{4}$ and let $x = x(\phi) = \sqrt{\cos 2\phi} \geq 0$. Then the arc of the lemniscate determined by x is understood to mean that part of the lemniscate which lies between the origin in $\mathbf{R}^2$ and $(r, \alpha) = (x(\phi),\ \phi)$.

(i) Prove that the length of the arc of the lemniscate determined by $x = x(\phi)$ is given by

$$\int_\phi^{\frac{\pi}{4}} \frac{1}{\sqrt{\cos 2\alpha}}\,d\alpha = \int_0^x \frac{1}{\sqrt{1-t^4}}\,dt.$$

Hint: Substitute $\cos 2\alpha = t^2$.

Exercise 3.44 or 4.34.(iii) now implies the following. If x and y are chosen positive and sufficiently close to 0, then the sum of the lengths of the arcs determined by x and y equals the length of the arc determined by $a(x, y)$, in the notation of the latter exercise. It is a remarkable fact that $a(x, y)$ can be constructed from x and y by means of ruler and compasses.

Let $\frac{\varpi}{2}$ be the length of the arc determined by 1, that is

$$\frac{\varpi}{2} = \int_0^{\frac{\pi}{4}} \frac{1}{\sqrt{\cos 2\alpha}}\, d\alpha = \int_0^1 \frac{1}{\sqrt{1-t^4}}\, dt = \frac{1}{4\sqrt{2\pi}}\Gamma\Big(\frac{1}{4}\Big)^2,$$

see Exercise 6.50.(vi). Then the length of the lemniscate equals 2ϖ. Note that $\frac{\pi}{2} = \int_0^1 \frac{1}{\sqrt{1-t^2}}\, dt$ and that the length of the unit circle equals 2π.

(ii) Use the substitution $\cos 2\alpha = \cos^2\phi$ to show that

$$\varpi = \sqrt{2}\int_0^{\frac{\pi}{2}} \frac{1}{\sqrt{1-\frac{1}{2}\sin^2\phi}}\, d\phi = \sqrt{2}K\Big(\frac{1}{\sqrt{2}}\Big),$$

where

$$K(k) := \int_0^{\frac{\pi}{2}} \frac{1}{\sqrt{1-k^2\sin^2\phi}}\, d\phi \qquad (0 < k < 1).$$

$K(k)$ is said to be *Legendre's form of the complete elliptic integral of the first kind with modulus* k (compare with Example 7.4.1). Combination of parts (i) and (ii) now gives

$$\Gamma\Big(\frac{1}{4}\Big)^2 = 4\sqrt{\pi}\, K\Big(\frac{1}{\sqrt{2}}\Big), \qquad \Gamma\Big(\frac{1}{4}\Big) = 3.625\,609\,908\,221\cdots.$$

In this calculation the following result is used.

(iii) Verify that, in the notation of Exercise 6.69,

$$\begin{aligned} K(k) &= \frac{\pi}{2}F\Big(\frac{1}{2}, \frac{1}{2}, 1 : k^2\Big) \\ &= \frac{\pi}{2}\Big(1 + \sum_{n\in\mathbf{N}}\Big(\frac{(2n-1)(2n-3)\cdots 3\cdot 1}{2n(2n-2)\cdots 4\cdot 2}k^n\Big)^2\Big) \\ &= \frac{\pi}{2}\Big(1 + \frac{1}{4}k^2 + \frac{9}{64}k^4 + \frac{25}{256}k^6 + \cdots\Big). \end{aligned}$$

In particular,

$$\Gamma\Big(\frac{1}{4}\Big)^2 = 2\pi^{\frac{3}{2}}F\Big(\frac{1}{2}, \frac{1}{2}, 1 : \frac{1}{2}\Big).$$

Moreover, deduce that K satisfies a hypergeometric differential equation.

Exercise 7.6 (Elliptic integrals and Catalan's constant – sequel to Exercises 6.39 and 7.5). We employ the notation from Exercise 7.5. Prove by changing the order of integration that

$$\int_0^1 K(k)\,dk = \int_0^{\frac{\pi}{2}} \frac{\phi}{\sin\phi}\,d\phi.$$

Now use the substitution $\phi = 2\arctan x$ to show, see Exercise 6.39.(iv),

$$\int_0^1 K(k)\,dk = 2\int_0^1 \frac{\arctan x}{x}\,dx = 2\mathbf{G}.$$

Conclude that

$$\mathbf{G} = \frac{\pi}{4}\Big(1 + \sum_{n\in\mathbf{N}} \frac{1}{2n+1}\Big(\frac{(2n-1)(2n-3)\cdots3\cdot1}{2n(2n-2)\cdots4\cdot2}\Big)^2\Big).$$

Exercise 7.7. Define $\phi : \,]0,1[\, \to \mathbf{R}^3$ by $\phi(t) = (\cos 2\pi t^2,\ \sin 2\pi t^2,\ 2\pi t^2)$.

(i) Calculate the Euclidean length of the curve $V = \mathrm{im}(\phi)$.

(ii) Prove that $\psi : \,]0, 2\pi\sqrt{2}[\, \to \mathbf{R}^3$ with $\psi(t) = (\cos\frac{t}{\sqrt{2}},\ \sin\frac{t}{\sqrt{2}},\ \frac{t}{\sqrt{2}})$, is the parametrization for V with the arc length as parameter.

Exercise 7.8. Let K be the intersection of the solid cylinders $\{\, x \in \mathbf{R}^3 \mid x_1^2 + x_2^2 \leq 1\,\}$ and $\{\, x \in \mathbf{R}^3 \mid x_1^2 + x_3^2 \leq 1\,\}$ (compare with Example 6.5.2). Prove that $\mathrm{area}(\partial K) = 16$.

Exercise 7.9. Let $V = \{\, x \in \mathbf{R}^3 \mid \|x\| = 1\,\}$. Prove

$$\int_V x_i^2\,d_2x = \frac{4\pi}{3} \qquad (i = 1, 2, 3).$$

Exercise 7.10. Let $V \subset \mathbf{R}^3$ be the triangle with vertices $(1,0,0)$, $(0,1,0)$ and $(0,0,1)$. Prove

$$\int_V x_1\,d_2x = \frac{1}{6}\sqrt{3}.$$

Exercise 7.11. Let $V \subset \mathbf{R}^3$ be the graph of $f : \,]0,1[\, \times \,]-1,1[\, \to \mathbf{R}$ given by $f(x) = x_1^2 + x_2$. Prove

$$\int_V x_1\,d_2x = \sqrt{6} - \frac{1}{3}\sqrt{2}.$$

Exercise 7.12. Let $a > 0$. Consider that part of the cylinder $\{ x \in \mathbf{R}^3 \mid x_1^2 + x_3^2 = a^2 \}$ that lies inside the cylinder $\{ x \in \mathbf{R}^3 \mid x_1^2 + x_2^2 = 2ax_2 \}$ and inside the octant $\{ x \in \mathbf{R}^3 \mid x_1 > 0,\ x_2 > 0,\ x_3 > 0 \}$. Prove that its area equals $2a^2$.

Exercise 7.13 (Girard's formula). A subset $D \subset S^2$ is said to be a *spherical diangle* with angle α if D is bounded by two half great circles whose tangent vectors at a point of intersection include an angle α. These great circles are said to be the sides of D.

(i) Prove that $\text{area}(D) = 2\alpha$.

A subset $\Delta \subset S^2$ is said to be a *spherical triangle* if Δ is the intersection of three spherical diangles that pairwise have a part of a side in common (compare with Exercise 5.27). Assume that the spherical diangles have angles α_1, α_2 and α_3, respectively.

(ii) Show that $\text{area}(\Delta) = \sum_{1 \le i \le 3} \alpha_i - \pi$.

Exercise 7.14. Assume $m > 0$ and $h > 0$. Let $V \subset \mathbf{R}^3$ be that part of the conical surface

$$\{ x \in \mathbf{R}^3 \mid x_3^2 = m^2(x_1^2 + x_2^2) \}$$

that lies between the planes $\{ x \in \mathbf{R}^3 \mid x_3 = 0 \}$ and $\{ x \in \mathbf{R}^3 \mid x_3 = h \}$.

(i) Show that the Euclidean area of V in $\mathbf{R}^3$ equals $\pi \frac{h^2}{m^2} \sqrt{m^2 + 1}$.

Next, consider the surface $V' \subset \mathbf{R}^3$ that is formed when one straight line segment lying on V is removed from the conical surface V. Then unroll this surface V' into a plane without stretching, shrinking or tearing it; thus is obtained a circular sector in $\mathbf{R}^2$.

(ii) Calculate the Euclidean area of this circular sector in $\mathbf{R}^2$.

Let W be a plane in $\mathbf{R}^3$ with the following two properties: W is parallel to a tangent plane (at $(0, 1, m)$, for example) to the conical surface V, and W goes through $(0, 0, h)$, the center of the top circle of V.

(iii) Prove that the length of the conic $V \cap W$ equals

$$\frac{h}{m} \int_{-1}^{1} \sqrt{1 + (m^2 + 1)t^2}\, dt.$$

Exercise 7.15 (Viviani's solid). This is the set $L \subset \mathbf{R}^3$ formed as the intersection of the unit ball in $\mathbf{R}^3$ and a solid cylinder, more precisely $L = \{ x \in \mathbf{R}^3 \mid \|x\| \le 1,\ x_1^2 + x_2^2 \le x_1 \}$. Then $\partial L = V_1 \cup V_2$, with

$$V_1 = \{ x \in \mathbf{R}^3 \mid \|x\| = 1,\ x_1^2 + x_2^2 \le x_1 \},$$
$$V_2 = \{ x \in \mathbf{R}^3 \mid \|x\| \le 1,\ x_1^2 + x_2^2 = x_1 \}.$$

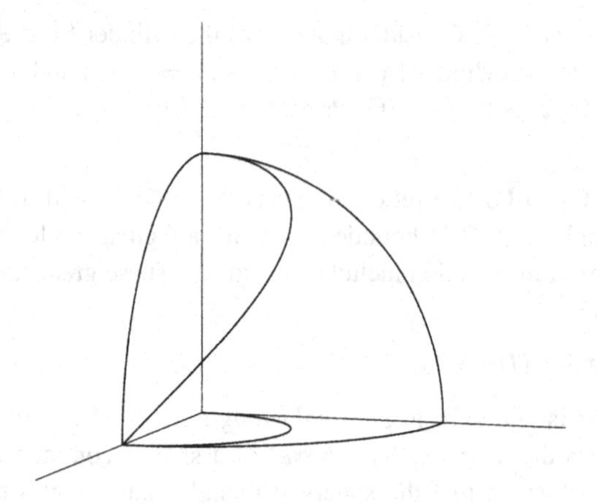

Illustration for Exercise 7.15: Viviani's solid

(i) Verify that $V_1 \cap V_2 = \mathrm{im}(\phi)$, where $\phi : \,]-\pi,\, \pi\,[\, \to \mathbf{R}^3$ is defined by

$$\phi(\alpha) = (\cos^2\alpha,\ \cos\alpha\sin\alpha,\ \sin\alpha) \qquad (-\pi < \alpha < \pi).$$

(ii) Prove that the restriction of ϕ to $]-\pi, 0\,[$ and to $]\,0, \pi\,[$, respectively, are C^∞ embeddings.

(iii) Show that the length of $V_1 \cap V_2$ equals the following complete elliptic integral of the second kind:

$$4\sqrt{2}\int_0^{\frac{\pi}{2}} \sqrt{1 - \tfrac{1}{2}\sin^2\alpha}\, d\alpha.$$

(iv) Demonstrate that the orthogonal projection of V_1 onto the plane $\{x \in \mathbf{R}^3 \mid x_3 = 0\}$ equals the set $D \times \{0\} \subset \mathbf{R}^3$ with

$$\begin{aligned} D &= \{\, y \in \mathbf{R}^2 \mid y_1^2 + y_2^2 \le y_1 \,\} \\ &= \{\, r(\cos\alpha,\ \sin\alpha) \in \mathbf{R}^2 \mid -\frac{\pi}{2} < \alpha \le \frac{\pi}{2},\ 0 \le r \le \cos\alpha \,\}. \end{aligned}$$

Verify $L = \{\, (y,\ x_3) \in \mathbf{R}^3 \mid y \in D,\ |x_3| \le \sqrt{1 - \|y\|^2} \,\}$.

(v) Calculate $\mathrm{vol}_3(L) = \frac{2}{3}\pi - \frac{8}{9}$.
Hint: $\int_0^{\frac{\pi}{2}} \sin^3\alpha\, d\alpha = \frac{2}{3}$.

(vi) Prove that the area of V_1 equals $2\pi - 4$ (see Example 7.4.10).

(vii) Demonstrate that the area of V_2 equals 4. The area of the boundary of Viviani's solid thus equals that of the half unit sphere.

Exercise 7.16. Let the notation be that of Example 7.4.8. Prove, for all $1 \leq j \leq 3$ and $x \in \mathbf{R}^3 \setminus A$,

$$\phi_{A,\,j}(x) = -\frac{1}{4\pi}\int_A \frac{y_j}{R}\,\frac{1}{\|x-y\|}\,d_2y = \begin{cases} -\dfrac{1}{3}x_j, & \|x\| < R; \\[2mm] -\dfrac{1}{3}\dfrac{R^3}{\|x\|^3}x_j, & \|x\| > R. \end{cases}$$

Verify that $\phi_{A,\,j}$ can be continuously continued over A.
Hint: Show

$$\frac{1}{4\pi}\int_A \|x-y\|\,d_2y = \begin{cases} \dfrac{1}{3}R\,(\|x\|^2 + 3R^2), & \|x\| < R; \\[2mm] \dfrac{1}{3}\dfrac{R^2}{\|x\|}(3\|x\|^2 + R^2), & \|x\| > R. \end{cases}$$

Then note that $\frac{\partial}{\partial x_j}\|x-y\| = \frac{x_j - y_j}{\|y-x\|}$ according to Example 2.4.8, and conclude that

$$\frac{1}{4\pi R}\int_A \frac{y_j - x_j}{\|y-x\|}\,d_2y = \begin{cases} -\dfrac{2}{3}x_j, & \|x\| < R; \\[2mm] \dfrac{x_j R}{\|x\|}\Big(\dfrac{R^2}{3\|x\|^2} - 1\Big), & \|x\| > R. \end{cases}$$

Background. In the theory of dielectrics from *electromagnetism* one encounters the potential of a charged sphere where the charge density is proportional to $\frac{y_3}{R} = \sin\theta$, varying from $+1$ at the north pole to -1 at the south pole.

Exercise 7.17 (Unrolling a cone – needed for Exercise 7.18). Let $I = \,]\,0, l\,[$ and let $\gamma : I \to \mathbf{R}^3$ be a C^1 mapping. Assume that $\gamma(s)$ and $\gamma'(s)$ are linearly independent, for all $s \in I$. Next define

$$\phi : D := I \times \,]\,0,1\,[\, \to \mathbf{R}^3 \qquad \text{by} \qquad \phi(s,t) = t\,\gamma(s),$$

and consider the cone $C = \mathrm{im}(\phi)$ in $\mathbf{R}^3$ determined by γ.

(i) Show that $\phi : D \to \mathbf{R}^3$ is a C^1 immersion.

(ii) From now on assume ϕ to be an C^1 embedding, which implies that C is a C^1 manifold. Then prove that $\mathrm{area}(C) = \frac{1}{2}\int_I \|\gamma(s) \times \gamma'(s)\|\,ds$.

Introduce the angle function $\alpha : I \to \mathbf{R}$ by $\alpha(s) = \int_0^s \frac{\|\gamma(\sigma)\times\gamma'(\sigma)\|}{\|\gamma(\sigma)\|^2}\,d\sigma$. By shrinking I if necessary, we may arrange that $\alpha(l) < \pi$. Furthermore, define

$$\Upsilon : D \to V := \mathbf{R}_+ \times \,]-\pi, \pi\,[\qquad \text{by} \qquad \Upsilon(s,t) = (t\|\gamma(s)\|,\ \alpha(s)).$$

(iii) Prove that the angle function is monotonically increasing and denote by β its inverse. Deduce that Υ is a C^1 diffeomorphism onto its image $\{ (r, \alpha) \in V \mid 0 < \alpha < \alpha(l),\ 0 < r < r(\alpha) \}$, where we write $r(\alpha) = \|\gamma \circ \beta(\alpha)\|$. Define $\Psi : V \to U$ with $\Psi(r, \alpha) = r(\cos\alpha, \sin\alpha)$ as in Example 3.1.1 and show that

$$\upsilon := \Psi \circ \Upsilon \circ \phi^{-1} : C \to \upsilon(C) \subset U$$

with

$$\upsilon(t\gamma(s)) = t\|\gamma(s)\| \,(\cos\alpha(s),\ \sin\alpha(s))$$

is a bijection that maps a ruling on C to a line segment in $\mathbf{R}^2$ of the same length that issues from the origin.

(iv) On the strength of Example 6.6.4 verify

$$\text{area}(C) = \text{area}\,(\upsilon(C)) = \frac{1}{2}\int_0^{\alpha(l)} r(\alpha)^2\, d\alpha.$$

In other words, υ is an *unrolling* of the cone $C \subset \mathbf{R}^3$ onto $\upsilon(C) \subset \mathbf{R}^2$ without distortion of area.

Exercise 7.18 (Area of tangent cluster is area of tangent sweep – sequel to Exercise 7.17). Let $J = \,]\,0, l\,[$, and let $f \in C(J)$ be given. Suppose that $\gamma : J \to \mathbf{R}^3$ is a C^2 parametrization by arc length of the curve $\text{im}(\gamma)$ and define, for $c \in \mathbf{R}$,

$$\phi_c : D := \{ (s, t) \in \mathbf{R}^2 \mid s \in J,\ 0 < t < f(s) \} \to \mathbf{R}^3$$

by

$$\phi_c(s, t) = c\,\gamma(s) + t\,\gamma'(s).$$

We refer to the conical surface $C := \text{im}(\phi_0) \subset \mathbf{R}^3$ as the *tangent cluster* determined by γ and f, and to the surface $S := \text{im}(\phi_1) \subset \mathbf{R}^3$ as the corresponding *tangent sweep*. Note that the difference between C and S corresponds precisely to that between tangent vectors and geometric tangent vectors. Denote by $\kappa(s) = \|\gamma''(s)\|$ the curvature of $\text{im}(\gamma)$ at $\gamma(s)$ as in Definition 5.8.1 and assume that $\kappa(s) > 0$, for all $s \in J$.

(i) Show that $\phi_c : D \to \mathbf{R}^3$ is a C^2 immersion.

(ii) From now on suppose that ϕ_0 and ϕ_1 are embeddings. Prove (compare with Exercise 7.17.(ii))

$$\text{area}(C) = \text{area}(S) = \frac{1}{2}\int_J f(s)^2 \kappa(s)\, ds.$$

(iii) Verify that the angle function α from Exercise 7.17 equals $\alpha(s) = \int_0^s \kappa(\sigma)\, d\sigma$ in this case. And using part (iv) from that exercise show $\text{area}(S) = \text{area}(C) = \text{area}\,(\upsilon(C))$, where υ is the unrolling of C into $\mathbf{R}^2$.

Exercise 7.19 (Pseudosphere and non-Euclidean geometry – sequel to Exercise 5.51). Consider the tractrix and the pseudosphere from Exercise 5.51.

(i) Prove that the length of the part of the tractrix lying between $(x, \pm f(x))$ and $(1, 0)$ equals $\log\left(\frac{1}{x}\right)$, for all $x \in\,]\,0, 1\,]$.

(ii) Conclude that the tractrix does not have finite length.

(iii) Calculate the area of the pseudosphere. Why does your answer not surprise?

Let V^+ be the upper half of the pseudosphere.

(iv) Verify that we obtain a parametrization of an open part of V^+ by means of $\widetilde{\phi} : \,]-\pi, \pi\,[\times [\,1, \infty\,[\,\to \mathbf{R}^3$ satisfying

$$\widetilde{\phi}(u, v) = \left(\frac{\cos u}{v}, \frac{\sin u}{v}, \log(v + \sqrt{v^2 - 1}) - \frac{\sqrt{v^2 - 1}}{v}\right).$$

Hint: Set $x = \frac{1}{v}$ in Exercise 5.51.(i) and rotate.

(v) Assume u and $v : I \to \mathbf{R}$ are two C^1 functions of range such that $\gamma : I \to V^+$ is a well-defined C^1 curve if $\gamma(t) = \widetilde{\phi}(u(t), v(t))$. Verify that the length of γ is given by (assuming convergence of the integral)

$$\int_I \frac{1}{v(t)} \sqrt{u'(t)^2 + v'(t)^2}\, dt.$$

Background. The properties (iv) and (v) imply that this open part of V^+ can in fact be regarded as subset of the upper half-plane

$$\mathcal{H} = \{\, x \in \mathbf{R}^2 \mid x_2 > 0 \,\} \simeq \{\, z \in \mathbf{C} \mid \operatorname{Im} z > 0 \,\}.$$

Here $\mathcal{H}$ is endowed with the *metric*, that is, the concept of distance, which assigns to a C^1 curve $\gamma : I \to \mathcal{H}$ the length

$$L(\gamma) = \int_I \frac{1}{\gamma_2(t)} \sqrt{\gamma_1'(t)^2 + \gamma_2'(t)^2}\, dt = \int_I \frac{|\gamma'(t)|}{\operatorname{Im} \gamma(t)}\, dt.$$

The set $\mathcal{H}$ with this metric is a standard model of a *non-Euclidean geometry* (more precisely, a two-dimensional one, of constant curvature -1).

(vi) Verify that, with respect to this metric, the arcs in $\mathcal{H}$ of concentric circles centered on the x_1-axis that lie within a (Euclidean) fixed angle from the center are all of the same length.

This makes it plausible that in this geometry these arcs play the role of parallel line segments. We shall now prove this. Define $J : \mathcal{H} \to \mathbf{C}$ by $J(z) = 1/\overline{z}$.

(vii) Prove that $J : \mathcal{H} \to \mathcal{H}$, and that J is an *involution*, that is, $J^2 = I$. Verify that J is an *isometry* or a *metric-preserving* mapping of $\mathcal{H}$ into itself, that is, $L(\gamma) = L(J\gamma)$, for every C^1 curve $\gamma : I \to \mathcal{H}$.

A straight line in $\mathbf{R}^2$ is the set of fixed points of the reflection in that straight line; consequently, straight lines in $\mathbf{R}^2$ are one-dimensional sets of fixed points of nontrivial involutive isometries of $\mathbf{R}^2$. We now **define** straight lines in $\mathcal{H}$ to be those one-dimensional submanifolds that occur as sets of fixed points of nontrivial involutive isometries of $\mathcal{H}$.

(viii) Deduce from part (vii) that the segment in $\mathcal{H}$ of the unit circle in $\mathbf{R}^2$ about 0 is a straight line in $\mathcal{H}$. Verify that the mappings $\mathcal{H} \to \mathcal{H}$, with $t_a(z) = z + a$ for $a \in \mathbf{R}$, and $d_r(z) = rz$ for $r > 0$, are isometries of $\mathcal{H}$. Conclude that the segment in $\mathcal{H}$ of the circle in $\mathbf{R}^2$ about $(a, 0)$ of radius r is a straight line in $\mathcal{H}$ (consider the involutive isometry $t_a d_r J d_r^{-1} t_a^{-1}$).

Background. The semicircles in $\mathcal{H}$ associated with circles in $\mathbf{R}^2$ centered on the x_1-axis are therefore straight lines in this geometry on $\mathcal{H}$. It is now obvious that *Euclid's parallel postulate* is violated in $\mathcal{H}$; indeed, given a point in $\mathcal{H}$ and a straight line in $\mathcal{H}$ which does not contain that point, there exist many straight lines in $\mathcal{H}$ that contain the given point but do not intersect the given straight line (and are therefore "parallel" to it).

Exercise 7.20 (Identity by spherical coordinates). Suppose $f \in C^1(\mathbf{R}^n)$ has compact support.

(i) Show (see Exercise 7.21 for $\text{hyperarea}_{n-1}(S^{n-1})$)

$$f(x) = \frac{1}{\text{hyperarea}_{n-1}(S^{n-1})} \int_{\mathbf{R}^n} \frac{\langle \operatorname{grad} f(x-y), y \rangle}{\|y\|^n} \, dy \qquad (x \in \mathbf{R}^n).$$

Hint: We have $f(x) = -\int_0^\infty f'(x - r\omega)\, dr = \int_0^\infty \langle \operatorname{grad} f(x - r\omega), \omega \rangle \, dr$, for every $\omega \in S^{n-1}$. Next integrate this equality over $\omega \in S^{n-1}$ and change from spherical to rectangular coordinates in $\mathbf{R}^n$ (see Example 7.4.12).

(ii) Deduce

$$|f(x)| \le \frac{1}{\text{hyperarea}_{n-1}(S^{n-1})} \int_{\mathbf{R}^n} \frac{\|\operatorname{grad} f(x-y)\|}{\|y\|^{n-1}} \, dy \qquad (x \in \mathbf{R}^n).$$

Exercise 7.21 (Spherical coordinates in $\mathbf{R}^n$ and hyperarea of $(n-1)$-sphere – sequel to Exercises 3.18 and 6.50 – needed for Exercises 7.22, 7.23, 7.24, 7.25, 7.26, 7.28, 7.29, 7.30 and 7.53). Consider the $(n-1)$-dimensional unit sphere $S^{n-1} = \{ x \in \mathbf{R}^n \mid \|x\| = 1 \}$ and the n-dimensional unit ball $B^n = \{ x \in \mathbf{R}^n \mid \|x\| \le 1 \}$.

(i) Prove that Formula (7.26) is also valid for $f : \mathbf{R}^n \to \mathbf{R}$ with $f(x) = e^{-\|x\|^2}$.

(ii) Prove

$$\text{hyperarea}_{n-1}(S^{n-1}) = \frac{2\pi^{\frac{n}{2}}}{\Gamma(\frac{n}{2})} \qquad (n \in \mathbf{N}).$$

Hint: Calculate $\int_{\mathbf{R}^n} e^{-\|x\|^2}\,dx$ by the use of Cartesian coordinates and Example 6.10.8, and then also by part (i). Subsequently apply Exercise 6.50.(i).

(iii) Verify that (ii) is consistent with Example 7.4.11.

(iv) Prove $n\,\text{vol}_n(B^n) = \text{hyperarea}_{n-1}(S^{n-1})$, for $n \in \mathbf{N}$.
Hint: A ball is a union of concentric spheres.

(v) Calculate $\text{vol}_n(B^n)$, and verify that the answer is consistent with that found in Exercise 6.50.(viii).

(vi) Prove

$$\frac{d}{dr}\,\text{vol}_n\,(B^n(r)) = \frac{d}{dr}\frac{\pi^{\frac{n}{2}}}{\Gamma(\frac{n}{2}+1)}r^n = \frac{2\pi^{\frac{n}{2}}}{\Gamma(\frac{n}{2})}r^{n-1} = \text{hyperarea}_{n-1}\,(S^{n-1}(r)).$$

Here $B^n(r)$ and $S^{n-1}(r)$ are the ball and the sphere, respectively, in $\mathbf{R}^n$ about the origin and of radius r.

(vii) Verify that Exercise 3.18.(iv) implies that the mapping

$$\phi : \,]-\pi, \pi[\, \times \,\Big]-\frac{\pi}{2}, \frac{\pi}{2}\Big[^{n-2} \to S^{n-1} \setminus \{x \in \mathbf{R}^n \mid x_1 \le 0,\; x_2 = 0\}$$

is a C^∞ embedding, if

$$\phi\begin{pmatrix}\alpha\\ \theta_1\\ \theta_2\\ \vdots\\ \theta_{n-2}\end{pmatrix} = \begin{pmatrix} \cos\alpha\,\cos\theta_1\,\cos\theta_2\cdots\cos\theta_{n-3}\cos\theta_{n-2}\\ \sin\alpha\,\cos\theta_1\,\cos\theta_2\cdots\cos\theta_{n-3}\cos\theta_{n-2}\\ \sin\theta_1\,\cos\theta_2\cdots\cos\theta_{n-3}\cos\theta_{n-2}\\ \vdots\\ \sin\theta_{n-3}\,\cos\theta_{n-2}\\ \sin\theta_{n-2}\end{pmatrix}.$$

(viii) Use the hint from part (ii) and Exercise 3.18.(iii) to conclude that

$$\omega_\phi(\alpha,\,\theta_1,\,\dots,\,\theta_{n-2}) = \cos\theta_1\cos^2\theta_2\cdots\cos^{n-2}\theta_{n-2}.$$

Further, verify that by means of Exercise 6.50.(iii) one obtains

$$\text{hyperarea}_{n-1}(S^{n-1}) = 2\pi\,\Gamma\Big(\frac{1}{2}\Big)^{n-2}\prod_{j=1}^{n-2}\frac{\Gamma(\frac{j+1}{2})}{\Gamma(\frac{j+2}{2})}.$$

The result from part (viii) can be generalized as follows. Set $S_+^{n-1} = \{ y \in \mathbf{R}_+^n \mid \|y\| = 1 \}$. By analogy with Exercise 6.50 we then define the *generalized Beta function* $\mathrm{B} : \mathbf{R}_+^n \to \mathbf{R}$ by

$$\mathrm{B}(p_1, \dots, p_n) := \frac{\Gamma(p_1) \cdots \Gamma(p_n)}{\Gamma(p_1 + \cdots + p_n)}.$$

(ix) Prove that we now have the following generalization of the formulae from Exercise 6.50.(iii) and Exercise 6.64.(iii) (see also Example 7.4.10), for $p \in \mathbf{R}_+^n$:

$$\begin{aligned} \int_{S_+^{n-1}} \prod_{1 \le j \le n} y_j^{p_j} \, d_{n-1} y &= \frac{1}{2^{n-1}} \mathrm{B}\Big(\frac{p_1 + 1}{2}, \dots, \frac{p_n + 1}{2} \Big) \\ &= \frac{\prod_{1 \le j \le n} \Gamma(\frac{p_j + 1}{2})}{2^{n-1} \Gamma\big(\frac{n + \sum_{1 \le j \le n} p_j}{2} \big)}. \end{aligned}$$

Consider this formula in the particular case that $p_1 = \cdots = p_n = 0$, and verify that the result is consistent with that from part (ii).
Hint: Imitate the technique from part (ii), or use part (viii).

(x) Suppose that $p_j \in 2\mathbf{N}_0$, for $1 \le j \le n$. Deduce that the average of the monomial function $y \mapsto y^p = \prod_{1 \le j \le n} y_j^{p_j}$ over S^{n-1} satisfies, in the notation of Exercise 6.50.(vii) and with $(-1)!! = 1$,

$$\frac{1}{\mathrm{hyperarea}_{n-1}(S^{n-1})} \int_{S^{n-1}} y^p \, d_{n-1} y = \frac{\prod_{1 \le j \le n} (p_j - 1)!!}{\prod_{0 \le l < \frac{|p|}{2}} (n + 2l)} \in \mathbf{Q}.$$

Prove this equality also by application of the technique from part (ii) to the function $x \mapsto x^p e^{-\|x\|^2}$ on $\mathbf{R}^n$.

(xi) Using spherical coordinates and part (iv), prove the following identity, which also is a direct consequence of part (x) and Exercise 6.65.(vi)

$$\frac{n + |p|}{\mathrm{vol}_n(B^n)} \int_{B^n} x^p \, dx = \frac{n}{\mathrm{hyperarea}_{n-1}(S^{n-1})} \int_{S^{n-1}} y^p \, d_{n-1} y.$$

Exercise 7.22 (Pizzetti's formula – sequel to Exercises 2.52, 6.66 and 7.21). Suppose $f \in C^\infty(\mathbf{R}^n)$ is equal to its MacLaurin series. Verify for $r \in \mathbf{R}_+$ and $y \in S^{n-1}$

$$f(ry) = \sum_{k \in \mathbf{N}_0} r^k \sum_{|p| = k} \frac{y^p}{p!} D^p f(0),$$

where $p \in \mathbf{N}_0^n$. Using Exercise 7.21.(x) and the Multinomial Theorem from Exercise 2.52.(ii) prove *Pizzetti's formula* for the *spherical mean* of f over the sphere of center 0 and radius r

$$\begin{aligned}
&\frac{1}{\text{hyperarea}_{n-1}(S^{n-1})} \int_{S^{n-1}} f(ry)\, d_{n-1}y \\
&= \sum_{k \in \mathbf{N}_0} \frac{r^{2k}}{2^k \prod_{0 \le l < k}(n+2l)} \sum_{|p|=k} \frac{D^{2p} f(0)}{p!} \\
&= \sum_{k \in \mathbf{N}_0} \frac{r^{2k}}{2^k \prod_{0 \le l < k}(n+2l)} \frac{(\sum_{1 \le j \le n} D_j^2)^k}{k!} f(0) \\
&= \sum_{k \in \mathbf{N}_0} \frac{r^{2k}}{2^k k! \prod_{0 \le l < k}(n+2l)} \Delta^k f(0) \\
&= \Gamma\Big(\frac{n}{2}\Big) \sum_{k \in \mathbf{N}_0} \frac{\Delta^k f(0)}{k!\, \Gamma(\frac{n}{2}+k)} \Big(\frac{r}{2}\Big)^{2k} \\
&= \Gamma\Big(\frac{n}{2}\Big) \Big(\frac{r\sqrt{-\Delta}}{2}\Big)^{1-\frac{n}{2}} J_{\frac{n}{2}-1}(r\sqrt{-\Delta}) f(0).
\end{aligned}$$

Here Δ denotes the Laplace operator and $J_{\frac{n}{2}-1}$ the Bessel function of order $\frac{n}{2}-1$ as in Exercise 6.66. Furthermore, the last equality is obtained by a formal substitution in the power series for the Bessel function from part (ix) of the latter exercise. See Exercise 7.54 for a different proof.

Exercise 7.23 (Sequel to Exercises 6.50 and 7.21). Prove (compare with Exercise 6.99.(iii))

$$\int_{\mathbf{R}^n} \frac{1}{(1+\|x\|^2)^{\frac{n+1}{2}}}\, dx = \frac{1}{2} \text{hyperarea}_n(S^n).$$

Hint: Substitute spherical coordinates first, and subsequently $r = \tan \beta$.
More generally, prove by means of Exercises 6.50.(iv) and 7.21.(ii)

$$\int_{\mathbf{R}^n} \frac{1}{(1+\|x\|^2)^p}\, dx = \pi^{\frac{n}{2}} \frac{\Gamma(p-\frac{n}{2})}{\Gamma(p)} \qquad (p > \frac{n}{2}).$$

Exercise 7.24 (Sequel to Exercises 6.50 and 7.21). Using Exercise 6.50.(iv) and Legendre's duplication formula from Exercise 6.50.(vi), show

$$\int_{\mathbf{R}^{2n}} \int_{\mathbf{R}} (t^2 + (1+\|x\|^2)^2)^{-(n+1)}\, dt\, dx = \frac{\pi^{n+1}}{4^n\, n!} \qquad (n \in \mathbf{N}).$$

Exercise 7.25 (Sequel to Exercise 7.21). Let $B^n \subset \mathbf{R}^n$ be the unit ball and let $S^{n-1} \subset \mathbf{R}^n$ be the unit sphere. In physics the *moment of inertia I_i of B^n about the x_i-axis* is defined by

$$I_i = \int_{B^n} \sum_{1\le j\le n,\ j\ne i} x_j^2\,dx \qquad (1 \le i \le n).$$

(i) Show that I_i is in fact independent of i.

(ii) Using Exercise 7.21.(iv) prove, for $1 \le i \le n$,

$$I_i = \frac{n-1}{n(n+2)}\,\text{hyperarea}_{n-1}(S^{n-1}) = \frac{n-1}{n+2}\,\text{vol}_n(B^n).$$

Hint: One has $\sum_{1\le i\le n}\sum_{1\le j\le n,\ j\ne i} x_j^2 = (n-1)\sum_{1\le k\le n} x_k^2$.

Exercise 7.26 (Generalized Beta function and standard $(n-1)$-simplex – sequel to Exercises 6.64 and 7.21 – needed for Exercise 7.52). Prove, for $p \in \mathbf{R}^n_+$,

$$\int_0^1\int_0^{1-y_1}\cdots\int_0^{1-\sum_{1\le j\le n-2} y_j} \prod_{1\le j<n} y_j^{p_j-1}\Big(1-\sum_{1\le j<n} y_j\Big)^{p_n-1} dy_{n-1}\cdots dy_2\,dy_1$$

$$= \mathrm{B}(p_1,\dots,p_n) = \frac{\prod_{1\le j\le n}\Gamma(p_j)}{\Gamma(\sum_{1\le j\le n} p_j)}.$$

Hint: Substitute $y_{n-1} = (1-\sum_{1\le j\le n-2} y_j)\,t$ in the innermost integral, and apply mathematical induction over $n \in \mathbf{N}$. Alternatively, apply Dirichlet's formula from Exercise 6.64 to the two outermost integrals, and proceed by induction in this case also. A third possibility is to apply Exercise 3.19.
The *standard $(n-1)$-simplex* Σ^{n-1} in $\mathbf{R}^n$ is defined by

$$\Sigma^{n-1} = \{\, y \in \mathbf{R}^n \mid \sum_{1\le j\le n} y_j = 1,\ y_j \ge 0 \text{ for } 1 \le j \le n \,\}.$$

Prove, for $p \in \mathbf{R}^n_+$,

$$\int_{\Sigma^{n-1}} \prod_{1\le j\le n} y_j^{p_j-1}\,d_{n-1}y = \sqrt{n}\,\mathrm{B}(p_1,\dots,p_n).$$

In particular, therefore,

$$\text{hyperarea}_{n-1}(\Sigma^{n-1}) = \frac{\sqrt{n}}{(n-1)!}.$$

Exercise 7.27 (Simplex coordinates in $\mathbf{R}^n$ – sequel to Exercise 3.19 – needed for Exercises 7.28 and 7.38). The *standard $(n-1)$-simplex* Σ^{n-1} in $\mathbf{R}^n$ is defined by

$$\Sigma^{n-1} = \{ x \in \mathbf{R}^n \mid \sum_{1 \le j \le n} x_j = 1,\ x_j \ge 0 \text{ for } 1 \le j \le n \}.$$

Let $\phi : D \to \Sigma^{n-1}$, with $D \subset \mathbf{R}^{n-1}$ open, be a C^1 parametrization of an open part of Σ^{n-1} with negligible complement (see part (vi) for explicit formulae).

(i) Verify that the mapping $\Psi : \mathbf{R}_+ \times D \to \mathbf{R}^n_+$ given by $\Psi(c, y) = c\,\phi(y)$ is a C^1 diffeomorphism on an open dense subset in $\mathbf{R}^n_+$. To verify the injectivity of Ψ, use $\langle \phi(y), (1, \dots, 1) \rangle = 1$. Show that

$$|\det D\Psi(c, y)| = c^{n-1} |\det (D\phi(y) \mid \phi(y))|.$$

(ii) Define $e = \frac{1}{\sqrt{n}}(1, \dots, 1) \in \mathbf{R}^n$. Demonstrate

$$(\star) \qquad \langle \phi(y), e \rangle = \frac{1}{\sqrt{n}}, \qquad \langle D_j\phi(y), e \rangle = 0 \qquad (y \in D,\ 1 \le j < n).$$

Conclude that e is a unit vector in $\mathbf{R}^n$ that is orthogonal to Σ^{n-1} at every point of Σ^{n-1}. Verify that the vectors e and $D_j\phi(y)$, for $1 \le j < n$, together form a basis for $\mathbf{R}^n$; and prove by means of $(\star)$ that there exist numbers $c_j \in \mathbf{R}$, for $1 \le j < n$, with

$$\phi(y) = \frac{1}{\sqrt{n}} e + \sum_{1 \le j < n} c_j\, D_j\phi(y).$$

(iii) Now prove by means of parts (i) and (ii)

$$|\det D\Psi(c, y)| = \frac{1}{\sqrt{n}} c^{n-1} \omega_\phi(y),$$

where ω is the Euclidean $(n-1)$-dimensional density on Σ^{n-1}. Conclude, for $f \in C_c(\mathbf{R}^n_+)$, that

$$\begin{aligned}\int_{\mathbf{R}^n_+} f(x)\, dx &= \frac{1}{\sqrt{n}} \int_{\mathbf{R}_+} c^{n-1} \int_{\Sigma^{n-1}} f(cy)\, d_{n-1}y\, dc \\ &= \frac{1}{\sqrt{n}} \int_{\Sigma^{n-1}} \int_{\mathbf{R}_+} c^{n-1} f(cy)\, dc\, d_{n-1}y.\end{aligned}$$

(iv) Verify that the formula from part (iii) also applies to the function $f(x) = e^{-\sum_{1 \le j \le n} x_j}$, for $x \in \mathbf{R}^n_+$. Conclude that the formula for hyperarea$_{n-1}(\Sigma^{n-1})$ from Exercise 7.26 follows immediately.

(v) Verify that $\Sigma^2 \subset \mathbf{R}^3$ is congruent with a regular triangle in $\mathbf{R}^2$ having edges of length $\sqrt{2}$. Give a geometrical proof of

$$\mathrm{area}(\Sigma^2) = \frac{1}{2} \cdot \frac{1}{2}\sqrt{6} \cdot \sqrt{2} = \frac{1}{2}\sqrt{3}.$$

Likewise, verify that $\Sigma^3 \subset \mathbf{R}^4$ is congruent with a regular *tetrahedron* in $\mathbf{R}^3$ having edges of length $\sqrt{2}$. Give a geometrical proof of

$$\mathrm{vol}(\Sigma^3) = \frac{1}{3} \cdot \frac{1}{3}\sqrt{12} \cdot \frac{1}{2}\sqrt{3} = \frac{1}{3}.$$

Finally, we prove the existence of an embedding ϕ as above.

(vi) Verify that Exercise 3.19.(i) implies that the mapping

$$\phi : \,]0, 1\,[^{n-1} \to \{\, y \in \mathbf{R}^n_+ \mid \sum_{1 \le j \le n} y_j = 1 \,\}$$

is a C^∞ embedding, if we define $\phi(y)$ as

$$(y_1\, y_2\, y_3 \cdots y_{n-1},\ (1-y_1)\, y_2\, y_3 \cdots y_{n-1},\ \cdots,\ (1-y_{n-2})\, y_{n-1},\ (1-y_{n-1})).$$

Prove, by part (ii) and Exercise 3.19.(iii),

$$\omega_\phi(y) = \sqrt{n}\, y_2\, y_3^2 \cdots y_{n-1}^{n-2} \qquad (y \in \,]0, 1\,[^{n-1}).$$

Exercise 7.28 (Feynman's formulae – sequel to Exercises 6.50, 7.21 and 7.27). Define S^{n-1}_+ as in Exercise 7.21. Let $a \in \mathbf{R}^n_+$.

(i) Prove

$$\int_{S^{n-1}_+} \frac{1}{\langle a, y\rangle^n}\, d_{n-1}y = \frac{1}{(n-1)!\, \prod_{1 \le j \le n} a_j}.$$

Hint: One has

$$\int_{\mathbf{R}_+} e^{-a_j x_j}\, dx_j = \frac{1}{a_j} \qquad (1 \le j \le n),$$

use Formula (7.26) and properties of the Gamma function from Exercise 6.50.

(ii) Now show, for $p \in \mathbf{R}^n_+$,

$$\int_{S^{n-1}_+} \frac{\prod_{1 \le j \le n} y_j^{p_j - 1}}{\langle a, y\rangle^{\sum_{1 \le j \le n} p_j}}\, d_{n-1}y = \frac{\mathrm{B}(p_1,\, p_2,\, \dots,\, p_n)}{\prod_{1 \le j \le n} a_j^{p_j}},$$

where B stands for the generalized Beta function from Exercise 7.21.
Hint: One has

$$\int_{\mathbf{R}_+} e^{-a_j x_j} x_j^{p_j - 1}\, dx_j = \frac{\Gamma(p_j)}{a_j^{p_j}} \qquad (1 \le j \le n),$$

Next let Σ^{n-1} be as in Exercise 7.27.

(iii) Verify, for $p \in \mathbf{R}^n_+$,

$$\int_{\Sigma^{n-1}} \frac{\prod_{1\le j\le n} y_j^{p_j-1}}{\langle a, y\rangle^{\sum_{1\le j\le n} p_j}}\, d_{n-1}y = \sqrt{n}\, \frac{\mathrm{B}(p_1, p_2, \dots, p_n)}{\prod_{1\le j\le n} a_j^{p_j}}.$$

Hint: See Exercise 7.27.(iii).

Background. In quantum electrodynamics the *Feynman formulae* above are important tools in making calculations concerning Feynman diagrams.

Exercise 7.29 (Sequel to Exercise 7.21 – needed for Exercise 7.30). Let $S^n \subset \mathbf{R}^{n+1}$ be the unit sphere, let f in $C(\mathbf{R})$ be continuous and let $x \in \mathbf{R}^{n+1}$. One then has the following, known as *Poisson's formula*:

$$\int_{S^n} f(\langle x, y\rangle)\, d_n y = \text{hyperarea}_{n-1}(S^{n-1}) \int_{-1}^{1} f(\|x\|t)\,(1-t^2)^{\frac{n}{2}-1}\, dt.$$

Hint: The expression on the left–hand side is independent of the direction of $x \in \mathbf{R}^{n+1}$; therefore choose $x = (0, \dots, 0, \|x\|)$. Now use Exercise 7.21.(viii).

Exercise 7.30 (Fourier transform of radial function and Hankel's formula – sequel to Exercises 0.8, 6.51, 6.66, 6.99, 6.102, 7.21 and 7.29 – needed for Exercises 7.31 and 8.20). Assume that $f \in \mathcal{S}(\mathbf{R}^n)$ has the property that there exists a function $f_0 \in \mathcal{S}(\mathbf{R})$ with $f(x) = f_0(\|x\|)$, for all $x \in \mathbf{R}^n$.

(i) Make use of spherical coordinates $x = ry$, where $r \in \mathbf{R}_+$ and $y \in S^{n-1}$, to prove

$$\widehat{f}(\xi) = \int_{\mathbf{R}_+} f_0(r) r^{n-1} \int_{S^{n-1}} e^{-ir\langle y, \xi\rangle}\, d_{n-1}y\, dr \qquad (\xi \in \mathbf{R}^n).$$

(ii) Using Exercise 7.29, show that

$$\widehat{f}(\xi) = \int_{\mathbf{R}_+} f_0(r) r^{n-1}\, \text{hyperarea}_{n-2}(S^{n-2}) \int_{-1}^{1} e^{-ir\|\xi\|t}\,(1-t^2)^{\frac{n-3}{2}}\, dt\, dr.$$

(iii) Use Exercises 7.21.(ii) and 6.66 to prove that

$$\|\xi\|^{\frac{n}{2}-1} \widehat{f}(\xi) = (2\pi)^{\frac{n}{2}} \int_{\mathbf{R}_+} r^{\frac{n}{2}} f_0(r) J_{\frac{n}{2}-1}(\|\xi\| r)\, dr \qquad (\xi \in \mathbf{R}^n).$$

(iv) Conclude that $\widehat{f} \in \mathcal{S}(\mathbf{R}^n)$ also is a radial function.

(v) Demonstrate, by means of Example 6.11.4, that part (iii) implies

$$\rho^n e^{-\frac{1}{2}\rho^2} = \int_{\mathbf{R}_+} r^{\frac{n}{2}} e^{-\frac{r^2}{2\rho^2}} J_{\frac{n}{2}-1}(r)\,dr \qquad (\rho \in \mathbf{R}_+).$$

Prove that one obtains in particular, using Exercise 6.66.(x) and (v), for $\rho \in \mathbf{R}$,

$$\sqrt{\frac{\pi}{2}} e^{-\frac{1}{2}\rho^2} = \int_{\mathbf{R}_+} e^{-\frac{1}{2}r^2} \cos(\rho r)\,dr,$$

$$\sqrt{\frac{\pi}{2}} \rho e^{-\frac{1}{2}\rho^2} = \int_{\mathbf{R}_+} r e^{-\frac{1}{2}r^2} \sin(\rho r)\,dr.$$

See also Exercise 6.83.

(vi) Prove by the Fourier Inversion Theorem, for $x \in \mathbf{R}^n$,

$$\|x\|^{\frac{n}{2}-1} f(x) = \int_{\mathbf{R}_+} \rho\, J_{\frac{n}{2}-1}(\|x\|\rho) \int_{\mathbf{R}_+} r^{\frac{n}{2}} f_0(r) J_{\frac{n}{2}-1}(\rho r)\,dr\,d\rho.$$

Use Exercise 6.66.(x) to prove that, in the case $n = 1$, this formula takes the following form:

$$f(x) = \frac{2}{\pi} \int_{\mathbf{R}_+} \cos(x\xi) \int_{\mathbf{R}_+} \cos(\xi r) f(r)\,dr\,d\xi \qquad (x \in \mathbf{R}).$$

Now also prove the latter directly from the Fourier Inversion Theorem.

In what follows we write x instead of $\|x\|$ and λ instead of $\frac{n}{2} - 1$; and, finally $g(x) = \|x\|^{\frac{n}{2}-1} f_0(\|x\|)$.

(vii) Now verify that, using (vi), one obtains *Hankel's formula*, known from Exercise 6.102,

$$g(x) = \int_{\mathbf{R}_+} \int_{\mathbf{R}_+} g(r) J_\lambda(r\rho) r\,dr\, J_\lambda(x\rho)\rho\,d\rho \qquad (x > 0).$$

Background. Here we have proved Hankel's formula for $\lambda \in \{-\frac{1}{2}, 0, \frac{1}{2}, 1, \dots\}$. Actually, the formula is valid for all $\lambda \geq -\frac{1}{2}$ and suitably chosen functions g on $\mathbf{R}_+$, as was shown in Exercise 6.102.

(viii) Combining part (iii) and Exercise 6.99.(ii), prove the following, known as *Gegenbauer's formula*:

$$\int_{\mathbf{R}_+} e^{-tr} r^{\frac{n}{2}} J_{\frac{n}{2}-1}(\rho r)\,dr = \frac{2^{\frac{n}{2}}\Gamma(\frac{n+1}{2})}{\sqrt{\pi}} \frac{t\rho^{\frac{n}{2}-1}}{(t^2+\rho^2)^{\frac{n+1}{2}}} \qquad (t > 0,\ \rho > 0).$$

In particular (compare with Exercise 0.8),

$$\int_{\mathbf{R}_+} e^{-tr} \cos \rho r\,dr = \frac{t}{t^2+\rho^2}, \qquad \int_{\mathbf{R}_+} r e^{-tr} J_0(\rho r)\,dr = \frac{t}{(t^2+\rho^2)^{\frac{3}{2}}}.$$

(ix) Using Exercise 6.66.(vi), show

$$-\frac{2}{r}\frac{d}{ds}\Big(\frac{J_\lambda(\sqrt{s}\,r)}{s^{\frac{\lambda}{2}}}\Big) = \frac{J_{\lambda+1}(\sqrt{s}\,r)}{s^{\frac{\lambda+1}{2}}} \qquad \Big(s > 0,\ \lambda > -\frac{1}{2}\Big).$$

Deduce, for $k \in \mathbf{N}$ with $\lambda - k \geq -\frac{1}{2}$,

$$\frac{J_\lambda(\sqrt{s}\,r)}{s^{\frac{\lambda}{2}}} = \Big(-\frac{2}{r}\Big)^k \Big(\frac{d}{ds}\Big)^k \Big(\frac{J_{\lambda-k}(\sqrt{s}\,r)}{s^{\frac{\lambda-k}{2}}}\Big).$$

Apply this identity with $\lambda = \frac{n}{2} - 1$ and $k = \lambda + \frac{1}{2}$, or $k = \lambda$, if n is odd or even, respectively. Using furthermore Exercise 6.66.(x), prove

$$\frac{1}{(\sqrt{s})^{\frac{n}{2}-1}} J_{\frac{n}{2}-1}(\sqrt{s}\,r)$$

$$= \begin{cases} \Big(-\frac{2}{r}\Big)^{\frac{n-1}{2}} \Big(\frac{2}{\pi r}\Big)^{\frac{1}{2}} \Big(\frac{d}{ds}\Big)^{\frac{n-1}{2}} \cos(\sqrt{s}\,r), & n \text{ odd;} \\ \Big(-\frac{2}{r}\Big)^{\frac{n}{2}-1} \Big(\frac{d}{ds}\Big)^{\frac{n}{2}-1} J_0(\sqrt{s}\,r), & n \text{ even.} \end{cases}$$

Now deduce from part (iii), that $\widehat{f}(\xi)$ equals, for $\xi \in \mathbf{R}^n$,

$$= \begin{cases} (-1)^{\frac{n-1}{2}} 2^n \pi^{\frac{n-1}{2}} \Big(\frac{d}{ds}\Big|_{s=\|\xi\|^2}\Big)^{\frac{n-1}{2}} \displaystyle\int_{\mathbf{R}_+} f_0(r)\cos(\sqrt{s}\,r)\,dr, & n \text{ odd;} \\ (-1)^{\frac{n}{2}-1} 2^{n-1} \pi^{\frac{n}{2}} \Big(\frac{d}{ds}\Big|_{s=\|\xi\|^2}\Big)^{\frac{n}{2}-1} \displaystyle\int_{\mathbf{R}_+} r f_0(r)\, J_0(\sqrt{s}\,r)\,dr, & n \text{ even.} \end{cases}$$

Conversely, knowing the values of the integrals in part (viii), we can obtain the identity in Exercise 6.99.(ii) by means of the formula above.

Exercise 7.31 (Sequel to Exercises 6.62, 6.86 and 7.30). For $0 < s < n$, define

$$f_s : U := \mathbf{R}^n \setminus \{0\} \to \mathbf{R} \qquad \text{by} \qquad f_s = \frac{2^{\frac{s}{2}} \|\cdot\|^{-s}}{\Gamma(\frac{n-s}{2})}.$$

Then we have the following claim:

$$\widehat{f_s} = (2\pi)^{\frac{n}{2}} f_{n-s}.$$

(i) Verify that $\widehat{f_s}$ is a well-defined function, use Exercise 7.30.(iv) to show that $\widehat{f_s}$ is radial, and prove it to be of degree of homogeneity $s - n$. Deduce there exists a constant $c(n, s) \in \mathbf{C}$ such that $\widehat{f_s}(\xi) = c(n, s)\|\xi\|^{s-n}$, for $\xi \in U$. Next use the Parseval–Plancherel identity from Exercise 6.86.(iv) to evaluate $\int_{\mathbf{R}^n} \widehat{f_s}(x) e^{-\frac{1}{2}\|x\|^2}\, dx$ and prove the claim.

(ii) Suppose $n = 1$ and recall the functional equation for the zeta function from Exercise 6.62 or 6.89. Show that $g_s : U \to \mathbf{R}$ defined by

$$g_s = \frac{(2\pi)^{\frac{s}{2}} \|\cdot\|^{-s}}{\zeta(s)} \qquad \text{satisfies} \qquad \widehat{g_s} = \sqrt{2\pi}\, g_{1-s} \qquad (0 < s < 1).$$

Exercise 7.32. Assume, for $i = 1, 2$, that V_i are compact C^k submanifolds in $\mathbf{R}^{n_i}$ of dimension d_i, where $d_i < n_i$.

(i) Prove that $V := V_1 \times V_2$ is a compact C^k submanifold in $\mathbf{R}^{n_1+n_2}$ of dimension $d := d_1 + d_2$.

Let $f : V \to \mathbf{R}$ be a continuous function.

(ii) Prove that $x_1 \mapsto \int_{V_2} f(x_1, x_2)\, d_{d_2}x_2$, for $x_1 \in V_1$, defines a continuous function on V_1, and that

$$\int_V f(x)\, d_d x = \int_{V_1} \int_{V_2} f(x_1, x_2)\, d_{d_2}x_2\, d_{d_1}x_1.$$

In particular, verify that $\mathrm{vol}_d(V) = \mathrm{vol}_{d_1}(V_1)\, \mathrm{vol}_{d_2}(V_2)$.

Define the *n-dimensional torus* T^n in $\mathbf{R}^{2n}$ by $T^n = \{ x \in \mathbf{R}^{2n} \mid x_{2k-1}^2 + x_{2k}^2 = 1 \ (1 \le k \le n) \}$.

(iii) Verify that T^n is a compact C^∞ submanifold in $\mathbf{R}^{2n}$ of dimension n, and calculate the Euclidean n-dimensional volume of T^n.

Exercise 7.33. (See Exercise 8.27.) Let V be a C^1 submanifold in $\mathbf{R}^3$ of dimension d. Let $\rho : V \to \mathbf{R}$ be a continuous function. In electrostatics one defines the *field* determined by the *charge density* ρ on V as the mapping $E : \mathbf{R}^3 \setminus V \to \mathbf{R}^3$ with

$$E_i(x) = \frac{1}{4\pi} \int_V \frac{x_i - y_i}{\|x - y\|^3}\, \rho(y)\, d_d y \qquad (1 \le i \le n).$$

(i) Let V be the plane $x_1 = 0$ in $\mathbf{R}^3$ and let $\rho \equiv 1$ on V. Prove $E(x) = \frac{1}{2}(\operatorname{sgn} x_1, 0, 0)$, for $x \notin V$, and conclude that the field is perpendicular to V and of constant magnitude $\frac{1}{2}$.

(ii) Let V be the x_3-axis in $\mathbf{R}^3$ and let $\rho \equiv 1$ on V. Prove

$$E(x) = \frac{1}{2\pi(x_1^2 + x_2^2)}(x_1, x_2, 0) \qquad (x \notin V),$$

and conclude that, with cylindrical symmetry about V, the field is perpendicular to V, and of magnitude equal to the reciprocal of 2π times the distance from x to V.

Hint: Use a substitution of the form $u = \tan v$.

Now assume that V is compact, and define (compare with Example 7.4.8 and Exercise 8.28.(iv)) the *potential* of V with density ρ as the function $\phi : \mathbf{R}^3 \setminus V \to \mathbf{R}$ with

$$\phi(x) = \frac{1}{4\pi} \int_V \frac{\rho(y)}{\|x - y\|} \, d_d y.$$

(iii) Use the Differentiation Theorem 2.10.4 to prove $E = -\operatorname{grad} \phi$.

Exercise 7.34. Let $f \in C_c(\mathbf{R}^n)$ be a C^k function, for $k \in \mathbf{N}$, and define $f_t = f * \psi_t$, for $t > 0$, similarly as in Formula (7.55). For each $\alpha \in \mathbf{N}_0^n$ with $|\alpha| \leq k$, prove that the functions $D^\alpha f_t$ converge uniformly in $\mathbf{R}^n$ to $D^\alpha f$, as $t \downarrow 0$.

Exercise 7.35 (Hyperarea as derivative of volume – sequel to Exercise 5.11). Assume $V \subset \mathbf{R}^n$ to be a C^2 hypersurface. According to Exercise 5.11.(iii) there exist for every $x \in V$ a number $\delta > 0$, an open set $D \subset \mathbf{R}^{n-1}$, a C^2 embedding $\phi : D \to \mathbf{R}^n$ with $\operatorname{im}(\phi) \subset V$, and an open neighborhood $U(x)$ of x in $\mathbf{R}^n$, such that

$$\Psi_\phi : \,]-\delta, \delta[\, \times D \to U(x) \qquad \text{with} \qquad \Psi_\phi(s, y) = \phi(y) + s\,\nu(\phi(y)),$$

is a C^1 diffeomorphism (see Sequel to Example 5.3.11); here $\nu(\phi(y))$ is as in Formula (7.37).

(i) Assume that V is compact. Then prove that there exists a number $t > 0$ such that

$$\Psi : \,]-t, t[\, \times V \to V_t \qquad \text{with} \qquad \Psi(s, x) = x + s\,\nu(x),$$

is a C^1 diffeomorphism onto an open neighborhood V_t of V in $\mathbf{R}^n$. Prove

$$V_t = \{\, z \in \mathbf{R}^n \mid \text{there exists } x \in V \text{ with } \|x - z\| < t \,\},$$

that is, the open tubular neighborhood V_t of V of radius t equals the open neighborhood of V consisting of the points whose distance to V is less than t.

Let $f : V \to \mathbf{R}$ be continuous. Then define a continuation $\widetilde{f} : V_t \to \mathbf{R}$ of f to V_t by

$$\widetilde{f}(x + s\,\nu(x)) = f(x) \qquad (x \in V, \ |s| < t).$$

(ii) Prove that $\widetilde{f}$ is well-defined on V_t and continuous.

(iii) Using that $|\det D\Psi_\phi(0, y)| = \omega_\phi(y)$ show

$$\frac{1}{2}\frac{d}{dt}\bigg|_{t=0} \int_{V_t} \widetilde{f}(x)\,dx = \int_V f(x)\,d_{n-1}x,$$

whence

$$\frac{1}{2}\frac{d}{dt}\bigg|_{t=0} \mathrm{vol}_n(V_t) = \mathrm{hyperarea}_{n-1}(V).$$

Background. The results above explain the formula from Exercise 7.21.(vi).

Exercise 7.36 (Integration over tubular neighborhood of level hypersurface – needed for Exercises 7.37, 7.38 and 7.39). Let $U_0 \subset \mathbf{R}^n$ be an open subset, and let $g \in C^k(U_0)$, for $k \in \mathbf{N}$. Assume that $x^0 \in U_0$ and $g(x^0) = c^0$ (that is, $x^0 \in N(c^0) = \{ x \in U_0 \mid g(x) = c^0 \}$), and assume g to be a submersion at x^0. Finally, let $f \in C(U_0)$. We now prove that there exist a number $\gamma > 0$ and an open neighborhood U of x^0 in U_0 such that

$$\int_{\{x \in U_0 \mid |g(x)-c^0|<\gamma\} \cap U} f(x)\,dx = \int_{c^0-\gamma}^{c^0+\gamma} \int_{N(c)\cap U} \frac{f(\widetilde{x})}{\|\,\mathrm{grad}\, g(\widetilde{x})\|}\,d_{n-1}\widetilde{x}\,dc.$$

Heuristic argument: Let $x \in N(c)$ be fixed for the moment, and assume that $\mathrm{grad}\, g(x)$ points in the direction of the x_n-axis. Then $T_x N(c)$ is spanned by the set of standard basis vectors $e_j \in \mathbf{R}^n$, with $1 \le j < n$. If $g(x + dx) = c + dc$, then

$$c + dc = g(x) + \langle\, \mathrm{grad}\, g(x),\, dx\,\rangle + \cdots = c + \|\,\mathrm{grad}\, g(x)\|\,dx_n + \cdots,$$

and therefore

$$dc = \|\,\mathrm{grad}\, g(x)\|\,dx_n, \qquad \text{that is,} \qquad dx_n = \frac{dc}{\|\,\mathrm{grad}\, g(x)\|}.$$

Indicating a point in $N(c)$ by $\widetilde{x}$, we obtain from the identity $dx = dx_1 \cdots dx_n = (dx_1 \cdots dx_{n-1})\,dx_n$ the equality

$$dx = d_{n-1}\widetilde{x}\,dx_n = \frac{1}{\|\,\mathrm{grad}\, g(\widetilde{x})\|}\,d_{n-1}\widetilde{x}\,dc.$$

The following observation is an additional argument to show that the formula above must contain the reciprocal of $\|\,\mathrm{grad}\, g(\widetilde{x})\|$. If $\|\,\mathrm{grad}\, g(\widetilde{x})\|$ is large, the volume of a rectangular domain between the level hypersurfaces $N(c)$ and $N(c+dc)$ is small. But this volume can be calculated by multiplication of $d_{n-1}\widetilde{x}$, the hyperarea (which has a normal value) of the basis, and $\frac{dc}{\|\,\mathrm{grad}\, g(\widetilde{x})\|}$, the height (which is small).

(i) Prove by the Submersion Theorem 4.5.2.(iv) that we have an open neighborhood U of x^0 in U_0 and a C^k diffeomorphism $\Psi : V \to U$ of open sets in $\mathbf{R}^n$ such that

$$(\star) \qquad (g \circ \Psi)(x_1, \ldots, x_{n-1}, c) = c.$$

With the notation

$$x' = (x_1, \dots, x_{n-1}) \in \mathbf{R}^{n-1},$$
$$V((x^0)'; \eta) = \{ x' \in \mathbf{R}^{n-1} \mid \|x' - (x^0)'\| < \eta \},$$

and the Remark at the end of the proof of the Submersion Theorem one has, for suitably chosen numbers η and $\gamma > 0$,

$$V = \{ y = (x', c\ ) \in \mathbf{R}^{n-1} \times \mathbf{R} \mid x' \in V((x^0)'; \eta),\ |c - c^0| < \gamma \},$$
$$U = \{ x = (x', x_n) \in \mathbf{R}^{n-1} \times \mathbf{R} \mid x' \in V((x^0)'; \eta),\ |g(x) - c^0| < \gamma \}.$$

(ii) Assume $|c - c^0| < \gamma$. Verify that $V((x^0)'; \eta) \ni x' \mapsto \Psi_c(x') := \Psi(x', c)$ is a parametrization of $N(c) \cap U$; that is,

$$N(c) \cap U = \operatorname{im}(\Psi_c).$$

(iii) Write $x = \Psi(y)$, for $y \in V$. Conclude by the chain rule that $(\star)$ implies $Dg(x)D\Psi(y) = (0, \dots, 0, 1)$. Use this to prove that there exist numbers $c_j \in \mathbf{R}$ such that, for $1 \le j < n$,

$$\operatorname{grad} g(x) \perp D_j\Psi(y);$$

$$D_n\Psi(y) = \frac{1}{\|\operatorname{grad} g(x)\|^2} \operatorname{grad} g(x) + \sum_{1 \le j < n} c_j D_j\Psi(y).$$

(iv) Conclude that the vectors $\operatorname{grad} g(x)$ and $(D_1\Psi \times \cdots \times D_{n-1}\Psi)(y)$ are linearly dependent in $\mathbf{R}^n$, and that therefore

$$\begin{aligned} &|\det D\Psi(y)| \\ &= \frac{1}{\|\operatorname{grad} g(x)\|^2} |\det (D_1\Psi(y) \ \cdots \ D_{n-1}\Psi(y) \ \ \operatorname{grad} g(x))| \\ &= \frac{1}{\|\operatorname{grad} g(x)\|^2} |\langle (D_1\Psi \times \cdots \times D_{n-1}\Psi)(y),\ \operatorname{grad} g(x) \rangle| \\ &= \frac{1}{\|\operatorname{grad} g(x)\|} \|(D_1\Psi \times \cdots \times D_{n-1}\Psi)(y)\|. \end{aligned}$$

(v) Derive from parts (ii) – (iv)

$$|\det D\Psi(x', c)| = \frac{\omega_{\Psi_c}(x')}{\|\operatorname{grad} g(\Psi(x', c))\|} \qquad ((x', c) \in V).$$

Now prove by the Change of Variables Theorem 6.6.1 and Corollary 6.4.3

$$\begin{aligned}\int_U f(x)\,dx &= \int_V (f\circ\Psi)(x',c)|\det D\Psi(x',c)|\,dx'\,dc\\ &= \int_{c^0-\gamma}^{c^0+\gamma}\int_{V((x^0)';\eta)} f(\Psi_c(x'))\frac{\omega_{\Psi_c}(x')}{\|\operatorname{grad} g(\Psi_c(x'))\|}\,dx'\,dc\\ &= \int_{c^0-\gamma}^{c^0+\gamma}\int_{N(c)\cap U}\frac{f(\widetilde{x})}{\|\operatorname{grad} g(\widetilde{x})\|}\,d_{n-1}\widetilde{x}\,dc.\end{aligned}$$

The inner integral is said to be the $(n-1)$-dimensional integral of f over the level hypersurface $N(c)\cap U$ with respect to $|\frac{dx}{dg}|$, the *Gel'fand–Leray density* associated with g on $N(c)$. One therefore has

$$\int_U f(x)\,dx = \int_{c^0-\gamma}^{c^0+\gamma}\int_{N(c)\cap U} f(\widetilde{x})\left|\frac{dx}{dg}\right|(\widetilde{x})\,d\widetilde{x}\,dc.$$

The result above is a local one. In order to arrive at global assertions we now assume the following: $N(c^0)$ is compact and g is a submersion at every point of $N(c^0)$.

(vi) Prove that there exists a $\gamma>0$ such that, for every $c\in\mathbf{R}$ with $|c-c^0|<\gamma$, the set $N(c)$ is compact and g is a submersion at every point of $N(c)$.

(vii) Use a partition of unity to prove that, for every $c\in\mathbf{R}$ with $|c-c^0|<\gamma$,

$$\int_{N(c)} f(\widetilde{x})\left|\frac{dx}{dg}\right|(\widetilde{x})\,d\widetilde{x},$$

the $(n-1)$-dimensional integral of f over $N(c)$ with respect to the density $\left|\frac{dx}{dg}\right|$ on $N(c)$, is well-defined and is a continuous function of c.

(viii) Prove

$$\int_{\{x\in U_0\,|\,|g(x)-c^0|<\gamma\}} f(x)\,dx = \int_{c^0-\gamma}^{c^0+\gamma}\int_{N(c)} f(\widetilde{x})\left|\frac{dx}{dg}\right|(\widetilde{x})\,d\widetilde{x}\,dc.$$

(ix) Conclude that, for all $c\in\mathbf{R}$ with $|c-c^0|<\gamma$,

$$\lim_{\delta\downarrow 0}\frac{1}{2\delta}\int_{\{x\in U_0\,|\,|g(x)-c^0|<\delta\}} f(x)\,dx = \int_{N(c)} f(\widetilde{x})\left|\frac{dx}{dg}\right|(\widetilde{x})\,d\widetilde{x}.$$

Exercise 7.37 (Sequel to Exercises 6.64 and 7.36). Using Exercise 7.36, we now give another proof of Exercise 6.64.(iv). That is, let $\Delta^2 = \{ x \in \mathbf{R}^2_+ \mid x_1 + x_2 < 1 \}$, assume $p_1, p_2 \in \mathbf{R}_+$, and let $h : [0, 1] \to \mathbf{R}$ be a continuous function. We then have the following, known as *Dirichlet's formula*:

$$\int_{\Delta^2} x_1^{p_1-1} x_2^{p_2-1} h(x_1 + x_2)\, dx = \frac{\Gamma(p_1)\Gamma(p_2)}{\Gamma(p_1 + p_2)} \int_0^1 x^{p_1+p_2-1} h(x)\, dx.$$

From this point onward the notation is that of Exercise 7.36. Verify the correctness of the following assertions. Defining $g : \mathbf{R}^2 \to \mathbf{R}$ by $g(x) = x_1 + x_2$ one has, for all $x \in \mathbf{R}^2$, that $\| \operatorname{grad} g(x) \| = \sqrt{2}$, and $\Delta^2 = \{ x \in \mathbf{R}^2_+ \mid 0 < g(x) < 1 \}$. For $0 < c < 1$ the following holds:

$$N(c) \cap \Delta^2 = \{ \widetilde{x} \in \mathbf{R}^2_+ \mid \widetilde{x}_1 + \widetilde{x}_2 = c \} = \operatorname{im}(\phi_c),$$

where $\phi_c : \,]0, c[\, \to \mathbf{R}^2$ with $\phi_c(y) = (y, c - y)$. Consequently, $\omega_{\phi_c}(y) = \| \frac{d\phi_c}{dy}(y) \| = \sqrt{2}$, and therefore

$$\int_{N(c)\cap\Delta^2} \frac{f(\widetilde{x})}{\| \operatorname{grad} g(\widetilde{x}) \|}\, d_1\widetilde{x} = \int_0^c f(y, c - y)\, dy.$$

It therefore follows, with $f(x) = x_1^{p_1-1} x_2^{p_2-1} h(x_1 + x_2)$,

$$\begin{aligned} \int_{\Delta^2} x_1^{p_1-1} x_2^{p_2-1} h(x_1 + x_2)\, dx &= \int_0^1 \int_0^c y^{p_1-1} (c - y)^{p_2-1} h(c)\, dy\, dc \\ &= \int_0^1 h(c)\, c^{p_1+p_2-1}\, dc \int_0^1 t^{p_1-1} (1 - t)^{p_2-1}\, dt. \end{aligned}$$

Exercise 7.38 (Sequel to Exercises 7.27 and 7.36). Using Exercise 7.36, we now give a different proof of the formula from Exercise 7.27.(iii). That is, for every $f \in C_c(\mathbf{R}^n_+)$,

$$\begin{aligned} \int_{\mathbf{R}^n_+} f(x)\, dx &= \frac{1}{\sqrt{n}} \int_{\mathbf{R}_+} c^{n-1} \int_{\Sigma^{n-1}} f(cy)\, d_{n-1}y\, dc \\ &= \frac{1}{\sqrt{n}} \int_{\Sigma^{n-1}} \int_{\mathbf{R}_+} c^{n-1} f(cy)\, dc\, d_{n-1}y. \end{aligned}$$

Define $g : \mathbf{R}^n_+ \to \mathbf{R}$ by $g(x) = x_1 + \cdots + x_n$, then

$$\| \operatorname{grad} g(x) \| = \sqrt{n} \quad (x \in \mathbf{R}^n_+), \qquad \mathbf{R}^n_+ = \bigcup_{0<c<\infty} N(c).$$

For $c > 0$ we have

$$N(c) = \{ \widetilde{x} \in \mathbf{R}^n_+ \mid \sum_{1\le j\le n} \widetilde{x}_j = c \} = \{ c\widetilde{x} \in \mathbf{R}^n_+ \mid \widetilde{x} \in N(1) \} = \operatorname{im}(\phi_c),$$

with

$$\phi_c :]\,0, 1\,[^{n-1} \to \mathbf{R}^n_+ \qquad \text{given by} \qquad \phi_c(y) = c\,\phi_1(y).$$

Note that $e := \frac{1}{\sqrt{n}}(1, \dots, 1)$ is a unit vector in $\mathbf{R}^n$ which at every point of $N(c)$ is perpendicular to $N(c)$, for all $c > 0$. Therefore we obtain, for $y \in]\,0, 1\,[^{n-1}$,

$$\omega_{\phi_c}(y) = \det(D\phi_c(y) \mid e) = c^{n-1}\det(D\phi_1(y) \mid e) = c^{n-1}\omega_{\phi_1}(y),$$

where ω is the Euclidean $(n-1)$-dimensional density on $N(c)$. As a result

$$\begin{aligned}\int_{\mathbf{R}^n_+} f(x)\,dx &= \int_{\mathbf{R}_+}\int_{]\,0,1\,[^{n-1}} f(c\phi_1(y))\frac{c^{n-1}\omega_{\phi_1}(y)}{\sqrt{n}}\,dy\,dc \\ &= \frac{1}{\sqrt{n}}\int_{\mathbf{R}_+} c^{n-1}\int_{\Sigma^{n-1}} f(cy)\,d_{n-1}y\,dc.\end{aligned}$$

Exercise 7.39 (Intersection of hypercube – sequel to Exercises 6.109 and 7.36). Let $g : \mathbf{R}^n \to \mathbf{R}$ be defined by $g(x) = \sum_{1\le j\le n} x_j$, and let $N(c)$ be the level hypersurface in $\mathbf{R}^n$ for g determined by $c \in \mathbf{R}$. The intersection of the *hypercube* $[-\frac{1}{2}, \frac{1}{2}]^n$ with $N(c)$ is a bounded subset $\Xi^{n-1}(c)$ of the hypercube, perpendicular to the diagonal in the direction $(1, \dots, 1)$

$$\Xi^{n-1}(c) = [-\frac{1}{2}, \frac{1}{2}]^n \bigcap N(c).$$

As in Exercise 6.109 we write χ for the characteristic function of the interval $[-\frac{1}{2}, \frac{1}{2}] \subset \mathbf{R}$, and $\chi_n : \mathbf{R} \to \mathbf{R}$ for the n-fold convolution of χ with itself. Finally we define the function $\chi \otimes \cdots \otimes \chi : \mathbf{R}^n \to \mathbf{R}$ by

$$\chi \otimes \cdots \otimes \chi(x) = \chi(x_1)\cdots\chi(x_n) \qquad (x \in \mathbf{R}^n).$$

(i) Calculate the $(n-1)$-dimensional integral of $\chi \otimes \cdots \otimes \chi$ over $N(c)$ with respect to the density $|\frac{dx}{dg}|$ on $N(c)$, and conclude that the following formula holds for the $(n-1)$-dimensional area of $\Xi^{n-1}(c)$:

$$\text{hyperarea}_{n-1}(\Xi^{n-1}(c)) = \sqrt{n}\,\chi_n(c) \qquad (c \in \mathbf{R}).$$

We recall the result from Exercise 7.27.(iv) that $\text{hyperarea}_{n-1}(\Sigma^{n-1}) = \frac{\sqrt{n}}{(n-1)!}$.

(ii) Now demonstrate, for $c \in \mathbf{R}$,

$$\begin{aligned}&\text{hyperarea}_{n-1}(\Xi^{n-1}(c)) \\ &\quad = \text{hyperarea}_{n-1}(\Sigma^{n-1}) \sum_{0\le j\le [c+\frac{n}{2}]} (-1)^j\binom{n}{j}(c + \frac{n}{2} - j)^{n-1}.\end{aligned}$$

In particular, for $c \in \mathbf{R}$ satisfying $c + \frac{n}{2} \in \mathbf{N}_0$,

$$\text{hyperarea}_{n-1}(\Xi^{n-1}(c)) = \text{hyperarea}_{n-1}(\Sigma^{n-1})\, A(n-1,\, c + \frac{n}{2}),$$

where the *Euler numbers* $A(n,\, k)$ known from combinatorics, not to be confused with the Euler numbers E_n from Exercise 6.29, are given by

$$A(n,\, k) = \sum_{0 \le j \le k} (-1)^j \binom{n+1}{j} (k-j)^n.$$

(iii) Verify that $\Xi^2(0) \subset \mathbf{R}^3$ is congruent with a regular *hexagon* (ἕξ = six) in $\mathbf{R}^2$ with edges of length $\frac{1}{2}\sqrt{2}$, and that $\Xi^2(0)$ is the disjoint union of six regular triangles, all of them congruent with $\frac{1}{2}\Sigma^2$. Likewise, verify that $\Xi^3(0) \subset \mathbf{R}^4$ is congruent with a regular *octahedron* (ὀκτώ = eight) in $\mathbf{R}^3$ with edges of length $\sqrt{2}$.

Exercise 7.40. Let $S = \{\, x \in \mathbf{R}^3 \mid \|x\| = 1 \,\}$ and let $f : \mathbf{R}^3 \to \mathbf{R}^3$ be defined by $f(x) = (2x_1,\, x_2^2,\, x_3^2)$. Prove that

$$\int_S \langle f,\, \nu\rangle(y)\, d_2 y = \frac{8\pi}{3}.$$

Exercise 7.41. Let $S = \{\, x \in \mathbf{R}^3 \mid \|x\| = 1 \,\}$ and let $g : \mathbf{R}^3 \to \mathbf{R}$ be defined by $g(x) = x_1^2 + x_2 + x_3$. Prove by Gauss' Divergence Theorem (compare with Exercise 7.9)

$$\int_S g(y)\, d_2 y = \frac{4\pi}{3}.$$

Exercise 7.42. Consider the cylinder $\Omega = \{\, x \in \mathbf{R}^3 \mid x_1^2 + x_2^2 < 1,\; -1 < x_3 < 1 \,\}$ and $f : \mathbf{R}^3 \to \mathbf{R}^3$ with $f(x) = (x_1 x_2^2,\, x_1^2 x_2,\, x_2)$. Prove

$$\int_{\partial\Omega} \langle f,\, \nu\rangle(y)\, d_2 y = \pi.$$

Exercise 7.43. The mapping $\Psi : \mathbf{R}^3 \to \mathbf{R}^3$ is given by

$$\Psi(y) = ((3 + y_3 \sin y_2) \cos y_1,\; (3 + y_3 \sin y_2) \sin y_1,\; y_3 \cos y_2).$$

Furthermore,

$$C = \{\, y \in \mathbf{R}^3 \mid 0 \le y_1 \le \pi,\; 0 \le y_2 \le 2\pi,\; 0 \le y_3 \le 1 \,\},$$
$$D = \{\, y \in \mathbf{R}^3 \mid 0 < y_1 < \pi,\; 0 \le y_2 \le 2\pi,\; y_3 = 1 \,\}.$$

Finally, define the vector field $f : \mathbf{R}^3 \to \mathbf{R}^3$ by $f(x) = (x_3 - x_1,\, 2x_2 + 2,\, x_1 x_2)$.

(i) Calculate the volume of $\Psi(C)$.

(ii) Describe the boundary of $\Psi(C)$.

(iii) Calculate $\int_{\Psi(D)} \langle f, \nu \rangle(y)\, d_2 y$, where ν is the outer normal to $\Psi(D)$ with respect to $\Psi(C)$.

Exercise 7.44. Let $\Omega \subset \mathbf{R}^3$ be the bounded open subset bounded by

the unit sphere	$\{ x \in \mathbf{R}^3 \mid \|x\| = 1 \}$;
the cylindrical surface	$\{ x \in \mathbf{R}^3 \mid x_1^2+{}_2^2 = 1 \}$;
the two horizontal planes	$\{ x \in \mathbf{R}^3 \mid x_3 = h_i \} \quad (1 \le i \le 2)$,

where $-1 < h_1 < h_2 < 1$. Thus the boundary $\partial\Omega$ is the union of a shell S_1 on the sphere, a shell S_2 on the cylindrical surface, and two plane regions. We want to prove in two different ways that the areas of the shells S_1 and S_2 are equal. (This result does not generalize to dimensions $n \ne 3$.)

(i) Show this by applying Gauss' Divergence Theorem to the vector field $f : \mathbf{R}^3 \to \mathbf{R}^3$ with

$$f(x) = \Big(\frac{x_1}{x_1^2 + x_2^2}, \frac{x_2}{x_1^2 + x_2^2}, 0 \Big).$$

(ii) Now give the proof by calculating the two areas.

Exercise 7.45. Let $\Omega \subset \mathbf{R}^n$ be as in Theorem 7.6.1.

(i) Prove in two different ways

$$n \operatorname{vol}_n(\Omega) = \int_{\partial\Omega} \langle y, \nu(y) \rangle \, d_{n-1} y.$$

Verify that Formula (8.26) is a special case of the formula above.
Hint: Assume $0 \in \Omega$ and let U be an open neighborhood of y in $\mathbf{R}^n$. Consider the solid cone with apex at 0 and base $\partial\Omega \cap U$.

(ii) Prove $n \operatorname{vol}_n(B^n) = \operatorname{hyperarea}_{n-1}(S^{n-1})$ (compare with Exercise 7.21.(iv)).

Exercise 7.46 (Sequel to Exercise 2.32). Let $f : \mathbf{R}^n \to \mathbf{R}$ be positively homogeneous of degree $d \in \mathbf{R}$, that is $f(tx) = t^d f(x)$, for all $t \in \mathbf{R}_+$ and $x \in \mathbf{R}^n$. Using Exercise 2.32.(ii) prove

$$\int_{B^n} \Delta f(x)\, dx = d \int_{S^{n-1}} f(y)\, d_{n-1} y.$$

Deduce (compare with Exercise 7.21.(x))

$$\int_{S^{n-1}} y_j^2 \, d_{n-1} y = \operatorname{vol}_n(B^n) \qquad (1 \le j \le n).$$

Exercise 7.47 (Generalization of Example 7.9.4 – needed for Exercise 8.24). Let V be a bounded C^1 hypersurface in $\mathbf{R}^n$ that occurs as an image under one parametrization. Assume that $0 \notin \overline{V}$, and that every half-line originating from 0 has at most one point in common with V, but is not tangent to V at that point. Let $f : x \mapsto \frac{1}{|S^{n-1}|\,\|x\|^n} x : \mathbf{R}^n \setminus \{0\} \to \mathbf{R}^n$ be the Newton vector field from Example 7.9.4. Prove

$$\int_V \langle f, \nu \rangle(y)\, d_{n-1}y = \frac{\text{solid angle subtended by } V \text{ from } 0}{\text{total solid angle from } 0}.$$

Hint: There exists an $\epsilon > 0$ such that $\|x\| \le \epsilon$ implies $x \notin V$. Consider

$$\Omega = \{\, \lambda x \mid x \in V \text{ and } \frac{\epsilon}{\|x\|} < \lambda < 1 \,\},$$

and note that this determines an orientation of V, while $\int_{\partial\Omega} \langle f, \nu \rangle(y)\, d_{n-1}y = 0$.

Exercise 7.48. Let $g \in C^2(\mathbf{R}^n)$ and assume

$$\Omega = \{\, x \in \mathbf{R}^n \mid g(x) < 0 \,\} \neq \emptyset; \qquad \partial\Omega = \{\, x \in \mathbf{R}^n \mid g(x) = 0 \,\};$$
$$\overline{\Omega} \text{ is bounded}; \qquad \|\operatorname{grad} g(x)\| = 1 \quad (x \in \partial\Omega).$$

(i) Prove the following: $\partial\Omega$ is a compact C^1 submanifold in $\mathbf{R}^n$ of dimension $n-1$; Ω is Jordan measurable and $\operatorname{vol}_n(\Omega) > 0$; Ω lies at one side of $\partial\Omega$.

(ii) Show that $\operatorname{hyperarea}_{n-1}(\partial\Omega) = \int_\Omega \Delta g(x)\, dx$.

Exercise 7.49. Let $B(R)$ and $S(R)$ be the open ball and the sphere, respectively, in $\mathbf{R}^n$ about 0 and of radius R, and let $f : B(R) \to \mathbf{R}$ be a differentiable function such that f and $D_j f$, for $1 \le j \le n$, can be extended to continuous functions on $\overline{B(R)}$. Prove

$$\int_{B(R)} (Df(x)\,x + nf(x))\, dx = R \int_{S(R)} f(y)\, d_{n-1}y.$$

Exercise 7.50. Let $\Omega \subset \mathbf{R}^n$ be as in Theorem 7.6.1, let $f \in C^3(\mathbf{R}^n)$ and let $\operatorname{grad} f = 0$ on $\partial\Omega$. Prove that

$$\int_\Omega (\Delta f)^2(x)\, dx = \int_\Omega \sum_{1 \le i,j \le n} (D_i D_j f(x))^2\, dx.$$

Exercise 7.51. Let $\Omega \subset \mathbf{R}^n$ be as in Theorem 7.6.1 and let $k \ge 0$.

(i) Prove

$$(n+k+1)\int_\Omega x_n\|x\|^k\,dx = \int_{\partial\Omega} y_n\|y\|^k\langle\, y,\ \nu(y)\,\rangle\,d_{n-1}y.$$

Let $a > 0$ and assume next that $\Omega = \Omega(a)$ is given by $\Omega(a) = \{\, x \in \mathbf{R}^n \mid \|x\| < a,\ 0 < x_n \,\}$. Then note that $\partial\Omega = V(a) \cup V'(a)$ is a disjoint union if

$$V(a) = \{\, y \in \mathbf{R}^n \mid \|y\| = a,\ 0 < y_n \,\},$$
$$V'(a) = \{\, y \in \mathbf{R}^n \mid \|y\| \le a,\ 0 = y_n \,\}.$$

(ii) Show

$$\int_{\Omega(a)} x_n\|x\|^k\,dx = \frac{a^{k+2}}{n+k+1}\int_{V(a)} \nu_n(y)\,d_{n-1}y.$$

(iii) Let B^{n-1} be the unit ball in $\mathbf{R}^{n-1}$. Verify

$$\int_{\Omega(a)} x_n\|x\|^k\,dx = \frac{a^{n+k+1}}{n+k+1}\,\mathrm{vol}_{n-1}(B^{n-1}).$$

Hint: See Illustration for 7.4.III: Hyperarea.

Exercise 7.52 (Sequel to Exercises 6.65 and 7.26). Let the notation be as in those exercises, but assume that $p_i > 1$, for $1 \le i \le n$. Verify that $\Sigma^{n-1} \subset \partial\Delta^n$, and that the function $y \mapsto \prod_{1\le j\le n} y_j^{p_j-1}$ vanishes at the points of $\partial\Delta^n \setminus \Sigma^{n-1}$. Now proceed to prove, by Theorem 7.7.3, that in this case the formulae from Exercise 6.65.(iv) and Exercise 7.26.(ii) are equivalent.

Exercise 7.53 (Spherical mean, Mean Value Theorem for a harmonic function, Darboux's equation, and Liouville's Theorem – sequel to Exercise 7.21 – needed for Exercises 7.54, 7.67 and 8.33). Let $S(x;r)$ and $B(x;r)$ be the sphere and the ball, respectively, in $\mathbf{R}^n$ about $x \in \mathbf{R}^n$ and of radius $r > 0$. Let $\Omega \subset \mathbf{R}^n$ be as in Theorem 7.6.1. For every $f \in C(\Omega)$, $x \in \Omega$ and every $r > 0$ for which $S(x;r) \subset \Omega$, we define the *spherical mean* $m_f(x,r)$ of f over the sphere in $\mathbf{R}^n$ about $x \in \mathbf{R}^n$ and of radius $r > 0$, by

$$m_f(x,r) = \frac{1}{\mathrm{hyperarea}_{n-1}(S(x;r))}\int_{S(x;r)} f(y)\,d_{n-1}y.$$

(i) Verify that

$$m_f(x,r) = \frac{1}{|S^{n-1}|}\int_{S^{n-1}} f(x+ry)\,d_{n-1}y.$$

In this connection, see Exercise 7.21.(ii) for $|S^{n-1}| := \mathrm{hyperarea}_{n-1}(S^{n-1})$.

(ii) For all $f \in C^2(\Omega)$, $x \in \Omega$ and $r > 0$ for which $S(x; r) \subset \Omega$, prove that

$$\frac{\partial m_f}{\partial r}(x, r) = \frac{1}{|S^{n-1}|} \int_{S^{n-1}} \langle \operatorname{grad} f(x + ry),\ \nu(y) \rangle \, d_{n-1}y.$$

(iii) Let B^n be the unit ball in $\mathbf{R}^n$, and $g : B^n \to \mathbf{R}^n$ the vector field $x' \mapsto \operatorname{grad} f(x + rx')$. Check that

$$\operatorname{div} g(x') = r\, \Delta f(x + rx') \qquad (x' \in B^n).$$

(iv) Then conclude that

$$\frac{\partial m_f}{\partial r}(x, r) = \frac{r}{|S^{n-1}|} \int_{B^n} \Delta f(x + rx') \, dx'.$$

We say that $f \in C(\Omega)$ possesses the *mean value property* on Ω if, for all $x \in \Omega$ and all $r > 0$ for which $S(x; r) \subset \Omega$,

$$f(x) = m_f(x, r).$$

The *Mean Value Theorem for harmonic functions* then asserts that f satisfies the mean value property on Ω if and only if f is harmonic on Ω (that is, $\Delta f = 0$ on Ω).

(v) Using part (iv), prove the Mean Value Theorem for harmonic functions (see Exercise 7.69.(iv) for another proof).

We now want to prove that the function $m_f : \Omega \times \mathbf{R}_+ \supset\!\!\to \mathbf{R}$ satisfies the following, known as *Darboux's equation*:

$$(\star) \qquad \frac{\partial^2 m_f}{\partial r^2}(x, r) + \frac{n-1}{r} \frac{\partial m_f}{\partial r}(x, r) = \Delta_x m_f(x, r).$$

Here Δ_x is the Laplace operator with respect to the variable $x \in \Omega$.

(vi) Verify

$$r^n \int_{B^n} f(x + rx') \, dx' = |S^{n-1}| \int_0^r \rho^{n-1} \, m_f(x, \rho) \, d\rho.$$

(vii) Now show by means of part (iv)

$$\frac{\partial m_f}{\partial r}(x, r) = \frac{1}{r^{n-1}} \Delta_x \left(\int_0^r \rho^{n-1} \, m_f(x, \rho) \, d\rho \right),$$

and

$$\frac{\partial}{\partial r}\left(r^{n-1} \frac{\partial m_f}{\partial r}(x, r) \right) = \Delta_x (r^{n-1} \, m_f(x, r)).$$

Then show that the differential equation $(\star)$ holds.

To conclude, we prove *Liouville's Theorem* which asserts that a bounded harmonic function on $\mathbf{R}^n$ is constant. (See Exercise 7.70.(viii) for a different proof.)

(viii) Assume f is bounded and harmonic on $\mathbf{R}^n$. Prove by part (vi), for every $x \in \mathbf{R}^n$ and $r \in \mathbf{R}_+$,

$$f(x) = \frac{n}{|S^{n-1}|}\int_{B^n} f(x + rx')\,dx' = \frac{n}{r^n|S^{n-1}|}\int_{B(x,r)} f(x')\,dx'.$$

Use this to demonstrate, for every $x \in \mathbf{R}^n$ and $r \in \mathbf{R}_+$, with $\|f\|$ the supremum of f on $\mathbf{R}^n$,

$$\begin{aligned}
&|f(x) - f(0)| \\
&= \frac{n}{r^n|S^{n-1}|}\left|\int_{B(x,r)} f(x')\,dx' - \int_{B(0,r)} f(x'')\,dx'\right| \\
&\le \frac{n}{r^n|S^{n-1}|}\|f\|\left(\int_{\|x'-x\|<r,\ \|x'\|>r} dx' + \int_{\|x'\|<r,\ \|x'-x\|>r} dx'\right) \\
&\le \frac{n}{r^n|S^{n-1}|}\|f\|\int_{r-\|x\|<\|x'\|<r+\|x\|} dx' = \frac{n}{r^n}\|f\|\int_{r-\|x\|}^{r+\|x\|}\rho^{n-1}\,d\rho \\
&= \frac{1}{r^n}\|f\|((r+\|x\|)^n - (r-\|x\|)^n) = \mathcal{O}\Big(\frac{1}{r}\Big), \qquad r \to \infty.
\end{aligned}$$

Verify that this proves Liouville's Theorem.

Exercise 7.54 (Pizzetti's formula – sequel to Exercise 7.53). Suppose $f \in C^\infty(\mathbf{R}^n)$ is equal to its MacLaurin series. We will give another proof of *Pizzetti's formula* from Exercise 7.22

$$\frac{1}{\text{hyperarea}_{n-1}(S^{n-1})}\int_{S^{n-1}} f(ry)\,d_{n-1}y = \sum_{k\in\mathbf{N}_0}\frac{r^{2k}}{2^k k!\prod_{0\le l<k}(n+2l)}\Delta^k f(0).$$

To this end, note that $m_f(r)$, the *spherical mean* of f over the sphere of center 0 and radius $|r|$, is an odd function of $r \in \mathbf{R}$. Deduce from the MacLaurin series for f the existence of $c_k \in \mathbf{R}$ such that $m_f(r) = \sum_{k\in\mathbf{N}_0} c_k r^{2k}$. Substituting this series in *Darboux's equation* from Exercise 7.53 prove

$$m_{\Delta f}(r) = \sum_{p\in\mathbf{N}_0} 2(p+1)(2p+n)c_{p+1}r^{2p}.$$

Next show by mathematical induction over $k \in \mathbf{N}$

$$m_{\Delta^k f}(r) = \sum_{p\in\mathbf{N}_0} 2^k\frac{(p+k)!}{p!}\prod_{0\le l<k}(2p+n+2l)c_{p+k}r^{2p},$$

and deduce Pizzetti's formula by taking $r = 0$.

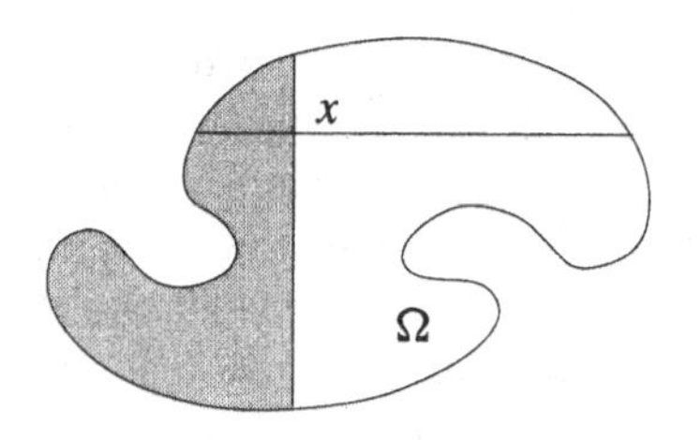

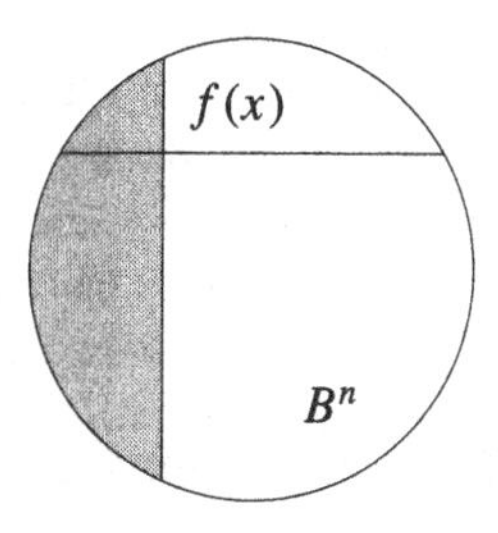

Illustration for Exercise 7.55: Isoperimetric inequality

Exercise 7.55 (Isoperimetric inequality – sequel to Exercise 5.39). As was shown in Example 7.9.1,

$$n \operatorname{vol}_n(\Omega) = \operatorname{hyperarea}_{n-1}(\partial\Omega),$$

if $\Omega = B^n$. We now want to prove that, in general, one has for a set $\Omega \subset \mathbf{R}^n$ as in Theorem 7.6.1 the following, known as the *isoperimetric inequality*:

$$\Big(\frac{\operatorname{vol}_n(\Omega)}{\operatorname{vol}_n(B^n)}\Big)^{n-1} \le \Big(\frac{\operatorname{hyperarea}_{n-1}(\partial\Omega)}{\operatorname{hyperarea}_{n-1}(S^{n-1})}\Big)^n.$$

That is, among all permissible sets Ω for which $\operatorname{hyperarea}_{n-1}(\partial\Omega)$ has a prescribed value, the ball (of suitable radius) has the largest volume. Assume for the moment $\operatorname{vol}_n(\Omega) = \operatorname{vol}_n(B^n)$.

(i) Prove that

$$t \mapsto \operatorname{vol}_n(B^n \cap \{\, y \in \mathbf{R}^n \mid y_1 < t \,\})$$

is a monotonically increasing continuous function on $\mathbf{R}$.

(ii) Use the Intermediate Value Theorem 1.9.5 and mathematical induction over $j = 1, 2, \dots, n$ to find numbers

$$f_j(x) = f_j(x_1, \dots, x_j) \in \mathbf{R} \qquad (1 \le j \le n)$$

such that

$$f(x) := (f_1(x), \dots, f_n(x)) \in B^n;$$

$$\begin{aligned}\operatorname{vol}_{n+1-j}(B^n \cap \{\, y \in \mathbf{R}^n \mid y_1 = f_1(x), \dots, y_{j-1} = f_{j-1}(x),\ y_j < f_j(x) \,\})\\ = \operatorname{vol}_{n+1-j}(\Omega \cap \{\, y \in \mathbf{R}^n \mid y_1 = x_1, \dots, y_{j-1} = x_{j-1},\ y_j < x_j \,\}).\end{aligned}$$

This can be worded as follows. The hyperplane $H_{n-1}(x)$ through x orthogonal to the x_1-axis divides Ω into two parts, with volumes $v_n^{(1)}$ and $v_n^{(2)}$. Let $\widetilde{H}_{n-1}(x)$ be the hyperplane through $(f_1(x), 0, \dots, 0)$ orthogonal to the x_1-axis, such that $\widetilde{H}_{n-1}(x)$ divides B^n into two parts, with volumes $v_n^{(1)}$ and $v_n^{(2)}$, respectively. Now let $H_{n-2}(x)$ be the intersection of $\Omega \cap H_{n-1}(x)$ with the hyperplane through x orthogonal to the x_2-axis. Then $H_{n-2}(x)$ divides $\Omega \cap H_{n-1}(x)$ into two parts, with $(n-1)$-dimensional volumes $v_{n-1}^{(1)}$ and $v_{n-1}^{(2)}$, respectively. Let $\widetilde{H}_{n-2}(x)$ be the $(n-2)$-dimensional affine submanifold through $(f_1(x),\ f_2(x), 0, \dots, 0)$ orthogonal to the x_1-axis and the x_2-axis, such that $\widetilde{H}_{n-2}(x)$ divides $B^n \cap \widetilde{H}_{n-1}(x)$ into two parts, with $(n-1)$-dimensional volumes $v_{n-1}^{(1)}$ and $v_{n-1}^{(2)}$, respectively. We continue in this way, until we have found lines $H_1(x)$ and $\widetilde{H}_1(x)$, and finally points $H_0(x) = \{x\}$ and $\widetilde{H}_0(x) = \{f(x)\}$. This then constitutes the definition of $f(x)$.

We assume that the mapping $f : \Omega \to B^n$ thus defined is differentiable (problems may occur here, for example, if $\partial\Omega$ contains "plane" regions). Moreover, almost by definition, f is volume-preserving, that is, one has $|\det Df(x)| = 1$.

(iii) Verify that, with $\lambda_j(x) = D_j f_j(x) > 0$,

$$Df(x) = \begin{pmatrix} \lambda_1(x) & \star & \cdots & \star \\ 0 & \lambda_2(x) & \ddots & \vdots \\ \vdots & \ddots & \ddots & \star \\ 0 & \cdots & 0 & \lambda_n(x) \end{pmatrix}.$$

(iv) Prove by means of Exercise 5.39.(ii)

$$n = n\Big(\prod_{1\le j\le n} \lambda_j(x)\Big)^{1/n} \le \sum_{1\le j\le n} \lambda_j(x) = \operatorname{div} f(x).$$

(v) Conclude that

$$n \operatorname{vol}_n(\Omega) \le \int_{\partial\Omega} \langle f, \nu\rangle(y)\, d_{n-1}y \le \int_{\partial\Omega} d_{n-1}y = \operatorname{hyperarea}_{n-1}(\partial\Omega).$$

(vi) Prove the isoperimetric inequality in the general case.

Exercise 7.56 (Normal derivative in polar coordinates – needed for Exercise 7.57). Define $\Psi : \mathbf{R}^2 \to \mathbf{R}^2$ by $\Psi(r,\alpha) = r(\cos\alpha,\ \sin\alpha)$. Let $r > 0$ and $dr > 0$, and assume that α and $\alpha + d\alpha$ satisfy α and $\alpha + d\alpha \in\ ]-\pi, \pi\,[$. Let $\Omega \subset \mathbf{R}^2$ be given by

$$\Omega = \{\,\Psi(r',\alpha') \mid r < r' < r + dr,\ \alpha < \alpha' < \alpha + d\alpha\,\}.$$

(i) Verify that $\partial\Omega$ is the union of the following four sets:

$$\{\,\Psi(r,\alpha') \mid \alpha < \alpha' < \alpha + d\alpha\,\}, \qquad \{\,\Psi(r',\alpha) \mid r < r' < r + dr\,\},$$
$$\{\,\Psi(r + dr,\alpha') \mid \alpha < \alpha' < \alpha + d\alpha\,\}, \qquad \{\,\Psi(r',\alpha + d\alpha) \mid r < r' < r + dr\,\}.$$

(ii) Show that

$$\begin{aligned}
\nu(\Psi(r,\alpha')) &= (-\cos\alpha', & -\sin\alpha'),\\
\nu(\Psi(r',\alpha)) &= (\ \ \sin\alpha, & -\cos\alpha),\\
\nu(\Psi(r+dr,\alpha')) &= (\ \ \cos\alpha', & \sin\alpha'),\\
\nu(\Psi(r',\alpha+d\alpha)) &= (-\sin(\alpha+d\alpha), & \cos(\alpha+d\alpha)).
\end{aligned}$$

(iii) Prove by analogy with Exercise 3.8.(iv)

$$\begin{pmatrix} D_1 f \\ D_2 f \end{pmatrix} \circ \Psi = \begin{pmatrix} \cos\alpha & -\sin\alpha \\ \sin\alpha & \cos\alpha \end{pmatrix} \begin{pmatrix} \dfrac{\partial(f\circ\Psi)}{\partial r} \\ \dfrac{1}{r}\dfrac{\partial(f\circ\Psi)}{\partial\alpha} \end{pmatrix}.$$

(iv) Verify

$$\begin{aligned}
\frac{\partial f}{\partial\nu}(\Psi(r,\alpha')) &= -\cos\alpha'\, D_1 f(\Psi(r,\alpha')) - \sin\alpha'\, D_2 f(\Psi(r,\alpha'))\\
&= -\frac{\partial(f\circ\Psi)}{\partial r}(r,\alpha'),\\
\frac{\partial f}{\partial\nu}(\Psi(r',\alpha)) &= -\frac{1}{r}\frac{\partial(f\circ\Psi)}{\partial\alpha}(r',\alpha),\\
\frac{\partial f}{\partial\nu}(\Psi(r+dr,\alpha')) &= \frac{\partial(f\circ\Psi)}{\partial r}(r+dr,\alpha'),\\
\frac{\partial f}{\partial\nu}(\Psi(r',\alpha+d\alpha)) &= \frac{1}{r}\frac{\partial(f\circ\Psi)}{\partial\alpha}(r',\alpha+d\alpha).
\end{aligned}$$

Exercise 7.57 (Laplacian in polar coordinates – sequel to Exercise 7.56). (See Exercise 3.8.(v).) Let $x \in \mathbf{R}^n$ and assume that the sets Ω below satisfy the conditions of Theorem 7.6.1, and $x \in \Omega$. Let the function f be as in Example 7.9.6.

(i) Prove

$$\Delta f(x) = \lim_{\Omega\downarrow\{x\}} \frac{1}{\mathrm{vol}_n(\Omega)} \int_\Omega \Delta f(x)\, dx = \lim_{\Omega\downarrow\{x\}} \frac{1}{\mathrm{vol}_n(\Omega)} \int_{\partial\Omega} \frac{\partial f}{\partial\nu}(y)\, d_{n-1}y.$$

Define $\Psi : \mathbf{R}^2 \to \mathbf{R}^2$ and $\Omega \subset \mathbf{R}^2$ as in Exercise 7.56.

(ii) Prove by means of Exercise 7.56.(iv) that, for small values of $dr > 0$ and $d\alpha > 0$, modulo terms of higher order in dr and $d\alpha$,

$$\begin{aligned}\int_{\partial\Omega} \frac{\partial f}{\partial \nu}(y)\, d_{n-1}y &\equiv -\frac{\partial(f\circ\Psi)}{\partial r}(r,\alpha)\, r\, d\alpha - \frac{1}{r}\frac{\partial(f\circ\Psi)}{\partial\alpha}(r,\alpha)\, dr \\ &\quad + \frac{\partial(f\circ\Psi)}{\partial r}(r+dr,\alpha)\,(r+dr)\, d\alpha + \frac{1}{r}\frac{\partial(f\circ\Psi)}{\partial\alpha}(r,\alpha+d\alpha)\, dr \\ &= \Big(\frac{\partial}{\partial r}\Big(r\,\frac{\partial(f\circ\Psi)}{\partial r}\Big)(r,\alpha) + \frac{1}{r}\frac{\partial^2(f\circ\Psi)}{\partial\alpha^2}(r,\alpha)\Big)\, dr\, d\alpha.\end{aligned}$$

(iii) Prove the following formula for the Laplacian in polar coordinates:

$$(\Delta f)\circ\Psi = \frac{1}{r^2}\Big(\Big(r\frac{\partial}{\partial r}\Big)^2 + \frac{\partial^2}{\partial\alpha^2}\Big)(f\circ\Psi).$$

Exercise 7.58 (Divergence in spherical coordinates). Let

$$\Psi : V = \mathbf{R}_+ \times\ ]-\pi,\ \pi\,[\ \times \Big]-\frac{\pi}{2},\ \frac{\pi}{2}\Big[\to \mathbf{R}^3$$

be the substitution of variables $x = \Psi(r,\alpha,\theta) = r(\cos\alpha\cos\theta,\ \sin\alpha\cos\theta,\ \sin\theta)$. Let e_r, e_α and e_θ be the orthonormal vectors as defined in Exercise 3.8.(iv), and let $f : \mathbf{R}^3 \to \mathbf{R}^3$ be a vector field. Then on V we define the component functions f_r, f_α and f_θ of f in spherical coordinates by

$$(f\circ\Psi)(r,\alpha,\theta) = f_r(r,\alpha,\theta)\, e_r + f_\alpha(r,\alpha,\theta)\, e_\alpha + f_\theta(r,\alpha,\theta)\, e_\theta.$$

Consider numbers $r > 0$ and $dr > 0$; α and $d\alpha > 0$, with α and $\alpha + d\alpha \in\ ]-\pi,\pi\,[$; and θ and $d\theta > 0$ with θ and $\theta + d\theta \in \big]-\frac{\pi}{2},\frac{\pi}{2}\big[$. Let $\Omega \subset \mathbf{R}^3$ be defined by

$$\Omega = \{\,\Psi(r',\alpha',\theta') \mid r < r' < r+dr,\ \alpha < \alpha' < \alpha + d\alpha,\ \theta < \theta' < \theta + d\theta\,\}.$$

(i) Verify that $\partial\Omega$ is the union of six smooth surfaces, given by the condition that precisely one of the r', α' or θ' is constant. Prove that the areas of these respective surfaces, for small values of $dr > 0$, $d\alpha > 0$ and $d\theta > 0$, modulo terms of higher order in dr, $d\alpha$ and $d\theta$, are given by (see the Illustration for Example 7.4.11)

$$\begin{array}{llll} (r+dr)^2\cos\theta\, d\alpha\, d\theta & (r' = r+dr), & r^2\cos\theta\, d\alpha\, d\theta & (r' = r), \\ r\, dr\, d\theta & (\alpha' = \alpha + d\alpha), & r\, dr\, d\theta & (\alpha' = \alpha), \\ r\cos(\theta + d\theta)\, dr\, d\alpha & (\theta' = \theta + d\theta), & \cos\theta\, dr\, d\alpha & (\theta' = \theta). \end{array}$$

(ii) Now apply Gauss' Divergence Theorem to Ω, and conclude that, for small values of $dr > 0$, $d\alpha > 0$ and $d\theta > 0$, modulo terms of higher order in dr, $d\alpha$ and $d\theta$,

$$
\begin{aligned}
&(\operatorname{div} f)(\Psi(r,\alpha,\theta))\, r^2\cos\theta\, dr\, d\alpha\, d\theta \\
&\quad\equiv\ (f_r(r+dr,\ \alpha,\ \theta)(r+dr)^2 - f_r(r,\alpha,\theta)r^2)\cos\theta\, d\alpha\, d\theta \\
&\quad\ +(f_\alpha(r,\ \alpha+d\alpha,\ \theta) - f_\alpha(r,\alpha,\theta))\, r\, dr\, d\theta \\
&\quad\ +(f_\theta(r,\ \alpha,\ \theta+d\theta)\cos(\theta+d\theta) - f_\theta(r,\alpha,\theta)\cos\theta)\, r dr\, d\alpha \\
&\quad = \Big(\cos\theta\,\frac{\partial(r^2 f_r)}{\partial r}(r,\alpha,\theta) + r\,\frac{\partial f_\alpha}{\partial\alpha}(r,\alpha,\theta) \\
&\quad\ + r\,\frac{\partial(\cos\theta\, f_\theta)}{\partial\theta}(r,\alpha,\theta)\Big)\, dr\, d\alpha\, d\theta.
\end{aligned}
$$

(iii) Conclude, by taking a limit, that one has the following formula for the divergence in spherical coordinates:

$$(\operatorname{div} f)\circ\Psi = \frac{1}{r^2}\frac{\partial(r^2 f_r)}{\partial r} + \frac{1}{r\cos\theta}\frac{\partial f_\alpha}{\partial\alpha} + \frac{1}{r\cos\theta}\frac{\partial(\cos\theta\, f_\theta)}{\partial\theta}.$$

(iv) Calculating in two different ways, demonstrate that for the identity mapping $f : \mathbf{R}^3 \to \mathbf{R}^3$ one has $\operatorname{div} f = 3$.

(v) Define on V the component functions f^r, f^α and f^θ of f in spherical coordinates by

$$f\circ\Psi = f^r\frac{\partial\Psi}{\partial r} + f^\alpha\frac{\partial\Psi}{\partial\alpha} + f^\theta\frac{\partial\Psi}{\partial\theta}.$$

Verify that $f^r = f_r$, $f^\alpha = \frac{1}{r\cos\theta} f_\alpha$ and $f^\theta = \frac{1}{r} f_\theta$, and conclude that

$$(\operatorname{div} f)\circ\Psi = \frac{1}{\det D\Psi}\Big(\frac{\partial(f^r \det D\Psi)}{\partial r} + \frac{\partial(f^\alpha \det D\Psi)}{\partial\alpha} + \frac{\partial(f^\theta \det D\Psi)}{\partial\theta}\Big).$$

See Exercise 7.60 for a generalization of this result.

Exercise 7.59 (Outer normal on the boundary of an image – needed for Exercise 7.60). Let $\Psi : V \to U$ be a C^1 diffeomorphism of open subsets of $\mathbf{R}^n$. Assume a and $b \in V$ are such that the rectangle $B = B(a,b) \subset V$, where

$$B(a,b) = \{\, y \in \mathbf{R}^n \mid a_j \le y_j \le b_j\ (1 \le j \le n)\,\}.$$

Define $\partial_{i,\pm}B$, the "smooth parts of the $(n-1)$-dimensional faces of B" by

$$\partial_{i,\pm}B = \{\, y \in \mathbf{R}^n \mid a_j < y_j < b_j\ (j \ne i),\ y_i = {}^{b_i}_{a_i}\,\} \qquad (1 \le i \le n).$$

Let $\Omega = \Psi(B) \subset U$. We now give a formula for the outer normal $\nu(x)$ on $\partial\Omega$, for x in the "smooth part" of $\partial\Omega$.

(i) Verify that $S := \partial B \setminus \bigcup_{i,\pm} \partial_{i,\pm} B$ is an $(n-1)$-dimensional negligible set in $\mathbf{R}^n$. Prove that $\Psi S \subset \Omega$ is an $(n-1)$-dimensional negligible set in $\mathbf{R}^n$, and that $\partial\Omega \setminus \Psi S = \Psi(\partial B \setminus S) = \bigcup_{i,\pm} \Psi(\partial_{i,\pm} B)$.

Let $1 \le i \le n$, and parametrize the $(n-1)$-dimensional manifold $\Psi(\partial_{i,\pm} B)$ in $\mathbf{R}^n$ by the embedding, defined on an open subset of $\mathbf{R}^{n-1}$,

$$\Psi_{i,\pm} : (y_1, \dots, y_{i-1}, y_{i+1}, \dots, y_n) \mapsto \Psi(y_1, \dots, y_{i-1}, \begin{matrix} b_i \\ a_i \end{matrix}, y_{i+1}, \dots, y_n).$$

(ii) Verify that Ω lies at one side of $\Psi(\partial_{i,\pm} B)$ everywhere. Next, assume $n \ge 3$. Prove that on $\partial_{i,\pm} B$ one has the following equality of vector fields in $\mathbf{R}^n$:

$$\begin{aligned} &\omega_{\Psi_{i,\pm}} \, \nu \circ \Psi_{i,\pm} \\ &\quad = \pm \operatorname{sgn}(\det D\Psi)(-1)^{i-1} D_1\Psi \times \cdots \times D_{i-1}\Psi \times D_{i+1}\Psi \times \cdots \times D_n\Psi. \end{aligned}$$

To do so, use properties of the cross product, Formula (7.25), and

$$\begin{aligned} &\langle \pm D_i\Psi, \pm \operatorname{sgn}(\det D\Psi)(-1)^{i-1} D_1\Psi \times \cdots \times D_{i-1}\Psi \times D_{i+1}\Psi \times \cdots \times D_n\Psi \rangle \\ &\qquad = \operatorname{sgn}(\det D\Psi)(-1)^{i-1} \det(D_i\Psi \; D_1\Psi \cdots D_{i-1}\Psi \; D_{i+1}\Psi \cdots D_n\Psi) \\ &\qquad = |\det D\Psi| > 0. \end{aligned}$$

(iii) Show in a sketch, with B a rectangle in $\mathbf{R}^3$ and $\Psi = I$, that the formulae from part (ii) do indeed yield the outer normals.

Exercise 7.60 (Divergence in arbitrary coordinates – sequel to Exercises 3.14 and 7.59). Let the notation be as in Exercise 3.14. In the exercise we derived the following formula for $(\operatorname{div} f) \circ \Psi : V \to \mathbf{R}$, in terms of the component functions with respect to the moving frame determined by Ψ:

$$(\star) \qquad (\operatorname{div} f) \circ \Psi = \frac{1}{\sqrt{g}} \operatorname{div}(\sqrt{g}\, \Psi^* f) = \frac{1}{\sqrt{g}} \sum_{1 \le i \le n} D_i(\sqrt{g}\, f^{(i)}).$$

Now we obtain $(\star)$ by means of integration, see Exercise 8.40 for another proof. To verify $(\star)$, we choose $y \in V$ and $dy_i > 0$, for $1 \le i \le n$, such that the rectangle $B = B(y, y + dy) \subset V$, in the notation of Exercise 7.59. We also employ the other notations and results from that exercise.

(i) Prove by successive application of the Change of Variables Theorem 6.6.1 and Gauss' Divergence Theorem 7.8.5 that, for small values of the dy_i, modulo terms of higher order in the dy_i,

$$\begin{aligned} &(\operatorname{div} f) \circ \Psi(y) \sqrt{g}(y)\, dy_1 \cdots dy_n \equiv \int_B (\operatorname{div} f) \circ \Psi(y') |\det D\Psi(y')|\, dy' \\ &\quad = \int_\Omega \operatorname{div} f(x')\, dx' = \int_{\partial\Omega} \langle f, \nu \rangle(u')\, d_{n-1}u'. \end{aligned}$$

(ii) Conclude from Exercise 7.59.(i) that

$$\int_{\partial\Omega} \langle f, \nu\rangle(u')\, d_{n-1}u' = \sum_{i,\pm} \int_{\Psi(\partial_{i,\pm}B)} \langle f, \nu\rangle(u')\, d_{n-1}u'.$$

(iii) Let $1 \leq i \leq n$ and let $\Psi_{i,\pm}$ be the parametrization of $\Psi(\partial_{i,\pm}B)$ from Exercise 7.59; verify that the definition of integration over a hypersurface implies

$$\int_{\Psi(\partial_{i,\pm}B)} \langle f, \nu\rangle(u')\, d_{n-1}u' = \int_{\partial_{i,\pm}B} \langle\, f \circ \Psi_{i,\pm},\ \nu \circ \Psi_{i,\pm}\,\rangle(y')\, \omega_{\Psi_{i,\pm}}(y')\, dy'.$$

(iv) Conclude by Exercise 7.59.(ii) that the following equality of functions holds on $\partial_{i,\pm}B$:

$$\langle\, f \circ \Psi_{i,\pm},\ \nu \circ \Psi_{i,\pm}\,\rangle\, \omega_{\Psi_{i,\pm}} = \pm f^{(i)}\, |\det D\Psi|.$$

(v) Now prove, for small values of dy_i, and modulo terms of higher order in dy_i,

$$\begin{aligned}
&\sum_{\pm} \int_{\partial_{i,\pm}B} \langle\, f \circ \Psi_{i,\pm},\ \nu \circ \Psi_{i,\pm}\,\rangle(y')\, \omega_{\Psi_{i,\pm}}(y')\, dy' \\
&\qquad = \sum_{\pm} \int_{\partial_{i,\pm}B} \pm f^{(i)}(y')\, |\det D\Psi(y')|\, dy' \\
&\qquad \equiv \frac{(\sqrt{g}\, f^{(i)})(y + dy_i) - (\sqrt{g}\, f^{(i)})(y)}{dy_i}\, dy_1 \cdots dy_n.
\end{aligned}$$

Here we write $y + dy_i$ for $(y_1, \dots, y_{i-1}, y_i + dy_i, y_{i+1}, \dots, y_n) \in \partial_{i,+}B$.

(vi) Finally, prove $(\star)$ by taking the limit for $dy_i \downarrow 0$, for $1 \leq i \leq n$.

Exercise 7.61 (Laplacian in arbitrary coordinates – sequel to Exercises 3.8 and 3.16). In Exercise 3.16 we obtained the following formula for Δ on U in the new coordinates in V, valid for every $f \in C^2(U)$:

$$(\Delta f) \circ \Psi = \frac{1}{\sqrt{g}}\, \mathrm{div}\,(\sqrt{g}\, G^{-1}\, \mathrm{grad}(f \circ \Psi)) = \frac{1}{\sqrt{g}} \sum_{1 \leq i, j \leq n} D_i(\sqrt{g}\, g^{ij}\, D_j)(f \circ \Psi).$$

We now prove this identity by means of integration.

(i) Verify that, for every $h \in C_c(U)$,

$$\int_U h(x)\, dx = \int_V (h \circ \Psi)(y)\, \sqrt{g}(y)\, dy.$$

(ii) Prove by means of formula ($\star$) from Exercise 3.8.(ii) that, for f and $h \in C^1(U)$,

$$\langle\, (\operatorname{grad} f) \circ \Psi,\ (\operatorname{grad} h) \circ \Psi \,\rangle(y) = D(f \circ \Psi)(y) \circ G(y)^{-1} \circ \operatorname{grad}(h \circ \Psi)(y).$$

Show that writing out into matrix coefficients one finds the identity of functions on V

$$\langle\, (\operatorname{grad} f) \circ \Psi,\ (\operatorname{grad} h) \circ \Psi \,\rangle = \sum_{1 \le i, j \le n} g^{ij} D_i(f \circ \Psi)\, D_j(h \circ \Psi).$$

(iii) Prove that for every $x \in U$ there exists an open set $\Omega \subset \mathbf{R}^n$ that satisfies the conditions from Theorem 7.6.1, and additionally $x \in \Omega \subset \overline{\Omega} \subset U$.

(iv) Let $f \in C^2(U)$ and $h \in C^1_c(\Omega)$ be chosen arbitrarily. Apply Green's first identity from Example 7.9.6 to Ω, and conclude by parts (i) – (iii) that

$$\int_V (\Delta f) \circ \Psi(y)\, (h \circ \Psi)(y) \sqrt{g}(y)\, dy = \int_U (\Delta f)(x)\, h(x)\, dx$$

$$= -\int_U \langle\, \operatorname{grad} f,\ \operatorname{grad} h \,\rangle(x)\, dx$$

$$= -\int_V \langle\, (\operatorname{grad} f) \circ \Psi,\ (\operatorname{grad} h) \circ \Psi \,\rangle(y) \sqrt{g}(y)\, dy$$

$$= -\sum_{1 \le j \le n} \int_V \Big(\sum_{1 \le i \le n} \sqrt{g}\, g^{ij}\, D_i(f \circ \Psi) \Big)(y)\, D_j(h \circ \Psi)(y)\, dy$$

$$= \int_V \sum_{1 \le i, j \le n} D_j\Big(\sqrt{g}\, g^{ij}\, D_i(f \circ \Psi) \Big)(y)\, (h \circ \Psi)(y)\, dy.$$

In the last equality Corollary 7.6.2 has been used.

(v) Finally prove the desired identity, making use of the continuity of the integrands in (iv) and of the freedom in the choice of the function h.

Exercise 7.62 (Hamilton's equation – sequel to Exercise 2.32). In this exercise we write $x \in \mathbf{R}^{2n}$ as

$$x = (q, p) = (q_1, \dots, q_n, p_1, \dots, p_n) \in \mathbf{R}^n \times \mathbf{R}^n.$$

Let $\Omega \subset \mathbf{R}^{2n}$ be an open subset and $H : \Omega \to \mathbf{R}$ a C^1 submersion. Here H is said to be a *Hamiltonian*, while the $(2n-1)$-dimensional C^1 submanifold $N(c) = \{\, x \in \Omega \mid H(x) = c \,\}$ is said to be the *energy surface* for H with energy $c \in \mathbf{R}$.

Furthermore, $v_H : \Omega \to \mathbf{R}^{2n}$ is said to be the *Hamiltonian vector field* determined by H if

$$v_H(x) = \Big(\frac{\partial H}{\partial p_1}(x), \dots, \frac{\partial H}{\partial p_n}(x), -\frac{\partial H}{\partial q_1}(x), \dots, -\frac{\partial H}{\partial q_n}(x)\Big).$$

(i) Prove that $\langle \operatorname{grad} H(x), v_H(x) \rangle = 0$, for all $x \in \Omega$, and conclude that

$$v_H(x) \in T_x N(H(x)) \qquad (x \in \Omega).$$

That is, at every point of Ω the Hamiltonian vector field v_H is tangent to the energy surface for H containing that point.

Let $x : J \to \Omega$ with $x : t \mapsto x(t)$ be a C^1 curve with $x(0) = x^0 \in \Omega$. Assume that $t \mapsto x(t)$ is an integral curve of the following (ordinary) differential equation, known as *Hamilton's equation*:

$$x'(t) = v_H(x(t)) \qquad (t \in J),$$

that is,

$$q_j'(t) = \frac{\partial H}{\partial p_j}(x(t)), \qquad p_j'(t) = -\frac{\partial H}{\partial q_j}(x(t)) \qquad (1 \leq j \leq n,\ t \in J).$$

(ii) Now show that the image of the curve x lies in the energy surface $N(H(x^0))$ by proving

$$\frac{dH(x(t))}{dt} = \langle \operatorname{grad} H(x(t)),\ x'(t) \rangle = 0 \qquad (t \in J).$$

(iii) Let $n = 1$ and $H(x) = \frac{1}{2}(q^2 + p^2)$ for $x \in \mathbf{R}^2 \setminus \{0\}$. Verify that an integral curve of Hamilton's equation which satisfies $x(0) = (1, 0)$ is uniquely determined, and is given by

$$x(t) = (\cos t, -\sin t) \qquad (t \in \mathbf{R}).$$

Now let $n = 1$, and assume that the function H is positively homogeneous of degree m. In the notation of Example 6.6.8 we write $P_J \subset \mathbf{R}^2$ for the area swept out during the interval of time J.

(iv) **(Generalization of Exercise 0.5).** Prove $\operatorname{area}(P_J) = \frac{1}{2} m H(x^0) \operatorname{length}(J)$.
Hint: Use Euler's identity from Exercise 2.32.(ii).

Exercise 7.63 (Archimedes' law). At a point x in a homogeneous and incompressible liquid at rest a *hydrostatic pressure* $p(x) \in [\,0,\ \infty\,[$ exists, caused by the weight of the "column" of liquid above x. In more precise formulation, the rate of increase of this pressure is highest in the direction of gravity; we therefore have the equality of vectors in $\mathbf{R}^3$

$$\nabla p(x) = \rho(x) g,$$

where $\rho(x)$ is the mass density of the liquid at x, and g the acceleration due to gravity.

Assume a body is submerged into the liquid and that, after it has come to rest, it occupies the set $\Omega \subset \mathbf{R}^3$. According to *Pascal's law,* at the point $y \in \partial\Omega$ a force of magnitude $p(y)$ and direction $-\nu(y)$ is exerted on the body, in particular, the magnitude of this force is independent of the direction of the surface $\partial\Omega$ in y.

(i) Prove that the total force experienced by the body is described by the equality of vectors

$$-\int_{\partial\Omega} p(y)\nu(y)\, d_2 y = -g \int_{\Omega} \rho(x)\, dx.$$

(ii) Conclude that one has *Archimedes' law*: the force experienced by a body at rest in a liquid is the opposite of the gravitational force acting on that mass of the liquid that could fill the space taken up by the body.

Exercise 7.64 (Self-adjointness of Laplacian). Assume that the set $\Omega \subset \mathbf{R}^n$ satisfies the conditions of Theorem 7.6.1. Write $C_c^k(\Omega)$, with $0 \le k \le \infty$, for the linear space of the C^k functions $f : \Omega \to \mathbf{R}$ satisfying $\operatorname{supp}(f) \subset \Omega$.

(i) Verify that the integral inner product

$$\langle\, f,\ g\,\rangle = \int_{\Omega} f(x)\, g(x)\, dx \qquad (f,\ g \in C_c^0(\Omega))$$

is well-defined on the linear space $C_c^0(\Omega)$.

(ii) Show that functions f and $g \in C_c^2(\Omega)$ satisfy the conditions from Theorem 7.6.1. Demonstrate that the Laplace operator $\Delta \in \operatorname{Lin}(C_c^2(\Omega),\ C_c^0(\Omega))$ is a self-adjoint operator with respect to this inner product on $C_c^0(\Omega)$, that is, for f and $g \in C_c^2(\Omega)$,

$$\langle\, \Delta f,\ g\,\rangle = \int_{\Omega} \Delta f(x)\, g(x)\, dx = \int_{\Omega} f(x)\, \Delta g(x)\, dx = \langle\, f,\ \Delta g\,\rangle.$$

(iii) Prove that, for every $f \in C_c^2(\Omega)$,

$$\int_{\Omega} f(x)\, \Delta f(x)\, dx = -\int_{\Omega} \|\operatorname{grad} f(x)\|^2\, dx.$$

Assume that there exist a nontrivial $f \in C_c^2(\Omega)$ and a number $\lambda \in \mathbf{C}$ with $-\Delta f = \lambda f$. Then conclude that $\lambda = \mu^2$ with $\mu \ge 0$.

Exercise 7.65 (Dirichlet's principle). Let Ω be as in Theorem 7.6.1, and denote by F the linear space consisting of the functions $f \in C(\overline{\Omega})$ such that all $D_j f$, for $1 \leq j \leq n$, exist and satisfy the conditions of Theorem 7.6.1. Given $k \in C(\partial\Omega)$, write F_k for the subset of all $f \in F$ with $f|_{\partial\Omega} = k$; in particular, consider F_0.

(i) Using Green's first identity show that $f \in F$ is harmonic on Ω if and only if we have

$$(\star) \qquad \int_\Omega \langle\, \mathrm{grad}\, f(x),\ \mathrm{grad}\, g(x)\,\rangle\, dx = 0 \qquad (g \in F_0).$$

Suppose $k \in C(\partial\Omega)$ is fixed.

(ii) Consider $f \in F_k$. Prove that f is harmonic on Ω (in other words, f solves the Dirichlet problem $\Delta f = 0$ and satisfies the prescribed boundary condition) if and only if

$$(\star\star) \qquad \int_\Omega \|\,\mathrm{grad}\, f(x)\|^2\, dx \leq \int_\Omega \|\,\mathrm{grad}\, g(x)\|^2\, dx \qquad (g \in F_k).$$

Hint: Suppose the inequality holds. Then $g \in F_0$ implies $f + t\, g \in F_k$, for all $t \in \mathbf{R}$. It follows that the function

$$D : \mathbf{R} \to \mathbf{R} \qquad \text{given by} \qquad D(t) = \int_\Omega \|\,\mathrm{grad}(f + t\, g)(x)\|^2\, dx$$

attains its minimum at 0. Then verify by differentiation that $(\star)$ is satisfied and conclude on the strength of part (i) that f is harmonic. Conversely, suppose f is harmonic on Ω. If $g \in F_k$ is arbitrary, then $h = g - f \in F_0$. From the equality

$$\int_\Omega \|\,\mathrm{grad}\, g(x)\|^2\, dx$$

$$= \int_\Omega (\|\,\mathrm{grad}\, f(x)\|^2 + \|\,\mathrm{grad}\, h(x)\|^2 + 2\langle\, \mathrm{grad}\, f(x),\ \mathrm{grad}\, h(x)\,\rangle)\, dx$$

in conjunction with part (i) we obtain that $(\star\star)$ holds.

Exercise 7.66 (Orthogonality of spherical harmonic functions of different degree – sequel to Exercise 2.32). Let $l \in \mathbf{N}_0$ and let $\mathcal{H}_l$ be the linear space over $\mathbf{R}$ of the harmonic polynomials on $\mathbf{R}^n$ that are homogeneous of degree l. Consider $p_l \in \mathcal{H}_l$ and $p_m \in \mathcal{H}_m$ with $l, m \in \mathbf{N}_0$. Let $S^{n-1} = \{\, x \in \mathbf{R}^n \mid \|x\| = 1\,\}$. Apply Green's second identity to show

$$\int_{S^{n-1}} \Big(p_l \frac{\partial p_m}{\partial \nu} - p_m \frac{\partial p_l}{\partial \nu}\Big)(y)\, d_{n-1}y = 0.$$

Deduce from Euler's identity in Exercise 2.32.(ii)

$$\frac{\partial p_m}{\partial \nu}(y) = \langle \operatorname{grad} p_m(y),\ y \rangle = m\, p_m(y) \qquad (y \in S^{n-1}).$$

Conclude that

$$\int_{S^{n-1}} p_l(y) p_m(y)\, d_{n-1}y = 0 \qquad (l, m \in \mathbf{N}_0,\ l \neq m).$$

Background. See Exercises 3.17 and 5.61 for more properties of the spaces $\mathcal{H}_l$.

Exercise 7.67 (Newton's potential and Poisson's equation – sequel to Exercise 7.53 – needed for Exercises 7.68, 7.69, 7.70, 8.28 and 8.35). Define, for every $x \in \mathbf{R}^n$, *Newton's potential* $p = p_x : \mathbf{R}^n \setminus \{x\} \to \mathbf{R}$ of x by

$$p(x') = p_x(x') = \begin{cases} \dfrac{1}{2\pi} \log \|x - x'\|, & n = 2; \\[2ex] \dfrac{1}{(2-n)|S^{n-1}|} \dfrac{1}{\|x - x'\|^{n-2}}, & n \neq 2. \end{cases}$$

In this connection, see Exercise 7.21.(ii) for $|S^{n-1}| := \text{hyperarea}_{n-1}(S^{n-1})$. In what follows all differentiations of p are with respect to the variable $x' \in \mathbf{R}^n \setminus \{x\}$.

(i) For all $x \in \mathbf{R}^n$, verify that p_x is a C^∞ function on $\mathbf{R}^n \setminus \{x\}$. Verify that, as a consequence of Example 7.8.4, p_x is harmonic on $\mathbf{R}^n \setminus \{x\}$, that is, for all $x \in \mathbf{R}^n$,

$$\Delta p_x = 0 \qquad \text{on} \qquad \mathbf{R}^n \setminus \{x\}.$$

Let $S(x; r)$ be the sphere in $\mathbf{R}^n$ about x and of radius r, with outer normal.

(ii) Use Example 7.9.4 to prove that, for all $x \in \mathbf{R}^n$,

$$\lim_{r \downarrow 0} \int_{S(x;r)} p(x, y)\, d_{n-1}y = 0, \qquad \lim_{r \downarrow 0} \int_{S(x;r)} \frac{\partial p}{\partial \nu}(x, y)\, d_{n-1}y = 1.$$

(iii) Verify, for all $x \in \mathbf{R}^n$, that p_x and $D_j p_x$ are locally Riemann integrable on all of $\mathbf{R}^n$, but that this is not true of $D_j^2 p_x$, for $1 \leq j \leq n$.
Hint: Use spherical coordinates with respect to x for x'.

Let $\Omega \subset \mathbf{R}^n$ be as in Theorem 7.6.1. Let $f \in C^2(\Omega)$, and assume f and Df can be extended to continuous mappings on $\overline{\Omega}$.

(iv) Using Green's second identity, and applying parts (i) and (ii), prove that, with p_x being Newton's potential of x,

$$f(x) = \int_\Omega (p_x\, \Delta f)(x')\, dx' + \int_{\partial\Omega} \Big(f \frac{\partial p_x}{\partial \nu} - p_x \frac{\partial f}{\partial \nu} \Big)(y)\, d_{n-1}y \qquad (x \in \Omega).$$

In other words, the value of f at the point x can be expressed in terms of the values of Δf on Ω, and those of f and $\frac{\partial f}{\partial \nu}$ on $\partial\Omega$.
Hint: See Example 7.9.4.

(v) Assume that, in addition, f is harmonic on Ω; then prove

$$f(x) = \int_{\partial\Omega} \Big(f \frac{\partial p_x}{\partial \nu} - p_x \frac{\partial f}{\partial \nu} \Big)(y)\, d_{n-1}y \qquad (x \in \Omega).$$

Conclude by part (i) that $f \in C^\infty(\Omega)$, that is, a harmonic function is infinitely differentiable.

Next, let $f \in C_c^2(\mathbf{R}^n)$ and define *Newton's potential* ϕ of f by (see Example 6.11.5)

$$\phi = p(0, \cdot) * f : \mathbf{R}^n \to \mathbf{R}, \qquad \text{that is,} \qquad \phi(x) = \int_{\mathbf{R}^n} p(x, x') f(x')\, dx'.$$

(vi) Prove by part (iii) that ϕ is well-defined on all of $\mathbf{R}^n$, that $\phi \in C^2(\mathbf{R}^n)$, and that ϕ on $\mathbf{R}^n$ satisfies the following, known as *Poisson's equation*:

$$(\star) \qquad \Delta\phi = f.$$

Hint: Show that $\Delta\phi$ may be calculated by differentiation under the integral sign, that is

$$\Delta\phi(x) = \int_{\mathbf{R}^n} p(0, x')\, \Delta f(x - x')\, dx' = \int_{\mathbf{R}^n} (p_x\, \Delta f)(x')\, dx'.$$

Then choose $\Omega \subset \mathbf{R}^n$ as in Theorem 7.6.1 such that $x \in \Omega$ and $\operatorname{supp}(f) \subset \Omega$; then $f|_{\partial\Omega} = 0$ and $\frac{\partial f}{\partial \nu}\big|_{\partial\Omega} = 0$. Conclude by part (iv) that $(\star)$ is true.

(vii) Use Liouville's Theorem from Exercise 7.53.(viii) or from Exercise 7.70.(viii) to show that ϕ is the unique C^2 solution of $(\star)$ that satisfies the *boundary condition at infinity*

$$\lim_{\|x\| \to \infty} \phi(x) = 0.$$

Exercise 7.68 (Another computation of Newton's potential of a ball – sequel to Exercises 3.10 and 7.67). Let $A \subset \mathbf{R}^3$ denote the closed ball about the origin, of radius $R > 0$. Without integration we shall compute Newton's potential ϕ_A from Example 6.6.7,

$$\phi_A(x) = -\frac{1}{4\pi} \int_A \frac{1}{\|x - y\|}\, dy \qquad (\|x\| > R).$$

(i) Deduce from Exercise 7.67.(vi) that ϕ_A is harmonic on $\mathbf{R}^n \setminus A$.

(ii) Using Exercise 3.10.(i) or (iii) establish the existence of a and $b \in \mathbf{R}$ with

$$\phi_A(x) = \frac{a}{\|x\|} + b \qquad (\|x\| > R).$$

Verify $\lim_{\|x\| \to \infty} \phi_A(x) = 0$, and deduce $b = 0$.

(iii) Prove $\lim_{\|x\|\to\infty} \frac{\|x\|}{\|x-y\|} = 1$ uniformly for $y \in A$, and show

$$a = \lim_{\|x\|\to\infty} -\frac{1}{4\pi} \int_A \frac{\|x\|}{\|x-y\|}\, dy = -\frac{\mathrm{vol}(A)}{4\pi}.$$

Background. There are still other ways to arrive at this result. Verify that the definition of $\phi_A(x)$ makes sense for $x \in A$, and use Exercise 3.10.(iii) to prove that there exist p and $q \in \mathbf{R}$ with $\phi_A(x) = \frac{1}{6}\|x\|^2 - \frac{p}{\|x\|} + q$, for $0 < \|x\| < R$. Show $\phi_A(0) = -\frac{1}{2}R^2$ using an elementary integration over A by means of spherical coordinates. Deduce

$$\phi_A(x) = -\frac{1}{6}(3R^2 - \|x\|^2) \qquad (\|x\| < R).$$

Finally use the continuity of $\phi_A(x)$ for $\|x\| = R$.

(iv) Prove similarly the result from Example 7.4.8.

Exercise 7.69 (Green's function, Poisson's kernel and Mean Value Theorem for harmonic function – sequel to Exercise 7.67 – needed for Exercise 7.70). We employ the notation from Exercise 7.67. Let $\Omega \subset \mathbf{R}^n$ be as in Theorem 7.6.1. Let $p = p_x$ be Newton's potential of $x \in \Omega$, and make the assumption that for every $x \in \Omega$ there exists a function $q = q_{\Omega,x} \in C^2(\Omega)$ such that q_x and Dq_x can be continuously continued to $\partial\Omega$, while

$$\Delta q_x = 0 \quad \text{on} \quad \Omega, \qquad q_x|_{\partial\Omega} = -p_x|_{\partial\Omega}.$$

According to Example 7.9.7, this uniquely determines q_x. The *diagonal* $\Delta\Omega$ of Ω is the set $\{\, (x, x) \mid x \in \Omega \,\}$. *Green's function G for* Ω is defined by

$$G = G_\Omega : (\Omega \times \overline{\Omega}) \setminus \Delta\Omega \to \mathbf{R}, \qquad G(x, x') = G_{\Omega,x}(x') = p_x(x') + q_x(x').$$

Poisson's kernel P for Ω is defined by

$$P = P_\Omega : \Omega \times \partial\Omega \to \mathbf{R} \qquad \text{with} \qquad P(x, y) = \frac{\partial G_x}{\partial \nu}(y).$$

(i) Verify that G_x has the properties which p_x was shown to possess in parts (i) and (ii) of Exercise 7.67, and that $G_x|_{\partial\Omega} = 0$. Conclude that, for every f as in Exercise 7.67.(iv),

$$f(x) = \int_\Omega G(x, x')\, \Delta f(x')\, dx' + \int_{\partial\Omega} P(x, y)\, f(y)\, d_{n-1}y \qquad (x \in \Omega).$$

(ii) Let $B(x; r)$ be the closed ball in $\mathbf{R}^n$ about x and of radius $r > 0$. Assume x, $x' \in \Omega$, and let Ω' be obtained from Ω by leaving out $B(x; r)$ and $B(x'; r)$, for r sufficiently small. Apply the identity from part (i), with Ω replaced by Ω', and f by $z \mapsto G(x, z)$ and $z \mapsto G(x', z)$, respectively. Thus demonstrate the following symmetry of Green's function:

$$G(x, x') = G(x', x) \qquad ((x, x') \in (\Omega \times \Omega) \setminus \Delta\Omega).$$

(iii) Verify that in this case the Dirichlet problem $\Delta f = g$ on Ω and $f = h$ on $\partial\Omega$ from Example 7.9.7 can be solved by means of part (i)

$$f(x) = \int_{\Omega} G(x, x')\, g(x')\, dx' + \int_{\partial\Omega} P(x, y)\, h(y)\, d_{n-1}y \qquad (x \in \Omega).$$

(iv) Prove the *Mean Value Theorem for harmonic functions* (see Exercise 7.53 for its formulation).
Hint: Let $x \in \Omega$ and $r > 0$, and apply (i) with $\Omega = \{ x' \in \mathbf{R}^n \mid \|x' - x\| < r \}$. In this case, Green's function G for Ω differs by a constant from Newton's potential of x. Therefore, with m_f the spherical mean of f,

$$f(x) = m_f(x, r) + \int_{\Omega} G(x, x')\, \Delta f(x')\, dx'.$$

Exercise 7.70 (Green's function and Poisson's kernel for half-space and ball, Poisson's integral, Schwarz' Theorem and Liouville's Theorem – sequel to Exercises 1.3 and 7.69). Consider the special case of Exercise 7.69 where we defined, violating our standard convention,

$$\Omega_\pm = \mathbf{R}^{n+1}_\pm := \{ (x, t) \in \mathbf{R}^{n+1} \mid x \in \mathbf{R}^n,\ \pm t > 0 \}.$$

Then $\partial\Omega_\pm = \{ (x, 0) \in \mathbf{R}^{n+1} \}$, and the outer normal at every point of $\partial\Omega_\pm$ is given by $\mp(0, 1)$.

(i) Verify that, for $((x, t),\ (x', t')) \in (\Omega_\pm \times \overline{\Omega_\pm}) \setminus \Delta\Omega_\pm$, Green's function $G_\pm$ has the value $G_\pm((x, t),\ (x', t'))$ given by

$$\frac{1}{2\pi} \big(\log \|(x - x',\ t - t')\| - \log \|(x - x',\ t + t')\| \big) \qquad (n = 1);$$

$$\frac{1}{(1 - n)\, |S^n|} \left(\frac{1}{\|(x - x',\ t - t')\|^{n-1}} - \frac{1}{\|(x - x',\ t + t')\|^{n-1}} \right) \qquad (n > 1).$$

(ii) Now prove by differentiation that one has for Poisson's kernel $P_\pm$, for $(x, t) \in \Omega_\pm$ and $(y, 0) \in \partial\Omega_\pm$,

$$P_\pm((x, t), (y, 0)) = \frac{2}{|S^n|} \frac{\pm t}{\|(x - y, t)\|^{n+1}}.$$

Conclude (compare with Exercises 6.99.(iii) and 7.23) that

$$\int_{\mathbf{R}^n} \frac{1}{\|(x - y, t)\|^{n+1}}\, dy = \frac{1}{|t|} \frac{\pi^{\frac{n+1}{2}}}{\Gamma(\frac{n+1}{2})} \qquad (x \in \mathbf{R}^n,\ t \neq 0).$$

Let $r > 0$. Then consider the special case of Exercise 7.69 where

$$\Omega = B^n(r) := \{\, x \in \mathbf{R}^n \mid \|x\| < r \,\},$$

$$\text{and thus} \qquad \partial\Omega = S^{n-1}(r) := \{\, y \in \mathbf{R}^n \mid \|y\| = r \,\}.$$

(iii) Let $x \in B^n(r) \setminus \{0\}$. Using the symmetry identity from Exercise 1.3 prove, for all $y \in S^{n-1}(r)$,

$$\|x - y\| = \left\| \frac{r}{\|x\|} x - \frac{\|x\|}{r} y \right\| = \frac{\|x\|}{r} \left\| \frac{r^2}{\|x\|^2} x - y \right\|.$$

Verify $\check{x} := \frac{r^2}{\|x\|^2} x \notin B^n(r)$. (Actually, $\check{x}$ is the *inverse of x with respect to the sphere $S^{n-1}(r)$*, that is, $\check{x} \in \mathbf{R}^n$ is the vector in the direction of x for which $\|x\|\,\|\check{x}\| = r^2$.)

(iv) For $(x, x') \in (B^n(r) \times \overline{B^n(r)}) \setminus \Delta B^n(r)$, if $x \neq 0$, show that $G(x, x')$ equals

$$\begin{cases} \dfrac{1}{2\pi} \left(\log \|x - x'\| - \log \left\| \dfrac{r}{\|x\|} x - \dfrac{\|x\|}{r} x' \right\| \right) & (n = 2); \\[2ex] \dfrac{1}{(2 - n)\,|S^{n-1}|} \left(\dfrac{1}{\|x - x'\|^{n-2}} - \dfrac{1}{\left\| \frac{r}{\|x\|} x - \frac{\|x\|}{r} x' \right\|^{n-2}} \right) & (n \neq 2); \end{cases}$$

and further, if $x = 0$,

$$G(0, x') = \begin{cases} \dfrac{1}{2\pi} \log \dfrac{\|x'\|}{r} & (n = 2); \\[2ex] \dfrac{1}{(2 - n)\,|S^{n-1}|} \left(\dfrac{1}{\|x'\|^{n-2}} - \dfrac{1}{r^{n-2}} \right) & (n \neq 2). \end{cases}$$

Now prove by differentiation

$$P(x, y) = \frac{r^2 - \|x\|^2}{|S^{n-1}|\, r} \frac{1}{\|x - y\|^n} \qquad (x \in B^n(r),\ y \in S^{n-1}(r)).$$

Assume that $n = 2$ and let $y \in S^1(r)$. Prove that the level curves of the function $x \mapsto P(x, y) : B^2(r) \to \mathbf{R}$ are circles tangent to $S^1(r)$ at the point y.

Given $h \in C(S^{n-1}(r))$, define *Poisson's integral* $\mathcal{P}h : B^n(r) \to \mathbf{R}$ of h by

$$(\mathcal{P}h)(x) = \frac{r^2 - \|x\|^2}{|S^{n-1}|\,r} \int_{S^{n-1}(r)} \frac{h(y)}{\|x-y\|^n}\, d_{n-1}y \qquad (x \in B^n(r)).$$

(v) Prove that $\mathcal{P}h \in C^\infty(B^n(r))$. Check that a solution f of the Dirichlet problem $\Delta f = 0$ on $B^n(r)$ with $f|_{S^{n-1}(r)} = h$, is given by $f = \mathcal{P}h$, known as *Poisson's integral formula.*

(vi) Conclude that

$$\int_{S^{n-1}(r)} \frac{1}{\|x-y\|^n}\, d_{n-1}y = \frac{2\pi^{\frac{n}{2}}}{\Gamma(\frac{n}{2})} \frac{r}{r^2 - \|x\|^2} \qquad (x \in B^n(r)).$$

(vii) Prove *Schwarz' Theorem*, which asserts

$$\lim_{x \in B^n(r),\, x \to y} (\mathcal{P}h)(x) = h(y) \qquad (y \in S^{n-1}(r)).$$

Hint: Let $y^0 \in S^{n-1}(r)$ be fixed. Prove by part (vi), for $x \in B^n(r)$,

$$(\mathcal{P}h)(x) - h(y^0) = \frac{r^2 - \|x\|^2}{|S^{n-1}|\,r} \int_{S^{n-1}(r)} \frac{h(y) - h(y^0)}{\|x-y\|^n}\, d_{n-1}y.$$

Let $\epsilon > 0$ be chosen arbitrarily. By virtue of the continuity of h, among other arguments, there exist a partition of $S^{n-1}(r)$ into subsets S_1 and S_2, and a number $c > 0$ such that

$$y \in S_1 \quad\Longrightarrow\quad |h(y) - h(y^0)| < \frac{\epsilon}{2}, \qquad y \in S_2 \quad\Longrightarrow\quad \|y - y^0\| \geq 2c.$$

Hence, for all $x \in B^n(r)$,

$$\frac{r^2 - \|x\|^2}{|S^{n-1}|\,r} \int_{S_1} \frac{|h(y) - h(y^0)|}{\|x-y\|^n}\, d_{n-1}y < \frac{\epsilon}{2}.$$

Assume next that $x \in B^n(r)$ satisfies $\|x - y^0\| < c$. Then, for $y \in S_2$,

$$\|y - x\| \geq \|y - y^0\| - \|y^0 - x\| \geq 2c - c = c.$$

Let $\|h\|$ be the supremum of h on $S^{n-1}(r)$; then $|h(y) - h(y^0)| < 2\|h\|$, for all $y \in S^{n-1}(r)$. Ergo, there exists $0 < \delta < c$ such that for all $x \in B^n(r)$ with $\|x - y^0\| < \delta$, and for all $y \in S_2$,

$$\frac{r^2 - \|x\|^2}{r} \frac{r^{n-1}}{\|x-y\|^n} \leq (r - \|x\|)\frac{2r^{n-1}}{c^n} < \frac{\epsilon}{4\|h\|}.$$

This gives, for all $x \in B^n(r)$ with $\|x - y^0\| < \delta$,

$$\frac{r^2 - \|x\|^2}{|S^{n-1}|\,r} \int_{S_2} \frac{|h(y) - h(y^0)|}{\|x-y\|^n}\, d_{n-1}y < \frac{\epsilon}{2}.$$

(viii) Prove *Liouville's Theorem* from Exercise 7.53.(viii).
Hint: Let $x \in \mathbf{R}^n$. Apply Poisson's integral formula from part (v), with $r > \|x\|$ arbitrary, and with f itself as the function h on the boundary. Calculate the partial derivatives $D_j f$, for $1 \le j \le n$, by differentiation under the integral sign. One obtains, with $\|f\|$ the supremum of f on $\mathbf{R}^n$,

$$|D_j f(x)| \le 2\|f\|\frac{\|x\|}{r^2 - \|x\|^2} + n\|f\|\frac{r^2 - \|x\|^2}{|S^{n-1}|r}\int_{S^{n-1}(r)} \frac{1}{\|x - y\|^{n+1}}\, d_{n-1}y.$$

Subsequently, use $\|x - y\| \ge r - \|x\|$, for $y \in S^{n-1}(r)$, to estimate the integral, and show

$$|D_j f(x)| = \mathcal{O}\Big(\frac{1}{r}\Big), \qquad r \to \infty.$$

Exercise 7.71 (Uniqueness of solution of heat equation). Let $\Omega \subset \mathbf{R}^n$ be as in Theorem 7.6.1 and let $T > 0$. Let $C \subset \mathbf{R}^{n+1}$ be the open cylinder $\Omega \times \,]\,0, T\,[$. Let $f : \overline{C} \to \mathbf{R}$ be a continuous function for which the following continuous derivatives exist on C:

$$\frac{\partial f}{\partial x_j} \quad (1 \le j \le n), \qquad \frac{\partial^2 f}{\partial x_i \partial x_j} \quad (1 \le i, j \le n), \qquad \frac{\partial f}{\partial t}.$$

Assume that f satisfies on C the *heat equation* from Example 7.9.5, with $k = 1$.

(i) Prove, for every $t \in \,]\,0, T\,[$,

$$0 = \int_\Omega \Big(\big(\frac{\partial f}{\partial t} - \Delta_x f\big) f\Big)(x, t)\, dx = \frac{1}{2}\frac{\partial}{\partial t}\int_\Omega f(x, t)^2\, dx$$

$$+ \int_\Omega \|\operatorname{grad}_x f(x, t)\|^2\, dx - \int_{\partial\Omega} \Big(f\,\frac{\partial f}{\partial \nu}\Big)(y, t)\, d_{n-1}y.$$

Here the gradient is calculated with respect to the variable $x \in \mathbf{R}^n$.

Now assume $f(x, 0) = 0$, for $x \in \Omega$, and $f(y, t) = 0$, for $y \in \partial\Omega$ and $t \in \,]\,0, T\,[$.

(ii) Prove, for all $0 < t < T$,

$$0 \ge \frac{\partial}{\partial t}\int_\Omega f(x, t)^2\, dx; \qquad \text{and conclude that} \qquad \int_\Omega f(x, t)^2\, dx = 0.$$

Show $f(x, t) = 0$, for $x \in \Omega$ and $t \in [\,0, T\,]$.

We derive some additional results, under the assumption that f satisfies the heat equation and the *Neumann boundary condition*

$$\frac{\partial f}{\partial \nu}(y, t) = 0 \qquad (y \in \partial\Omega,\ t \in \,]\,0, T\,[\,).$$

(iii) Prove, for $0 \le t \le T$,

$$\int_\Omega f(x,t)\,dx = \int_\Omega f(x,0)\,dx, \qquad \int_\Omega f(x,t)^2\,dx \le \int_\Omega f(x,0)^2\,dx.$$

Exercise 7.72 (Needed for Exercise 7.73). Let $x^0 \in \mathbf{R}^3$ be chosen arbitrarily, and let $f : \mathbf{R}^3 \setminus \{x^0\} \to \mathbf{R}$ be defined by

$$f(x) = \log \|x - x^0\|.$$

(i) Using Example 2.4.8 prove, for $x \in \mathbf{R}^3 \setminus \{x^0\}$ (see also Exercise 3.8),

$$\operatorname{grad} f(x) = \frac{1}{\|x - x^0\|^2}(x - x^0), \qquad \Delta f(x) = \frac{1}{\|x - x^0\|^2}.$$

(ii) Prove that Δf is absolutely Riemann integrable over $\mathbf{R}^3 \setminus \{x^0\}$.

Now let $g \in C^1(\mathbf{R}^3)$ be a function with compact support. Part (ii) then leads to

$$\int_{\mathbf{R}^3} \frac{g(x)^2}{\|x - x^0\|^2}\,dx < \infty.$$

In fact, we have the following inequality: for every $x^0 \in \mathbf{R}^3$ and every $g \in C^1(\mathbf{R}^3)$ with compact support one has

$$(\star) \qquad \int_{\mathbf{R}^3} \frac{g(x)^2}{\|x - x^0\|^2}\,dx \le 4 \int_{\mathbf{R}^3} \|\operatorname{grad} g(x)\|^2\,dx.$$

We now prove $(\star)$.

(iii) Use Green's first identity to prove

$$\int_{\mathbf{R}^3} \frac{g(x)^2}{\|x - x^0\|^2}\,dx = -2 \int_{\mathbf{R}^3} \frac{g(x)}{\|x - x^0\|^2}\langle\, x - x^0,\ \operatorname{grad} g(x)\,\rangle\,dx.$$

Then derive from this

$$\int_{\mathbf{R}^3} \frac{g(x)^2}{\|x - x^0\|^2}\,dx \le 2\Big(\int_{\mathbf{R}^3} \frac{g(x)^2}{\|x - x^0\|^2}\,dx\Big)^{1/2}\Big(\int_{\mathbf{R}^3} \|\operatorname{grad} g(x)\|^2\,dx\Big)^{1/2};$$

and conclude that $(\star)$ holds.

Exercise 7.73 (Stability of an atom – sequel to Exercise 7.72). According to *quantum physics* an atom consists of an atomic nucleus with a positive electric charge, and negatively charged electrons moving around that nucleus. Assume that the magnitude of the electric charge of the nucleus is a multiple $Z \in \mathbf{N}$ of the electronic charge. An electron moving around the nucleus then finds itself in an electric field, caused by the atomic nucleus, which is described (to within a constant) by *Newton's potential* (see Exercise 7.67)

$$V(x) = -\frac{Z}{\|x\|} \qquad (x \in \mathbf{R}^3 \setminus \{0\}).$$

It has been found that an electron moves in certain orbits only. Furthermore, an electron can spontaneously jump from such an orbit to another orbit lying closer to the atomic nucleus, under simultaneous emission of an amount of energy. According to Schrödinger these orbits are parametrized by numbers $\lambda \in \mathbf{R}$ that are **eigenvalues** of the *Schrödinger operator with Newton's potential*; that is, those λ for which there exists a function $g \in C^2(\mathbf{R}^3)$ with $\int_{\mathbf{R}^3} g(x)^2\,dx = 1$ satisfying the following, known as *Schrödinger's equation*:

$$(\star) \qquad -\frac{1}{2}\Delta g(x) - Z\frac{g(x)}{\|x\|} = \lambda g(x) \qquad (x \in \mathbf{R}^3 \setminus \{0\}).$$

Then the amount of energy emitted upon the transition from an orbit with parameter λ to one with parameter μ is proportional to $\lambda - \mu$. If the collection of eigenvalues λ has a lower bound, this opens up the possibility for the electrons not to continually emit energy, but instead to remain in a ground state of minimal energy; this is then referred to as the *stability of the atom*. We now prove this. To avoid convergence problems we restrict ourselves to $g \in C^2(\mathbf{R}^3)$ with compact support.

(i) Multiply both sides of $(\star)$ by $g(x)$, to deduce

$$\frac{1}{2}\int_{\mathbf{R}^3} \|\operatorname{grad} g(x)\|^2\,dx - Z\int_{\mathbf{R}^3} \frac{g(x)^2}{\|x\|}\,dx = \lambda\int_{\mathbf{R}^3} g(x)^2\,dx.$$

(ii) Using $(\star)$ from Exercise 7.72, prove that

$$\begin{aligned}\int_{\mathbf{R}^3} \frac{g(x)^2}{\|x\|}\,dx &\leq \left(\int_{\mathbf{R}^3} \frac{g(x)^2}{\|x\|^2}\,dx \int_{\mathbf{R}^3} g(x)^2\,dx\right)^{1/2}\\ &\leq \frac{1}{2}\left(\frac{1}{4Z}\int_{\mathbf{R}^3} \frac{g(x)^2}{\|x\|^2}\,dx + 4Z\int_{\mathbf{R}^3} g(x)^2\,dx\right)\\ &\leq \frac{1}{2Z}\int_{\mathbf{R}^3} \|\operatorname{grad} g(x)\|^2 + 2Z\int_{\mathbf{R}^3} g(x)^2\,dx.\end{aligned}$$

(iii) Conclude that a number $\lambda \in \mathbf{R}$ for which there exists a nontrivial $g \in C^2(\mathbf{R}^3)$ with compact support satisfying $(\star)$ has the property $\lambda \geq -2Z^2$.

Exercises for Chapter 8

Exercise 8.1. Let $I \subset \mathbf{R}$ be a closed interval and $\gamma : I \to \mathbf{R}^n$ a C^1 mapping.

(i) Prove that there exists a C^1 mapping $\delta : [k_-, k_+] \to \mathbf{R}^n$ with $\mathrm{im}(\gamma) = \mathrm{im}(\delta)$ and also $\delta'(k_-) = \delta'(k_+) = 0$. (This mapping δ is not an immersion everywhere, and a fortiori not an embedding.)
Hint: Define the C^1 mapping $\psi : J := [0, 1] \to \mathbf{R}$ by $\psi(t) = t^2(2-t)^2$. Then $\psi(0) = 0$ and $\psi(1) = 1$, and thus $J \subset \mathrm{im}(\psi)$ according to the Intermediate Value Theorem 1.9.5. Moreover $\psi'(t) = 4t(1-t)(2-t) > 0$ for $0 < t < 1$, and therefore ψ is monotonically strictly increasing on J. Hence $\psi : J \to J$ is a C^1 mapping that preserves the order of the endpoints for which $\psi'(0) = \psi'(1) = 0$.

(ii) Using part (i) deduce that the image under a piecewise C^1 mapping can be written as the image under a C^1 mapping.

Exercise 8.2. Let $U \subset \mathbf{R}^n$ be an open rectangle with $0 \in U$, and let $f : U \to \mathbf{R}^n$ be a C^1 vector field with $Af = 0$. Then $g : U \to \mathbf{R}$ is a scalar potential for f if

$$g(x) = \sum_{1 \le i \le n} \int_0^{x_i} f_i(x_1, x_2, \dots, x_{i-1}, t, 0 \dots, 0)\, dt.$$

Prove this by direct calculation, and also by using that $g(x) = \int_{\gamma_x} \langle f(s), d_1 s \rangle$, where $\gamma_x : [0, n] \to \mathbf{R}^n$ is the curve from 0 to x successively following the directions of the standard basis vectors e_i, for $1 \le i \le n$,

$$\gamma_x(t) = (x_1, x_2, \dots, x_{i-1}, (t-i+1)x_i, 0 \dots, 0) \quad (1 \le i \le n,\ 0 \le t-i+1 \le 1).$$

Exercise 8.3. Define $f : \mathbf{R}^2 \to \mathbf{R}^2$ by $f(x) = (x_1 x_2^2,\ x_1 + x_2)$, and let $\Omega \subset \mathbf{R}^2$ be the open subset in the first quadrant bounded by the curves $\{ x \in \mathbf{R}^2 \mid x_1 = x_2 \}$ and $\{ x \in \mathbf{R}^2 \mid x_1^2 = x_2 \}$. Prove by integration over Ω, and also by using Green's Integral Theorem

$$\int_\Omega \mathrm{curl}\, f(x)\, dx = \frac{1}{12}.$$

Exercise 8.4 (Astroid – sequel to Exercise 5.19). This is the curve A defined as the zero-set in $\mathbf{R}^2$ of $g(x) = x_1^{2/3} + x_2^{2/3} - 1$.

(i) Verify that the length of A equals 6.

(ii) Prove that the area of the bounded set in $\mathbf{R}^2$ bounded by A equals $\frac{3}{8}\pi$.

Exercise 8.5 (Quadrature of parabola). Define $\phi : \mathbf{R} \to \mathbf{R}^2$ by $\phi(t) = (t, t^2)$; then $P = \mathrm{im}(\phi)$ is a parabola. Consider arbitrary t_+ and $t_- \in \mathbf{R}$ with $t_+ > t_-$, and define

$$\delta = \frac{t_+ - t_-}{2}, \qquad t_0 = \frac{t_+ + t_-}{2}.$$

(i) Let $t \in \mathbf{R}$ with $t_+ > t > t_-$. Demonstrate that the area of the triangle in $\mathbf{R}^2$ with vertices $\phi(t_+)$, $\phi(t)$ and $\phi(t_-)$ equals

$$\frac{1}{2}\left|\begin{array}{cc} t_+ - t_- & t_+ - t \\ t_+^2 - t_-^2 & t_+^2 - t^2 \end{array}\right|.$$

Calculate the unique t such that the corresponding triangle $\Delta(t_+, t_-)$ has maximal area, and show that this area equals δ^3.

(ii) Find the value of t for which the direction of the tangent space to P at $\phi(t)$ is the same as that of the straight line $l(t_+, t_-)$ through $\phi(t_+)$ and $\phi(t_-)$.

(iii) Prove in three different ways, including the use of successive integration and of Green's Integral Theorem, that area $(S(t_+, t_-)) = \frac{4}{3}\delta^3$, if $S(t_+, t_-)$ is the sector of the parabola P with base $l(t_+, t_-)$, that is, the bounded part of $\mathbf{R}^2$ which is bounded by the parabola P and the straight line $l(t_+, t_-)$.
Hint: Check that

$$S(t_+, t_-) = \{ x \in \mathbf{R}^2 \mid t_- \le x_1 \le t_+,\ x_1^2 \le x_2 \le (t_0 - \delta)^2 + 2t_0(x_1 - t_-) \}.$$

Further, use $\frac{1}{3}(t_+^3 - t_-^3) = \frac{2}{3}\delta^3 + 2t_0^2\delta$.

The *quadrature of the parabola* according to Archimedes follows from parts (i) and (iii): for all t_+ and $t_- \in \mathbf{R}$ with $t_+ > t_-$ one has

$$\text{area}\,(S(t_+, t_-)) = \frac{4\,\text{area}\,(\Delta(t_+, t_-))}{3}.$$

The area of a sector of a parabola therefore equals four thirds the area of the inscribed triangle having the same base as the sector of the parabola, and whose apex is the point at which the tangent line to the parabola runs parallel to that base.

We give another, direct proof of this result. Define $\Psi : \mathbf{R}^2 \to \mathbf{R}^2$ by

$$\Psi(y) = \begin{pmatrix} t_0 + \quad \delta\, y_1 \\ t_0^2 + 2t_0\delta\, y_1 + \delta^2 y_2 \end{pmatrix} = \phi(t_0) + \delta \begin{pmatrix} 1 & 0 \\ 2t_0 & \delta \end{pmatrix} y.$$

(iv) Demonstrate that Ψ is a C^∞ diffeomorphism. Verify $\Psi \circ \phi(t) = \phi(t_0 + \delta\, t)$, for all $t \in \mathbf{R}$. Now conclude that Ψ maps the parabola P into itself; and that the triangle $\Delta(t_+, t_-)$ from part (i) is the image under Ψ of the triangle with vertices $(1, 1)$, $(0, 0)$ and $(-1, 1)$.

(v) Prove that $\det D\Psi(y) = \delta^3$, for every $y \in \mathbf{R}^2$. Then conclude by part (iv) that the quadrature of the parabola has been reduced to a special case.

Exercise 8.6 (Steiner's hypocycloid and Kakeya's needle problem – sequel to Exercise 5.35). Let $b > 0$; let $\phi : \mathbf{R} \to \mathbf{R}^2$ and Steiner's hypocycloid $H \subset \mathbf{R}^2$ be defined by, respectively,

$$\phi(\alpha) = b\begin{pmatrix} 2\cos\alpha + \cos 2\alpha \\ 2\sin\alpha - \sin 2\alpha \end{pmatrix} \qquad \text{and} \qquad H = \mathrm{im}(\phi).$$

(i) Prove that the length of H equals $16b$, that is, 16 times the radius of the incircle of H.

(ii) Prove that area of the bounded set in $\mathbf{R}^2$ bounded by H equals $2\pi b^2$, that is, twice the area of the incircle of H.

Background. Consider the special case where $b = \frac{1}{4}$. Exercise 5.35.(vii) then implies that a needle of length 1 can be continuously rotated in the interior of H by an angle 2π; moreover, during the rotation the needle is always tangent to H. The area of that interior of H equals $\frac{\pi}{8}$, while that of a circle of diameter 1 equals $\frac{\pi}{4}$. Thus arises *Kakeya's needle problem*: what is the minimal area of a subset of $\mathbf{R}^2$ within which a needle of length 1 (lying in $\mathbf{R}^2$) can be continuously rotated. For a long time it was thought that $\frac{\pi}{8}$ would be the answer. It has been shown by Besicovitch, however, that there are sets in $\mathbf{R}^2$ of arbitrarily small area > 0 that have the desired property.[1]

Exercise 8.7. Let $C \subset \mathbf{R}^3$ be the intersection of the cylinder $\{ x \in \mathbf{R}^3 \mid x_1^2 + x_2^2 = 1 \}$ and the plane $\{ x \in \mathbf{R}^3 \mid x_1 + x_2 + x_3 = 1 \}$, and let C be oriented by the requirement that the tangent vector at the point $(1, 0, 0)$ have a positive x_3-component. Let $f : \mathbf{R}^3 \to \mathbf{R}^3$ be defined by $f(x) = (-x_2^3,\ x_1^3,\ -x_3^3)$.

(i) Prove

$$\int_C \langle f(s), d_1 s \rangle = 3 \int_{\{ x \in \mathbf{R}^2 \mid \|x\| \le 1 \}} \|x\|^2 \, dx = \frac{3\pi}{2}.$$

(ii) Also calculate $\int_C \langle f(s), d_1 s \rangle$ by means of a parametrization of C.
Hint: $\cos^4\alpha + \sin^4\alpha = \frac{1}{4}(3 + \cos 4\alpha)$.

[1] See Section 3.5 in Krantz, S. G.: *A Panorama of Harmonic Analysis.* Mathematical Association of America, Washington 1999.

Exercise 8.8. Let the surface $\Xi \subset \mathbf{R}^3$ be the union of the cylinder $\{ x \in \mathbf{R}^3 \mid x_1^2 + x_2^2 = 1,\ 0 < x_3 < 1 \}$ and the hemisphere $\{ x \in \mathbf{R}^3 \mid x_1^2 + x_2^2 + (x_3 - 1)^2 = 1,\ 1 \le x_3 \}$. Let Ξ be oriented by the requirement that the normal at $(0, 0, 2)$ point away from the origin. Define $f : \mathbf{R}^3 \to \mathbf{R}^3$ by $f(x) = (x_1 + x_1x_3 + x_2x_3^2,\ x_2 + x_1x_2x_3^3,\ x_1^2x_3^4)$. Prove

$$\int_\Xi \langle \operatorname{curl} f(x),\ d_2x \rangle = 0.$$

Exercise 8.9. Let $\phi : \mathbf{R}^2 \to \mathbf{R}^3$ be the mapping given by

$$\phi(r, \alpha) = (r\cos 4\alpha,\ r \sin 4\alpha,\ \cos\alpha).$$

Further, let

$$\begin{aligned} \Omega &= \{ (r, \alpha) \in \mathbf{R}^2_+ \mid \alpha < \pi,\ r < \sin\alpha \}, \qquad \Xi = \phi(\Omega), \\ \gamma_1 &= \{ (\sin\alpha\cos 4\alpha,\ \sin\alpha \sin 4\alpha,\ \cos\alpha) \in \mathbf{R}^3 \mid 0 < \alpha < \pi \}, \\ \gamma_2 &= \{ x \in \mathbf{R}^3 \mid x_1 = x_2 = 0,\ -1 < x_3 < 1 \}. \end{aligned}$$

Let γ_2 be oriented by the requirement that the tangent vector at $(0, 0, 0)$ have a positive x_3-component.

(i) Show that $\partial\Xi = \gamma_1 \cup \gamma_2$.

Assume one is given the function $g : \mathbf{R}^3 \to \mathbf{R}$ and the vector field $h : \mathbf{R}^3 \to \mathbf{R}^3$ defined by

$$g(x) = 2(x_3 + x_1x_2)\sqrt{3 + \|x\|^2},$$

$$h(x) = ((4 + x_1^2)x_2 + x_1x_3,\ (4 + x_2^2)x_1 + x_2x_3,\ 4 + x_3^2 + x_1x_2x_3).$$

(ii) Demonstrate that $\operatorname{grad} g$ restricted to the unit sphere $S^2 \subset \mathbf{R}^3$ equals the restriction of h to S^2, and conclude that

$$\int_{\gamma_1} \langle \operatorname{grad} g(s),\ d_1s \rangle = \int_{\gamma_1} \langle h(s),\ d_1s \rangle.$$

(iii) Calculate $\int_{\gamma_2} \langle \operatorname{grad} g(s),\ d_1s \rangle$.
Hint: Use the fact that integration by parts with respect to x_3 commutes with setting x_1 and x_2 equal to 0, so that the third component of $\operatorname{grad} g$ can easily be integrated.

(iv) Prove that $\operatorname{curl}\operatorname{grad} g = 0$ and conclude $\int_{\gamma_1} \langle h(s),\ d_1s \rangle = -8$.
Hint: It may be taken for granted that Stokes' Integral Theorem also holds for integration over Ξ.

Exercise 8.10. Let $g : U := \mathbf{R}^3 \setminus \{ (0,0,x_3) \mid x_3 \in \mathbf{R} \} \to \mathbf{R}^3$ be the vector field satisfying

$$g(x) = -x_3(f(x_1,x_2),\ 0) = \frac{x_3}{x_1^2+x_2^2}(x_2,\ -x_1,\ 0);$$

here f is the gradient vector field of the argument function: $\mathbf{R}^2 \setminus (\,]-\infty, 0] \times \{0\}) \to\ ]-\pi, \pi[$ (see Example 8.2.4). Let $\psi \in [-\frac{\pi}{2}, \frac{\pi}{2}]$ be fixed and suppose γ is the parametrization of the parallel on the unit sphere $S^2 \subset \mathbf{R}^3$ determined by ψ, given by

$$\gamma : \alpha \mapsto (\cos\alpha\cos\psi,\ \sin\alpha\cos\psi,\ \sin\psi) \in S^2 \qquad (-\pi < \alpha < \pi).$$

(i) Prove $\int_\gamma \langle\, g(s),\ d_1 s\,\rangle = -2\pi\sin\psi$ by explicit computation as well as by application of Formula (8.18).

(ii) Demonstrate

$$\operatorname{curl} g(x) = \frac{1}{x_1^2+x_2^2}(x_1, x_2, 0) \qquad \text{and} \qquad \langle\, \operatorname{curl} g(x),\ x\,\rangle = 1 \qquad (x \in U).$$

Let Ξ be the cap of the sphere S^2 determined by ψ as in Examples 7.4.6 and 8.5.1.

(iii) Verify by a direct computation using part (ii) and Example 7.4.6

$$\int_\Xi \langle\, \operatorname{curl} g(x),\ d_2 x\,\rangle = 2\pi(1-\sin\psi).$$

(iv) The results from parts (i) and (iii) seem to contradict Stokes' Integral Theorem. Prove this is not the case.
Hint: The vector field g is not defined in $(0,0,1) \in S^2$, and a limit argument based on the result from Example 8.2.4 gives the missing line integral having the value 2π.

Background. The line integral in part (i) gives the angle of daily rotation of *Foucault's pendulum* from Exercise 5.57. See Exercise 8.45 for the same computation in terms of differential forms.

Exercise 8.11 (Cauchy's Integral Theorem by differentiation under integral sign). As in Cauchy's original proof, verify Cauchy's Integral Theorem 8.3.12 by means of differentiation under the integral sign.
Hint: Combine Formula (8.27), Formula (8.11) and Lemma 8.3.10.(iii).
Also perform this computation working over $\mathbf{C}$. That is, consider

$$\int_{\gamma_{z_1}} f(z_2)\,dz_2 = \int_I (f \circ \Gamma)(z_1,z_2) D_2\Gamma(z_1,z_2)\,dz_2,$$

and differentiate under the integral sign. Next integrate by parts and finally use

$$D_1(f\circ\Gamma)\,D_2\Gamma - D_2(f\circ\Gamma)\,D_1\Gamma = (f'\circ\Gamma)\,D_1\Gamma\,D_2\Gamma - (f'\circ\Gamma)\,D_2\Gamma\,D_1\Gamma = 0.$$

Exercise 8.12 (Equivalence of holomorphic and complex-analytic – needed for Exercises 8.13, 8.16, 8.17, 8.21 and 8.22). Assume that $\Omega \subset \mathbf{C}$ and the holomorphic function $f : \Omega \to \mathbf{C}$ meet the conditions of Cauchy's Integral Theorem 8.3.11. For every $z \in \Omega$ there exists a number $R^0 > 0$ such that the circle $S(z; R)$ about z of radius $0 < R \leq R^0$ is contained in Ω. Let $\Omega' = \Omega'(R) \subset \mathbf{C}$ be the open set bounded by $S(z; R)$ and $\partial\Omega$.

(i) Apply Cauchy's Integral Theorem to Ω' in order to conclude that, for $0 < R \leq R^0$,

$$\int_{\partial\Omega} \frac{f(w)}{w - z}\, dw = \int_{S(z;R)} \frac{f(w)}{w - z}\, dw;$$

here $S(z; R)$ has positive orientation.

Now consider the parametrization $\alpha : \,]-\pi, \pi[\, \to S(z; R)$ with $\alpha : t \mapsto w(t) = z + Re^{it}$.

(ii) Verify that, for all $0 < R \leq R^0$,

$$\int_{S(z;R)} \frac{f(w)}{w - z}\, dw = i \int_{-\pi}^{\pi} f(z + Re^{it})\, dt.$$

(iii) Using the Continuity Theorem 2.10.2, prove the following, which is known as *Cauchy's integral formula*:

$$f(z) = \frac{1}{2\pi i} \int_{\partial\Omega} \frac{f(w)}{w - z}\, dw.$$

Let $z^0 \in \Omega$ be fixed for the moment. Then, for every $z \in \Omega$ and $w \in \partial\Omega$,

$$\frac{1}{w - z} = \frac{1}{w - z^0} + \frac{z - z^0}{w - z^0}\,\frac{1}{w - z} = \sum_{0 \leq k < n} \frac{(z - z^0)^k}{(w - z^0)^{k+1}} + \left(\frac{z - z^0}{w - z^0}\right)^n \frac{1}{w - z}.$$

It follows that, for every $z \in \Omega$,

$$f(z) = \sum_{0 \leq k < n} (z - z^0)^k \frac{1}{2\pi i} \int_{\partial\Omega} \frac{f(w)}{(w - z^0)^{k+1}}\, dw + R_n(z),$$

with

$$R_n(z) = \frac{1}{2\pi i} \int_{\partial\Omega} \left(\frac{z - z^0}{w - z^0}\right)^n \frac{f(w)}{w - z}\, dw.$$

(iv) Note that $\partial\Omega$ is compact; use this fact to prove the following. There exist numbers $\rho > 0$, $0 < p < 1$ and $q > 0$ such that, for every $z \in \mathbf{C}$ with $|z - z^0| < \rho$ and all $w \in \partial\Omega$,

$$z \in \Omega; \qquad \left|\frac{z - z^0}{w - z^0}\right| \leq p; \qquad |w - z| \geq q.$$

(v) Prove that, for every $z \in \mathbf{C}$ with $|z - z^0| < \rho$, we have $\lim_{n\to\infty} R_n(z) = 0$; then conclude that

$$f(z) = \sum_{n\in\mathbf{N}_0} (z - z^0)^n \frac{1}{2\pi i} \int_{\partial\Omega} \frac{f(w)}{(w - z^0)^{n+1}}\, dw.$$

That is, the holomorphic function f is *complex-analytic*, which means that f can be written as a complex power series near $z^0 \in \Omega$. The differentiability of a power series on its disk of convergence gives the reverse of this assertion.

(vi) Use n-fold differentiation of the power series in part (v) to conclude that

$$f^{(n)}(z^0) = \frac{n!}{2\pi i} \int_{\partial\Omega} \frac{f(w)}{(w - z^0)^{n+1}}\, dw \qquad (n \in \mathbf{N}_0).$$

Exercise 8.13 (Fundamental Theorem of Algebra – sequel to Exercise 8.12). Let $p : \mathbf{C} \to \mathbf{C}$ be a complex polynomial function of degree $n \in \mathbf{N}$, that is $p(z) = \sum_{0\le k\le n} c_k z^k$ with $c_k \in \mathbf{C}$ and $c_n \neq 0$. Suppose that $p(z) \neq 0$, for all $z \in \mathbf{C}$. Prove that $\check{p} : \mathbf{C} \to \mathbf{C}$ is a nowhere vanishing complex polynomial function if

$$\check{p}(z) := \sum_{0\le k\le n} c_{n-k} z^k = z^n p\Big(\frac{1}{z}\Big), \qquad \text{and deduce} \qquad p(z) = z^n \check{p}\Big(\frac{1}{z}\Big).$$

Let $\Omega = \{ z \in \mathbf{C} \mid |z| < 1 \}$. Using Cauchy's integral formula from Exercise 8.12.(iii), the substitution $z = \frac{1}{w}$ and Cauchy's Integral Theorem 8.3.11, prove

$$0 \neq \frac{2\pi i}{p(0)} = \int_{\partial\Omega} \frac{1}{z\, p(z)}\, dz = \int_{\partial\Omega} \frac{1}{z^{n+1}\, \check{p}(\frac{1}{z})}\, dz = -\int_{\partial\Omega} \frac{w^{n-1}}{\check{p}(w)}\, dw = 0.$$

From this contradiction obtain the *Fundamental Theorem of Algebra*, which states that p must have a zero in $\mathbf{C}$ (see Example 8.11.5 and Exercise 3.48 for other proofs).

Exercise 8.14 (Equivalence of holomorphic and orientation-preserving conformal). Let $U \subset \mathbf{C}$ be open; we identify U with an open set in $\mathbf{R}^2$ via $U \ni z = x_1 + i x_2 \leftrightarrow x$. Let $f : U \to \mathbf{C}$ be a complex-valued function, to be identified with the vector field $f : U \to \mathbf{R}^2$. Prove that the following four assertions are equivalent.

(i) $f : U \to \mathbf{C}$ is a holomorphic function at z with $f'(z) \neq 0$.

(ii) $f : U \to \mathbf{R}^2$ satisfies the Cauchy–Riemann equation at x and $Df(x) \neq 0$.

(iii) $\det Df(x) > 0$ and there exists $R = R(f, x) \in \mathbf{SO}(2, \mathbf{R})$ with $Df(x) = \sqrt{\det Df(x)}\, R = |f'(z)|\, R$.

(iv) $\det Df(x) > 0$ and there exists $c = c(f, x) \in \mathbf{R}$ with $\langle Df(x)u, Df(x)v \rangle = c\,\langle u, v\rangle$, for all $u, v \in \mathbf{R}^2$. In other words, f is an orientation-preserving conformal mapping at x (see Exercise 5.29).

Hint: First prove by means of $Df(x)^t Df(x) = c\,I$ that $c = \det Df(x)$, and show that (iv) $\Longrightarrow$ (ii).

Exercise 8.15. Let $U \subset \mathbf{C}$ be open and let $f : U \to \mathbf{C}$ be holomorphic. We identify f with a C^1 vector field $f : U \to \mathbf{R}^2$.

(i) Prove that for every C^1 curve $\gamma : I \to U$

$$\text{length}\,(f(\gamma)) = \int_I |f'(\gamma(t))|\,|\gamma'(t)|\,dt.$$

(ii) Let $L \in \mathcal{J}(U)$, and assume that $f|_L$ is an injection with $f'(x_1 + i x_2) \neq 0$, if $x \in L$. Demonstrate that

$$\text{area}\,((f(L)) = \int_L |f'(x_1 + i x_2)|^2\,dx.$$

Exercise 8.16 (Winding number and Residue Theorem – sequel to Exercise 8.12). Let $f : \mathbf{R}^2 \setminus \{0\} \to \mathbf{R}^2$ be given by $f(x) = \frac{1}{\|x\|^2} Jx$, with J as in Formula (8.20). Let $\Omega \subset \mathbf{R}^2$ be as in Green's Integral Theorem, and $a \in \Omega$.

(i) Prove

$$\frac{1}{2\pi} \int_{\partial\Omega} \langle\, f(s - a), d_1 s\,\rangle = 1.$$

Hint: There are various conceivable methods:

(a) use Example 7.9.4;

(b) prove $f = \operatorname{grad} \arg$ with $\arg : \mathbf{R}^2 \setminus (\,]-\infty, 0] \times \{0\}) \to \mathbf{R}$ the argument function;

(c) use Kronecker's integral from Example 8.11.9.

Let $U \subset \mathbf{C}$ be open and $a \in U$, and let γ be a closed C^1 curve with $\operatorname{im}(\gamma) \subset U \setminus \{a\}$.

(ii) Prove that there exists a number $w = w(\gamma, a) \in \mathbf{Z}$, the *winding number* of γ about a, with

$$\frac{1}{2\pi i} \int_\gamma \frac{1}{z - a}\,dz = w(\gamma, a).$$

Hint: Reduce to part (i), in particular Kronecker's integral, or use the following argument. We may assume $\gamma : I = [0, 1] \to U \setminus \{a\}$, and we introduce

$$g(x) = \int_0^x \frac{\gamma'(t)}{\gamma(t) - a}\,dt \qquad (x \in I).$$

Then $\gamma' - (\gamma - a)g' = 0$ on I, and hence $((\gamma - a)e^{-g})' = 0$. As a result, for $x \in I$,

$$e^{g(x)} = \frac{\gamma(x) - a}{\gamma(0) - a}, \qquad \text{in particular} \qquad e^{g(1)} = 1,$$
$$\text{and so} \qquad g(1) = 2\pi i w.$$

(iii) Verify that *Cauchy's integral formula* from Exercise 8.12.(iii) under the present conditions takes the following form. For $f : U \to \mathbf{C}$ holomorphic,

$$w(\gamma, a) f(a) = \frac{1}{2\pi i} \int_\gamma \frac{f(z)}{z - a}\, dz.$$

(iv) Now assume γ in $U \setminus \{a\}$ to be homotopic with the mapping $\gamma_1 : t \mapsto a + re^{2\pi i t}$, where $r > 0$ has been chosen sufficiently small to ensure $\operatorname{im}(\gamma_1) \subset U$. Prove that $w(\gamma, a) = w(\gamma_1, a) = 1$.

Let $f : U \setminus \{a\} \to \mathbf{C}$ be holomorphic. The *residue* $\operatorname{Res}_{z=a} f(z)$ of f at a is the unique number $r \in \mathbf{C}$ such that $z \mapsto f(z) - \frac{r}{z-a}$ has an antiderivative on a sufficiently small neighborhood of a in $U \setminus \{a\}$.

(v) By means of part (ii), prove

$$\frac{1}{2\pi i} \int_\gamma f(z)\, dz = w(\gamma, a) \operatorname*{Res}_{z=a} f(z).$$

(vi) Assume $\operatorname{im}(\gamma) \subset U \setminus \{a_1, \dots, a_m\}$ and $f : U \setminus \{a_1, \dots, a_m\} \to \mathbf{C}$ to be holomorphic. Prove the following *Residue Theorem*:

$$\frac{1}{2\pi i} \int_\gamma f(z)\, dz = \sum_{1 \le i \le m} w(\gamma, a_i) \operatorname*{Res}_{z=a_i} f(z).$$

Background. In *complex analysis* methods are developed for the efficient calculation of residues.

Exercise 8.17 (Generalization of Cauchy's integral formula – sequel to Exercise 8.12 – needed for Exercise 8.18). Let $i = \sqrt{-1}$. Identify $x = (x_1, \dots, x_{2n}) \in \mathbf{R}^{2n}$ with $z = (z_1, \dots, z_n) \in \mathbf{C}^n$, where $z_j = x_{2j-1} + i\, x_{2j}$; while $\overline{z}_j = x_{2j-1} - i\, x_{2j}$, for $1 \le j \le n$. We then have the real $2n$-dimensional vector space $T_x\mathbf{R}^{2n} \simeq T_z\mathbf{C}^n$, with the partial differentiations D_j, for $1 \le j \le 2n$, as basis vectors (see Exercise 5.75). Let $(T_x\mathbf{R}^{2n})_{\mathbf{C}}$ be the complexification of $T_x\mathbf{R}^{2n}$.

(i) Show that the following vectors form a basis over $\mathbf{C}$ for $(T_x\mathbf{R}^{2n})_{\mathbf{C}}$, with $1 \le j \le n$:

$$(\star) \qquad \frac{\partial}{\partial z_j} = \frac{1}{2}(D_{2j-1} - i\, D_{2j}), \qquad \frac{\partial}{\partial \overline{z}_j} = \frac{1}{2}(D_{2j-1} + i\, D_{2j}).$$

As usual, we write $T_x^*\mathbf{R}^{2n}$ for the dual vector space over $\mathbf{R}$ of $T_x\mathbf{R}^{2n}$. Then the dx_j, for $1 \le j \le 2n$, are basis vectors for $T_x^*\mathbf{R}^{2n}$ over $\mathbf{R}$. Let $(T_x^*\mathbf{R}^{2n})_\mathbf{C}$ be the dual of $(T_x\mathbf{R}^{2n})_\mathbf{C}$.

(ii) Prove that the basis over $\mathbf{C}$ for $(T_x^*\mathbf{R}^{2n})_\mathbf{C}$, dual to that in $(\star)$, is given by

$$dz_j = dx_{2j-1} + i\,dx_{2j}, \qquad d\overline{z}_j = dx_{2j-1} - i\,dx_{2j} \qquad (1 \le j \le n).$$

(iii) Prove

$$dx_1 \wedge dx_2 \wedge \cdots \wedge dx_{2n-1} \wedge dx_{2n} = \Big(\frac{i}{2}\Big)^n dz_1 \wedge d\overline{z}_1 \wedge \cdots \wedge dz_n \wedge d\overline{z}_n.$$

Now let $f : \mathbf{C}^n \to \mathbf{C}$ be a C^1 function, and consider $df(x) \in (T_x^*\mathbf{R}^{2n})_\mathbf{C}$.

(iv) Show that

$$df = \sum_{1\le j\le n} \Big(\frac{\partial f}{\partial z_j} dz_j + \frac{\partial f}{\partial \overline{z}_j} d\overline{z}_j\Big) =: \frac{\partial f}{\partial z} dz + \frac{\partial f}{\partial \overline{z}} d\overline{z} =: \partial f + \overline{\partial} f;$$

that is,

$$d = \partial + \overline{\partial}.$$

The function f is said to be *holomorphic* or *complex-differentiable* if it satisfies the *Cauchy–Riemann equation*

$$\overline{\partial} f = 0, \qquad \text{that is,} \qquad i\,D_{2j-1} f = D_{2j} f \qquad (1 \le j \le n).$$

(v) Let $n = 1$, and assume $f : \mathbf{C} \to \mathbf{C}$ to be holomorphic. Prove

$$df = \frac{\partial f}{\partial z} dz = \partial f.$$

Use this to show that the differential 1-form $f\,dz$ is closed on $\mathbf{C}$ (this is where the restriction $n = 1$ is important). Next, let $a \in \mathbf{C}$. Prove, using $\frac{\partial}{\partial \overline{z}}\big(\frac{1}{z-a}\big) = 0$, that the following differential 1-form is closed on $\mathbf{C} \setminus \{a\}$:

$$\frac{f}{z-a}\,dz.$$

Let $a \in \mathbf{C}$ and assume $f : \mathbf{C} \to \mathbf{C}$ to be an arbitrary C^1 function.

(vi) Conclude that on $\mathbf{C} \setminus \{a\}$

$$d\Big(\frac{f}{z-a}\,dz\Big) = -\frac{1}{z-a}\frac{\partial f}{\partial \overline{z}}\,dz \wedge d\overline{z}.$$

Let $\Omega \subset \mathbf{C}$ be a bounded open subset having a C^1 boundary $\partial\Omega$ and lying at one side of $\partial\Omega$. Let $f : \Omega \to \mathbf{C}$ be a C^1 function such that f and the total derivative Df can be extended to continuous functions on $\overline{\Omega}$.

(vii) Conclude by part (iii), and using polar coordinates (r, α) for $z - a \in \mathbf{C}$, that

$$dz \wedge d\overline{z} = -2i\, dx_1 \wedge dx_2 = -2i\, r dr \wedge d\alpha.$$

Use this to prove that $z \mapsto \frac{1}{z-a}\frac{\partial f}{\partial \overline{z}}(z)$ is absolutely Riemann integrable over Ω.

(viii) Now prove, analogously to Exercise 8.12.(iii), the following generalization of *Cauchy's integral formula*:

$$f(a) = \frac{1}{2\pi i}\int_{\partial\Omega} \frac{f(z)}{z-a}\, dz + \frac{1}{2\pi i}\int_{\Omega} \frac{1}{z-a}\frac{\partial f}{\partial \overline{z}}(z)\, dz \wedge d\overline{z} \qquad (a \in \Omega).$$

Exercise 8.18 (Generalization of Cauchy's Integral Theorem 8.3.12 – sequel to Exercise 8.17). Let Φ_0 and $\Phi_1 : \mathbf{C} \to \mathbf{C}$ be two homotopic C^1 mappings. Suppose $f : \mathbf{C} \to \mathbf{C}$ is a holomorphic function and let $\omega = f\, dz$ be the corresponding closed differential 1-form on $\mathbf{C}$ as in Exercise 8.17.(v). Apply the Homotopy Lemma 8.9.5 to find a C^1 function g on $\mathbf{C}$ such that $\Phi_1^*\omega - \Phi_0^*\omega = dg$. Next suppose that $\gamma : [\,0, 1\,] \to \mathbf{C}$ is a closed C^1 curve, thus, in particular, $\gamma(0) = \gamma(1)$. Now prove the following generalization of Cauchy's Integral Theorem 8.3.12:

$$\int_{\Phi_1\circ\gamma} f\, dz = \int_{\Phi_0\circ\gamma} f\, dz.$$

Exercise 8.19 (Sequel to Exercise 6.57). In four steps we shall prove (see Exercise 6.60.(iii) for another demonstration), for $s \in \mathbf{C}$,

$$(\star) \qquad \int_{\mathbf{R}_+} x^{s-1} \left\{ \begin{matrix} \sin \\ \cos \end{matrix} \right. x\, dx = \Gamma(s) \left\{ \begin{matrix} \sin \\ \cos \end{matrix} \right. (s\frac{\pi}{2}) \qquad \begin{matrix} (-1 < \operatorname{Re} s < 1); \\ (0 < \operatorname{Re} s < 1). \end{matrix}$$

(i) Let $a > 0$ and apply Cauchy's Integral Theorem to the function $f(z) = e^{-z}z^{s-1}$ and the set Ω which equals the (open) square with vertices 0, a, $a + ia$ and ia, of which the vertex 0, however, is cut off by a small quarter-circle of radius ϵ with $0 < \epsilon < a$. Show, for $0 < \operatorname{Re} s < 1$,

$$0 = \int_\epsilon^a e^{-x}x^{s-1}\, dx + \int_a^{a+ia} f(z)\, dz + \int_{a+ia}^{ia} f(z)\, dz$$

$$+ \int_a^\epsilon e^{-iy}(iy)^{s-1}\, d(iy) + \int_{\frac{\pi}{2}}^0 e^{-\epsilon e^{i\phi}}\epsilon^{s-1}e^{i(s-1)\phi}\, d(\epsilon e^{i\phi}) =: \sum_{1\le j\le 5} I_j.$$

(ii) Verify

$$\begin{aligned}|I_2 + I_3| &\le e^{-a+|\operatorname{Im} s|\frac{\pi}{4}} \int_0^a (a^2 + y^2)^{\frac{\operatorname{Re} s-1}{2}}\, dy \\ &\quad + e^{|\operatorname{Im} s|\frac{\pi}{2}} \int_0^a e^{-x}(x^2 + a^2)^{\frac{\operatorname{Re} s-1}{2}}\, dx \\ &\le e^{-a+|\operatorname{Im} s|\frac{\pi}{4}} 2a^{\operatorname{Re} s} + e^{|\operatorname{Im} s|\frac{\pi}{2}} 2a^{\operatorname{Re} s-1}.\end{aligned}$$

Furthermore,

$$|I_5| \le \frac{\pi}{2} \epsilon^{\operatorname{Re} s} e^{|\operatorname{Im} s|\frac{\pi}{2}}.$$

(iii) Conclude by taking limits for $\epsilon \downarrow 0$ and $a \to \infty$, for $0 < \operatorname{Re} s < 1$, that

$$\Gamma(s) = \int_{\mathbf{R}_+} e^{-x} x^{s-1}\, dx = e^{is\frac{\pi}{2}} \int_{\mathbf{R}_+} e^{-iy} y^{s-1}\, dy.$$

(iv) If we carry out the same reasoning as in parts (i) – (iii) but using the square of vertices 0, a, $a - ia$ and $-ia$, the square again being indented at 0 by a quarter-circle of radius ϵ, we obtain

$$\Gamma(s) = e^{-is\frac{\pi}{2}} \int_{\mathbf{R}_+} e^{iy} y^{s-1}\, dy.$$

Deduce the formulae in $(\star)$ by addition and subtraction, for $0 < \operatorname{Re} s < 1$.

(v) Verify that the first equation in $(\star)$ is valid for $-1 < \operatorname{Re} s < 1$.

Exercise 8.20 (Asymptotics of Bessel function – sequel to Exercises 6.66 and 7.30). We employ the notations from Exercise 6.66. Let $\lambda > -\frac{1}{2}$. Under the substitution $v(x) = \sqrt{x}\, u(x)$ Bessel's equation takes the form

$$(\star) \qquad v''(x) + \Big(1 - \frac{4\lambda^2 - 1}{4x^2}\Big) v(x) = 0 \qquad (x \in \mathbf{R}_+).$$

Neglecting terms $\mathcal{O}(\frac{1}{x^2})$, for $x \to \infty$, we obtain the harmonic equation $w'' + w = 0$, with $w(x) = a\cos(x - \mu)$, for constants a and $\mu \in \mathbf{R}$, as the general solution. This makes it plausible that a solution v of the equation $(\star)$ has the form

$$v(x) = a\cos(x - \mu) + \mathcal{O}\Big(\frac{1}{x}\Big), \qquad x \to \infty.$$

Indeed, for the Bessel function J_λ we shall prove

$$(\star\star) \qquad J_\lambda(x) = \sqrt{\frac{2}{\pi}} \frac{1}{\sqrt{x}} \cos\Big(x - \frac{\pi}{2}\lambda - \frac{\pi}{4}\Big) + \mathcal{O}\Big(\frac{1}{x\sqrt{x}}\Big), \qquad x \to \infty.$$

(i) Verify

$$f_\lambda(x) = \int_{-1}^{1} e^{ixt}(1-t^2)^{\lambda-\frac{1}{2}}\,dt \qquad (x \in \mathbf{R}).$$

Let $Z = \mathbf{C} \setminus (\,]-\infty, -1\,] \cup [\,1, \infty\,[\,)$.

(ii) Check that

$$g : z \mapsto e^{ixz}(1-z^2)^{\lambda-\frac{1}{2}} : Z \to \mathbf{C}$$

is a well-defined holomorphic function if we require that $(1-z^2)^{\lambda-\frac{1}{2}} > 0$, for $z \in\,]-1, 1\,[$.

(iii) Let $a > 0$ and apply Cauchy's Integral Theorem to the function g and the set $\Omega \subset Z$ which equals the (open) rectangle with vertices -1, 1, $1+ia$ and $-1+ia$. Conclude, for $x \in \mathbf{R}_+$, that

$$\begin{aligned} 0 &= f_\lambda(x) + i\int_0^a e^{ix(1+iy)}(y^2-2iy)^{\lambda-\frac{1}{2}}\,dy \\ &\quad + i\int_a^0 e^{ix(-1+iy)}(y^2+2iy)^{\lambda-\frac{1}{2}}\,dy + R(a), \end{aligned}$$

where $\lim_{a\to\infty} R(a) = 0$. Verify that then

$$f_\lambda(x) = I_+(x) + I_-(x), \qquad I_\pm(x) = \pm i e^{\mp ix}\int_{\mathbf{R}_+} e^{-xy}(y^2 \pm 2iy)^{\lambda-\frac{1}{2}}\,dy.$$

(iv) Show

$$(y^2 \pm 2iy)^{\lambda-\frac{1}{2}} = (\pm 2i)^{\lambda-\frac{1}{2}} y^{\lambda-\frac{1}{2}} + \begin{cases} \mathcal{O}(y^{\lambda+\frac{1}{2}}) & (0 \le y < 1); \\ \mathcal{O}(y^{2\lambda-1}) & (1 \le y < \infty). \end{cases}$$

Prove that $I_\pm(x)$ then equals, for $x \to \infty$,

$$\begin{aligned} &\frac{1}{2}(\pm 2i)^{\lambda+\frac{1}{2}} e^{\mp ix}\int_{\mathbf{R}_+} e^{-xy} y^{\lambda-\frac{1}{2}}\,dy + \mathcal{O}\Big(\int_0^1 e^{-xy} y^{\lambda+\frac{1}{2}}\,dy\Big) \\ &\quad + \mathcal{O}\Big(\int_1^\infty e^{-xy} y^{2\lambda-1}\,dy\Big) \\ &= \frac{1}{2}\Big(\pm\frac{2i}{x}\Big)^{\lambda+\frac{1}{2}} e^{\mp ix}\Gamma\big(\lambda+\frac{1}{2}\big) + \mathcal{O}(x^{-\lambda-\frac{3}{2}}) + \mathcal{O}(e^{-x}). \end{aligned}$$

(v) Now prove ($\star\star$).

(vi) Conclude, by part (v) and Exercise 6.66.(viii), that the Hankel transform $\mathcal{H}_\lambda f$ of a function f from Exercises 6.102 and 7.30 is well-defined for f having the property that $r \mapsto \sqrt{r}\,f(r)$ is absolutely Riemann integrable over $\mathbf{R}_+$.

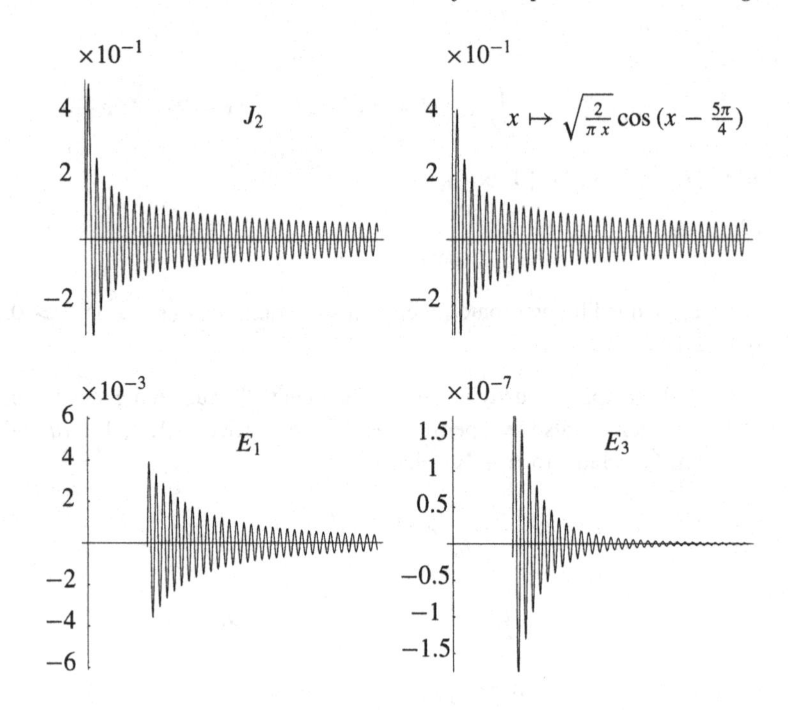

Illustration for Exercise 8.20: Asymptotics of Bessel function

$$E_1(x) = J_2(x) - \sqrt{\frac{2}{\pi x}} \cos\left(x - \frac{5\pi}{4}\right),$$

$$E_3(x) = J_2(x) - \sqrt{\frac{2}{\pi x}} \left(\left(1 - \frac{105}{128x^2}\right) \cos\left(x - \frac{5\pi}{4}\right) - \frac{15}{8x} \sin\left(x - \frac{5\pi}{4}\right)\right)$$

on $[\,50,\ 250\,]$

(vii) It is possible to formulate a stronger version of the result from part (iv). To show this, we write

$$I_\pm(x) = \frac{1}{2}(\pm 2i)^{\lambda+\frac{1}{2}} e^{\mp ix} \frac{1}{x^{\lambda+\frac{1}{2}}} \int_{\mathbf{R}_+} e^{-y} y^{\lambda-\frac{1}{2}} \left(1 \mp \frac{iy}{2x}\right)^{\lambda-\frac{1}{2}} dy.$$

Applying Taylor's formula for $z \mapsto (1+z)^{\lambda-\frac{1}{2}}$ at 0 with the remainder according to Lagrange (this is obtained from the integral formula for the remainder by application of the Intermediate Value Theorem 1.9.5), we find, for $m \in \mathbf{N}$,

$$
\begin{aligned}
&\Big(1 \mp \frac{iy}{2x}\Big)^{\lambda-\frac{1}{2}} \\
&= \sum_{0 \le n \le m} \binom{\lambda - \frac{1}{2}}{n} \Big(\mp\frac{iy}{2x}\Big)^n + \binom{\lambda-\frac{1}{2}}{m+1}\Big(\mp\frac{iy}{2x}\Big)^{m+1}\Big(1 \mp \frac{iy}{2x}t\Big)^{\lambda-\frac{3}{2}-m},
\end{aligned}
$$

for a $t \in [\,0, 1\,]$. We now note that $\big|1 \mp \frac{iy}{2x}t\big| \ge 1$ for the present x and y; and so the absolute value of the remainder, for $m \ge \lambda$ and those x and y, is dominated by

$$
\binom{\lambda-\frac{1}{2}}{m+1}\Big(\frac{y}{2x}\Big)^{m+1}.
$$

The integration with respect to y over $\mathbf{R}_+$ can now be carried out, and we find the following asymptotic expansion, for $x \to \infty$:

$$
J_\lambda(x) \sim \sqrt{\frac{2}{\pi}} \sum_{n \in \mathbf{N}_0} \frac{1}{n!\,2^n}\, \frac{\Gamma(\lambda+n+\frac{1}{2})}{\Gamma(\lambda-n+\frac{1}{2})}\, \frac{1}{x^{n+\frac{1}{2}}} \left\{ \begin{array}{l} (-1)^{\frac{n}{2}} \;\cos\Big(x - \dfrac{\pi}{2}\lambda - \dfrac{\pi}{4}\Big) \\[2ex] (-1)^{\frac{n+1}{2}} \sin\Big(x - \dfrac{\pi}{2}\lambda - \dfrac{\pi}{4}\Big) \end{array} \right. ,
$$

where we take the upper or the lower expression behind the brace according to whether n is even or odd, respectively. This asymptotic expansion can also be put into another form. To do so, we introduce

$$
\begin{aligned}
\phi(\lambda, x) &= x - \frac{\pi}{2}\lambda - \frac{\pi}{4}, \\
a(\lambda, n) &= \frac{(4\lambda^2 - 1^2)(4\lambda^2 - 3^2)\cdots((4\lambda^2 - (2n-1)^2)}{n!\,8^n}.
\end{aligned}
$$

One then has

$$
\begin{aligned}
J_\lambda(x) \sim \sqrt{\frac{2}{\pi\, x}}\bigg(&\cos\phi(\lambda, x) \sum_{n \in \mathbf{N}_0} (-1)^n \frac{a(\lambda, 2n)}{x^{2n}} \\
&- \sin\phi(\lambda, x) \sum_{n \in \mathbf{N}_0} (-1)^n \frac{a(\lambda, 2n+1)}{x^{2n+1}}\bigg), \qquad x \to \infty.
\end{aligned}
$$

Exercise 8.21 (Laplace's formula for Legendre polynomial – sequel to Exercises 0.9 and 8.12). Let $z \in \mathbf{C}$ and choose $\Omega \subset \mathbf{C}$ such that the conditions of Cauchy's Integral Theorem 8.3.11 are met, and such that $z \in \Omega$.

(i) Use Exercise 8.12.(vi) to prove the following, known as *Schläfli's formula* for the Legendre polynomial P_l, for $l \in \mathbf{N}_0$, from Exercise 0.4:

$$
P_l(z) = \frac{1}{2\pi i 2^l} \int_{\partial\Omega} \frac{(w^2-1)^l}{(w-z)^{l+1}}\, dw.
$$

Now let $x \in \mathbf{R}$ with $|x| > 1$ and choose $\Omega = \{ w \in \mathbf{C} \mid |w - x| < \sqrt{x^2 - 1} \}$.

(ii) Check that $t \mapsto w(t) = x + \sqrt{x^2 - 1}\, e^{it}$ is a parametrization of $\partial\Omega$. Then prove the following, known as *Laplace's formula*:

$$P_l(x) = \frac{1}{\pi} \int_0^\pi (x + \sqrt{x^2 - 1} \cos t)^l \, dt \qquad (x \in \mathbf{R},\ |x| > 1).$$

The choice $a = x$ and $b = \sqrt{x^2 - 1}$ leads to a special case of the integral studied in Exercise 0.9.

(iii) Prove by using that exercise

$$P_l(x) = \frac{1}{\pi} \int_0^\pi (x - \sqrt{x^2 - 1} \cos t)^{-l-1} \, dt \qquad (x \in \mathbf{R},\ |x| > 1).$$

Also prove, using the identity $P_l(x) = (-1)^l P_l(-x)$, for $x \in \mathbf{R}$ with $|x| > 1$,

$$P_l(x) = \frac{1}{\pi} \int_0^\pi (x - \sqrt{x^2 - 1} \cos t)^l \, dt = \frac{1}{\pi} \int_0^\pi (x + \sqrt{x^2 - 1} \cos t)^{-l-1} \, dt.$$

(iv) Analogously prove

$$P_l(x) = \frac{1}{\pi} \int_0^\pi (x + i\sqrt{1 - x^2} \cos t)^l \, dt \qquad (x \in \mathbf{R},\ |x| < 1).$$

Conclude that the zonal spherical function Y_l^0 from Exercise 3.17 satisfies

$$Y_l^0(\alpha, \theta) = \frac{1}{\pi} \int_0^\pi (\sin\theta + i \cos\theta \cos t)^l \, dt \qquad (|\alpha| < \pi,\ |\theta| < \frac{\pi}{2}).$$

Exercise 8.22 (Real and imaginary parts of a holomorphic function are harmonic – sequel to Exercise 8.12 – needed for Exercise 8.23). Let f and Ω be as in Exercise 8.12, and consider f_1 and f_2, with $f = f_1 + i f_2$, as functions on an open subset of $\mathbf{R}^2$.

(i) Prove by means of the Cauchy–Riemann equation that f_1 and f_2 are harmonic functions. Check that the vector fields $\operatorname{grad} f_1$ and $\operatorname{grad} f_2$ on Ω are both harmonic, and mutually orthogonal at every point of Ω.

Conversely, let $f_1 \in C^2(\Omega)$ be given, with $\Omega \subset \mathbf{R}^2$ open. We want to find $f_2 \in C^2(\Omega)$ such that $f := f_1 + i f_2$ is a complex-analytic function on $\Omega \subset \mathbf{C}$.

(ii) Prove that the Cauchy–Riemann equation for f gives the following condition on f_2:

$$\operatorname{grad} f_2 = J \operatorname{grad} f_1.$$

Show that the integrability condition $\operatorname{curl} \operatorname{grad} f_2 = 0$ implies the identity $\operatorname{div} J^t J \operatorname{grad} f_1 = \Delta f_1 = 0$, that is, the function f_1 has to be harmonic on Ω.

(iii) Assume f_1 to be harmonic on Ω and Ω to be simply connected. Check that a scalar potential $f_2 : \Omega \to \mathbf{R}$ exists for the vector field J grad f_1, and conclude that the f thus constructed has the desired property.

Exercise 8.23 (Poisson's integral formula and Schwarz' Theorem – sequel to Exercise 8.22). Let $\Omega = \{ z \in \mathbf{C} \mid |z| < 1 \}$ and define, for every $z \in \Omega$,

$$\Psi_z : \overline{\Omega} \to \mathbf{C} \qquad \text{by} \qquad \Psi_z(w) = \frac{w+z}{\overline{z}w+1}.$$

(i) Verify that for all $z \in \Omega$ the mapping $\Psi_z : \Omega \to \Omega$ is a C^1 diffeomorphism with inverse Ψ_{-z}; and also that $\Psi_z : \partial\Omega \to \partial\Omega$. Prove that we have $\lim_{z\in\Omega,\, z\to e^{i\alpha}} \Psi_z(e^{i\beta}) = e^{i\alpha}$, for all $\alpha, \beta \in \,]-\pi, \pi\,]$.

Now let $h \in C(\partial\Omega)$ be given, and define the *Poisson integral* $\mathcal{P}h : \Omega \to \mathbf{R}$ of h by

$$(\mathcal{P}h)(z) = \frac{1}{2\pi}\int_{-\pi}^{\pi} h(\Psi_z(e^{i\beta}))\, d\beta.$$

(ii) Prove by part (i) and Arzelà's Dominated Convergence Theorem 6.12.3 that

$$\lim_{z\in\Omega,\, z\to e^{i\alpha}} (\mathcal{P}h)(z) = h(e^{i\alpha}).$$

(iii) For all $z \in \Omega$, make the (z-dependent) substitution of variables on $]-\pi, \pi\,]$ given by $\alpha = \alpha(\beta)$, with $e^{i\alpha} = \Psi_z(e^{i\beta})$, and show that

$$\frac{d\beta}{d\alpha}(\alpha) = \frac{1-|z|^2}{|e^{i\alpha}-z|^2}.$$

Use this to derive the following, known as *Poisson's integral formula*:

$$(\mathcal{P}h)(z) = \frac{1-|z|^2}{2\pi}\int_{-\pi}^{\pi} \frac{h(e^{i\alpha})}{|e^{i\alpha}-z|^2}\, d\alpha.$$

(iv) Demonstrate that

$$\frac{1-|z|^2}{|e^{i\alpha}-z|^2} = \mathrm{Re}\left(\frac{e^{i\alpha}+z}{e^{i\alpha}-z}\right);$$

and conclude by Exercise 8.22 that $\mathcal{P}h$ is a harmonic function on Ω.

Background. On $\Omega = \{ x \in \mathbf{R}^2 \mid \|x\| < 1 \}$ a solution f of the Dirichlet problem $\Delta f = 0$, with $f|_{\partial\Omega} = h$, is obviously given by $f = \mathcal{P}h$, where (compare with Exercise 7.70.(v))

$$(\mathcal{P}h)(x) = \frac{1 - \|x\|^2}{|S^1|} \int_{S^1} \frac{h(y)}{\|x-y\|^2}\, d_1 y \qquad (x \in \Omega).$$

Part (ii) now tells us that $(\mathcal{P}h)(x)$ converges to $h(y)$, for $x \in \Omega$ approaching $y \in \partial\Omega$; this is *Schwarz' Theorem* (see Exercise 7.70.(vii)).

Exercise 8.24 (Sequel to Exercise 7.47). Demonstrate that the proof in that exercise comes down to the fact that f is solenoidal on $\mathbf{R}^n \setminus \{0\}$.

Exercise 8.25 (Sequel to Exercise 5.80 – needed for Exercises 8.26 and 8.29). Assume that f and $g : \mathbf{R}^3 \to \mathbf{R}^3$ both are C^2 vector fields.

(i) Prove, analogously to Grassmann's identity from Exercise 5.26.(ii),

$$\nabla \times (\nabla \times f) = \nabla \langle \nabla, f \rangle - \langle \nabla, \nabla \rangle f.$$

Conclude that

$$\operatorname{curl}(\operatorname{curl} f) = \operatorname{grad}(\operatorname{div} f) - \Delta f;$$

here the Laplacian Δ acts by components on f. Deduce that the component functions of a harmonic vector field on $\mathbf{R}^3$ are harmonic functions.

(ii) Prove from the antisymmetry of the cross product operator with respect to the inner product, that

$$\langle \nabla, f \times g \rangle = \langle \nabla \times f, g \rangle - \langle \nabla \times g, f \rangle,$$

and conclude that (see Exercise 8.39.(iv) for a different proof)

$$\operatorname{div}(f \times g) = \langle \operatorname{curl} f, g \rangle - \langle f, \operatorname{curl} g \rangle.$$

(iii) Prove, analogously to Grassmann's identity,

$$\begin{aligned} \nabla \times (f \times g) &= \langle \nabla, g \rangle f - \langle \nabla, f \rangle g \\ &= f \langle \nabla, g \rangle + \langle g, \nabla \rangle f - g \langle \nabla, f \rangle - \langle f, \nabla \rangle g, \end{aligned}$$

and conclude, using the formula from Exercise 5.80 for the commutator $[\cdot, \cdot]$ (for a different proof see Exercise 8.39.(vi)) that

$$\begin{aligned} \operatorname{curl}(f \times g) &= (\operatorname{div} g) f + \langle g, \operatorname{grad} \rangle f - (\operatorname{div} f)\, g - \langle f, \operatorname{grad} \rangle\, g \\ &= (\operatorname{div} g) f - (\operatorname{div} f)\, g - [f, g]. \end{aligned}$$

Here the differential operator $\langle g, \operatorname{grad} \rangle = g_1 D_1 + \cdots + g_3 D_3$ acts by components on the vector field f.

(iv) Let $\Omega \subset \mathbf{R}^3$ be as in Theorem 7.6.1. By means of part (ii), prove

$$\int_\Omega (\langle \operatorname{curl} f, g \rangle - \langle f, \operatorname{curl} g \rangle)(x)\,dx = \int_{\partial\Omega} \langle f \times g, \nu \rangle(y)\,d_2y$$
$$= \int_{\partial\Omega} \langle f, g \times \nu \rangle(y)\,d_2y.$$

Write $V_0^k(\Omega)$, with $k \in \mathbf{N}_0$, for the linear space of the C^k vector fields $f : \Omega \to \mathbf{R}^3$ satisfying $\operatorname{supp}(f) \subset \Omega$.

(v) Deduce that $\operatorname{curl} : V_0^1(\Omega) \to V_0^0(\Omega)$ is a self-adjoint linear operator with respect to the integral inner product on $V_0^0(\Omega)$, that is,

$$\langle \operatorname{curl} f, g \rangle = \langle f, \operatorname{curl} g \rangle := \int_\Omega \langle f, \operatorname{curl} g \rangle(x)\,dx \qquad (f, g \in V_0^1(\Omega)).$$

(vi) Show (see Exercise 8.39.(viii) for another proof)

$$\operatorname{grad}\langle f, g\rangle = \langle f, \operatorname{grad} \rangle g + \langle g, \operatorname{grad} \rangle f + f \times \operatorname{curl} g + g \times \operatorname{curl} f.$$

Exercise 8.26 (Maxwell's equations – sequel to Exercise 8.25 – needed for Exercises 8.27, 8.28, 8.31 and 8.33). In the theory of *electromagnetism* three time-dependent C^1 vector fields on $\mathbf{R}^3$ play a role:

the *electric field*	E	$: \mathbf{R}^3 \times \mathbf{R} \to \mathbf{R}^3,$	$(x, t) \mapsto E(x, t);$
the *magnetic field*	B	$: \mathbf{R}^3 \times \mathbf{R} \to \mathbf{R}^3,$	$(x, t) \mapsto B(x, t);$
the *current (density)*	j	$: \mathbf{R}^3 \times \mathbf{R} \to \mathbf{R}^3,$	$(x, t) \mapsto j(x, t);$

together with a time-dependent C^1 function on $\mathbf{R}^3$:

the *charge (density)* $\rho : \mathbf{R}^3 \times \mathbf{R} \to \mathbf{R}, \qquad (x, t) \mapsto \rho(x, t).$

These entities are mutually correlated by *Maxwell's equations*, which in addition contain several constants with physical meaning. If these constants are equated to 1 for simplicity, the equations read, in the absence of matter,

$$\langle \nabla, E \rangle = \rho, \qquad \nabla \times E = -\frac{\partial B}{\partial t}, \qquad \langle \nabla, B \rangle = 0, \qquad \nabla \times B = j + \frac{\partial E}{\partial t}.$$

Here the divergence $\langle \nabla, \cdot \rangle$ and the curl $\nabla \times$ are calculated with respect to the variable in $\mathbf{R}^3$.

(i) Prove that $\left\langle \nabla, j + \frac{\partial E}{\partial t} \right\rangle = 0$, and that this leads to the *continuity equation*

$$\langle \nabla, j \rangle + \frac{\partial \rho}{\partial t} = 0.$$

Let $\Omega \subset \mathbf{R}^3$ be an open set as in Theorem 7.6.1, and let $\Xi \subset \mathbf{R}^3$ be as in Stokes' Integral Theorem 8.4.4, that is, Ξ is an oriented surface having the closed curve $\partial \Xi$ with corresponding orientation as its boundary. Now prove the following assertions, under the assumption that the conditions of the theorems used are met.

(ii) **(Gauss' law).** The flux of E across the closed surface $\partial\Omega$ equals the charge inside Ω, that is

$$\int_{\partial\Omega} \langle E(y,t),\, d_2 y \rangle = \int_{\Omega} \rho(x,t)\, dx.$$

(iii) **(Faraday's law).** The circulation of E along the closed curve $\partial \Xi$ equals the negative of the rate of change of the flux of B across the surface Ξ, that is

$$\int_{\partial \Xi} \langle E(s,t),\, d_1 s \rangle = -\frac{\partial}{\partial t} \int_{\Xi} \langle B(y,t),\, d_2 y \rangle.$$

(iv) **(Absence of magnetic monopoles).** The flux of B across the closed surface $\partial\Omega$ vanishes, that is

$$\int_{\partial\Omega} \langle B(y,t),\, d_2 y \rangle = 0.$$

(v) **(Ampère–Maxwell law).** The circulation of B along the closed curve $\partial \Xi$ equals the flux of j across the surface Ξ plus the rate of change of the flux of E across the surface Ξ, that is

$$\int_{\partial \Xi} \langle B(s,t),\, d_1 s \rangle = \int_{\Xi} \langle j(y,t),\, d_2 y \rangle + \frac{\partial}{\partial t} \int_{\Xi} \langle E(y,t),\, d_2 y \rangle.$$

(vi) **(Law of conservation of charge).** The flux of j across the closed surface $\partial\Omega$ equals the negative of the rate of change of the charge inside Ω, that is

$$\int_{\partial\Omega} \langle j(y,t),\, d_2 y \rangle = -\frac{\partial}{\partial t} \int_{\Omega} \rho(x)\, dx.$$

The *electromagnetic energy* F and the *Poynting vector field* P are defined by, respectively,

$$F = \frac{1}{2}\langle E, E \rangle + \frac{1}{2}\langle B, B \rangle \qquad \text{and} \qquad P = E \times B.$$

(vii) **(Law of conservation of energy).** Prove by Exercise 8.25.(ii)

$$-\frac{\partial}{\partial t}\int_{\Omega} F(x,t)\,dx = \int_{\partial\Omega}\langle\, P(y,t),\, d_2y\,\rangle + \int_{\Omega}\langle\, E(x,t),\ j(x,t)\,\rangle\,dx,$$

that is, the flux of P across the closed surface $\partial\Omega$ equals the fraction due to dissipation across $\partial\Omega$ of minus the rate of change of the energy inside Ω.

We speak of *Maxwell's equations in vacuum* if $\rho = 0$ and $j = 0$, that is, if

$$\langle\, \nabla, E\,\rangle = 0, \qquad \langle\, \nabla, B\,\rangle = 0, \qquad \nabla\times E = -\frac{\partial B}{\partial t}, \qquad \nabla\times B = \frac{\partial E}{\partial t}.$$

Assume that E and B both are C^2 vector fields.

(viii) Prove by Exercise 8.25.(i) that in this case

$$\nabla\times(\nabla\times E) = -\Delta E, \qquad \nabla\times(\nabla\times B) = -\Delta B.$$

(ix) Prove

$$\Box E = 0, \qquad \Box B = 0, \qquad \text{where} \qquad \Box := D_t^2 - \Delta_x := D_t^2 - \sum_{1\le j\le 3} D_j^2.$$

That is, both E and B obey the *wave equation* for a time-dependent vector field G on $\mathbf{R}^3$

$$\Box\, G_i(x,t) = 0 \qquad (1 \le i \le 3,\ (x,t)\in \mathbf{R}^3\times\mathbf{R}).$$

Background. This prediction, in 1864/5, on theoretical grounds of the existence of *electromagnetic waves in vacuum* is one of the great triumphs of Maxwell's theory. The existence of radio waves was experimentally verified by Hertz in 1887.

Exercise 8.27 (Sequel to Exercise 8.26). Use Exercise 8.26.(ii) to give another proof of parts (i) and (ii) from Exercise 7.33.
Hint: Let $\partial\Omega$ in Exercise 8.26.(ii) be a straight circular cylinder with axis perpendicular to the plane $\{\, x\in\mathbf{R}^3 \mid x_1 = 0\,\}$, in Exercise 7.33.(i), or coinciding with the x_3-axis, in the case of Exercise 7.33.(ii).

Exercise 8.28 (Maxwell's equations: time-independent case – sequel to Exercises 7.67 and 8.26 – needed for Exercise 8.29 and 8.51). Under the assumption that the vector fields E and B on $\mathbf{R}^3$ are time-independent, one obtains Maxwell's laws in the following form:

$$\langle\, \nabla, E\,\rangle = \rho, \qquad \nabla\times E = 0, \qquad \langle\, \nabla, B\,\rangle = 0, \qquad \nabla\times B = j.$$

In this case E is curl-free and B is divergence-free on $\mathbf{R}^3$. Therefore, one may try to find solutions E and B of the form

$$E = -\nabla\phi, \qquad B = \nabla \times A,$$

with the function $\phi : \mathbf{R}^3 \to \mathbf{R}$ a *scalar potential* for E, and the vector field $A : \mathbf{R}^3 \to \mathbf{R}^3$ a *vector potential* for B. (In physics, the minus sign for $\nabla\phi$ is customary.) To limit the analytical complications we assume that $\rho \in C_c^2(\mathbf{R}^3)$ and $j \in C_c^2(\mathbf{R}^3, \mathbf{R}^3)$.

(i) Verify that ϕ has to satisfy *Poisson's equation*

$$(\star) \qquad \Delta\phi = -\rho,$$

while quite obviously $\nabla \times (\nabla\phi) = 0$.

(ii) Demonstrate that A has to satisfy $\nabla \times (\nabla \times A) = j$, while naturally we have $\langle \nabla, \nabla \times A \rangle = 0$.

(iii) Use Exercise 8.25.(i) to prove that, in addition, under the *Coulomb gauge condition*

$$\langle \nabla, A \rangle = 0,$$

the vector potential A has to satisfy Poisson's equation by components

$$(\star\star) \qquad \Delta A = -j.$$

We now inquire about solutions ϕ of $(\star)$ and A of $(\star\star)$ that satisfy the following *boundary condition at infinity*:

$$\lim_{\|x\|\to\infty} \phi(x) = 0, \qquad \lim_{\|x\|\to\infty} A(x) = 0.$$

(iv) Apply Exercise 7.67.(vii) and conclude (note that, contrary to our usual conventions for potentials, the minus sign is missing; this is because in electromagnetism the forces between like charges are repulsive)

$$\phi(x) = \frac{1}{4\pi} \int_{\mathbf{R}^3} \frac{\rho(x')}{\|x - x'\|}\, dx' \qquad (x \in \mathbf{R}^3).$$

Verify that now the electric field E is described by *Coulomb's law*

$$E(x) = \frac{1}{4\pi} \int_{\mathbf{R}^3} \frac{\rho(x')}{\|x - x'\|^3}(x - x')\, dx' \qquad (x \in \mathbf{R}^3).$$

(v) Prove, in similar fashion as in (iv),

$$A(x) = \frac{1}{4\pi} \int_{\mathbf{R}^3} \frac{1}{\|x - x'\|} j(x')\, dx' \qquad (x \in \mathbf{R}^3).$$

Verify that now the magnetic field B is described by the *Biot–Savart law*

$$B(x) = \frac{1}{4\pi} \int_{\mathbf{R}^3} \frac{1}{\|x - x'\|^3} j(x') \times (x - x')\, dx' \qquad (x \in \mathbf{R}^3).$$

(vi) Verify that the Coulomb gauge condition $\langle \nabla, A \rangle = 0$ is met.
Hint: Use Corollary 7.6.2 and the continuity equation $\langle \nabla, j \rangle = 0$ (see Exercise 8.26.(i)).

Exercise 8.29 (Helmholtz–Weyl decomposition – sequel to Exercises 8.25 and 8.28 – needed for Exercise 8.30). Let N be the Newton vector field on $\mathbf{R}^n$ from Example 7.8.4, and let $*$ be the convolution from Example 6.11.5. Demonstrate that the results from Exercise 8.28.(iv) and (v) can be generalized as follows.

(i) A C^3 vector field f on $\mathbf{R}^n$ with $Af = 0$ is uniquely determined by $\operatorname{div} f$, if this function has compact support on $\mathbf{R}^n$, via

$$f = (\operatorname{div} f) * N.$$

Here the integration is carried out by components of N.

(ii) A C^3 vector field f on $\mathbf{R}^n$ with $\operatorname{div} f = 0$ is uniquely determined by Af, if this vector field has compact support on $\mathbf{R}^n$, via

$$f = (Af) * N;$$

in more explicit notation, for $1 \leq i \leq n$ and $x \in \mathbf{R}^n$,

$$f_i(x) = \sum_{1 \leq j \leq n} \int_{\mathbf{R}^n} (D_j f_i - D_i f_j)(x')\, N_j(x - x')\, dx'.$$

(iii) Let f be a C^2 vector field on $\mathbf{R}^3$ with compact support. Verify there exist a C^1 function $g : \mathbf{R}^3 \to \mathbf{R}$ and a C^1 vector field $h : \mathbf{R}^3 \to \mathbf{R}^3$ such that we have the following *Helmholtz–Weyl decomposition*:

$$f = \operatorname{grad} g + \operatorname{curl} h.$$

Prove that g and h are solutions if

$$g(x) = \frac{1}{4\pi} \int_{\mathbf{R}^3} \frac{\langle f(x'), (x - x') \rangle}{\|x - x'\|^3}\, dx',$$

$$h(x) = \frac{1}{4\pi} \int_{\mathbf{R}^3} \frac{1}{\|x - x'\|^3} f(x') \times (x - x')\, dx'.$$

Deduce $f = \operatorname{grad} g + f_2$ with f_2 divergence-free on $\mathbf{R}^3$.
Hint: Write $f = \Delta(f * p)$, where the actions of convolution and Δ are according to components of f and $p(x) = -\frac{1}{4\pi}\frac{1}{\|x\|}$, for $x \in \mathbf{R}^3 \setminus \{0\}$, and use the identity from Exercise 8.25.(i).

Exercise 8.30 (Hodge decomposition – sequel to Exercise 8.29). We want to determine conditions for the uniqueness of the summands f_1 and f_2 occurring in the decomposition $f = f_1 + f_2 = \operatorname{grad} g + f_2$ from Exercise 8.29.(iii). And we would like to generalize this decomposition to $\mathbf{R}^n$. Therefore we consider a set $U \subset \mathbf{R}^n$ and vector fields f_1 and $f_2 : U \to \mathbf{R}^n$ satisfying the conditions from Theorem 7.6.1. Moreover, we assume that f_1 possesses a potential g on U, that f_2 is divergence-free on U, and that f_2 is parallel to ∂U, which means $\langle f_2, \nu \rangle(y) = 0$ for $y \in \partial U$, with ν as in Theorem 7.6.1.

(i) Prove $\operatorname{div}(g f_2) = \langle f_1, f_2 \rangle$ and use Gauss' Divergence Theorem 7.8.5 to conclude that

$$\int_U \langle f_1, f_2 \rangle(x)\, dx = 0.$$

(ii) Now assume $\widetilde{f_1}$ and $\widetilde{f_2}$ satisfy the same conditions as f_1 and f_2, respectively, and $f_1 + f_2 = \widetilde{f_1} + \widetilde{f_2}$. Prove $\int_U \|(f_1 - \widetilde{f_1})(x)\|^2\, dx = 0$, and deduce that $f_1 = \widetilde{f_1}$ and $f_2 = \widetilde{f_2}$ on U.

As to the existence of f_1 and f_2, we note that $f = \operatorname{grad} g + f_2$, with f_2 divergence-free on U and parallel to ∂U, implies $\operatorname{div} f = \operatorname{div} \operatorname{grad} g = \Delta g$ on U as well as $\langle f, \nu \rangle = \langle \operatorname{grad} g, \nu \rangle = \frac{\partial g}{\partial \nu}$ on ∂U. Given a C^1 vector field f on U it is therefore sufficient to determine a C^2 function g on U with

$$(\star) \qquad \Delta g = \operatorname{div} f \quad \text{on} \quad U, \qquad \frac{\partial g}{\partial \nu} = \langle f, \nu \rangle \quad \text{on} \quad \partial U,$$

where we also need that $\frac{\partial g}{\partial \nu}$ is well-defined on ∂U. Indeed, $f_1 = \operatorname{grad} g$ and $f_2 = f - \operatorname{grad} g$ then form a solution. A partial differential equation, together with a boundary condition

$$\Delta g = p \quad \text{on} \quad U, \qquad \frac{\partial g}{\partial \nu} = q \quad \text{on} \quad \partial U,$$

for given functions p and q, is said to be a *Neumann problem* on U.

(iii) Using Green's first identity, verify that the following condition is necessary for the solvability of the Neumann problem:

$$\int_U p(x)\, dx = \int_{\partial U} q(y)\, d_{n-1}y.$$

(iv) Verify that the condition from part (iii) is satisfied in our problem ($\star$).

We state without proof that, for sufficiently well-behaved ∂U, the condition from part (iii) is also sufficient for the solution of the Neumann problem.

(v) Assume the vector field f on U satisfies the integrability conditions. Prove that f_2 is a harmonic vector field on U, and that we have the direct sum decomposition

$$f = \operatorname{grad} g \oplus f_2 \qquad \text{with } f_2 \text{ harmonic on } U \text{ and parallel to } \partial U.$$

Assume $n = 3$. Let $\omega = \flat_1 f$ be the differential 1-form on U associated with f according to Example 8.8.2, and similarly ω_1 with f_1, and ω_2 with f_2.

(vi) Show ω_1 to be exact. Assume ω is closed. Then the cohomology class of ω in the first de Rham cohomology $H^1(U)$ has a harmonic representative, namely ω_2, satisfying

$$[\omega] = [\omega_2] \in H^1(U), \qquad d\omega_2 = 0, \qquad d^*\omega_2 = 0.$$

Here the *Hodge operator* $* : \Omega^1(U) \to \Omega^2(U)$ is defined by $*\flat_1 = \flat_2$, furthermore $* : \Omega^3(U) \to \Omega^0(U)$ by $*(f\,dx) = f$, and finally $d^* = *d* : \Omega^1(U) \to \Omega^0(U)$. (A more intrinsic definition is possible but is not discussed here for lack of space.)

Background. The sum decomposition of the closed form $\omega = \omega_1 + \omega_2$ into an exact form ω_1 and a harmonic form ω_2 is called a *Hodge decomposition*[2] of ω. It is used to investigate under what conditions on U the de Rham cohomology $H^k(U)$ is a finite-dimensional vector space, for $k \in \mathbf{N}_0$.

Exercise 8.31 (Maxwell's equations in covariant form – sequel to Exercise 8.26). We employ the notation, and make the assumptions, from Exercise 8.26. Note that $E(\cdot, t)$ and $B(\cdot, t)$ both are C^1 vector fields on $\mathbf{R}^3$ dependent on a parameter $t \in \mathbf{R}$. In this exercise, the operators $\flat$, curl, div, grad, Δ and the differential form $dx = dx_1 \wedge dx_2 \wedge dx_3$ are associated with $\mathbf{R}^3$. The operators d and $D_0 := \frac{\partial}{\partial t}$ are associated with $\mathbf{R}^4$. Define, for $(x, t) \in \mathbf{R}^4$,

$$\mathcal{E}(x,t) = \flat_1(E(\cdot,t))(x) \in \Omega^1(\mathbf{R}^4), \qquad \mathcal{B}(x,t) = \flat_2(B(\cdot,t))(x) \in \Omega^2(\mathbf{R}^4).$$

(i) Taking the indices i modulo 3, verify

$$\mathcal{E} = \sum_{1 \le i \le 3} E_i\,dx_i, \qquad \mathcal{B} = \sum_{1 \le i \le 3} B_i\,dx_{i+1} \wedge dx_{i+2}.$$

[2] See for more details in the case of $n = 3$: Cantarella, J., DeTurck, D., Gluck, H.: Vector calculus and the topology of domains in 3-space. Amer. Math. Monthly 109 (2002), 409 – 442.

Introduce the *Faraday form*

$$\mathcal{F} = \mathcal{E} \wedge dt + \mathcal{B} \in \Omega^2(\mathbf{R}^4).$$

(ii) Demonstrate, using Formula (8.54),

$$d\mathcal{F} = d\mathcal{E} \wedge dt + d\mathcal{B} = \flat_2(\operatorname{curl} E + D_0 B) \wedge dt + (\operatorname{div} B)\, dx \in \Omega^3(\mathbf{R}^4). \quad (1)$$

Define the *Hodge operator* $* \in \operatorname{Lin}(\Omega^2(\mathbf{R}^4), \Omega^2(\mathbf{R}^4))$ by, for $1 \leq i \leq 3$,

$$*(dx_{i+1} \wedge dx_{i+2}) = dx_i \wedge dt, \qquad *(dx_i \wedge dt) = -dx_{i+1} \wedge dx_{i+2}.$$

(More intrinsic definitions are possible but are not discussed here for lack of space.) Further, introduce

$$\mathcal{D} = \flat_2 E \in \Omega^2(\mathbf{R}^4), \qquad \mathcal{H} = \flat_1 B \in \Omega^1(\mathbf{R}^4).$$

(iii) Verify

$$*\mathcal{F} = \mathcal{H} \wedge dt - \mathcal{D} \in \Omega^2(\mathbf{R}^4). \quad (2)$$

Let

$$\mathcal{J} = -\rho\, dt + \flat_1 j \in \Omega^1(\mathbf{R}^4).$$

Introduce the *Hodge operator* $* \in \operatorname{Lin}(\Omega^1(\mathbf{R}^4), \Omega^3(\mathbf{R}^4))$ by

$$*(dx_i) = dx_{i+1} \wedge dx_{i+2} \wedge dt, \qquad *(dt) = dx.$$

Then $*$ is a bijection since it takes a basis into a basis. Therefore, define $* \in \operatorname{Lin}(\Omega^3(\mathbf{R}^4), \Omega^1(\mathbf{R}^4))$ as the inverse of the mapping just defined.

(iv) Now prove

$$\begin{aligned} d(*\mathcal{F}) &= d\mathcal{H} \wedge dt - d\mathcal{D} = \flat_2(\operatorname{curl} B - D_0 B) \wedge dt - (\operatorname{div} E)\, dx \\ &= *\mathcal{J} \in \Omega^3(\mathbf{R}^4). \end{aligned} \quad (3)$$

Thus, using (1) and (3), one may formulate Maxwell's equations as the following system of equations for the Faraday form on $\mathbf{R}^4$:

$$d\mathcal{F} = 0 \qquad \text{and} \qquad d^*\mathcal{F} = \mathcal{J},$$

where

$$d^* = *d* : \Omega^2(\mathbf{R}^4) \to \Omega^1(\mathbf{R}^4).$$

The Hodge operators can be shown to be independent of the choice of a basis in $\mathbf{R}^4$, but they do depend on the choice of an orientation. Consequently, the formulation of Maxwell's equations given above is independent of the choice of coordinates in $\mathbf{R}^4$. In physics an equation is said to be *covariant* if its form is independent of the choice of coordinates used to write the equation.

(v) Prove that, in terms of the exterior derivative d_3 associated with $\mathbf{R}^3$, Maxwell's equations take the form of identities between differential forms on $\mathbf{R}^3$ with additional dependence on a parameter in $\mathbf{R}$, as follows:

$$d_3\mathcal{D} = \rho\, dx, \qquad d_3\mathcal{E} + D_0\mathcal{B} = 0, \qquad d_3\mathcal{B} = 0, \qquad d_3\mathcal{H} - D_0\mathcal{D} = \flat_2 j.$$

(vi) Show that the integral theorems from Exercise 8.26.(ii) – (v) immediately follow, by application of Stokes' Theorem 8.6.10.

Since $\mathcal{F} \in \Omega^2(\mathbf{R}^4)$ is a closed C^1 differential form, it follows from Poincaré's Lemma 8.10.2 that there exists a C^2 differential form $\mathcal{G} \in \Omega^1(\mathbf{R}^4)$ with $\mathcal{F} = d\mathcal{G}$.

(vii) Demonstrate the existence of C^2 potentials $\phi : \mathbf{R}^4 \to \mathbf{R}$ and $A : \mathbf{R}^4 \to \mathbf{R}^3$ with

$$\mathcal{G} = -\phi\, dt + \flat_1 A \in \Omega^1(\mathbf{R}^4).$$

The equation $\mathcal{F} = d\mathcal{G}$ now leads to expressions for E and B in terms of ϕ and A

$$E = -\nabla\phi - D_0 A, \qquad B = \nabla \times A. \tag{4}$$

Use (2) and (4) to show that the equation $d^*\mathcal{F} = \mathcal{J}$ is equivalent to

$$d \sum_{1\le i\le 3} ((D_{i+1}A_{i+2} - D_{i+2}A_{i+1})\, dx_i \wedge dt + (D_i\phi + D_0 A_i)\, dx_{i+1} \wedge dx_{i+2})$$
$$= *\mathcal{J}. \tag{5}$$

Note that $\mathcal{G}$ is not completely determined by $\mathcal{J}$, and that, consequently, (5) does not completely determine ϕ and A; it follows that we may impose another condition.

(viii) Try to find $\mathcal{G}$ such that the *Lorentz gauge condition* $d^*\mathcal{G} = 0$ is satisfied. Then verify that

$$\langle \nabla, A \rangle + D_0\phi = 0.$$

Demonstrate that under this assumption (5) is equivalent to the following equations for ϕ and A, for given ρ and j, respectively:

$$\Box\phi = \rho, \qquad \Box A = j, \qquad \text{where} \qquad \Box = D_0^2 - \Delta \tag{6}$$

is the *wave operator* or *D'Alembertian*. In general, $\mathcal{G} + df$ will satisfy the Lorentz gauge condition if $\Box f = 0$.

In Exercise 8.34 we prove that solutions of (6) are given by the *retarded potentials*, for $(x, t) \in \mathbf{R}^4$ with $t > 0$,

$$\phi(x, t) = \frac{1}{4\pi} \int_{\|x-x'\|\le t} \frac{\rho(x', t - \|x - x'\|)}{\|x - x'\|}\, dx',$$

$$A(x, t) = \frac{1}{4\pi} \int_{\|x-x'\|\le t} \frac{j(x', t - \|x - x'\|)}{\|x - x'\|}\, dx'.$$

Exercise 8.32 (Invariance of wave operator under Lorentz transformation and special relativity – sequel to Exercise 2.39 – needed for Exercises 5.70 and 5.71). Let J_{n+1} in $\mathrm{Mat}(n+1, \mathbf{R})$ be the diagonal matrix having $-1, 1, \dots, 1$ on the main diagonal. The mapping

$$(y, \widetilde{y}) \mapsto \lceil y, \widetilde{y} \rceil = y^t J_{n+1} \widetilde{y} : \mathbf{R}^{n+1} \times \mathbf{R}^{n+1} \to \mathbf{R}$$

is a nondegenerate symmetric bilinear form on $\mathbf{R}^{n+1}$. A *Lorentz transformation* of $\mathbf{R}^{n+1}$ is a linear mapping $L \in \mathrm{End}(\mathbf{R}^{n+1})$ leaving this form invariant, that is, one that satisfies $\lceil Ly, L\widetilde{y} \rceil = \lceil y, \widetilde{y} \rceil$, for all $y, \widetilde{y} \in \mathbf{R}^{n+1}$. The *Lorentz group* $\mathbf{Lo}(n+1, \mathbf{R})$ consists of all Lorentz transformations of $\mathbf{R}^{n+1}$.

(i) Consider $(t, x) \in \mathbf{R} \times \mathbf{R}^n \simeq \mathbf{R}^{n+1}$. Prove

$$\lceil (t, x), (\widetilde{t}, \widetilde{x}) \rceil = t\widetilde{t} - \langle x, \widetilde{x} \rangle, \qquad (t, \widetilde{t} \in \mathbf{R},\ x, \widetilde{x} \in \mathbf{R}^n).$$

Show $L \in \mathbf{Lo}(n+1, \mathbf{R})$ if and only if $L^t J_{n+1} L = J_{n+1}$. From this deduce $\det L = \pm 1$ for $L \in \mathbf{Lo}(n+1, \mathbf{R})$, furthermore that $\mathbf{Lo}(n+1, \mathbf{R})$ indeed satisfies the axioms of a group, and also that $L \in \mathbf{Lo}(n+1, \mathbf{R})$ if and only if $L^t \in \mathbf{Lo}(n+1, \mathbf{R})$. Let S_{n+1} be a diagonal matrix with $S_{n+1}^2 = J_{n+1}$. Then $L \in \mathbf{Lo}(n+1, \mathbf{R})$ if and only if $S_{n+1} L S_{n+1}^{-1}$ is an orthogonal linear mapping (with complex coefficients).

(ii) By means of part (i) and Exercise 2.39 verify that the *wave operator* or *D'Alembertian*

$$\Box = D_t^2 - \Delta_x = D_t^2 - \sum_{1 \le j \le n} D_j^2$$

in $\mathbf{R}^{n+1}$ is invariant under Lorentz transformation, that is

$$\Box (f \circ L) = (\Box f) \circ L \qquad (f \in C^2(\mathbf{R}^{n+1}),\ L \in \mathbf{Lo}(n+1, \mathbf{R})).$$

Background. The invariance under Lorentz transformations of the wave operator, and also of Maxwell's equations, played a role in the development of the *theory of special relativity* in physics.

(iii) Assume $n = 1$. Then $L \in \mathbf{Lo}(2, \mathbf{R})$, $\det L = 1$ and $\operatorname{tr} L > 0$ if and only if there exists a number $\zeta \in \mathbf{R}$ with

$$L = L_\zeta = \begin{pmatrix} \cosh\zeta & \sinh\zeta \\ \sinh\zeta & \cosh\zeta \end{pmatrix}.$$

Verify that $L_\zeta \circ L_{\zeta'} = L_{\zeta+\zeta'}$ and thus $L_\zeta^{-1} = L_{-\zeta}$, for all ζ and $\zeta' \in \mathbf{R}$. The mapping L_ζ is said to be the *hyperbolic screw* or *boost* in $\mathbf{R}^2$ with *rapidity* ζ. **Hint:** If $L(1, 0) = (t, x)$ then $\lceil (t, x), (t, x) \rceil = t^2 - x^2 = 1$, and therefore there exists a unique number $\zeta \in \mathbf{R}$ with $t = \cosh\zeta$ and $x = \sinh\zeta$. For the computation of $L(0, 1) = (\widetilde{t}, \widetilde{x})$ use $\lceil (t, x), (\widetilde{t}, \widetilde{x}) \rceil = 0$ and $\det L = t\widetilde{x} - \widetilde{t}x = 1$.

Next we define $-1 < v < 1$ by

$$\tanh \zeta = v,$$

then

$$\zeta = \frac{1}{2} \log \frac{1+v}{1-v}, \qquad \cosh \zeta = \frac{1}{\sqrt{1-v^2}} =: \gamma, \qquad \sinh \zeta = \gamma\, v.$$

In physics ζ is called the *rapidity* of the velocity v. Addition of rapidities corresponds to the following *relativistic law for addition of velocities*:

$$v := \tanh(\zeta_1 + \zeta_2) = \frac{v_1 + v_2}{1 + v_1 v_2} \qquad (v_i = \tanh \zeta_i).$$

(iv) Verify, if $(\widetilde{t}, \widetilde{x}) = L_{-\zeta}(t, x)$ for t and $x \in \mathbf{R}$, and

$$L_{-\zeta} = \gamma \begin{pmatrix} 1 & -v \\ -v & 1 \end{pmatrix}, \qquad \text{that} \qquad \begin{aligned} \widetilde{t} &= \gamma(t - xv), \\ \widetilde{x} &= \gamma(x - tv). \end{aligned}$$

(v) Next we generalize to $\mathbf{R}^{n+1}$ the Lorentz transformations having the form from part (iv). Let $v \in S^{n-1} = \{ x \in \mathbf{R}^n \mid \|x\| = 1 \}$ and $-1 < v_0 < 1$ be arbitrary and write $\gamma = (1 - v_0^2)^{-\frac{1}{2}}$. Prove that $L \in \mathbf{Lo}(n+1, \mathbf{R})$ if

$$L \begin{pmatrix} t \\ x \end{pmatrix} = \begin{pmatrix} \gamma(t - v_0 \langle x, v \rangle) \\ x + ((\gamma - 1)\langle x, v \rangle - \gamma v_0 t) v \end{pmatrix} \qquad (t \in \mathbf{R},\ x \in \mathbf{R}^n).$$

Hint: Direct computation, or else proceed as follows. Write $x = x_\| + x_\perp$ for the decomposition in $\mathbf{R}^n$ of x in components parallel and perpendicular to v. Application of part (iv) to the linear subspace in $\mathbf{R}^{n+1} \simeq \mathbf{R} \times \mathbf{R}^n$ spanned by e_0 and v then gives

$$\widetilde{t} = \gamma(t - v_0 \langle x_\|, v \rangle), \qquad \widetilde{x}_\| = \gamma(x_\| - v_0 t\, v), \qquad \widetilde{x}_\perp = x_\perp.$$

Since $x_\| = \langle x, v \rangle\, v$ and $x_\perp = x - \langle x, v \rangle\, v$ we obtain

$$\widetilde{x} = \widetilde{x}_\perp + \widetilde{x}_\| = x_\perp + \gamma(x_\| - v_0 t\, v) = x - \langle x, v \rangle\, v + \gamma(\langle x, v \rangle\, v - v_0 t\, v).$$

(vi) Let $v \in S^{n-1}$ be arbitrary and let 0_n denote $0 \in \mathrm{Mat}(n, \mathbf{R})$. Put

$$V = \begin{pmatrix} 0 & v^t \\ v & 0_n \end{pmatrix} \in \mathrm{Mat}(n+1, \mathbf{R}).$$

Prove $V^t J_{n+1} + J_{n+1} V = 0$. Note that $vv^t \in \mathrm{Mat}(n, \mathbf{R})$, and show by induction

$$V^{2j} = \begin{pmatrix} 1 & 0 \\ 0 & vv^t \end{pmatrix}, \qquad V^{2j+1} = V \qquad (j \in \mathbf{N}).$$

Demonstrate for $\zeta \in \mathbf{R}$ (see Example 2.4.10 for the definition of exp)

$$\begin{aligned} \exp \zeta\, V &= \sum_{j \in \mathbf{N}_0} \frac{1}{j!} (\zeta\, V)^j = \begin{pmatrix} \cosh \zeta & \sinh \zeta\ v^t \\ \sinh \zeta\ v & I_n + (-1 + \cosh \zeta)\, vv^t \end{pmatrix} \\ &=: B_{\zeta, v}. \end{aligned}$$

Thus, for $t \in \mathbf{R}$ and $x \in \mathbf{R}^n$,

$$B_{\zeta, v} \begin{pmatrix} t \\ x \end{pmatrix} = \begin{pmatrix} t \cosh \zeta \ + \langle x, v \rangle \sinh \zeta \\ t \sinh \zeta\ v + \langle x, v \rangle \cosh \zeta\ v + x - \langle x, v \rangle\, v \end{pmatrix}.$$

$B_{\zeta, v}$ is said to be the *hyperbolic screw* or *boost* in the direction $v \in S^{n-1}$ with *rapidity* ζ. See Exercise 5.70.(xii) and (xiii) for another characterization of a boost. Verify that $\langle x, v \rangle\, v$, the component of x along v, undergoes a hyperbolic screw in the plane spanned by e_0 and v with rapidity ζ, and that $x - \langle x, v \rangle\, v$, the component of x perpendicular to v, remains unchanged under the action of $B_{\zeta, v}$. Now define $-1 < v_0 < 1$, the speed of the boost, by

$$\tanh \zeta = v_0, \qquad \text{then} \qquad \cosh \zeta = \frac{1}{\sqrt{1 - v_0^2}} =: \gamma, \qquad \sinh \zeta = \gamma\, v_0.$$

Verify that $B_{-\zeta, v}$ equals the Lorentz transformation L given in (v).

(vii) Consider $(\widetilde{t}, \widetilde{x}) = L(t, x)$ as in (v). Prove that elimination of γt from the expression for $\widetilde{x}$ in (v) gives

$$\widetilde{x} = \sqrt{1 - v_0^2}\, \langle x, v \rangle v + x - \langle x, v \rangle v - v_0 \widetilde{t} v.$$

Deduce that for two different points with coordinates x and y, and $\widetilde{x}$ and $\widetilde{y}$, respectively,

$$\| \widetilde{x} - \widetilde{y} \| = (1 - v_0^2)\, \| (x - y)_{\parallel} \|^2 + \| (x - y)_{\perp} \|^2.$$

For a stationary observer objects that move with velocity $v_0 v \in \mathbf{R}^3$ contract by a factor $(1 - v_0^2)^{\frac{1}{2}}$ along the direction of motion while there is no contraction perpendicular to the direction of motion; this is the *FitzGerald–Lorentz contraction* of space.

(viii) Now assume $v = e_1 \in \mathbf{R}^3$. Then similarly elimination of γx_1 from the expression for $\widetilde{t}$ in (v) gives

$$\widetilde{t} = \sqrt{1 - v_0^2}\, t - v_0 \widetilde{x}_1.$$

Deduce that for two different moments with coordinates t and u, and $\widetilde{t}$ and $\widetilde{u}$, respectively,

$$\widetilde{t} - \widetilde{u} = \sqrt{1 - v_0^2}\,(t - u).$$

For a stationary observer clocks that move with velocity $v_0 v \in \mathbf{R}^3$ run slower by a factor $(1 - v_0^2)^{\frac{1}{2}}$; this is the *dilatation* of time. In particular, long journeys across cosmic distances would be instantaneous for an observer traveling with the speed 1 of light (in our usual normalization).

Exercise 8.33 (Wave equation in three space variables – sequel to Exercises 3.22, 7.53, 8.26 – needed for Exercise 8.34). Consider the *wave equation*, which we encountered in Maxwell's theory, in particular in Exercise 8.26.(ix)

$$(\star) \qquad \frac{1}{c^2} D_t^2 u(x, t) = \Delta_x u(x, t) = \sum_{1 \le j \le 3} D_j^2 u(x, t) \qquad (c > 0),$$

for a C^2 function $u : \mathbf{R}^3 \times \mathbf{R} \to \mathbf{R}$, with $(x, t) \mapsto u(x, t)$. We want to solve the *initial value problem* for this equation, that is, we look for solutions u of $(\star)$ which in addition satisfy the following initial conditions, for $t = 0$:

$$(\star\star) \qquad u(x, 0) = f(x), \qquad D_t u(x, 0) = g(x) \qquad (x \in \mathbf{R}^3),$$

for given functions $f \in C^3(\mathbf{R}^3)$ and $g \in C^2(\mathbf{R}^3)$.

Form the spherical means with respect to the space variable, as in Exercise 7.53, for the functions u, f and g, and write the resulting functions as

$$m_u : \mathbf{R}^3 \times \mathbf{R} \times \mathbf{R} \to \mathbf{R} \qquad \text{and} \qquad m_f,\ m_g : \mathbf{R}^3 \times \mathbf{R} \to \mathbf{R}, \qquad \text{respectively.}$$

In particular,

$$m_u(x, r, t) = \frac{1}{4\pi} \int_{\|y\|=1} u(x + ry,\ t)\, d_2 y.$$

(i) Prove

$$m_u(x, 0, t) = u(x, t), \qquad m_u(x, r, 0) = m_f(x, r),$$
$$D_t m_u(x, r, 0) = m_g(x, r).$$

(ii) Show

$$\frac{1}{c^2} D_t^2 m_u(x,\ r,\ t) = \Delta_x m_u(x,\ r,\ t).$$

(iii) Prove by means of Exercise 7.53 that, for $x \in \mathbf{R}^3$ fixed, the function $(r, t) \mapsto m_u(x, r, t)$ satisfies the following partial differential equation:

$$\frac{1}{c^2} D_t^2 m_u(x, r, t) = D_r^2 m_u(x, r, t) + \frac{2}{r} D_r m_u(x, r, t).$$

Conclude that $(r, t) \mapsto r m_u(x, r, t)$ satisfies the wave equation in one space variable

$$\frac{1}{c^2} D_t^2 (r m_u(x, r, t)) = D_r^2 (r m_u(x, r, t)),$$

$$r m_u(x, r, 0) = r m_f(x, r), \qquad D_t (r m_u(x, r, 0)) = r m_g(x, r).$$

(iv) Prove, by means of Exercise 3.22.(iii),

$$\begin{aligned} r m_u(x, r, t) &= \frac{1}{2}((r + ct) m_f(x, r + ct) + (r - ct) m_f(x, r - ct)) \\ &\quad + \frac{1}{2c} \int_{r-ct}^{r+ct} s m_g(x, s)\, ds. \end{aligned}$$

The definitions of $m_f(x, r)$ and $m_g(x, r)$ show that the functions $r \mapsto m_f(x, r)$ and $r \mapsto m_g(x, r)$ are well-defined on all of $\mathbf{R}$, and are even functions.

(v) On the basis of the foregoing observation, prove that

$$\begin{aligned} m_u(x, r, t) &= \frac{(ct + r) m_f(x, ct + r) - (ct - r) m_f(x, ct - r)}{2r} \\ &\quad + \frac{1}{2cr} \int_{ct-r}^{ct+r} s m_g(x, s)\, ds. \end{aligned}$$

Hint: $\int_{-(ct+r)}^{ct+r} s m_g(x, s)\, ds = 0$.

(vi) In the formula in (v), take the limit for $r \to 0$, and prove by (i)

$$\begin{aligned} u(x, t) &= D_p\big|_{p=ct} (p\, m_f(x, p)) + t\, m_g(x, ct) \\ &= D_t (t\, m_f(x, ct)) + t\, m_g(x, ct). \end{aligned}$$

That is, u is given by the following, known as *Kirchhoff's formula*:

$(\star\star\star)$ $\qquad u(x, t)$

$$= \frac{1}{4\pi} \int_{\|y\|=1} (f(x + cty) + t\, g(x + cty) + t\, D_t f(x + cty))\, d_2 y.$$

(vii) Conclude that formula $(\star\star\star)$ gives the **unique** solution of the initial value problem $(\star)$ and $(\star\star)$.

In the following we shall assume that there exists a bounded set $K \subset \mathbf{R}^3$ such that

$$\operatorname{supp}(f) \subset K, \qquad \operatorname{supp}(g) \subset K.$$

(viii) Let $t > 0$ and $u(x,t) \neq 0$. Prove that then there exists a $z \in K$ such that x lies on the sphere in $\mathbf{R}^3$ of center z and radius ct. Thus, in particular, there exists, for all $t > 0$, an open ball B_t in $\mathbf{R}^3$ about the origin and of t-dependent radius such that

$$x \notin B_t \qquad \Longrightarrow \qquad u(\cdot,\, t) = 0 \text{ in a neighborhood of } x.$$

Note that according to formula $(\star\star\star)$ the solution u may be one order less differentiable than the initial f and g. This is a "*focusing effect*": irregularities from various places in the initial data are focused, thus leading to *caustics*, that is, (smaller) sets of stronger irregularity. Nevertheless the solution u is "on average well-behaved", as becomes evident from the following. Define the *energy $E(t)$ of the solution u at time t* by

$$E(t) = \frac{1}{2}\int_{\mathbf{R}^3}\Big(\Big(\frac{1}{c}\,D_t u\Big)^2 + \|\operatorname{grad}_x u\|^2\Big)(x,\, t)\,dx,$$

with grad_x the gradient with respect to the variable $x \in \mathbf{R}^3$.

(ix) Prove that E is a conserved quantity, that is, $t \mapsto E(t)$ is a constant function. **Hint:** One has

$$\begin{aligned}\frac{dE}{dt}(t) &= \int_{B_t}\Big(\frac{1}{c^2}\,(D_t u)\,(D_t^2 u) + \langle\, \operatorname{grad}_x u,\ \operatorname{grad}_x(D_t u)\,\rangle\Big)(x,t)\,dx\\ &= \int_{B_t} D_t u\,\Big(\frac{1}{c^2}\,D_t^2 u - \Delta_x u\Big)(x,t)\,dx\\ &\quad + \int_{\partial B_t}\frac{\partial u}{\partial \nu}(y,t)\,D_t u(y,t)\,d_2 y = 0,\end{aligned}$$

by Green's first identity from Example 7.9.6, and part (viii).

Exercise 8.34 (Inhomogeneous wave equation – sequel to Exercises 2.74 and 8.33 – needed for Exercise 8.35). We want to find a C^2 function $u : \mathbf{R}^4 \to \mathbf{R}$ satisfying, for a given function $g \in C^2(\mathbf{R}^4)$, the *inhomogeneous wave equation*

$$(\star) \qquad \Box u = g.$$

Let $\tau \in \mathbf{R}$ be chosen arbitrarily, and let $(x, t) \mapsto v(x, t; \tau)$ be a solution of the initial value problem

$$\Box v = 0, \qquad v(x, \tau) = 0, \qquad D_0 v(x, \tau) := \frac{\partial v}{\partial t}(x, \tau) = g(x, \tau) \qquad (x \in \mathbf{R}^3).$$

On account of Kirchhoff's formula from Exercise 8.33.(vi), this is satisfied by

$$v(x, t; \tau) = \frac{t - \tau}{4\pi} \int_{\|y\|=1} g(x + (t - \tau)y; \tau)\, d_2 y \qquad ((x, t) \in \mathbf{R}^3 \times \mathbf{R}).$$

Now define, assuming convergence,

$$(\star) \qquad u(x, t) = \int_0^t v(x, t; \tau)\, d\tau \qquad ((x, t) \in \mathbf{R}^3 \times \mathbf{R}).$$

Then $u(x, 0) = 0$, and we find, by means of Exercise 2.74,

$$(\star\star) \qquad D_0 u(x, t) = v(x, t; t) + \int_0^t D_0 v(x, t; \tau)\, d\tau = \int_0^t D_0 v(x, t; \tau)\, d\tau,$$

because $v(x, t; t) = 0$ in view of the initial condition on v. Hence, $D_0 u(x, 0) = 0$. Furthermore, differentiation of $(\star\star)$ gives

$$D_0^2 u(x, t) = D_0 v(x, t; t) + \int_0^t D_0^2 v(x, t; \tau)\, d\tau = g(x, t) + \int_0^t D_0^2 v(x, t; \tau)\, d\tau.$$

And, from $(\star)$,

$$\Delta u(x, t) = \int_0^t \Delta v(x, t; \tau)\, d\tau = \int_0^t D_0^2 v(x, t; \tau)\, d\tau.$$

Upon subtracting these results we obtain

$$\Box u(x, t) = g(x, t), \qquad u(x, 0) = 0, \qquad D_0 u(x, 0) = 0.$$

Therefore

$$\begin{aligned} u(x, t) &= \frac{1}{4\pi} \int_0^t (t - \tau) \int_{\|y\|=1} g(x + (t - \tau)y; \tau)\, d_2 y\, d\tau \\ &= \frac{1}{4\pi} \int_0^t \tau \int_{\|y\|=1} g(x + \tau y; t - \tau)\, d_2 y\, d\tau. \end{aligned}$$

Substitution of $y = \Psi(x') = \frac{1}{\tau}(x' - x)$ leads to the *retarded potential* from Exercise 8.31

$$\begin{aligned} u(x, t) &= \frac{1}{4\pi} \int_0^t \frac{1}{\tau} \int_{\|x'-x\|=|\tau|} g(x'; t - \tau)\, d_2 x'\, d\tau \\ &= \frac{1}{4\pi} \int_{\|x-x'\|\leq|t|} \frac{g(x', t - \operatorname{sgn}(t)\|x - x'\|)}{\|x - x'\|}\, dx'. \end{aligned}$$

Background. The value of the potential u at the point $(x, t) \in \mathbf{R}^4$ with $t \geq 0$ is exclusively determined by the charge density g at points $(x', t') \in \mathbf{R}^4$ with $t \geq \|x - x'\| = t - t' \geq 0$. These points (x', t') lie on that part of a "rearward" conical surface in $\mathbf{R}^4$ which has apex (x, t) and lies in the "positive" half-space in $\mathbf{R}^4$. The method used here to solve an inhomogeneous equation is a special case of *Duhamel's principle*.

Exercise 8.35 (Fundamental solution – sequel to Exercises 6.49, 6.68, 6.92, 6.105, 7.67 and 8.34). Let $f \in C_c^2(\mathbf{R}^n)$ be given. Prove that the inhomogeneous partial differential equation $P(D)u = f$ on $\mathbf{R}^n$ has a solution $u = f * E \in C^2(\mathbf{R}^n)$, where $*$ denotes convolution and $E \in C^\infty(\mathbf{R}^n \setminus \{0\})$ satisfies the homogeneous partial differential equation $P(D)E = 0$ on $\mathbf{R}^n \setminus \{0\}$, in the following cases:

	variable	$P(D)$	$E(x)$	Exer.
(i)	$x \in \mathbf{R}$	$D^s \quad (s > 0)$	$\dfrac{x_+^{s-1}}{\Gamma(s)}$	6.105
(ii)	$x \in \mathbf{R}^2$	$\frac{1}{2}(D_1 + iD_2)$	$\dfrac{1}{\pi}\dfrac{1}{x_1 + ix_2}$	6.49
(iii)	$x \in \mathbf{R}^2$	Δ	$\dfrac{1}{2\pi}\log\|x\|$	7.67
(iv)	$x \in \mathbf{R}^n \quad (n \neq 2)$	Δ	$\dfrac{1}{(2-n)\|S^{n-1}\|}\dfrac{1}{\|x\|^{n-2}}$	7.67
(v)	$x \in \mathbf{R}^3$	$\Delta + \mu^2$	$-\dfrac{1}{4\pi}\dfrac{e^{\pm i\mu\|x\|}}{\|x\|}$	6.68
(vi)	$(x, t) \in \mathbf{R}^n \times \mathbf{R}$	$\dfrac{\partial}{\partial t} - k\,\Delta_x$	$\begin{cases} \dfrac{1}{(4\pi kt)^{n/2}} e^{-\frac{\|x\|^2}{4kt}} & (t > 0) \\ 0, & (t \leq 0) \end{cases}$	6.92

Such a solution E is said to be a *fundamental solution* for the partial differential operator $P(D)$. The operator in (i) is fractional differentiation; in (ii) it is the Cauchy–Riemann operator; in (iii) and (iv) the Laplace operator; in (v) the Helmholtz operator (see the technique of the Exercise 7.67); and in (vi) the heat operator.

Background. In the case of the wave operator $D_t^2 - \Delta_x$ from Exercise 8.34 the situation is more complicated; it turns out that E never is a differentiable function on $\mathbf{R}^{n+1} \setminus \{0\}$. For instance, for $n = 1$ a solution E is given by the function with the constant value $\frac{1}{2}$ on the forward cone $\{ (x, t) \in \mathbf{R}^2 \mid |x| < t \}$, and the value 0 elsewhere. And E even fails to be a function for larger values of n; nevertheless it always is a *distribution*, a generalization of the notion of a function. For that reason the retarded potential from Exercise 8.34 where $n = 3$ can not immediately be recognized as a convolution product.

Exercise 8.36 (Brouwer's Fixed-point Theorem – sequel to Exercise 6.103). The assertion from Example 8.8.3 holds for an arbitrary continuous mapping $f : U \to \mathbf{R}^n$ instead of a C^2 mapping f. We now prove this, leaving it to the reader to fill in the details.

Suppose $f(x) \neq x$, for all $x \in B^n$. The continuous function $x \mapsto \|f(x) - x\|$ then reaches a minimum of value $4m > 0$ on B^n. By application of Weierstrass' Approximation Theorem from Exercise 6.103 by components, for example, f can be approximated by means of a polynomial function $\widetilde{p} : B^n \to \mathbf{R}^n$ such that in the uniform norm $\|\cdot\|$ on B^n

$$\|f - \widetilde{p}\| < m.$$

This gives $\|\widetilde{p}\| \leq 1 + m$. We therefore have $p := \frac{1}{1+m}\widetilde{p} : B^n \to B^n$, while

$$\|f - p\| \leq \|f - \widetilde{p}\| + \Big(1 - \frac{1}{1+m}\Big)\|\widetilde{p}\| < m + \frac{m}{1+m}(1+m) = 2m.$$

Consequently, for all $x \in B^n$,

$$\begin{aligned}\|p(x) - x\| &= \|f(x) - x - (f - p)(x)\| \geq \|f(x) - x\| - \|f - p\| \\ &\geq 4m - 2m = 2m > 0.\end{aligned}$$

By Example 8.8.3, the polynomial function p does have a fixed point $x \in B^n$, and this implies a contradiction.

Exercise 8.37 (Sequel to Exercise 6.23). Use Exercise 6.23.(i) in order to show that Brouwer's Fixed-point Theorem is false for open balls.

Exercise 8.38. Prove that Formula (8.49) can be written as

$$\begin{aligned}&(\omega \wedge \eta)(v_1, \dots, v_{k+l}) \\ &\quad = \frac{1}{k!\,l!} \sum_{\sigma \in S_{k+l}} \operatorname{sgn}(\sigma)\,(\omega \circ \sigma)(v_1, \dots, v_k)\,(\eta \circ \sigma)(v_{k+1}, \dots, v_{k+l}).\end{aligned}$$

Hint: Note that in this formula $\{\sigma(1), \dots, \sigma(k)\}$ and $\{\sigma(k+1), \dots, \sigma(k+l)\}$ are not ordered.

Exercise 8.39 (Vector analysis in $\mathbf{R}^3$). We derive the formulae in Exercise 8.25.(ii) and (iii) using differential forms. If v is a vector field on $\mathbf{R}^3$, let $\flat_1 v \in \Omega^1(\mathbf{R}^3)$ and $\flat_2 v \in \Omega^2(\mathbf{R}^3)$, be the corresponding 1-form and 2-form, respectively, as in Example 8.8.2. Further, denote by i_v the contraction with the vector field v as in Formula (8.57), and by L_v the Lie derivative in the direction of v as in Formula (8.55).

(i) Recall that $i_v dx = \flat_2 v$, and deduce from Formula (5.29) that $L_v dx = \operatorname{div} v\, dx \in \Omega^3(\mathbf{R}^3)$.

(ii) Prove, for vector fields v_1 and v_2 on $\mathbf{R}^3$,

$$\langle v_1, v_2\rangle = i_{v_1}\flat_1 v_2 \in \Omega^0(\mathbf{R}^3), \qquad \langle v_1, v_2\rangle\, dx = \flat_1 v_1 \wedge \flat_2 v_2 \in \Omega^3(\mathbf{R}^3).$$

Further, show

$$\flat_1(v_1\times v_2) = -i_{v_1}\flat_2 v_2 = i_{v_2}\flat_2 v_1, \qquad \flat_1 v_1\wedge\flat_1 v_2 = \flat_2(v_1\times v_2) = i_{v_1\times v_2}\, dx.$$

(iii) Suppose v_3 is a vector field on $\mathbf{R}^3$. Compute $i_{v_1}(\flat_1 v_2\wedge\flat_1 v_3) \in \Omega^1(\mathbf{R}^3)$ using (ii) and the antiderivation property of i_{v_1}, and deduce Grassmann's formula from Exercise 5.26.(ii).

(iv) Compute $d(\flat_1 v_1 \wedge \flat_1 v_2) \in \Omega^3(\mathbf{R}^3)$ by means of (ii) and the results in Formula (8.54), and deduce the identity in Exercise 8.25.(ii).

(v) Prove $[L_{v_1}, i_{v_2}] = i_{[v_1, v_2]}$. To this end, note that $[L_{v_1}, i_{v_2}]$ is an antiderivation that takes k-forms to $(k-1)$-forms and that vanishes on $\Omega^0(\mathbf{R}^3)$. It is sufficient therefore to establish the identity on $\Omega^1(\mathbf{R}^3)$.

(vi) Apply the homotopy formula to $\flat_2(\operatorname{curl}(v_1 \times v_2)) = d(\flat_1(v_1 \times v_2)) = d i_{v_2}\flat_2 v_1$. Then, using part (i), note that $L_{v_2}\flat_2 v_1 = L_{v_2} i_{v_1} dx$ and apply part (v). Finally, deduce the identity in Exercise 8.25.(iii).

(vii) In the same way as above show $\operatorname{curl}(f v) = f \operatorname{curl} v + (\operatorname{grad} f) \times v$, for $f \in C^1(\mathbf{R}^3)$.

(viii) Prove the identity in Exercise 8.25.(vi). To do so, start with $\flat_1(\operatorname{grad}\langle v_1, v_2\rangle) = (d\circ i_{v_1})\flat_1 v_2$ and apply the homotopy formula. Further, use $L_{v_1} f = (Df)v_1$, for $f \in C^1(\mathbf{R}^3)$, and $[d, L_{v_1}] = 0$.

Exercise 8.40 (Divergence in arbitrary coordinates – sequel to Exercise 3.14). Using differential forms we give two different proofs of the following formula $(\star)$ from Exercise 3.14:

$$(\star) \qquad (\operatorname{div} f)\circ\Psi = \frac{1}{\det D\Psi}\sum_{1\le i\le n} D_i(f^{(i)}\, \det D\Psi).$$

Here U and V are open subsets of $\mathbf{R}^n$, while $f: U \to \mathbf{R}^n$ is a C^1 vector field and $\Psi: V\to U$ is a C^1 diffeomorphism, and $f\circ\Psi = \sum_{1\le i\le n} f^{(i)}\, D_i\Psi : V \to \mathbf{R}^n$.

(i) Consider $\flat_{n-1} f \in \Omega^{n-1}(U)$, and derive from Example 8.8.2 the following equality of differential forms in $\Omega^{n-1}(V)$:

$$\Psi^*(\flat_{n-1} f) = \sum_{1\le i\le n} (-1)^{i-1} f^{(i)} \det D\Psi\, dy_{\hat{i}};$$

deduce

$$d(\Psi^*(\flat_{n-1} f)) = \sum_{1\le i\le n} D_i(f^{(i)} \det D\Psi)\, dy \in \Omega^n(V).$$

Prove, for $g \in C(U)$,

$$(\star\star) \qquad \Psi^*(g\, dx) = g \circ \Psi \det D\Psi\, dy$$

and verify, as in Example 8.8.2,

$$\Psi^*(d(\flat_{n-1} f)) = (\operatorname{div} f) \circ \Psi \ \det D\Psi\, dy.$$

Using Theorem 8.6.12 deduce Formula ($\star$).

(ii) Using the homotopy formula from Lemma 8.9.1 deduce that $\operatorname{div} f\, dx = d \circ i_f\, dx \in \Omega^n(U)$. Prove, by applying Ψ^*, Formula ($\star\star$) and Theorem 8.6.12,

$$\begin{aligned}(\operatorname{div} f) \circ \Psi \ \det D\Psi\, dy &= d(\Psi^*(i_f\, dx)) = d(i_{\Psi^* f} \Psi^* dx)\\ &= d(\det D\Psi\, i_{\Psi^* f}\, dy).\end{aligned}$$

Here, in the notation of Exercise 3.14, we have the vector field $\Psi^* f : V \to \mathbf{R}^n$ satisfying $(\Psi^* f)(y) = D\Psi(y)^{-1}(f \circ \Psi)(y) = \sum_{1\le i\le n} f^{(i)}(y) e_i$. Deduce Formula ($\star$) using

$$d(\det D\Psi\, i_{\Psi^* f}\, dy) = \sum_{1\le i\le n} D_i(f^{(i)} \det D\Psi)\, dy.$$

Exercise 8.41 (Lie derivative of vector field and differential form – sequel to Exercise 3.14 – needed for Exercises 8.42, 8.43 and 8.46). Let U be open in $\mathbf{R}^n$. Suppose X is the vector field on U satisfying $X = \frac{d}{dt}\big|_{t=0} \Phi^t$, for a one-parameter group of C^1 diffeomorphisms $(\Phi^t)_{t\in\mathbf{R}}$. If ω is a C^1 differential form in $\Omega^k(U)$ and $X_1, \dots, X_k$ are C^1 vector fields on U, then $g := \omega(X_1, \dots, X_k)$ belongs to $C^1(U)$.

(i) Verify that Definition 8.6.7 of the pullback $(\Phi^t)^*$ acting on differential forms gives, for $x \in U$,

$$\begin{aligned}&(\Phi^t)^* g(x) = \omega(X_1, \dots, X_k)(\Phi^t(x))\\ &= \omega(\Phi^t(x))\big(D\Phi^t(x) D\Phi^t(x)^{-1} X_1(\Phi^t(x)), \dots, D\Phi^t(x) D\Phi^t(x)^{-1} X_k(\Phi^t(x))\big)\\ &= ((\Phi^t)^*\omega)(x)((\Phi^t)^* X_1(x), \dots, (\Phi^t)^* X_k(x)).\end{aligned}$$

Here we used that, in view of the definition of pullback of a vector field from Exercise 3.14,

$$D\Phi^t(x)^{-1}(X_i \circ \Phi^t)(x) = (\Phi^t)^* X_i(x) \qquad (1 \le i \le k).$$

Note that $L_X g = Xg := (Dg)X$. Now define, for a vector field Y on U,

$$L_X Y = \frac{d}{dt}\Big|_{t=0} (\Phi^t)^* Y.$$

(ii) Using Proposition 2.7.6 and the definition of Lie derivative of a differential form from Formula (8.55), deduce from part (i) the following derivation property for the Lie derivative L_X:

$$\begin{aligned} X(\omega(X_1, \dots, X_k)) &= (L_X\omega)(X_1, \dots, X_k) \\ &\quad + \sum_{1 \le i \le k} \omega(X_1, \dots, L_X X_i, \dots, X_k). \end{aligned}$$

Next we study $L_X Y$, for vector fields X and Y on U. The proper framework for studying vector fields is that of derivations; it is in this context that one obtains the correct functorial properties.

(iii) Prove, on account of the chain rule, for any $f \in C^1(U)$,

$$\begin{aligned} ((\Phi^t)^* Y)(f)(x) &= Df(x)((\Phi^t)^* Y)(x) \\ &= Df(x) D\Phi^{-t}(\Phi^t(x))(Y \circ \Phi^t)(x) = D(f \circ \Phi^{-t})(\Phi^t(x)) Y(\Phi^t(x)) \\ &= (\Phi^t)^*(Y((\Phi^{-t})^* f))(x). \end{aligned}$$

Deduce

$$\begin{aligned} (L_X Y) f &= \frac{d}{dt}\Big|_{t=0} ((\Phi^t)^* Y) f = \frac{d}{dt}\Big|_{t=0} (\Phi^t)^*(Y((\Phi^{-t})^* f)) \\ &= X(Y(f)) - Y(X(f)), \end{aligned}$$

which implies

$$L_X Y = XY - YX = [X, Y]; \qquad (L_X\omega)(X_1, \dots, X_k)$$

$$= X(\omega(X_1, \dots, X_k)) + \textstyle\sum_{1 \le i \le k} \omega(X_1, \dots, [X_i, X], \dots, X_k).$$

Exercise 8.42 (Homotopy formula and exterior derivative – sequel to Exercise 8.41 – needed for Exercise 8.43). Let U be open in $\mathbf{R}^n$, let $\omega \in \Omega^k(U)$ be a C^1 form, and let $X_1, \dots, X_{k+1}$ be C^1 vector fields on U. The homotopy formula from Lemma 8.9.1 implies

$$i_{X_1} d\omega = L_{X_1}\omega - d i_{X_1}\omega,$$

while the derivation property for L_{X_1} from Exercise 8.41 yields

$$\begin{aligned}(L_{X_1}\omega)(X_2,\dots,X_{k+1}) &= X_1(\omega(X_2,\dots,X_{k+1}))\\ &\quad - \sum_{2\le j\le k+1} \omega(X_2,\dots,[X_1,X_j],\dots,X_{k+1}).\end{aligned}$$

By combination of these two formulae derive

$$\begin{aligned}d\omega(X_1,\dots,X_{k+1}) &= X_1(\omega(X_2,\dots,X_{k+1}))\\ &\quad + \sum_{1<j\le k+1} (-1)^{j-1}\omega([X_1,X_j],X_2,\dots,\widehat{X_j},\dots,X_{k+1})\\ &\quad -d(i_{X_1}\omega)(X_2,\dots,X_{k+1}).\end{aligned}$$

More generally, one can tackle the last term, which involves $i_{X_1}\omega \in \Omega^{k-1}(U)$, by the same method. Verify the following formula by mathematical induction over $k \in \mathbf{N}_0$:

$$\begin{aligned}d\omega(X_1,\dots,X_{k+1}) &= \sum_{1\le i\le k+1} (-1)^{i-1}X_i(\omega(X_1,\dots,\widehat{X_i},\dots,X_{k+1}))\\ &\quad + \sum_{1\le i<j\le k+1} (-1)^{j-i}\omega([X_i,X_j],X_1,\dots,\widehat{X_i},\dots,\widehat{X_j},\dots,X_{k+1}).\end{aligned}$$

Background. In algebraic contexts, for instance in *Lie algebra cohomology*, the formula above is often adopted as the **definition** of the exterior derivative d. Furthermore, the result from Proposition 8.6.11 is a direct consequence. However, a direct proof of $d^2 = 0$ (compare with Theorem 8.7.2) on the basis of this definition is tedious and not illuminating; therefore we give a different argument in Exercise 8.43 under a mildly restrictive extra condition. (Using some more theory, one may get rid of this restriction.)

Exercise 8.43 (Proof by algebra of $d^2 = 0$ – sequel to Exercises 8.41 and 8.42). Let the notation be as in Exercise 8.42. Furthermore, let $\Phi : U \to U$ be a diffeomorphism and let Φ_* be the corresponding pushforward of vector fields on U as defined in Exercise 3.15. For a vector field X on U and $f \in C^1(U)$ we define $Xf \in C(U)$ by $Xf = (Df)X$.

(i) Verify $\Phi^*((\Phi_*X)f) = X(\Phi^* f)$, and conclude that

$$(\Phi_*X)f = (\Phi^{-1})^*(X(\Phi^* f)).$$

(ii) Let Y be a vector field on U. Deduce from part (i) that

$$\Phi_*[X,Y] = [\,\Phi_*X,\ \Phi_*Y\,].$$

(iii) Using Exercise 8.41.(i) prove, for a C^1 differential form $\omega \in \Omega^k(U)$ and C^1 vector fields $X_1, \dots, X_k$ on U,

$$(\Phi^*\omega)(X_1, \dots, X_k) = \Phi^*(\omega(\Phi_* X_1, \dots, \Phi_* X_k)).$$

(iv) Successively apply part (iii), Exercise 8.42, and parts (i) and (ii) to obtain

$$\Phi^*(d\omega) = d(\Phi^*\omega), \qquad \text{in other words} \qquad [\Phi^*, d] = 0.$$

Let X be as in Exercise 8.41 but otherwise arbitrary and deduce $[L_X, d] = 0$.

(v) Prove the homotopy formula $L_X\omega = d(i_X\omega) + i_X(d\omega)$ on the basis of the formula for $L_X\omega$ from Exercise 8.41.(iii), and for $d\omega$ from Exercise 8.42, respectively. Next conclude that $[i_X, d^2] = 0$ by means of part (iv) and the homotopy formula. Finally, use mathematical induction over $k \in \mathbf{N}_0$ to show $d^2\omega = 0$, for every $\omega \in \Omega^k(U)$.

Exercise 8.44 (Closed but not exact). Suppose $n \geq 2$. Define, as in Formula (8.73)

$$\sigma \in \Omega^{n-1}(\mathbf{R}^n \setminus \{0\}) \qquad \text{by} \qquad \sigma(x) = i_x\Big(\frac{1}{\|x\|^n}\,dx\Big) = \frac{1}{\|x\|^n} \sum_{1 \leq i \leq n} (-1)^{i-1} x_i \, d\widehat{x_i}.$$

(i) Demonstrate that the closed differential form σ is not exact, that is, there is no $\eta \in \Omega^{n-2}(\mathbf{R}^n \setminus \{0\})$ with $\sigma = d\eta$.
Hint: Recall that $\int_{S^{n-1}} \sigma = |S^{n-1}|$ and apply Stokes' Theorem, noting that $\partial S^{n-1} = \emptyset$.

(ii) Take $n = 2$ and let $\Psi : V \to U$ with $\Psi(r, \alpha) = r(\cos\alpha, \sin\alpha)$ be the substitution of polar coordinates from Example 3.1.1. Prove $\Psi^*\sigma = d\alpha$ on V.

Background. The angle function α is multi-valued on $\mathbf{R}^2 \setminus \{0\}$, and this is the obstruction why σ is not exact on all of $\mathbf{R}^2 \setminus \{0\}$. On the other hand, the summands involving multiples of 2π are annihilated when one applies d to α.

Exercise 8.45. Let $U = \mathbf{R}^3 \setminus \{ (0, 0, x_3) \mid x_3 \in \mathbf{R} \} \subset \mathbf{R}^3$ and let ω be the C^∞ differential form

$$\omega = \frac{x_3}{x_1^2 + x_2^2}(x_2\, dx_1 - x_1\, dx_2) \in \Omega^1(U).$$

(i) Verify

$$d\omega = \frac{1}{x_1^2 + x_2^2}(x_1\, dx_2 \wedge dx_3 + x_2\, dx_3 \wedge dx_1) \in \Omega^2(U) \qquad \text{and} \qquad d^2\omega = 0.$$

Fix $\psi \in [-\frac{\pi}{2}, \frac{\pi}{2}]$ and define the C^∞ embedding

$$\phi : D := \,]-\pi, \pi[\, \times \left]\psi, \frac{\pi}{2}\right[\to U$$

$$\text{by} \qquad \phi(\alpha, \theta) = (\cos\alpha\cos\theta,\ \sin\alpha\cos\theta,\ \sin\theta).$$

(ii) Prove $\phi^*\omega = -\sin\theta\, d\alpha \in \Omega^1(D)$. Deduce

$$d(\phi^*\omega) = \phi^*(d\omega) = \cos\theta\, d\alpha \wedge d\theta \in \Omega^2(D),$$

and verify the second identity also by a direct computation.

(iii) Check the identity $\int_\phi d\omega = \int_{\partial\phi} \omega$ from Stokes' Theorem by proving

$$\begin{aligned} \int_D \cos\theta\, d\alpha d\theta &= 2\pi(1 - \sin\psi) = -\sin\psi \int_{-\pi}^{\pi} d\alpha - \int_{\pi}^{-\pi} d\alpha \\ &= \int_{\partial D} -\sin\theta\, d\alpha. \end{aligned}$$

Background. The oriented line integral above gives the angle of daily rotation of *Foucault's pendulum* from Exercise 5.57. See Exercise 8.10 for the same computation in terms of vector fields.

Exercise 8.46 (Hamiltonian mechanics in terms of differential forms – sequel to Exercises 3.8, 3.15, 5.76 and 8.41). In mechanics the cotangent bundle $T^*Q \simeq Q \times \mathbf{R}^{d*}$ of a submanifold Q of dimension d, see Exercise 5.76, plays an important role. T^*Q arises as *momentum phase space* of a system: $q \in Q$ represents the *generalized coordinates* and $p \in \mathbf{R}^{d*}$ the *generalized momenta* for the system. The evolution in time of the system is described by Hamilton's equation in part (v) below, which is a system of $2d$ first-order ordinary differential equations. As in Exercise 5.17.(ii) one proves that T^*Q is a submanifold of dimension $2d$.

(i) Define $\pi : T^*Q \to Q$ as the projection onto the first factor. Then we have $D\pi(q, p) : T_{(q,p)}T^*Q \to T_qQ$, for all $(q, p) \in T^*Q$; and additionally $p : T_qQ \to \mathbf{R}$. Therefore the *tautological* 1-*form* τ on T^*Q may be introduced by

$$\tau(q, p) = p \circ D\pi(q, p) : T_{(q,p)}T^*Q \to \mathbf{R}.$$

Show

$$\tau = \sum_{1 \le i \le d} p_i\, dq_i \in \Omega^1(T^*Q).$$

The following explains the name of τ. Let $\eta \in \Omega^1(Q)$; in other words, let $\eta : Q \to T^*Q$ be a section of the cotangent bundle, that is, $\pi \circ \eta = I$ on Q; then $\eta^*\tau = \eta$. Indeed, use Definition 8.6.7 to prove $(\eta^*\tau)(q) = \tau(q, \eta(q)) \circ D\eta(q) = \eta(q) \circ D\pi(q, \eta(q)) \circ D\eta(q) = \eta(q)$.

(ii) Next introduce the *symplectic 2-form* σ on T^*Q by

$$\sigma = d\,\tau. \qquad \text{Verify} \qquad \sigma = \sum_{1\le i\le d} dp_i \wedge dq_i \in \Omega^2(T^*Q).$$

This implies that τ is all but closed; now prove that σ itself is closed. Verify for the d-fold exterior product

$$\sigma^d = \sigma \wedge \cdots \wedge \sigma = d!\,(-1)^{\frac{d(d+1)}{2}}\, dq_1 \wedge \cdots \wedge dq_d \wedge dp_1 \wedge \cdots \wedge dp_d.$$

In other words, $\frac{(-1)^{\frac{d(d+1)}{2}}}{d!}\sigma^d$ is the Euclidean volume form on T^*Q.

(iii) On the strength of Definition 8.6.2 verify, for vector fields v and $\widetilde{v}$ on T^*Q,

$$\sigma(v, \widetilde{v}) = \sum_{1\le i\le d} (v_{d+i}\widetilde{v}_i - v_i\widetilde{v}_{d+i}) = \langle\, v,\ J_d\widetilde{v}\,\rangle,$$

with

$$J_d = \begin{pmatrix} 0 & -I_d \\ I_d & 0 \end{pmatrix} \in \mathbf{GL}(2d, \mathbf{R}).$$

Prove that σ is a nondegenerate bilinear form.

(iv) A vector field v on T^*Q is said to be a *Hamilton vector field* corresponding to the *Hamiltonian* $H : T^*Q \to \mathbf{R}$ if $i_v\sigma = -dH$ (see (8.57)), that is, $i_v\sigma$ is an exact differential 1-form on T^*Q. Now deduce from (iii) that $\langle\, J_d v, \widetilde{v}\,\rangle = \langle\, \text{grad}\, H,\ \widetilde{v}\,\rangle$, and use the nondegeneracy of σ to conclude that $v = v_H := -J_d\, \text{grad}\, H$.

(v) Prove that a solution curve $x = (q, p) : J \to T^*Q$ of the Hamilton vector field v_H satisfies the following, known as *Hamilton's equation* (compare with Exercise 7.62), that is, for $1 \le j \le n$ and $t \in J$,

$$x'(t) = v_H(x(t)) \quad\Longleftrightarrow\quad q_j'(t) = \frac{\partial H}{\partial p_j}(x(t)), \qquad p_j'(t) = -\frac{\partial H}{\partial q_j}(x(t)).$$

(vi) Show that $i_{v_H}\sigma$ from part (iv) is exact, at least locally, if and only if $i_{v_H}\sigma$ is closed, that is, $di_{v_H}\sigma = L_{v_H}\sigma = 0$, on account of (ii) and the homotopy formula.

(vii) Assume that $(\Psi^t)_{t\in\mathbf{R}}$ is a one-parameter group of C^1 diffeomorphisms of T^*Q having a Hamilton vector field v_H as tangent vector field and that σ is the symplectic 2-form. By means of part (vi) show, for $t \in \mathbf{R}$,

$$\frac{d}{dt}(\Psi^t)^*\sigma = \frac{d}{dh}\bigg|_{h=0}(\Psi^{t+h})^*\sigma = (\Psi^t)^*\frac{d}{dh}\bigg|_{h=0}(\Psi^h)^*\sigma = (\Psi^t)^*L_{v_H}\sigma = 0.$$

Consequently, $(\Psi^t)^*\sigma = \sigma$; such diffeomorphisms are called *canonical transformations*. Using part (ii) deduce *Liouville's Theorem*, which asserts that the Euclidean volume form on T^*Q is invariant under $(\Psi^t)_{t\in\mathbf{R}}$.

(viii) For a canonical transformation Ψ prove, for all $y \in T^*Q$ and $v, \widetilde{v} \in T_y(T^*Q)$,

$$\sigma(\Psi(y))(D\Psi(y)v,\, D\Psi(y)\widetilde{v}) = \sigma(y)(v, \widetilde{v});$$

hence

$$v^\tau D\Psi(y)^\tau J_d D\Psi(y)\widetilde{v} = v^\tau J_d \widetilde{v},$$

on account of (iii). Here we have written the transpose as $^\tau$ instead of ^{t}, in order to avoid any confusion with the time variable t. Another way of saying this is that $G = D\Psi(y)$ belongs to the *symplectic group* $\mathbf{Sp}(d, \mathbf{R})$ defined by

$$\mathbf{Sp}(d, \mathbf{R}) = \{\, G \in \mathbf{GL}(2d, \mathbf{R}) \mid G^\tau J_d G = J_d \,\}.$$

Originally this group was called the linear complex group. This terminology was too confusing, so the Latin roots in com-plex (meaning "plaited together") were replaced by the Greek roots sym-plectic.

(ix) Prove the mapping $J_d \in \mathrm{End}(T^*\mathbf{R}^d)$ is canonical, in view of $J_d \in \mathbf{Sp}(d, \mathbf{R})$.

(x) A diffeomorphism $\psi : Q \to Q$ induces the mappings

$$TQ \to TQ \qquad \text{with} \qquad (q, q') \mapsto (\psi(q),\ D\psi(q)q');$$
$$\Psi : T^*Q \to T^*Q \qquad \text{with} \qquad \Psi(q, p) = (\psi(q),\ (D\psi(q)^{-1})^\tau p).$$

Prove $D\Psi(q, p)(\delta q, \delta p) = (D\psi(q)\delta q,\ (D\psi(q)^{-1})^\tau \delta p)$, for $(\delta q, \delta p) \in \mathbf{R}^{2d}$. Verify that the induced mapping $\Psi : T^*Q \to T^*Q$ is a canonical transformation.

(xi) Let $\Psi : T^*Q \to T^*Q$ be a canonical transformation and $H : T^*Q \to \mathbf{R}$ a Hamiltonian. Suppose x is as in part (v). Using Exercises 3.15.(i) and 3.8.(ii)) show, with $x = \Psi(y)$ and $t \in J$,

$$y'(t) = D\Psi(y(t))^{-1} J_d\, (D\Psi(y(t))^{-1})^\tau\ \mathrm{grad}(\Psi^*H)(y(t)) = v_{\Psi^*H}(y(t)).$$

Here we have used that $G \in \mathbf{Sp}(d, \mathbf{R})$ if and only if $G^{-1}J_d(G^{-1})^\tau = J_d$. In other words, the pullback Ψ^*v_H of the vector field v_H under the canonical transformation Ψ is the Hamiltonian vector field corresponding to the pullback Ψ^*H of H under Ψ, that is, $\Psi^*v_H = v_{\Psi^*H}$. Prove this also via

$$i_{\Psi^*v_H}\sigma = i_{\Psi^*v_H}\Psi^*\sigma = \Psi^*(i_{v_H}\sigma) = \Psi^*dH = d\Psi^*H = i_{v_{\Psi^*H}}\sigma.$$

(xii) Define, for functions f and $g \in C^1(T^*Q)$, the *Poisson brackets* $\{f, g\} \in C(T^*Q)$ of f and g by $\{f, g\} = i_{v_f} dg = dg(v_f)$. Using $dg = -i_{v_g}\sigma$ from part (iv) show

$$\{f, g\} = \sigma(v_f, v_g) = \sum_{1 \le j \le d} \Big(\frac{\partial f}{\partial p_j}\frac{\partial g}{\partial q_j} - \frac{\partial f}{\partial q_j}\frac{\partial g}{\partial p_j} \Big).$$

The definition of $\{f, g\}$ is independent of the choice of coordinates $(q, p) = \xi \in T^*Q$, in view of the invariance of σ under canonical transformations.

(xiii) Prove the relation $[\, v_f, v_g \,] = v_{\{f,g\}}$ between the Lie and the Poisson brackets. In fact, apply successively the formula from Exercise 8.41.(iii) for the Lie derivative of a differential form with $X = v_f$ and $\omega = i_{v_g}\sigma \in \Omega^1(T^*Q)$, part (vi), and Exercise 8.41.(iii) again, to obtain

$$\begin{aligned}(L_{v_f}(i_{v_g}\sigma))(v) &= v_f(i_{v_g}\sigma(v)) + (i_{v_g}\sigma)([\, v, v_f \,]) \\ &= -(L_{v_f}\sigma)(v_g, v) + v_f(\sigma(v_g, v)) + \sigma(v_g, [\, v, v_f \,]) \\ &= \sigma([\, v_f, v_g \,], v) = (i_{[\, v_f, v_g \,]}\sigma)(v),\end{aligned}$$

and deduce $i_{[\, v_f, v_g \,]}\sigma = L_{v_f}(-dg) = -d(L_{v_f} g) = -d\{f, g\} = i_{v_{\{f,g\}}}\sigma$ from part (iv), the homotopy formula and part (xii). Show that the Poisson brackets satisfy Jacobi's identity from Exercise 5.26.(iii), that is

$$\{f_1, \{f_2, f_3\}\} + \{f_2, \{f_3, f_1\}\} + \{f_3, \{f_1, f_2\}\} = 0.$$

(xiv) Suppose $x : J \to T^*Q$ is a solution curve of the Hamilton vector field v_H as in part (v), and let $f \in C^1(T^*Q)$. Verify by means of Formula (2.12)

$$(f \circ x)'(t) = \{H, f\}(x(t)), \qquad \text{in particular} \qquad (H \circ x)' = 0.$$

That is, the Hamiltonian H is a conserved quantity. Hamilton's equation itself takes the form

$$q_j' = \{H, q_j\} \qquad p_j' = \{H, p_j\} \qquad (1 \le j \le d), \qquad x' = \{H, x\}.$$

Here we have extended the definition of the Poisson brackets to vector-valued functions. Assuming convergence deduce that the solution is given, with $\delta_H x = \{H, x\}$, by

$$x(t) = e^{t\delta_H} x(0) = \sum_{n \in \mathbf{N}_0} \frac{t^n}{n!}\{H, \cdots \{H, \{H, x\}\} \cdots\}(0) \qquad (t \in \mathbf{R}).$$

This formula shows an analogy with descriptions of *quantum physics*.

(xv) The Hamiltonian of a classical point particle in $\mathbf{R}$ of mass m under the influence of gravity $F(q) = -g$, for $q \in \mathbf{R}$, is given by $H(q, p) = \frac{p^2}{2m} + mgq$. Prove

$$\{H, q\} = \frac{p}{m}, \qquad \{H, \{H, q\}\} = -g, \qquad \{H, \{H, \{H, q\}\}\} = 0;$$

and obtain the *law of free fall* as the solution, where $p = mq'$,

$$q(t) = q(0) + \frac{p(0)t}{m} - \frac{gt^2}{2} = q(0) + q'(0)t - \frac{gt^2}{2} \qquad (t \in \mathbf{R}).$$

Exercise 8.47 (Minimal hypersurface – needed for Exercise 8.48). Consider a compact oriented C^2 submanifold $V \subset \mathbf{R}^n$ of codimension 1, and a one-parameter group of C^2 diffeomorphisms $(\Phi^t)_{t\in\mathbf{R}}$ of $\mathbf{R}^n$ with C^1 tangent vector field $v : \mathbf{R}^n \to \mathbf{R}^n$. Then all the $V_t = \Phi^t(V)$, for $t \in \mathbf{R}$, are compact oriented C^2 hypersurfaces too. Select C^1 mappings $n : \mathbf{R} \times \mathbf{R}^n \to \mathbf{R}^n$ such that $n_t(x) = n(t, x)$, for $x \in V_t$, is the normal to V_t at x compatible with the orientation on V_t. According to Formula (7.15) the Euclidean hyperarea form on V_t is given by $\omega_t = i_{n_t}\,dx$. Note that $V_0 = V$, and write $\omega_0 = \omega$.

(i) Use the homotopy formula from Lemma 8.9.1 to verify

$$\begin{aligned}
\frac{d}{dt}\Big|_{t=0} \int_{V_t} \omega_t &= \frac{d}{dt}\Big|_{t=0} \int_V (\Phi^t)^* \omega_t \\
&= \int_V \Big(\frac{d}{dt}\Big|_{t=0} (\Phi^t)^* \Big) \omega_0 + \int_V \frac{d}{dt}\Big|_{t=0} \omega_t \\
&= \int_V (d \circ i_v + i_v \circ d)\,\omega + \int_V i_{\frac{d}{dt}\big|_{t=0} n_t}\, dx.
\end{aligned}$$

(ii) Deduce from $\|n_t\|^2 = 1$, for $t \in \mathbf{R}$, that

$$\Big\langle \frac{d}{dt}\Big|_{t=0} n_t,\, n_0 \Big\rangle = 0; \qquad \text{thus} \qquad \frac{d}{dt}\Big|_{t=0} n_t(x) \in T_x V;$$

accordingly

$$\int_V i_{\frac{d}{dt}\big|_{t=0} n_t}\, dx = 0,$$

since computing the integral involves evaluation of dx at the points $x \in V$ on n vectors belonging to the $(n-1)$-dimensional space $T_x V$.

(iii) Use Stokes' Theorem to show

$$\frac{d}{dt}\Big|_{t=0} \int_{V_t} \omega_t = \int_{\partial V} i_v \omega + \int_V i_v \circ d(i(n)\,dx) = \int_V i_v (d \circ i(n)\,dx).$$

Apply the equality $d \circ i_n \, dx = \operatorname{div} n \, dx$ from Example 8.9.2 and prove

$$\frac{d}{dt}\bigg|_{t=0} \int_{V_t} \omega_t = \int_V (\operatorname{div} n) \, i_v \, dx.$$

Background. We call $\operatorname{div} n = \operatorname{tr} Dn : V \to \mathbf{R}$ the *mean curvature* of V (see Section 5.7); it depends on the choice of orientation. A hypersurface V with fixed boundary and having smallest possible hyperarea is called a *minimal hypersurface*. The arguments above imply that the mean curvature of a minimal hypersurface vanishes identically, as one sees by taking the vector field v restricted to V equal to $f\,n$, for arbitrary C^1 functions f.

Exercise 8.48 (Catenoid and helicoid are minimal surfaces – sequel to Exercises 4.6 and 8.47). As in Exercise 4.6 we define $\phi : \mathbf{R}^2 \to \mathbf{R}^3$ by $x = \phi(s, t) = (\cosh s \cos t, \ \cosh s \sin t, \ s)$, and we call $C = \operatorname{im}(\phi)$ the catenoid.

(i) Show that the Gauss mapping $n : C \to S^2$ is given by

$$n(x) = \frac{1}{\cosh s}(-\cos t, -\sin t, \sinh s) \qquad ((s, t) \in \mathbf{R}^2).$$

(ii) Take $D_1\phi(s, t)$ and $D_2\phi(s, t)$ as basis vectors for T_xC. With respect to this basis compute, as in Example 5.7.2, the matrix of the Weingarten mapping $Dn(x) \in \operatorname{End}^+(T_xC)$ to be

$$\frac{1}{\cosh^2 s}\begin{pmatrix} 1 & 0 \\ 0 & -1 \end{pmatrix}.$$

Deduce from Exercise 8.47 that C is a minimal surface.

(iii) Fix $a \in \mathbf{R}_+$. Compute the area of the subset C_a of C consisting of the $x \in C$ with $|x_3| < a$ to be $2\pi(a + \cosh a \sinh a)$. On the other hand, the area of the two disks $D_a^{\pm} = \{ x \in \mathbf{R}^3 \mid x_1^2 + x_2^2 \leq \cosh^2 a, \ x_3 = \pm a \}$ equals $2\pi \cosh^2 a$. So the minimal surface C_a will not minimize the area among all surfaces with boundary the two circles $\partial D_a^{\pm}$ if $a + \cosh a \sinh a > \cosh^2 a$, that is, if $2a > 1 + e^{-2a}$, which is satisfied for a sufficiently large.

(iv) Prove that the helicoid from Exercise 4.8 is a minimal surface.

Exercise 8.49 (Special case of Gauss–Bonnet Theorem). Consider a compact oriented C^2 submanifold $V \subset \mathbf{R}^n$ of codimension 1. Extending the theory of Section 5.7 in a straightforward manner we say that the Gaussian curvature K of V is given by $K = \det Dn$, where $n : V \to S^{n-1}$ is the Gauss mapping. Let

$\omega \in \Omega^{n-1}(\mathbf{R}^n)$ be the differential form from Example 8.11.9 that computes the Euclidean $(n-1)$-dimensional hyperarea of V. Prove

$$n^*\omega_{S^{n-1}} = K\,\omega_V, \qquad \text{and deduce} \qquad (\star) \qquad \frac{1}{|S^{n-1}|}\int_V K(y)\,d_{n-1}y = \deg(n).$$

In particular, if $V \subset \mathbf{R}^3$ is a multi–donut with g holes, prove that $\deg(n) = 1 - g$ by means of Formula (8.68). The number g is called the *genus* of the multi–donut.
Background. The equality $(\star)$ above is half the assertion of the *Gauss–Bonnet Theorem.* The other half identifies $\deg(n) \in \mathbf{Z}$ as an invariant of V. Furthermore, in $\mathbf{R}^3$ the integer $2\deg(n)$ equals the *Euler characteristic* $\chi(V)$ of V: partition V into a finite number of triangles, then $\chi(V)$ equals the number of vertices minus the number of edges plus the number of faces of the triangles, irrespective of the chosen subdivision.

Exercise 8.50 (Zeros of a holomorphic function). As usual, write $\mathbf{C} \ni z = x_1 + ix_2 \leftrightarrow x = (x_1, x_2) \in \mathbf{R}^2$. Let $f = f_1 + if_2 : \mathbf{C} \to \mathbf{C}$ be a holomorphic function and set $U = \{\, z \in \mathbf{C} \mid f(z) \neq 0\,\}$.

(i) Let $\sigma \in \Omega^1(\mathbf{R}^2 \setminus \{0\})$ be as in Formula (8.73). Verify that we have on $\mathbf{R}^2 \setminus \{0\}$ and U, respectively,

$$\sigma(x) = \frac{-x_2\,dx_1 + x_1\,dx_2}{x_1^2 + x_2^2}, \qquad d\log f = \frac{df}{f} = \frac{1}{2}d\log\|f\|^2 + if^*\sigma.$$

In complex function theory it is shown that f has only isolated zeros (see Example 2.2.6) if $f \neq 0$.

(ii) Suppose that $f'(z) \neq 0$ if $f(z) = 0$. Use the Cauchy–Riemann equation to show $\det Df(x) = |f'(z)|^2 > 0$, and deduce that $\operatorname{sgn}(\det(Df(x)) = 1$, for every zero $x \in \mathbf{R}^2$ for f.

(iii) Let $\Omega \subset \mathbf{C}$ be as in Example 8.11.11. By means of parts (i) and (ii) prove

$$\frac{1}{2\pi i}\int_{\partial\Omega}\frac{df}{f} = \frac{1}{2\pi}\int_{\partial\Omega} f^*\sigma = n(f, \Omega),$$

where $n(f, \Omega)$ is the number of zeros of f that belong to Ω.

(iv) In the case of $f(a) = f'(a) = 0$, for some $a \in \Omega$, we argue as follows. In view of Exercise 8.12.(v) we may develop f in a power series about a, hence $f(z) = \sum_{n \geq n(f,a)} c_n(z-a)^n$, with $n(f, a) \in \mathbf{N}$ and $0 \neq c_{n(f,a)} \in \mathbf{C}$. This gives $f(z) = (z-a)^{n(f,a)}\,g(z)$, for z near a, with g holomorphic near a and $g(a) \neq 0$. Hence, for z near a,

$$\frac{f'(z)}{f(z)} = \frac{n(f,a)}{z-a} + \frac{g'(z)}{g(z)} = \frac{n(f,a)}{z-a} + h(z),$$

with h holomorphic near a. If $f^{-1}(\{0\}) \cap \Omega = \{\, a_i \mid 1 \le i \le m \,\}$, deduce

$$\frac{1}{2\pi i}\int_{\partial\Omega}\frac{df}{f} = \sum_{1\le i\le m} n(f, a_i) =: n(f, \Omega).$$

Exercise 8.51 (Linking number, Kronecker's integral and Biot–Savart's law – sequel to Exercise 8.28). Let V_i be connected compact and oriented C^2 submanifolds of $\mathbf{R}^n$ of dimension d_i, for $1 \le i \le 2$, where $d_1 + d_2 = n - 1$, and suppose these have no point in common. (Best example: two disjoint closed curves in $\mathbf{R}^3$.) In the notation of Example 8.11.9 define the *linking number* $L(V_1, V_2)$ as $W(\phi(V_1 \times V_2), 0)$, where $\phi : V_1 \times V_2 \to \mathbf{R}^n \setminus \{0\}$ is given by $\phi(x_1, x_2) = x_2 - x_1$.

(i) Prove

$$L(V_1, V_2) = \frac{1}{|S^{n-1}|}\int_{V_1\times V_2}\phi^*\Big(\frac{1}{\|x\|^n} i_x dx\Big).$$

(ii) We will compute the integral in (i) for a pair of closed curves $V_i = \mathrm{im}(\gamma_i)$ in $\mathbf{R}^3$, where $\gamma_i : [\,0, 1\,] \to V_i$. Verify, for the standard basis vectors $e_i \in \mathbf{R}^2$,

$$D(\phi \circ (\gamma_1 \times \gamma_2))(t_1, t_2)\, e_i = (-1)^i \gamma_i'(t_i) \qquad (1 \le i \le 2),$$

and show that this implies

$$L(V_1, V_2) = \frac{1}{4\pi}\int_0^1\int_0^1 \frac{\det\big(\gamma_1(t_1) - \gamma_2(t_2)\ \ \gamma_1'(t_1)\ \ \gamma_2'(t_2)\big)}{\|\gamma_1(t_1) - \gamma_2(t_2)\|^3}\, dt_1\, dt_2$$

$$= \int_0^1 \Big\langle \frac{1}{4\pi}\int_0^1 \frac{1}{\|\gamma_1(t_1) - \gamma_2(t_2)\|^3}\gamma_2'(t_2) \times (\gamma_1(t_1) - \gamma_2(t_2))\, dt_2,\ \gamma_1'(t_1)\Big\rangle\, dt_1.$$

Recognize the inner integral as the Biot–Savart law from Exercise 8.28.(v) describing the magnetic field at $\gamma_1(t_1)$ due to a steady unit electric current flowing around the closed loop V_2. Deduce that $L(V_1, V_2)$ is precisely the work done by this magnetic field on a unit magnetic pole which makes one circuit around V_1.

(iii) The curves $\gamma_1(t) = (\cos t, \sin t, 0)$ and $\widetilde{\gamma}_r(t) = r(-1 + \cos t, 0, -\sin t)$, with $r > 1$, define two disjoint oriented circles in $\mathbf{R}^3$ that are linked. Prove that $\widetilde{\gamma}_r$ converges to γ_2 with $\gamma_2(t) = (0, 0, -t)$, for $r \to \infty$. Now verify that $L(V_1, V_2) = 1$ by explicit evaluation of the integral in (ii). Note that $\phi(V_1 \times V_2)$ is the cylinder $\{\, x \in \mathbf{R}^3 \mid x_1^2 + x_2^2 = 1 \,\}$, which winds once around the origin in $\mathbf{R}^3$.
Hint: Compute $\int_{\mathbf{R}_+} \frac{1}{(1+t_2^2)^{3/2}}\, dt_2$ by means of Exercise 6.50.(iv) or the substitution $t_2 = \sinh u$.

Notation

^{c}	complement	8
$\widehat{\ }$	Fourier transform	466
$*$	convolution	468
$\circ$	composition	17
$\overline{\{\cdot\}}$	closure	8
$\flat$	mapping from vector fields to differential forms	583
∂	boundary	9
$\partial_V A$	boundary of A in V	11
$\dfrac{\partial}{\partial\nu}$	derivative in direction of outer normal	534
$\times$	cross product	147
$\wedge$	exterior multiplication	577
$\bigwedge^k V^*$	k-th exterior power of V^*	568
Δ	Laplace operator	470, 528
∇	nabla or del	59, 528
∇_X	covariant derivative in direction of X	407
$\Box$	wave operator	755
$\Vert\cdot\Vert$	Euclidean norm on $\mathbf{R}^n$	3
$\Vert\cdot\Vert_{\mathrm{Eucl}}$	Euclidean norm on $\mathrm{Lin}(\mathbf{R}^n, \mathbf{R}^p)$	39
$\langle\cdot,\cdot\rangle$	standard inner product	2
$\{\cdot,\cdot\}$	Poisson brackets	773
$[\cdot,\cdot]$	Lie brackets	169
1_A	characteristic function of A	34
$\int$	integral	426
$\overline{\int}$	upper Riemann integral	426
$\underline{\int}$	lower Riemann integral	426
$\int_\phi \omega$	integral of differential form ω along mapping ϕ	572
$\int_V f(x)\,d_d x$	integral of f over V w.r.t. Euclidean density	495
$\int_V f(x)\rho(x)\,dx$	integral of f over V w.r.t. density ρ	490
$\int_V \langle f, \nu\rangle(y)\,dy$	flux of f across V w.r.t. ν	530
$\int_\gamma \langle f(s), d_1 s\rangle$	oriented line integral of f along γ	537
$\int_\Xi \langle g(x), d_2 x\rangle$	oriented surface integral of vector field g over Ξ	560
$\int_{\partial\Omega} f(z)\,dz$	complex line integral of f along $\partial\Omega$	555
$(\alpha)_k$	shifted factorial	180

$\mathcal{J}(A)$	collection of compact and Jordan measurable subsets of A	431
$L^{\perp}$	orthocomplement of linear subspace L	201
L_v	Lie derivative in direction of vector field v	585
lim	limit	6, 12
$\mathrm{Lin}(\mathbf{R}^n, \mathbf{R}^p)$	linear space of linear mappings $\mathbf{R}^n \to \mathbf{R}^p$	38
$\mathrm{Lin}^k(\mathbf{R}^n, \mathbf{R}^p)$	linear space of k-linear mappings $\mathbf{R}^n \to \mathbf{R}^p$	63
$\mathbf{Lo}(n+1, \mathbf{R})$	Lorentz group of $\mathbf{R}^{n+1}$	756
$\mathbf{Lo}^{\circ}(4, \mathbf{R})$	proper Lorentz group	386
$\mathrm{Mat}(n, \mathbf{R})$	linear space of $n \times n$ matrices with coefficients in $\mathbf{R}$	38
$\mathrm{Mat}(p \times n, \mathbf{R})$	linear space of $p \times n$ matrices with coefficients in $\mathbf{R}$	38
$N(c)$	level set of c	15
$\mathcal{O} = \{ O_i \mid i \in I \}$	open covering of set	30
$\mathbf{O}(\mathbf{R}^n)$	orthogonal group, of orthogonal operators in $\mathrm{End}(\mathbf{R}^n)$	73
$\mathbf{O}(n, \mathbf{R})$	orthogonal group, of orthogonal matrices in $\mathrm{Mat}(n, \mathbf{R})$	124
$\mathbf{R}^n$	Euclidean space of dimension n	2
$\mathcal{R}(\mathbf{R}^n)$	linear space of Riemann integrable functions on $\mathbf{R}^n$ with compact support	428
$\overline{S}$	upper sum	425
$\underline{S}$	lower sum	425
Sf	symmetric part of Df	539
S^{n-1}	unit sphere in $\mathbf{R}^n$	124
$\mathcal{S}(\mathbf{R}^n)$	linear space of Schwartz functions on $\mathbf{R}^n$	466
$\mathfrak{sl}(n, \mathbf{C})$	Lie algebra of $\mathbf{SL}(n, \mathbf{C})$	385
$\mathbf{SL}(n, \mathbf{R})$	special linear group, of matrices in $\mathrm{Mat}(n, \mathbf{R})$ with determinant 1	303
$\mathbf{SO}(3, \mathbf{R})$	special orthogonal group, of orthogonal matrices in $\mathrm{Mat}(3, \mathbf{R})$ with determinant 1	219
$\mathbf{SO}(n, \mathbf{R})$	special orthogonal group, of orthogonal matrices in $\mathrm{Mat}(n, \mathbf{R})$ with determinant 1	302
$\mathbf{Sp}(n, \mathbf{R})$	symplectic group of $\mathbf{R}^n$	772
$\mathfrak{su}(2)$	Lie algebra of $\mathbf{SU}(2)$	385
$\mathbf{SU}(2)$	special unitary group, of unitary matrices in $\mathrm{Mat}(2, \mathbf{C})$ with determinant 1	377
sup	supremum	21
supp	support of function	427
$T_x V$	tangent space of submanifold V at point x	134
$T^* V$	cotangent bundle of submanifold V	399
$T_x^* V$	cotangent space of submanifold V at point x	399
tr	trace	39
$V(a; \delta)$	closed ball of center a and radius δ	8
vol_n	n-dimensional volume	429

Index

Printed in the United States
By Bookmasters